Biomaterials, Bioengineering and Sustainability

Volume 4

There is an urgent need to address the current paradigm shifts in bioengineering for Human Health aiming the creation of breakthrough tissue engineered products, manufacturing technologies and effective regenerative treatments for tackling different diseases/disorders in a personalized manner. Yet, the excessive costs and wastes related to the development and production of biomaterials and advanced therapy medicinal products, and the increasing use of plastics in cell culture methods and animal derived reagents has recognized the importance of decreasing the direct carbon footprint and thus, implement sustainable principles and solutions in the innovation ecosystem. The main goal of the volumes in Biomaterials, Bioengineering and Sustainability series is to catapult and consolidate new concepts and solutions towards the development of the next-generation of sustainable and eco-friendly biomaterials and tissue engineering and regenerative medicine approaches. Each volume will focus on the latest developments dealing with the identification of new sources of sustainable or recycled biomaterials, providing ideas for green technologies and methods that can be applied for biomaterials advanced processing and scaffolding strategies, and applications in biofabrication, tissue engineering, regenerative medicine, and drug delivery systems. It also aims to include the exploitation of renewable and sustainable source of human cells applied for cell therapies or in combination with sustainable biomaterials. The develop complex in vitro 3D/4D models and dynamic cell culture systems will be other subjects to be further explored from a sustainable perspective. This series aims to attract the contributions of leading experts in bioengineering, cell biology, materials engineering, and environmental sciences.

Md. Faiyazuddin • Meghraj Suryawanshi
Editors

Design and Processing of Green Materials

Exploring Sustainable Applications in Medical and Pharmaceuticals Sciences

Springer

Editors
Md. Faiyazuddin
Department of Pharmaceutics,
School of Pharmacy
Al-Karim University
Katihar, Bihar, India

Meghraj Suryawanshi
Department of Pharmaceutics
Sandip Institute of Pharmaceutical
Sciences (SIPS)
Affiliated To Savitribai Phule Pune
University (SPPU, Pune)
Nashik, Maharashtra, India

ISSN 2731-751X ISSN 2731-7528 (electronic)
Biomaterials, Bioengineering and Sustainability
ISBN 978-3-031-91789-9 ISBN 978-3-031-91790-5 (eBook)
https://doi.org/10.1007/978-3-031-91790-5

This Springer imprint is published by the registered company Springer Nature Switzerland AG
The registered company address is: Gewerbestrasse 11, 6330 Cham, Switzerland

This book is lovingly dedicated to my beloved mother, **Sabiha Khatoon**, who has been the guiding light of my life—my source of strength, compassion, and unwavering support. Her prayers have been my protection, her wisdom my direction, and her love my foundation. She is the heart of our family and the soul of our home. Every accomplishment I achieve is built upon her sacrifices and unconditional love.

"And We have enjoined upon man [care] for his parents. His mother carried him, [increasing her] in weakness upon weakness, and his weaning is in two years. Be grateful to Me and to your parents; to Me is the [final] destination."

—Surah Luqman (31:14)

"And lower to them the wing of humility out of mercy and say, 'My Lord, have mercy upon them as they brought me up [when I was] small.'"

—Surah Al-Isra (17:24)

May Allah grant her a long life filled with health, happiness, and continued blessings. Ameen.

—Professor (Dr.) Md. Faiyazuddin

Preface

The escalating environmental challenges and the growing global demand for sustainable development have urged researchers, scientists, and industries to adopt eco-conscious approaches across various disciplines. In the biomedical and pharmaceutical sciences, this shift is both a necessity and a responsibility—one that prioritizes reducing ecological footprints while enhancing human health outcomes. It is in this spirit that the present volume, *Design and Processing of Green Materials: Exploring Sustainable Applications in Medical and Pharmaceuticals Sciences*, has been developed.

This edited book presents a comprehensive collection of recent advancements, research insights, and innovative applications in the realm of green biomaterials. Authored by leading experts and emerging scholars, the chapters address key issues related to sustainability in material design, synthesis, processing, and biomedical application. The book is structured to provide a holistic perspective, beginning with fundamental concepts and progressing through applied technologies and regulatory considerations.

Topics covered include biodegradable polymers, green chemistry methods for biomaterials synthesis, eco-friendly fabrication techniques, biomimetic strategies, and novel approaches in tissue engineering, regenerative medicine, and drug delivery. Additionally, the volume explores life cycle assessments, intellectual property landscapes, and regulatory frameworks that shape the future of sustainable biomedical practices.

The book opens with Chap. 1, which introduces the concept of green biomaterials, offering a foundational understanding of their significance, development challenges, and potential applications in health and sustainability. This sets the stage for Chap. 2, where the focus shifts to polymeric materials, discussing their synthesis, biodegradation, and growing role in drug delivery and tissue engineering. In Chap. 3, the authors explore environmentally benign chemical methods for material synthesis, promoting the use of renewable resources, solvent-free reactions, and energy-efficient processing techniques. Chapter 4 presents innovative approaches that integrate ecological responsibility into tissue engineering, from sustainable scaffold production to low-energy bioprocessing systems. Chapter 5 delves into the fabrication of medical devices using green techniques such as 3D bioprinting and electrospinning, emphasizing the use of biodegradable inputs. Complementing this, Chap. 6 illustrates how regenerative medicine offers sustainable, long-term therapeutic solutions that reduce dependency on synthetic implants and prolonged pharmacological treatments. Nature-inspired design takes center stage in Chap. 7, where the authors discuss the design and functionality of biomaterials modeled after natural structures. Chapter 8 builds upon this by examining how these materials interact with biological systems, influencing cell behavior and contributing to functional tissue regeneration. In Chap. 9, the spotlight is on eco-designed scaffolds that not only support tissue regrowth but also minimize environmental burden through sustainable sourcing and degradation. This theme continues in Chap. 10, which surveys cutting-edge bioprinting strategies that allow precise control over architecture and composition, using biocompatible and biodegradable materials. Chapter 11 discusses green innovations in drug delivery, particularly biodegradable and stimuli-responsive systems that reduce systemic toxicity and environmental impact. Natural polymers take precedence in Chap. 12, where plant-derived materials are explored for their roles in wound healing, drug encapsulation, and tissue scaffolds. Chapter 13 presents clinical applications of bioactive natural materials in the treatment of chronic wounds, emphasizing the dual benefits of efficacy and biodegradability. Shifting focus to environmental and production considerations, Chap. 14 provides tools and methodologies for evaluating the environmental impact of biomaterials across their entire lifecycle. In Chap. 15, readers are introduced to the regulatory and innovation landscape, showcasing how green biomaterials are advancing through patent filings, clinical validation, and formulation development. Finally, Chap. 16 addresses the safety, standardization, and regulatory frameworks necessary to ensure the responsible adoption of biomaterials in healthcare.

Together, these chapters provide a comprehensive overview of sustainable biomaterials, addressing both advances and challenges in their development, processing, application, and regulation. The book bridges fundamental research with translational potential, offering insights that are critical for researchers, clinicians, policymakers, and industry professionals seeking to build a greener and more sustainable future in medicine and pharmaceuticals.

We extend our sincere thanks to all the contributors for their dedicated efforts, and to Springer Nature for supporting this initiative. We hope this volume serves as both a resource and an inspiration for future innovations in green material science.

Katihar, Bihar, India — Md. Faiyazuddin
Nashik, Maharashtra, India — Meghraj Suryawanshi

Acknowledgments

We thank Almighty, the All-Merciful, the All-Compassionate, for His countless blessings and guidance throughout this endeavor. We sincerely thank each and every contributor for participating in this beneficial work.

First and foremost, we extend our sincere thanks to all the contributing authors for their valuable research contributions and dedication to advancing sustainable biomaterials. We are also deeply grateful to the reviewers and subject experts who provided insightful feedback to enhance the quality of each chapter.

We would like to acknowledge **Jamia Hamdard**, **Sandip University**, and **Saveetha University**—our esteemed institutions—for providing strong foundations for our academic and research pursuits.

Our sincere appreciation goes to the editorial and production team at **Springer Nature**, especially **Ramya Ravichandran**, Senior Editor; **Eva Loerinczi**, Senior Production Editor; **Peggy Papandreou**, Publishing Manager; **Immaculate Jayanthi**, Assistant Editor; and **Gonzalo Cordova**, Senior Editorial Assistant. Their professionalism, patience, and support were invaluable throughout the publication process.

We also extend our thanks to **Dr. Miguel Oliveira** (i3Bs, University of Minho) for his invaluable guidance and support.

The volume editors would also like to express their gratitude to **Springer Nature** and all its employees for offering a professional environment that fosters effective collaboration with subject matter experts.

We are thankful to our colleagues, students, and collaborators for their encouragement and intellectual engagement during the development of this volume.

Lastly, we extend our deepest gratitude to our families for their unwavering love, prayers, and support, which have been our constant source of strength.

To all who contributed directly or indirectly, we offer our sincere thanks.

Contents

Green Biomaterials: Challenges and Opportunities

1

Bindiya Chauhan, Bhuvnesh Kumar Singh, Anil Kumar, Neelanchal Trivedi, Md. Faiyazuddin, and Thomas J. Webster

Abstract

Green biomaterials (GBs) are defined as materials derived from biodegradable, renewable, and environmental sources as eco-friendly implants. These emerging materials have great potential to withstand and achieve sustainable developments to enhance various healthcare and industrial applications. GBs have significant advantages over conventional materials as they are biocompatible, reduce environmental impact, and have the capability to replace petroleum-based materials. Plant-derived composites, microbial products, and bio-based polymers are GBs, presenting many opportunities for the world of GBs. Technological advancements in bio- and nano-engineering are providing the capability to create materials with enhanced functions and characteristics. The amalgam of green chemistry within the circular economy, further provides a sustainable lifecycle of products

B. Chauhan (✉)
SGT College of Pharmacy, SGT University, Gurugram, Haryana, India

B. K. Singh · A. Kumar
Faculty of Pharmacy, Moradabad Educational Trust Group of Institutions, Moradabad, Uttar Pradesh, India

N. Trivedi
Invertis Institute of Pharmacy, Invertis University, Bareilly, Uttar Pradesh, India

Md. Faiyazuddin
School of Pharmacy, Al–Karim University, Katihar, Bihar, India

T. J. Webster
School of Biomedical Engineering and Health Sciences, Hebei University of Technology, Tianjin, China

School of Engineering, Saveetha University, Kuthambakkam, Tamil Nadu, India

Division of Pre-College and Undergraduate Studies, Brown University, Providence, RI, USA

Md. Faiyazuddin, M. Suryawanshi (eds.), *Design and Processing of Green Materials*, Biomaterials, Bioengineering and Sustainability 4,
https://doi.org/10.1007/978-3-031-91790-5_1

with enhanced recyclability and less waste. GBs in medical applications are providing promising solutions for preparing safe and quality implants, enhancing drug delivery systems, and creating tissue-engineered scaffolds for eco-friendly healthcare practices. In the present era of advanced GB research, this review was prepared with a critical view of GBs as it provides a platform to explore emerging GB applications with a focus on providing relevant cross-disciplinary research and policy initiatives to foster innovation and practical-based applications for clinical applications. In summary, this chapter highlights that GBs are truncating the growth of material science and biotechnology to significantly contribute to global sustainability goals while enhancing medicine.

Keywords

Green biomaterials · Conventional material · Sustainability · Bio-based polymers · Biodegradable materials · Renewable resources · Eco-friendly · Nanotechnology · Biocompatibility

1.1 Introduction

The global focus on improving environmental protection and sustainability has significantly catalyzed green material development and utilization. Green biomaterials (GBs) are considered some of the most interesting materials toward environmental conservation due to their renewability and biodegradability. Some biomaterials (such as flexmosa, flex tiles, and similar other products) provide an alternative and green approach to replace non-biodegradable products, which have a hazardous impact on the environment and people's health. Various biodegradable polymers, composites, and other forms of GBs, possess low impact on the environment and life cycle operations in biomedical, environmental, and industrial contexts (Mohanty et al. 2018, 2002). "Green biomaterials" is the specific term derived from material engineering with environmentalism. Mostly these materials have been derived from renewable resources, such as cellulose, starch, and chitosan, and/or are biodegradable such as polylactic acid (Bonsignore et al. 2021; Khan and Alamry 2021). Due to their biodegradability, they disintegrate into harmless products, instead of materials that cause permanent harm to the environment and healthcare system through waste management and pollution control (Rathod et al. 2024; Kalia et al. 2011).

The principal characteristics of GBs are their bioactivity and biocompatibility proven to be have promising applications in numerous biomedical applications such as drug delivery, wound healing, and tissue engineering (Yi et al. 2020; Watkins et al. 2015). Examples include numerous natural polymers available in nature, one of which includes hydrogels which have been utilized in tissue regeneration because of its similarity to extracellular matrices (Hoffman 2012). In the era of current environmental challenges, GBs play an important role in resolving the following issues: microplastic particle presence in all consumables, carbon dioxide emissions, and replacing traditional materials with petroleum-based products (Song et al. 2009).

Despite their potential, various limiting factors are include scalability, functionality, and cost. Moreover, to proceed with the translation of GBs into a wider set of medical applications, regulatory agency approval and market acceptance is required (Wei 2011; Iavicoli et al. 2014).

1.1.1 GBs in Comparison to Conventional Materials

In terms of environmental compatibility, functional versatility, and sustainability, GBs have demonstrated a significant paradigm shift from conventional materials. Although traditional materials have long been the cornerstone of biomedical, industrial, and technological applications (such as for metals, synthetic polymers, and ceramics), because of their deteriorating impact on the environment and continuous reliance on finite fossil resources, there is a large desire for greener alternatives (Chaudhry et al. 2018; Eppler and Jewell 2019). GBs potentially align with the sustainable development goals based on their origin from natural sources and natural lifecycles (Saratale et al. 2018).

1.1.1.1 Origin and Resources Utilization

Most conventional materials are derived from petroleum-based sources and they are depleting resources and are leading to environmental deterioration as they are from non-renewable resources. In contrast, GBs are mainly sourced through renewable resources such as agricultural waste, plant biomass, and microbial fermentation products (Gholap et al. 2024a). Examples of biomaterials derived from bioprocesses and natural polymers are chitosan, polyhydroxyalkanoates, etc. (Hopewell et al. 2009).

1.1.1.2 Environmental Impact and Biodegradability

The major drawback of conventional materials, particularly plastics or polymers, is their long-term pollution persistence in the environment, leading to challenges to microplastic accumulation and declining soil strength (Thompson et al. 2009). On the other hand, GBs are designed with this aspect in mind to degrade under natural conditions, with minimal impact on the ecological system, and to degrade into harmless byproducts such as organic matter, water, and carbon dioxide (Tokiwa et al. 2009).

1.1.1.3 Energy and Carbon Footprint

The process of developing and manufacturing conventional materials are energy-intensive and substantially emit gas due to the involvement of high-temperatures. GBs are developed utilizing energy-efficient synthesis processes, involving fermentation, and enzymatic reactions, significantly reducing its carbon footprint (Shah et al. 2008). The development of GBs has been integrated with the principles of green chemistry to ensure the production of the least toxic reagents and solvents to the environment (Song et al. 2009).

1.1.1.4 Functional Attributes and Applications

In terms of the functional attributes of conventional materials, they excel in thermal stability and longevity, proven to be suitable for industrial applications. GBs are continuously evolving to meet these standards of high tensile strength and sustainability (Nair and Laurencin 2007).

1.1.1.5 Latest Innovations and Ongoing Trials on GBs

Table 1.1 highlights the dynamic advancements and experimental efforts in the field of GBs, showcasing their versatility and sustainability across various industries. Key innovations include bioplastics derived from agricultural waste, offering biodegradable solutions for packaging and medical implants, and plant-based textiles, which replace synthetic fabrics with renewable cellulose-based alternatives. Bioceramics from algae and bio-inks for 3D bioprinting are advancing applications in regenerative medicine and bone repair, demonstrating high biocompatibility and sustainable sourcing. Bio-composites (like hemp-reinforced polymers) are enhancing automotive and construction industries with lightweight and renewable materials. Innovations (such as seaweed-based edible packaging and soy protein adhesives) emphasize eco-friendly replacements for conventional plastics and formaldehyde-based adhesives. Further, plant oil-based lubricants and starch-derived hydrogels show promise in industrial and pharmaceutical applications, while bio-mineralized green concrete addresses the need for durable, low-carbon construction materials. These developments, many in various trial phases, underline the transformative potential of biomaterials in fostering sustainability and reducing ecological footprints.

This chapter delves into the properties of GBs, and their applications, with the main focus on fostering environmental sustainability and innovation across various sectors. Emerging challenges, trends, and future perspectives have been discussed to provide a comprehensive understanding of this evolving field.

1.2 Historical Evolution of Biomaterials

Biomaterial science has evolved from the use of passive, biocompatible, or at best bioadhesive materials to multiphased, multifunctional systems that interact and interface with tissues. Various materials like gold, ivory, and silk were used as implant materials because these materials were available and to a certain extent were biocompatible.

The twentieth century however witnessed a revolution with the development of polymers which were synthetic ceramics and metallic alloys designed for use in medicine. With the principles of biocompatibility, as well as further modification of implant surfaces, the concept of using materials to stimulate certain biological responses was introduced. In the nineteenth century, gold dental fillings and silk sutures were introduced. In the mid-twentieth century, the manufacturing of orthopedic implants with titanium and stainless steel was introduced. In the late twentieth

Table 1.1 Innovations and ongoing trials in GBs

Category	Innovation/Trial	Material	Applications	Key Benefits	Status
Bioplastics (Bhatia et al. 2021)	Development of polyhydroxyalkanoates (PHA) derived from agricultural waste.	PHA	Packaging, medical implants	Biodegradable, reduces agricultural waste	In pilot production
Plant-Based Textiles (Felgueiras et al. 2021)	Trial of cellulose-based fabrics using enzymatic processing to replace synthetic textiles.	Cellulose	Clothing, upholstery	Renewable, lower carbon footprint	Ongoing trials
Bio-Ceramics (Dorozhkin 2010)	Algae-derived bio-ceramics for dental and orthopedic implants.	Algae-based calcium phosphate	Bone repair, prosthetics	High compatibility, sustainable sourcing	Clinical trials (phase 1)
3D Bioprinting (Fatimi et al. 2022)	Using plant-based bio-inks for creating tissue scaffolds.	Alginate and cellulose derivatives	Tissue engineering, regenerative medicine	Non-toxic, customizable, abundant availability	Lab-scale development
Bio-Composites (Elfaleh et al. 2023)	Natural fiber-reinforced polymers (e.g., hemp, flax) for automotive components.	Hemp, flax	Automotive interiors, construction	Lightweight, strong, renewable	Commercial testing
Edible Packaging (Waseem et al. 2023)	Development of seaweed-based edible films as a replacement for plastic wraps.	Seaweed	Food packaging	Edible, biodegradable, waste-free	Large-scale trials
Green Adhesives (Waseem et al. 2023)	Soy protein-based adhesives for wood products.	Soy protein	Furniture, construction	Formaldehyde-free, eco-friendly	Commercial production
Bio-Lubricants (Berman 2024)	Formulation of plant oil-based lubricants for industrial and automotive use.	Castor oil, soybean oil	Machinery, automotive engines	Non-toxic, biodegradable	Early adoption
Pharmaceuticals (Lee and Hwang 2023; Gholap et al. 2024b)	Corn starch-derived hydrogels for controlled drug delivery systems.	Corn starch	Drug delivery, wound care	Biodegradable, precise release control	Preclinical testing
Green Concrete (Javeed et al. 2024)	Bacteria-mediated self-healing concrete using urease-producing microbes.	Bio-mineralized concrete	Infrastructure, construction	Increased durability, reduced CO_2 emissions	Field trials

century, the usage of bioactive glasses and hydroxyapatite in bone regeneration was introduced (Ratner et al. 2024; Park and Lakes 2007).

Biomaterials were used even before the scientific field of "biomaterials science" became common. Early biomaterial applications were predominantly empirical, driven by trial-and-error approaches as highlighted by the following examples:

- *Medicine*: Catgut sutures, teeth implants from shells, and trepanation equipment of bones and stones. Source materials tended to be organized according to their utility rather than their likely compatibility with other items.
- *Industry*: Silk and wool were used in production then transferred to medical applications and natural rubber as elastic bandages or surgical gloves, etc.

The first bioactive materials helped establish principles for today's therapeutic uses and prosthetics, as clinicians understood the role of material properties such as longevity and biocompatibility (Hench 1998; Donaruma 1987).

The millennials have a paradigm shift in thinking towards GBs. Particularly, the idea of GBs is currently in high demand as the world is moving towards increasing sustainability and environmental responsibility. GBs are produced from renewable sources or are developed to provide less harm to the environment when made or discarded. Major features include biodegradation, non-toxicity and recyclability as well as the use of natural resources (Susheel 2011; Mohanty et al. 2005).

1.3 Classification of GBs

GBs, or materials made of renewable sources, can transform the science of materials to have a tremendous influence reducing negative impacts on the environment (Xu et al. 2022). A large potential for such kind of materials can been claimed beneficial for biomedical, industrial, and environmental applications (Pesode et al. 2023).

These are then followed by the extensive classifications of biomaterials and subclasses, as mentioned in Fig. 1.1:

1.3.1 Biopolymers

Biopolymers, either naturally produced by or bioengineered by a living organism, are derived from renewable resources. They are essentially biodegradable, hence, are optimal for sustainable applications.

1.3.1.1 Polylactic Acid (PLA): Production, Properties, and Applications

PLA, referring to the product of lactic acid polymerization, is produced from renewable sources like corn starch or sugarcane. It is biodegradable, thermoplastic in nature, and its excellent mechanical properties are quite alike those of traditional plastics. This versatility is exploited in using it in packaging materials, biomedical devices like sutures, implants, and 3D printing materials, due to biocompatibility and versatility (DeStefano et al. 2020).

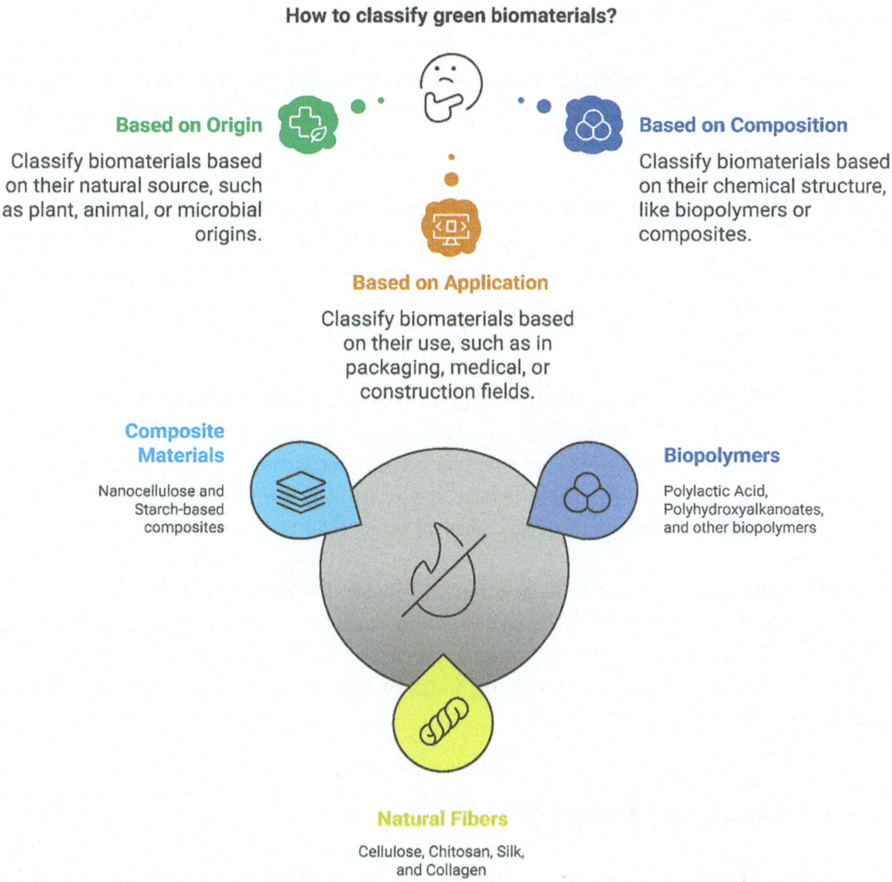

Fig. 1.1 Classification of GBs

1.3.1.2 Polyhydroxyalkanoates (PHA): Types, Synthesis, and Uses

PHAs is the family of polyesters which are produced by microorganisms as intracellular storage materials. The produced mechanical properties depend upon their monomer composition, dependent on renewable carbon sources which are utilized in fermentation processes. PHAs are biodegradable plastics, drug delivery systems, and agricultural films (Chen et al. 2013).

1.3.1.3 Other Notable Biopolymers: Polycaprolactone (PCL) and Polyglycolic Acid (PGA)

Polycaprolactone, or PCL, is a synthetic biodegradable polyester that is known for its low melting temperatures and is applied to tissue engineering and controlled drug release. Polyglycolic acid, or PGA, is a synthetic polymer that has high tensile strength and biodegradability, hence very often applied in surgical sutures and biodegradable packaging (Hasnain et al. 2018).

1.3.2 Natural Fibers

Originated in plants, animals, and/or microorganisms, natural fibers are flexible GBs giving biodegradability and renewability, along with functionality (Monroe et al. 2020).

1.3.2.1 Cellulose: Structure, Derivatives, and Applications in GBs

Cellulose is the most abundant natural polymer produced from plant cell walls and has derivatives, like cellulose acetate and carboxymethyl cellulose, applied in textiles, medicines, and biodegradable films. It also serves as reinforcement in some biocomposites such as cellulose nanofibers (Seddiqi et al. 2021).

1.3.2.2 Chitosan: Extraction, Modification, and Biomedical Uses

One such technique for extraction includes deacetylation of derivative chitin out of crustacean shells. Chitosan is another versatile derivative as it is biocompatible, antimicrobial, and biodegradable. Their application ranges from a wound dressing delivery system to water treatments, among other things (Pakizeh et al. 2021).

1.3.2.3 Silk and Collagen: Traditional Fibers in Modern GBs

Silk, obtained from silkworms or spiders, and collagen, found in animal tissues, are natural proteins with great mechanical properties and biocompatibility. They are heavily utilized in tissue engineering, sutures, and regenerative medicine (Vepari and Kaplan 2007).

1.3.3 Composite Materials

Composite materials are made up of two or more constituents, primarily to provide the superior property of combining strength, flexibility, and functionality with sustainability.

1.3.3.1 Nanocellulose Composites: Synthesis, Properties, and Applications

Nanocellulose composites are composites of cellulose fibers at the nanometer scale. They increase their strength while reducing the weight as well. These endow with great barrier properties along with mechanical properties that neoplasticizes lightweight constructions, biomedical scaffolding, and electronic applications (Iravani and Varma 2022).

1.3.3.2 Starch-Based Composites: Formulation, Advantages, and Biomedical Applications

Starch-based composite is derived from natural starch blended with biodegradable other materials. It is low in cost, nontoxic, and found to be used in a drug delivery system, a green packaging source, and agricultural applications (Onyeaka et al. 2022).

1.3.3.3 Hybrid Composites: Combining Strength, Biodegradability, and Functionality

Hybrid composites are produced based on the combination of natural and synthetic polymers, maximizing mechanical properties, thermal stability, and biodegradability. Automotive parts, medical implants, and sustainable packaging may be among some of these applications (Satchanska et al. 2024).

1.3.4 Emerging GBs

Novel GBs harness innovative sources and technologies to address certain specific challenges in sustainability and material performance as described below.

1.3.4.1 Algal Biomaterials: Harnessing Marine Resources

These biomaterials are algae-derived and include alginate, carrageenan, and agar. These natural resources find their application in food packaging, wound care, and bioplastics because of their gel-forming and biodegradable properties (Sabu Mathew et al. 2024; Faiyazuddin and Hakeem 2024).

1.3.4.2 Fungal Biomaterials: Mycelium-Based Innovations

Fungal mycelium is increasingly being researched for its sustainable alternative use in plastics and synthetic foams. The application areas vary from packaging to construction materials, even leather alternatives, all focused on biodegradability and low cost (Madusanka et al. 2024).

1.3.4.3 Engineered Proteins and Peptides: Precision Design for Biodegradability

Modern behavioral research aids in the development of protein and peptide constructs designed specifically for various uses, offering a potential high precision in terms of biodegradability and biocompatibility, thus, presenting applications from drug delivery to regenerative medicine and advanced coatings (Yin et al. 2017).

This classification declares the dynamics and broadness of GBs in their applications while revealing the chances of revolutionizing several industries towards sustainable progress (Mahmud et al. 2024).

1.4 Applications of GBs

GBs find diverse applications in several sectors because they offer the sustainable and environmental benefits associated with them. In the packaging industry, they increasingly displace traditional plastics by the biodegradable plastic ones such as polylactic acid (PLA) and cellulose-based films to present an environmentally friendly plastic waste solution (Moshood et al. 2022a). In construction, bio-based composites made from natural fibers or mycelium are being used for an eco-friendly insulation, panels, and even bricks with reduced dependence on non-renewable

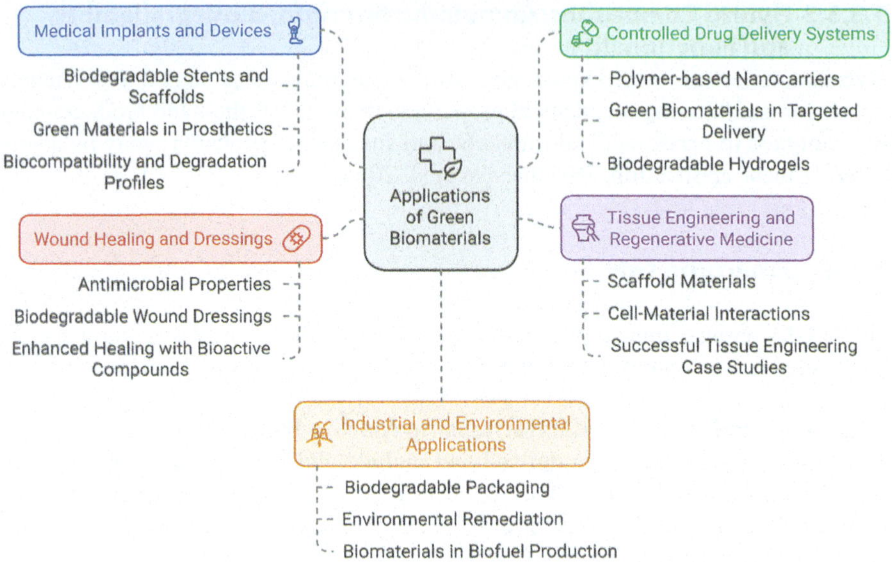

Fig. 1.2 Applications of GBs

materials (Alaneme et al. 2023). These applications contribute to both reducing environmental impact and promoting the development of the bio-based economy through the support of sustainable alternatives across different sectors, as seen in Fig. 1.2.

1.4.1 Medical Implants and Devices

The future promise of development lies in implant and devices using GBs for sustainable health technologies. On the other hand, researchers explore GBs as a part of drug delivery systems that involve controlled mechanisms minimizing environmental impact in comparison with plastic-based conventional devices (Trucillo 2024).

1.4.1.1 Biodegradable Stents and Scaffolds: Case Studies and Performance

Stents and scaffolds in biodegradable forms are key innovations developed to address some of the constraints associated with the more commonly used traditional nonbiodegradable materials in medical implants. These implants are usually produced from biopolymers, such as poly(lactic-co-glycolic acid) (PLGA) or polycaprolactone (PCL), and they degrade inside the body, so the need for surgical removal is completely eliminated (Modrák et al. 2023). Case studies prove that in cardiovascular applications, biodegradable stents help in the restoration of vascular function and facilitate healing, hence reducing complications that might arise from the use of such a device. Healing is slow as these stents degrade and stabilize the implantation site temporarily, thereby avoiding chronic inflammation or clotting. In performance testing, the results are encouraging; they have used the implant for repairing arteries and have high potential in orthopaedic and neural implants (Zong et al. 2022).

1.4.1.2 Green Materials in Prosthetics: Innovations and Challenges

Green materials in prosthetics represent a shift towards more sustainable, bio-friendly alternatives to traditional prosthetic materials like titanium or carbon fiber. The incorporation of polymers, such as PLA and PCL, along with composites of natural fibers to cut down on environmental damage is also being considered. Light and strong prosthetic limbs of the future will be the hallmark of green materials while presenting a further advantage through diminished weight and comfort (Hussain et al. 2024). Challenges remain in the optimization of their mechanical properties and long-term durability. Such challenges are being met by material science innovations with smart polymers that can sense environmental stimuli, improving both functionality and sustainability. The progress has still been hampered by challenges in scaling the green prosthetic materials for mass production (Virumbrales et al. 2024).

1.4.1.3 Biocompatibility and Degradation Profiles in Implants

Biocompatibility and degradation profiles are essential in the design of GBs for implants. Biocompatibility means that a material will perform appropriately with an appropriate host response when applied in the body, without causing any adverse reaction. Materials such as PLGA, PCL, and chitosan have been widely studied in terms of their ability to support cell growth, tissue regeneration, and immune rejection avoidance (Moghadasi et al. 2022). These materials have degradation profiles tailored to fit the healing process so that they break down at the optimal rate for the tissue they are supporting. Poor degradation profiles may result in premature failure or chronic inflammation, while over-degradation compromises implant function. Therefore, there is a delicate balance needed to make biodegradable implants work (Sheikh et al. 2015).

1.4.2 Controlled Drug Delivery Systems

GBs are used in DDS, not by making therapies patient-friendly, but by more sustainable health care practice, less reliance on synthetic, non-biodegradable polymers (Pires et al. 2023).

1.4.2.1 Polymer-Based Nanocarriers: Design and Mechanisms

Polymer-based nanocarriers, including micelles, nanoparticles, dendrimers, etc., have captured the attention in controlled drug delivery systems, as they can carry and protect drugs with the possibility of targeting their sustenance (Begines et al. 2020). The general characteristics of such nanocarriers include biomaterials such as poly(lactic acid) for poly(glycolic acid) and chitosan that are biodegradable and biocompatible. Characteristics include optimization in terms of size and surface charge and biomaterial polymer for improved availability and stability (Rocha et al. 2022). Mechanistically, polymer-based nanocarriers comprise the integration of drugs into its matrix and then provide an extended-release kinetics together with diminished systemic toxicity. The main advantage these systems can offer is the possibility to target a specific tissue, especially in cancer therapies, permitting localized treatment thereby reducing side effects (Dristant et al. 2023; Sharmila et al. 2024).

1.4.2.2 GBs in Targeted Delivery: Applications in Oncology

In terms of having drugs delivered directly to the sites of action, GBs are becoming prominent in oncology. In addition to being an environmentally sustainable alternative to treating cancer, biopolymer-based nanoparticles, liposomes, and hydrogels form GBs that encapsulate anticancer agents and initiate release at the targeted tumor site (Drabczyk et al. 2024). This localized drug delivery minimizes the systemic exposure and side effects, which is one of the significant disadvantages of traditional chemotherapy. Studies have shown that greens such as chitosan and alginate are effective in enhancing drug release specificity to cancer cells along with targeting ligands, hence increasing the treatment efficacy while at the same time reducing the environmental footprint of the drug delivery systems by using biodegradable materials (Imran et al. 2023).

1.4.2.3 Biodegradable Hydrogels: Sustained Release Formulations

Another intriguing GB for controlled drug delivery is the biodegradable hydrogel, which serves as a platform for a sustained localized release of a drug. Natural polymeric hydrogels, for example, chitosan, alginate, and gelatin, have very high-water content, biocompatibility, and tunable degradation rates. Such biodegradable hydrogels can be used particularly in wound healing and tissue regeneration (Teixeira et al. 2023). They may be loaded with growth factors or antimicrobial agents to effect healing over time. Their use in controlled release formulations is beneficial in reducing dosing frequency and side effects and thus leading to targeted therapeutic benefits (Adepu and Ramakrishna 2021).

1.4.3 Tissue Engineering and Regenerative Medicine

This could lead to the advancement of regenerative medicine with improved patient outcomes if these GBs have the possibility of being integrated into biological systems in a manner that encourages tissue regeneration yet at low risks of rejection (Dzobo et al. 2018).

1.4.3.1 Scaffold Materials: Designing for Biocompatibility and Function

Scaffold materials are crucial in tissue engineering by supporting the growth of cells and the development of tissues. The new generation of scaffolds with application in tissue engineering is currently being fabricated using GBs, mainly natural polymers (collagen, chitosan, silk fibroin) and biopolymers (PLA, PCL), which can both provide compatibility and biodegradability (Krishani et al. 2023). General idea is that such a material should mimic the extracellular matrix architecture, to provide the necessary mechanical support to cells that have to adhere, proliferate, and differentiate. This rate of degradation is engineered to keep up with the rate of tissue regeneration so that when tissue has been newly formed, it will degrade. Tissue-engineered constructs, as applications for regenerative medicine, rely heavily on the interaction of biocompatibility and optimal functionality (Golebiowska et al. 2024).

1.4.3.2 Cell-Material Interactions: Enhancing Regeneration with GBs

Cell-material interactions form the core of tissue engineering. It is, after all, the scaffold material that will support cell attachment, migration, and differentiation. GBs would improve such interactions by creating an environment in which cells may flourish. Materials like chitosan and gelatin have intrinsic bioactive properties and thus facilitate cellular activities, including growth and differentiation. In addition, surface properties of GBs may be engineered to provide interaction with specific cell types that enhance tissue regeneration. Examples include bone, skin, and cartilage regeneration. The bioactive compounds added in these materials, for instance, growth factors or peptides, can enhance their regenerative capacity (Wang et al. 2022).

1.4.3.3 Case Studies: Successful Tissue Engineering with GBs

Case studies have proved that GBs are very efficient in tissue engineering. For example, collagen-based scaffolds were used in skin regeneration, where promising results have been seen in the treatment of burns. PCL and PLA scaffolds have been used in bone tissue engineering for supporting osteoblast proliferation and mineralization (Dong and Lv 2016). When such GBs are associated with stem cells or bioactive molecules, they have proven to be capable of regenerating functional tissues in vitro and animal models. The success of these case studies clearly establishes the prospects of GBs in advancing the field of regenerative medicine, proposing sustainable and efficient alternatives compared to synthetic materials (Wu et al. 2018).

1.4.4 Wound Healing and Dressings

Biodegradable dressings made from such plant-based polymers or polymer composites are increasingly getting popular; they support wound healing besides degrading innocuously in the body eliminating the requirement for surgical removal. Such dressings could provide a more sustainable and effective patient-friendly approach to healing chronic wounds, burns, and surgical incisions than traditional wound care while contributing to reduced environmental burdens from the use of synthetic, non-biodegradable alternatives (Xu et al. 2023).

1.4.4.1 Antimicrobial Properties of GBs: Mechanisms and Applications

The most interesting green materials having antimicrobial properties which have been derived from natural origins include chitosan, alginate, and silk. This material can be used for the treatment of wound healing because of the existence of antimicrobial activity. For example, the disruption of bacterial cell walls, enzymatic activity, and the release of bioactive compounds such as antimicrobial peptides can delay the growth of pathogens (Matica et al. 2019). These materials in wound dressings not only prevent infections but also support regeneration of tissue and accelerate the healing process. GBs are of great benefit with chronic wounds, where infections are the main cause; they provide an eco-friendly and sustainable answer to wound care (Zhang et al. 2024).

1.4.4.2 Biodegradable Wound Dressings: Case Studies and Innovations

The most interesting green materials having antimicrobial properties which have been derived from natural origins include chitosan, alginate, and silk. This material can be used for the treatment of wound healing because of the existence of antimicrobial activity. For example, the disruption of bacterial cell walls, enzymatic activity, and the release of bioactive compounds such as antimicrobial peptides can delay the growth of pathogens (Dai et al. 2011). These materials in wound dressings not only prevent infections but also support regeneration of tissue and accelerate the healing process. GBs are of great benefit with chronic wounds, where infections are the main cause; they provide an eco-friendly and sustainable answer to wound care (Deng et al. 2022).

1.4.4.3 Enhanced Healing: Integration of Bioactive Compounds

So, the bioactive compounds integrated with the GBs promise very well in enhancing wound healing processes. Growth factors, peptides, and natural extracts are included into biodegradable material like aloe vera, honey for improving cell proliferation and formation of collagen and subsequent new angiogenesis. The biological agents are associated with accelerating the complete repair of tissues and improving the quality of wound closure. For example, topical treatment with these compounds through biodegradable dressings would guarantee prolonged liberation and, therefore, an increase in the therapeutic effect. This is part of what makes future advanced solutions in wound management effective and environmentally sustainable (Trinh et al. 2022).

1.4.5 Industrial and Environmental Applications

Due to their sustainability and eco-friendly properties, GBs have gained utmost attention in industrial and environmental applications. Bioplastics are alternatives to petroleum-based products applied in the industries of packing, textiles, and building, replacing non-renewable resources and reducing carbon imprint (Atiwesh et al. 2021).

1.4.5.1 Biodegradable Packaging: Innovations and Market Adoption

One of the most important industrial applications of GBs is biodegradable packaging. These materials are derived from renewable resources such as starch, PLA, and PHA which are developed to replace the petroleum-based plastics. Innovations in biodegradable packaging have developed flexible films, rigid containers, and food packaging, which break down naturally over time (Cheng et al. 2024). Market adoption is growing due to the shift of consumers and industries towards sustainable alternatives, but challenges remain in terms of cost-effectiveness and scaling up production. Material property improvement and wide acceptance in the marketplace will therefore determine the success of biodegradable packaging (Boz et al. 2020).

1.4.5.2 GBs in Environmental Remediation: Case Studies

GBs are also drawn upon in environmental remediation—notably in cleaning pollutants from water, soil, and air. Activated carbon, biochar, and biopolymer materials, for example, have been used to adsorb heavy metals, organic pollutants, and toxic substances. Case studies are seen in the detoxification of various wastewaters through chitosan-based material as well as air purification through algae-based composites (Liu et al. 2024). These green materials have an advantage as a more affordable and environmentally friendly substitute for conventional remediation strategies, thereby providing the double satisfaction of reducing pollutants while also recovering natural resources (Al et al. 2023).

1.4.5.3 Renewable Energy: Biomaterials in Biofuel Production

Biomaterials have emerged as sources from organic wastes, which include agricultural residues, algae, and other biomass sources, for the production of biofuel as a renewable energy source. Cellulose and lignocellulosic biomass are identified as GBs, which are serving as feedstocks for the production of bioethanol and biogas using microbial fermentation techniques (Clauser et al. 2021). To date, research had been done on optimizing the biomaterials into efficient production of biofuel with improved yield and optimum energy input and ensured sustainability of the entire process. The production of GBs contributes in fuelling the diversionary force away from fossil fuel dependency and results in more environmentally friendly substitute for generation of energy (El-Araby 2024).

1.5 Environmental and Economic Impacts of GBs

There is immense potential associated with using GBs in environmental and developmental problems (Pesode et al. 2023; Gholap et al. 2024c). These are renewable resources, which in turn, make them eco-friendly, as they minimize carbon emissions throughout their lifecycle-from raw materials such as plants, algae, and biowaste, through production, to disposal. These resources significantly minimize carbon emissions associated with producing and disposing of products using renewables (Jaiswal et al. 2022). Biodegradable materials would cause reduced long-term pollution in the environment, especially for packaging, construction, and healthcare industries. GBs further expand the bio-economy to other regions of the economy, making the employment numbers grow and facilitating innovations in sustainable technology development, which is shown in Fig. 1.3 (Xu et al. 2022; Priya et al. 2023).

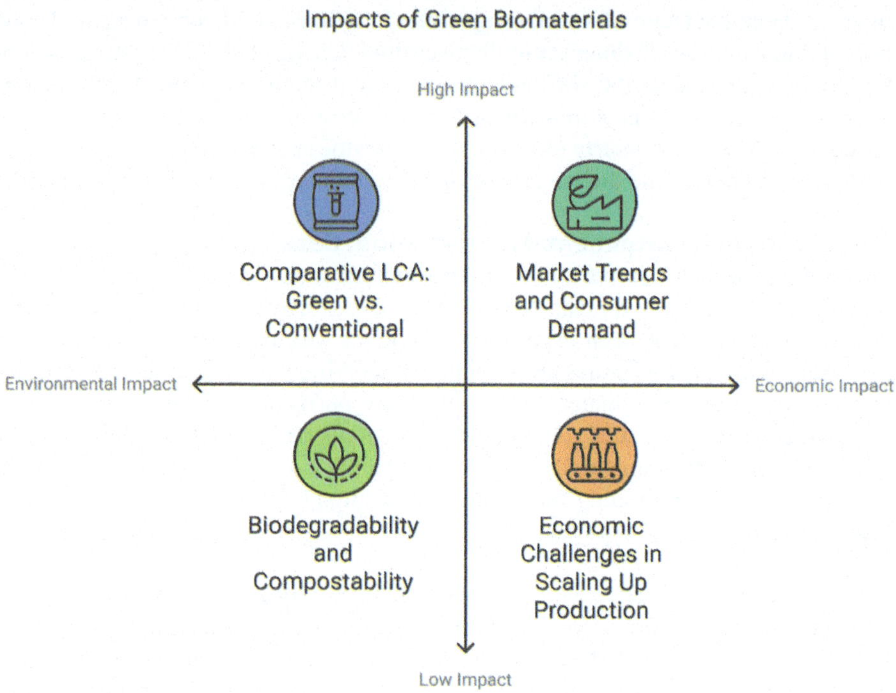

Fig. 1.3 Environmental and economic impacts of GBs

1.5.1 Life Cycle Assessment (LCA) of GBs

LCA is an all-encompassing technique that aims at examining the environmental effects of the GBs from the source to the final disposal or recycling. Such evaluation serves to assess the sustainability of these materials regarding their energy consumption, greenhouse gases emissions, water use, brute wastewater and solid waste generation (Khasreen et al. 2009; Choi et al. 2012). In the case of GBs, LCA generally portrays lesser adverse environmental impacts in comparison to that of its petroleum fought alternatives. Through delivering an intricate, cradle-to-grave perception of the environmental performance of GBs, LCA is an important resource that stakeholders in research, industry, and policymaking need to make decisions within global sustainability goals (Patel and Singh 2023; Hjuler and Hansen 2018).

1.5.1.1 Methodologies for LCA in Biomaterials

LCA is a key tool for evaluating the environmental impact of GBs from production to disposal. It considers all stages of a biomaterial's life cycle, including raw material extraction, manufacturing, transportation, usage, and end-of-life disposal. LCA methodologies involve data collection on resource consumption, emissions, waste generation, and energy use throughout these stages. Results of an LCA can determine the area in which reduction in environmental impact can be obtained in the designing of GBs (Jiao et al. 2024).

1.5.1.2 Comparative LCA: Green vs. Conventional Biomaterials

A comparison of life cycles for GBs to traditional materials, such as petroleum-based plastics or metals, would indicate a vast difference in terms of environmental benefits (Zia et al. 2021). Biomaterials generally create less carbon footprint since they are derived from renewable resources and will biodegrade, so waste will not end up in landfills or contribute to pollution (Pipitone et al. 2021). Also, GBs often require less energy and produce fewer harmful substances in the production process. These advantages position GBs at a more sustainable position in many industries, including packaging and biomedical devices (Kumar and Verma 2021).

1.5.1.3 Case Studies: Environmental Impact of Key GBs

The following case studies demonstrate some of the positive environmental impact of GBs (Atiwesh et al. 2021). For example, the manufacture of PLA from corn-based starch has been proven to emit fewer greenhouse gases than the traditional method of plastic production. I'm another bioplastics, like PHA and biocomposite products made from natural fibers, which has been found to be much less toxic to soil and water than plastics (Kumar et al. 2024). Such cases are proofs of the environmental effectiveness of GBs and their promotion of sustainable practices in varied industries (Synani et al. 2024).

1.5.2 Economic Considerations

Adoption of GBs entails several economic issues that weigh higher initial costs with long-term benefits (Mahmud et al. 2024). While production costs of GBs are often costlier than the conventional materials owing to factors such as availability of feedstock, the processing technology, and scale-up issues, these cost savings are offset by avoided environmental remediation costs, costs of complying with stringent regulations, and the growing market demand for sustainable products (Abolore et al. 2024).

1.5.2.1 Cost Analysis: Production and Market Pricing

GBs often tend to be expensive compared to the traditional materials as a result of using renewable raw materials, the complexity involved in their production processes, and special technologies (Thomas et al. 2023). With time, biotechnology and material science advancements improve production efficiency, which increases economies of scale and hence lowers costs. Other influences include consumer demand, the degree of regulatory support and supply chain of raw materials in terms of market pricing of GBs (Chen et al. 2024). Therefore, it is expected that costs associated with GBs will be lower, even more competitive than with the traditional materials as production becomes scalable, technology develops over time (Bradu et al. 2022).

1.5.2.2 Market Trends and Consumer Demand

The market for GBs has been rapidly moving forward because the consumer is environmentally aware of issues and the desire for sustainable products (Moshood et al. 2022a). The major drivers are packaging, textile, and biomedical devices; people seek alternatives to traditional plastics and metals (Moshkbid et al. 2024). Consumer preference for eco-friendly and regulatory pressure to reduce environmental impacts are significant factors driving these market trends (Yu and Zeng 2024). The shift toward sustainability is therefore continued as biomaterials made green become more and more adopted across various sectors (Verma et al. 2024).

1.5.2.3 Economic Challenges in Scaling Up Production

GBs are very promising from the environmental perspective, but the production is a serious economic challenge at this stage. Infrastructure, technology, and supply chain investment will be high when moving from the laboratory to commercial scale production (Bradu et al. 2022). Raw materials such as bio-based feedstocks can be expensive due to competition with food and availability due to agricultural yields (Arias et al. 2023). All this is an indication that quite a bit of research and innovation will be required to bring the production cost down, the efficiency up, and to increase the scalability of GBs (Vrchota et al. 2020).

1.5.3 Waste Management and Pollution Reduction

GBs are now playing an important role in waste management and pollution reduction with biodegradable alternatives from traditional plastics and synthetic materials (Atiwesh et al. 2021). As opposed to typical materials, which can continue to remain in the environment for decades, many GBs have been designed with the feature of natural decomposability, thus cutting down long-term environmental pollution and putting less pressure on waste management systems (Coppola et al. 2021; Singh et al. 2022).

1.5.3.1 The Role of GBs in a Circular Economy

This would therefore be crucial in the circular economy by reusing, recycling, and degrading materials with GBs (Ritzen et al. 2023). As they are developed from renewable sources, they could be developed to degrade to non-toxic by-products, and thus reduce landfill wastes (Das et al. 2022). They contribute to the establishment of a more sustainable and closed-loop circular economy wherein materials are recycled and regeneratively reused continuously (Mishra et al. 2023).

1.5.3.2 Biodegradability and Compostability: Environmental Benefits

Another environmental advantage is that of biodegradability and compostability in GBs, reducing long-term waste accumulation and pollution levels (Samir et al. 2022). PLA, PHA, and other biopolymers naturally break down to nontoxic substances because of microbial action (Chauhan et al. 2023). Biodegradable materials, unlike the traditional plastics that last for hundreds of years in the environment, are reducing waste and also preventing environmental pollution (Acharjee et al. 2022). Besides this, these materials can easily be composted and restore nutrients to the soil hence help in recycling organic waste (Moshood et al. 2022a).

1.5.3.3 Reducing Pollution: Case Studies of Effective Implementation

Such studies have shown that effective implementation of GBs to reduce pollution is possible (Rosenboom et al. 2022). In fact, biodegradable PLA packaging has been proved to reduce plastic waste in landfills and oceans (Moshood et al. 2022b). It has also been applied successfully in wastewater treatment by the removal of heavy metals and organic pollutants by chitosan (Islam et al. 2024). These case studies, therefore, underline a new contribution of GBs in pollution minimization and other alternative ways to the use of traditional materials toward cleaner and healthier environments (Chia et al. 2020).

1.6 Challenges in the Development and Adoption of GBs

Green materials derived from biological sources have shown great potential in their ability for various uses like tissue engineering and biomedical applications while minimizing environmental impacts (Brovold et al. 2018). GBs with the support of refined research and development usage in nanomaterials have the capability to greatly challenge conventional materials as they possess many limitations and most significantly as emerging micro-pollutants. In terms of the main factors, which play a key role in GBs usage are their functionality and sturdiness in comparison to various conventional materials (Mishra et al. 2018). GBs, degradation rate has to be controlled and requires adaptability according to the intended application. For instance, PLA-like biopolymers, are relatively stiff and heat-resistant, their operation under high-stress conditions is hindered due to low mechanical strength. An example of biomedical applications is medical implants, which require a parallel rate of tissue healing with the rate of material degradation, to achieve uniformity is difficult (Phiri et al. 2023).

Further challenges include the property to enhance the strength and flexibility while maintaining the biodegradable properties. Nanotechnology incorporation as nanostructures in GBs without modification in carbon footprint is a technological challenge and is costly (Bin et al. 2024). Economic constraints are major hurdles in development and there in subsequent market releases. Naturally derived polymers and additives are scarce in developing biomaterials. The utilization of superior GBs in technologies like 3D bio-printing and electrospinning, manufacturing requisite surplus capital and operating costs. In wind turbines, there are possibilities to reduce manufacturing costs and increase volume, with the support of government and regulatory authorities providing subsidized funds to promote GB usage at an economic scale (Shabbirahmed et al. 2023).

1.7 Conclusion

Bio-based green materials as part of the fourth generation of materials have immense potential to resolve environmental problems without compromising functionality in various uses. However, the reactivity of these materials can pose a problem in their widespread development and application to replace normal basic materials. Thus, it brings out a clear perspective based on the scientific realm. Among the factors considered when evaluating GBs, performance is the most important measure of applicability. Some difficulties are encountered in the replication or enhancement of the values of the conventional material counterparts. For instance, most biopolymers including PLA have low mechanical properties and unfavorable thermal stability thereby their application is limited especially under high stress conditions. Moreover, the biodegradation rate of GBs is sensitive to the application's needs, so it has to be adjusted appropriately. For example, in medical implants, materials should degrade at certain rates that should match the rates of healing tissues; this is hard to achieve systematically.

References

Abolore RS, Jaiswal S, Jaiswal AK (2024) Green and sustainable pretreatment methods for cellulose extraction from lignocellulosic biomass and its applications: a review. Carbohydr Polym Technol Appl 7:100396. https://doi.org/10.1016/J.CARPTA.2023.100396

Acharjee SA, Bharali P, Gogoi B, Sorhie V, Walling B, Alemtoshi. (2022) PHA-based bioplastic: a potential alternative to address microplastic pollution. Water Air Soil Pollut 234:21. https://doi.org/10.1007/S11270-022-06029-2

Adepu S, Ramakrishna S (2021) Controlled drug delivery systems: current status and future directions. Molecules 26:5905. https://doi.org/10.3390/MOLECULES26195905

Al MMA, Shin WS, Sarker A, Septian A, Das K, Deepo DM et al (2023) A critical review of sustainable application of biochar for green remediation: research uncertainty and future directions. Sci Total Environ 904:166813. https://doi.org/10.1016/J.SCITOTENV.2023.166813

Alaneme KK, Anaele JU, Oke TM, Kareem SA, Adediran M, Ajibuwa OA et al (2023) Mycelium-based composites: a review of their bio-fabrication procedures, material properties, and potential for green building and construction applications. Alex Eng J 83:234–250. https://doi.org/10.1016/J.AEJ.2023.10.012

Arias A, Feijoo G, Moreira MT (2023) Biorefineries as a driver for sustainability: key aspects, actual development and future prospects. J Clean Prod 418:137925. https://doi.org/10.1016/J.JCLEPRO.2023.137925

Atiwesh G, Mikhael A, Parrish CC, Banoub J, Le TAT (2021) Environmental impact of bioplastic use: a review. Heliyon 7:e07918. https://doi.org/10.1016/J.HELIYON.2021.E07918

Begines B, Ortiz T, Pérez-Aranda M, Martínez G, Merinero M, Argüelles-Arias F et al (2020) Polymeric nanoparticles for drug delivery: recent developments and future prospects. Nano 10:1403. https://doi.org/10.3390/NANO10071403

Berman D (2024) Plant-based oils for sustainable lubrication solutions—review. Lubricants 12:300. https://doi.org/10.3390/LUBRICANTS12090300

Bhatia SK, Otari SV, Jeon JM, Gurav R, Choi YK, Bhatia RK et al (2021) Biowaste-to-bioplastic (polyhydroxyalkanoates): conversion technologies, strategies, challenges, and perspectives. Bioresour Technol 326:124733. https://doi.org/10.1016/J.BIORTECH.2021.124733

Bin RA, Haque M, Islam SMM, Uddin Labib KMR (2024) Nanotechnology-enhanced fiber-reinforced polymer composites: recent advancements on processing techniques and applications. Heliyon 10:e24692. https://doi.org/10.1016/J.HELIYON.2024.E24692

Bonsignore G, Patrone M, Martinotti S, Ranzato E (2021) "Green" biomaterials: the promising role of honey. Journal of functional. Biomaterials 12:72. https://doi.org/10.3390/JFB12040072

Boz Z, Korhonen V, Sand CK (2020) Consumer considerations for the implementation of sustainable packaging: a review. Sustain For 12:2192. https://doi.org/10.3390/SU12062192

Bradu P, Biswas A, Nair C, Sreevalsakumar S, Patil M, Kannampuzha S et al (2022) RETRACTED ARTICLE: recent advances in green technology and industrial revolution 4.0 for a sustainable future. Environ Sci Pollut Res 30:124488–124519. https://doi.org/10.1007/S11356-022-20024-4

Brovold M, Almeida JI, Pla-Palacín I, Sainz-Arnal P, Sánchez-Romero N, Rivas JJ et al (2018) Naturally-derived biomaterials for tissue engineering applications. Adv Exp Med Biol 1077:421. https://doi.org/10.1007/978-981-13-0947-2_23

Chaudhry N, Dwivedi S, Chaudhry V, Singh A, Saquib Q, Azam A et al (2018) Bio-inspired nanomaterials in agriculture and food: current status, foreseen applications and challenges. Microb Pathog 123:196–200. https://doi.org/10.1016/J.MICPATH.2018.07.013

Chauhan B, Dodamani S, Malik S, Almalki WH, Haque S, Sayyed RZ (2023) Microbial approaches for pharmaceutical wastewater recycling and management for sustainable development: a multicomponent approach. Environ Res 237:116983. https://doi.org/10.1016/J.ENVRES.2023.116983

Chen Q, Liang S, Thouas GA (2013) Elastomeric biomaterials for tissue engineering. Prog Polym Sci 38:584–671. https://doi.org/10.1016/j.progpolymsci.2012.05.003

Chen L, Zhang Y, Chen Z, Dong Y, Jiang Y, Hua J et al (2024) Biomaterials technology and policies in the building sector: a review. Environ Chem Lett 22:715–750. https://doi.org/10.1007/S10311-023-01689-W

Cheng J, Gao R, Zhu Y, Lin Q (2024) Applications of biodegradable materials in food packaging: a review. Alex Eng J 91:70–83. https://doi.org/10.1016/J.AEJ.2024.01.080

Chia WY, Ying Tang DY, Khoo KS, Kay Lup AN, Chew KW (2020) Nature's fight against plastic pollution: algae for plastic biodegradation and bioplastics production. Environ Sci Ecotechnol 4:100065. https://doi.org/10.1016/J.ESE.2020.100065

Choi JK, Friley P, Alfstad T (2012) Implications of energy policy on a product systems dynamic life-cycle environmental impact: survey and model. Renew Sust Energ Rev 16:4744–4752. https://doi.org/10.1016/j.rser.2012.05.032

Clauser NM, González G, Mendieta CM, Kruyeniski J, Area MC, Vallejos ME (2021) Biomass waste as sustainable raw material for energy and fuels. Sustain For 13:794. https://doi.org/10.3390/SU13020794

Coppola G, Gaudio MT, Lopresto CG, Calabro V, Curcio S, Chakraborty S (2021) Bioplastic from renewable biomass: a facile solution for a greener environment. Earth Syst Environ 5:231–251. https://doi.org/10.1007/S41748-021-00208-7

Dai T, Tanaka M, Huang YY, Hamblin MR (2011) Chitosan preparations for wounds and burns: antimicrobial and wound-healing effects. Expert Rev Anti-Infect Ther 9:857–879. https://doi.org/10.1586/ERI.11.59

Das O, Babu K, Shanmugam V, Sykam K, Tebyetekerwa M, Neisiany RE et al (2022) Natural and industrial wastes for sustainable and renewable polymer composites. Renew Sust Energ Rev 158:112054. https://doi.org/10.1016/J.RSER.2021.112054

Deng X, Gould M, Ali MA (2022) A review of current advancements for wound healing: biomaterial applications and medical devices. J Biomed Mater Res B Appl Biomater 110:2542–2573. https://doi.org/10.1002/JBM.B.35086

DeStefano V, Khan S, Tabada A (2020) Applications of PLA in modern medicine. Eng Regen 1:76–87. https://doi.org/10.1016/J.ENGREG.2020.08.002

Donaruma LG (1987) Definitions in biomaterials. In: Williams DF (ed) . Elsevier, Amsterdam, p 72. Journal of Polymer Science Part C: Polymer Letters 1988;26:414–414. https://doi.org/10.1002/POL.1988.140260910

Dong C, Lv Y (2016) Application of collagen scaffold in tissue engineering: recent advances and new perspectives. Polymers 8:42. https://doi.org/10.3390/POLYM8020042

Dorozhkin SV (2010) Bioceramics of calcium orthophosphates. Biomaterials 31:1465–1485. https://doi.org/10.1016/J.BIOMATERIALS.2009.11.050

Drabczyk A, Kudłacik-Kramarczyk S, Jamroży M, Krzan M (2024) Biomaterials in drug delivery: advancements in cancer and diverse therapies—review. Int J Mol Sci 25:3126. https://doi.org/10.3390/IJMS25063126

Dristant U, Mukherjee K, Saha S, Maity D (2023) An overview of polymeric nanoparticles-based drug delivery system in cancer treatment. Technol Cancer Res Treat 22:15330338231152084. https://doi.org/10.1177/15330338231152083

Dzobo K, Thomford NE, Senthebane DA, Shipanga H, Rowe A, Dandara C et al (2018) Advances in regenerative medicine and tissue engineering: innovation and transformation of medicine. Stem Cells Int 2018:2495848. https://doi.org/10.1155/2018/2495848

El-Araby R (2024) Biofuel production: exploring renewable energy solutions for a greener future. Biotechnol Biofuels Bioprod 17:1–32. https://doi.org/10.1186/S13068-024-02571-9

Elfaleh I, Abbassi F, Habibi M, Ahmad F, Guedri M, Nasri M et al (2023) A comprehensive review of natural fibers and their composites: an eco-friendly alternative to conventional materials. Results Eng 19:101271. https://doi.org/10.1016/J.RINENG.2023.101271

Eppler HB, Jewell CM (2019) Biomaterials as tools to decode immunity. Adv Mater 32:e1903367. https://doi.org/10.1002/ADMA.201903367

Faiyazuddin Md, Hakeem KR (2024) Handbook of research in marine pharmaceutics. https://doi.org/10.1201/9781003537403

Fatimi A, Okoro OV, Podstawczyk D, Siminska-Stanny J, Shavandi A (2022) Natural hydrogel-based bio-inks for 3D bioprinting in tissue engineering: a review. Gels 8:179. https://doi.org/10.3390/GELS8030179

Felgueiras C, Azoia NG, Gonçalves C, Gama M, Dourado F (2021) Trends on the cellulose-based textiles: raw materials and technologies. Front Bioeng Biotechnol 9:608826. https://doi.org/10.3389/FBIOE.2021.608826

Gholap AD, Sawant SD, Pardeshi SR, Khuspe PR, Sayyad SF, Chavan M, et al. (2024a) Omics tools for determining microbial composition: a systematic review. In Microbiome engineering: the new dimension of biotechnology. pp 35–50

Gholap AD, Gupta J, Kamandar P, Bhowmik DD, Rojekar S, Faiyazuddin M et al (2024b) Harnessing Nanovaccines for effective immunization—a special concern on COVID-19: facts, Fidelity, and future prospective. ACS Biomater Sci Eng 10:271–297. https://doi.org/10.1021/ACSBIOMATERIALS.3C01247/ASSET/IMAGES/MEDIUM/AB3C01247_0006.GIF

Gholap AD, Pardeshi SR, Hatvate NT, Dhorkule N, Sayyad SF, Faiyazuddin M et al (2024c) Environmental implications and nanotechnological advances in octocrylene-enriched sunscreen formulations: a comprehensive review. Chemosphere 358:142235. https://doi.org/10.1016/J.CHEMOSPHERE.2024.142235

Golebiowska AA, Intravaia JT, Sathe VM, Kumbar SG, Nukavarapu SP (2024) Decellularized extracellular matrix biomaterials for regenerative therapies: advances, challenges, and clinical prospects. Bioact Mater 32:98–123. https://doi.org/10.1016/J.BIOACTMAT.2023.09.017

Hasnain MS, Ahmad SA, Chaudhary N, Hoda MN, Nayak AK (2018) Biodegradable polymer matrix nanocomposites for bone tissue engineering. In: Applications of nanocomposite materials in orthopedics. Elsevier, pp 1–37. https://doi.org/10.1016/B978-0-12-813740-6.00001-6

Hench LL (1998) Bioceramics. J Am Ceram Soc 81:1705–1728. https://doi.org/10.1111/J.1151-2916.1998.TB02540.X

Hjuler SV, Hansen SB (2018) LCA of biofuels and biomaterials. In: Life cycle assessment: theory and practice. Springer International Publishing, pp 755–782. https://doi.org/10.1007/978-3-319-56475-3_30

Hoffman AS (2012) Hydrogels for biomedical applications. Adv Drug Deliv Rev 64:18–23. https://doi.org/10.1016/J.ADDR.2012.09.010

Hopewell J, Dvorak R, Kosior E (2009) Plastics recycling: challenges and opportunities. Philos Trans R Soc B Biol Sci 364:2115–2126. https://doi.org/10.1098/RSTB.2008.0311

Hussain M, Khan SM, Shafiq M, Abbas N (2024) A review on PLA-based biodegradable materials for biomedical applications. Giant 18:100261. https://doi.org/10.1016/J.GIANT.2024.100261

Iavicoli I, Leso V, Ricciardi W, Hodson LL, Hoover MD (2014) Opportunities and challenges of nanotechnology in the green economy. Environ Health 13:1–11. https://doi.org/10.1186/1476-069X-13-78/FIGURES/1

Imran HM, Tang Y, Wang S, Yan X, Liu C, Guo L et al (2023) Optimized DOX drug deliveries via chitosan-mediated nanoparticles and stimuli responses in cancer chemotherapy: a review. Molecules 29:31. https://doi.org/10.3390/MOLECULES29010031

Iravani S, Varma RS (2022) Cellulose-based composites as scaffolds for tissue engineering: recent advances. Molecules 27:8830. https://doi.org/10.3390/MOLECULES27248830

Islam M, Xayachak T, Haque N, Lau D, Bhuiyan M, Pramanik BK (2024) Impact of bioplastics on the environment from its production to end-of-life. Process Saf Environ Prot 188:151–166. https://doi.org/10.1016/J.PSEP.2024.05.113

Jaiswal KK, Chowdhury CR, Yadav D, Verma R, Dutta S, Jaiswal KS et al (2022) Renewable and sustainable clean energy development and impact on social, economic, and environmental health. Energy Nexus 7:100118. https://doi.org/10.1016/J.NEXUS.2022.100118

Javeed Y, Goh Y, Mo KH, Yap SP, Leo BF (2024) Microbial self-healing in concrete: a comprehensive exploration of bacterial viability, implementation techniques, and mechanical properties. J Mater Res Technol 29:2376–2395. https://doi.org/10.1016/J.JMRT.2024.01.261

Jiao H, Ali SS, Alsharbaty MHM, Elsamahy T, Abdelkarim E, Schagerl M et al (2024) A critical review on plastic waste life cycle assessment and management: challenges, research gaps, and future perspectives. Ecotoxicol Environ Saf 271:115942. https://doi.org/10.1016/J.ECOENV.2024.115942

Kalia S, Dufresne A, Cherian BM, Kaith BS, Avérous L, Njuguna J et al (2011) Cellulose-based bio- and nanocomposites: a review. Int J Polym Sci 2011:837875. https://doi.org/10.1155/2011/837875

Khan A, Alamry KA (2021) Recent advances of emerging green chitosan-based biomaterials with potential biomedical applications: a review. Carbohydr Res 506:108368. https://doi.org/10.1016/J.CARRES.2021.108368

Khasreen MM, Banfill PFG, Menzies GF (2009) Life-cycle assessment and the environmental impact of buildings: a review. Sustain For 1:674–701. https://doi.org/10.3390/SU1030674

Krishani M, Shin WY, Suhaimi H, Sambudi NS (2023) Development of scaffolds from bio-based natural materials for tissue regeneration applications: a review. Gels 9:100. https://doi.org/10.3390/GELS9020100

Kumar B, Verma P (2021) Life cycle assessment: blazing a trail for bioresources management. Energy Convers Manag X 10:100063. https://doi.org/10.1016/J.ECMX.2020.100063

Kumar R, Lalnundiki V, Shelare SD, Abhishek GJ, Sharma S, Sharma D et al (2024) An investigation of the environmental implications of bioplastics: recent advancements on the development of environmentally friendly bioplastics solutions. Environ Res 241:117707. https://doi.org/10.1016/J.ENVRES.2023.117707

Lee CS, Hwang HS (2023) Starch-based hydrogels as a drug delivery system in biomedical applications. Gels 9:951. https://doi.org/10.3390/GELS9120951

Liu Y, Biswas B, Hassan M, Naidu R (2024) Green adsorbents for environmental remediation: synthesis methods, Ecotoxicity, and reusability prospects. PRO 12:1195. https://doi.org/10.3390/PR12061195

Madusanka C, Udayanga D, Nilmini R, Rajapaksha S, Hewawasam C, Manamgoda D et al (2024) A review of recent advances in fungal mycelium-based composites. Discov Mater 4:1–19. https://doi.org/10.1007/S43939-024-00084-8/FIGURES/5

Mahmud MZA, Mobarak MH, Hossain N (2024) Emerging trends in biomaterials for sustainable food packaging: a comprehensive review. Heliyon 10:e24122. https://doi.org/10.1016/J.HELIYON.2024.E24122

Matica MA, Aachmann FL, Tøndervik A, Sletta H, Ostafe V (2019) Chitosan as a wound dressing starting material: antimicrobial properties and mode of action. Int J Mol Sci 20:5889. https://doi.org/10.3390/IJMS20235889

Mishra RK, Ha SK, Verma K, Tiwari SK (2018) Recent progress in selected bio-nanomaterials and their engineering applications: an overview. J Sci Adv Mater Devices 3:263–288. https://doi.org/10.1016/J.JSAMD.2018.05.003

Mishra K, Siwal SS, Nayaka SC, Guan Z, Thakur VK (2023) Waste-to-chemicals: green solutions for bioeconomy markets. Sci Total Environ 887:164006. https://doi.org/10.1016/J.SCITOTENV.2023.164006

Modrák M, Trebuňová M, Balogová AF, Hudák R, Živčák J (2023) Biodegradable materials for tissue engineering: development, classification and current applications. J Funct Biomater 14:159. https://doi.org/10.3390/JFB14030159

Moghadasi K, Mohd Isa MS, Ariffin MA, Mohd Jamil MZ, Raja S, Wu B et al (2022) A review on biomedical implant materials and the effect of friction stir-based techniques on their mechanical and tribological properties. J Mater Res Technol 17:1054–1121. https://doi.org/10.1016/J.JMRT.2022.01.050

Mohanty AK, Misra M, Drzal LT (2002) Sustainable bio-composites from renewable resources: opportunities and challenges in the green materials world. J Polym Environ 10:19–26. https://doi.org/10.1023/A:1021013921916/METRICS

Mohanty AK, Misra M, Drzal LT (eds) (2005) Natural fibers, biopolymers, and biocomposites. CRC Press. https://doi.org/10.1201/9780203508206

Mohanty AK, Vivekanandhan S, Pin JM, Misra M (2018) Composites from renewable and sustainable resources: challenges and innovations. Science 362:536–542. https://doi.org/10.1126/SCIENCE.AAT9072

Monroe K, Kirk T, Hull V, Biswas E, Murawski A, Quirino RL (2020) Vegetable oil-based polymeric materials: synthesis, properties, and applications. In: Encyclopedia of renewable and sustainable materials: volume 1–5, 295–302. https://doi.org/10.1016/B978-0-12-803581-8.11521-8

Moshkbid E, Cree DE, Bradford L, Zhang W (2024) Biodegradable alternatives to plastic in medical equipment: current state, challenges, and the future. J Compos Sci 8:342. https://doi.org/10.3390/JCS8090342

Moshood TD, Nawanir G, Mahmud F, Mohamad F, Ahmad MH, AbdulGhani A (2022a) Sustainability of biodegradable plastics: new problem or solution to solve the global plastic pollution? Curr Res Green Sustain Chem 5:100273. https://doi.org/10.1016/J.CRGSC.2022.100273

Moshood TD, Nawanir G, Mahmud F, Mohamad F, Ahmad MH, AbdulGhani A (2022b) Biodegradable plastic applications towards sustainability: recent innovations in green products. Clean Eng Technol 6:100404. https://doi.org/10.1016/J.CLET.2022.100404

Nair LS, Laurencin CT (2007) Biodegradable polymers as biomaterials. Prog Polym Sci (Oxford) 32:762–798. https://doi.org/10.1016/J.PROGPOLYMSCI.2007.05.017

Onyeaka H, Obileke K, Makaka G, Nwokolo N (2022) Current research and applications of starch-based biodegradable films for food packaging. Polymers 14:1126. https://doi.org/10.3390/POLYM14061126

Pakizeh M, Moradi A, Ghassemi T (2021) Chemical extraction and modification of chitin and chitosan from shrimp shells. Eur Polym J 159:110709. https://doi.org/10.1016/J.EURPOLYMJ.2021.110709

Park J, Lakes RS (2007) Biomaterials: an introduction, 3rd edn, pp 1–561. https://doi.org/10.1007/978-0-387-37880-0/COVER

Patel K, Singh SK (2023) Environmental sustainability analysis of biofuels: a critical review of LCA studies. Clean Techn Environ Policy 25:2489–2510. https://doi.org/10.1007/S10098-023-02596-Y

Pesode P, Barve S, Wankhede SV, Ahmad A (2023) Sustainable materials and Technologies for Biomedical Applications. Adv Mater Sci Eng 2023:6682892. https://doi.org/10.1155/2023/6682892

Phiri R, Mavinkere Rangappa S, Siengchin S, Oladijo OP, Dhakal HN (2023) Development of sustainable biopolymer-based composites for lightweight applications from agricultural waste biomass: a review. Adv Ind Eng Polym Res 6:436–450. https://doi.org/10.1016/J.AIEPR.2023.04.004

Pipitone E, Caltabellotta S, Occhipinti L (2021) A life cycle environmental impact comparison between traditional, hybrid, and electric vehicles in the European context. Sustain For 13:10992. https://doi.org/10.3390/SU131910992

Pires PC, Mascarenhas-Melo F, Pedrosa K, Lopes D, Lopes J, Macário-Soares A et al (2023) Polymer-based biomaterials for pharmaceutical and biomedical applications: a focus on topical drug administration. Eur Polym J 187:111868. https://doi.org/10.1016/J.EURPOLYMJ.2023.111868

Priya AK, Alagumalai A, Balaji D, Song H (2023) Bio-based agricultural products: a sustainable alternative to agrochemicals for promoting a circular economy. RSC Sustain 1:746–762. https://doi.org/10.1039/D3SU00075C

Rathod S, Preetam S, Pandey C, Bera SP (2024) Exploring synthesis and applications of green nanoparticles and the role of nanotechnology in wastewater treatment. Biotechnol Rep 41:e00830. https://doi.org/10.1016/J.BTRE.2024.E00830

Ratner BD et al. Biomaterials science: an introduction to materials in medicine – Google Books. Available at: https://books.google.co.in/books/about/Biomaterials_Science.html?id=8hBq-dLLaxwC. Accessed 11 Dec 2024

Ritzen L, Sprecher B, Bakker C, Balkenende R (2023) Bio-based plastics in a circular economy: a review of recovery pathways and implications for product design. Resour Conserv Recycl 199:107268. https://doi.org/10.1016/J.RESCONREC.2023.107268

Rocha CV, Gonçalves V, da Silva MC, Bañobre-López M, Gallo J (2022) PLGA-based composites for various biomedical applications. Int J Mol Sci 23:2034. https://doi.org/10.3390/IJMS23042034

Rosenboom JG, Langer R, Traverso G (2022) Bioplastics for a circular economy. Nat Rev Mater 7:117–137. https://doi.org/10.1038/s41578-021-00407-8

Sabu Mathew S, Jaiswal AK, Jaiswal S (2024) Carrageenan-based sustainable biomaterials for intelligent food packaging: a review. Carbohydr Polym 342:122267. https://doi.org/10.1016/J.CARBPOL.2024.122267

Samir A, Ashour FH, Hakim AAA, Bassyouni M (2022) Recent advances in biodegradable polymers for sustainable applications. NPJ Mater Degrad 6:1–28. https://doi.org/10.1038/s41529-022-00277-7

Saratale RG, Karuppusamy I, Saratale GD, Pugazhendhi A, Kumar G, Park Y et al (2018) A comprehensive review on green nanomaterials using biological systems: recent perception and their future applications. Colloids Surf B Biointerfaces 170:20–35. https://doi.org/10.1016/J.COLSURFB.2018.05.045

Satchanska G, Davidova S, Petrov PD (2024) Natural and synthetic polymers for biomedical and environmental applications. Polymers 16:1159. https://doi.org/10.3390/POLYM16081159

Seddiqi H, Oliaei E, Honarkar H, Jin J, Geonzon LC, Bacabac RG et al (2021) Cellulose and its derivatives: towards biomedical applications. Cellulose 28:1893–1931. https://doi.org/10.1007/S10570-020-03674-W

Shabbirahmed AM, Sekar R, Gomez LA, Sekhar MR, Hiruthyaswamy SP, Basavegowda N et al (2023) Recent developments of silk-based scaffolds for tissue engineering and regenerative medicine applications: a special focus on the advancement of 3D printing. Biomimetics 8:16. https://doi.org/10.3390/BIOMIMETICS8010016

Shah AA, Hasan F, Hameed A, Ahmed S (2008) Biological degradation of plastics: a comprehensive review. Biotechnol Adv 26:246–265. https://doi.org/10.1016/J.BIOTECHADV.2007.12.005

Sharmila S, Saranya A, Arulprakasajothi M, Saranya R, Srimanickam B, Abel SK et al (2024) Expediting the bioactivity of zinc sulfide nanoparticles with copper oxide as a nanocomposite. BMC Chem 18:1–14. https://doi.org/10.1186/S13065-024-01320-1/FIGURES/5

Sheikh Z, Najeeb S, Khurshid Z, Verma V, Rashid H, Glogauer M (2015) Biodegradable materials for bone repair and tissue engineering applications. Materials 8:5744–5794. https://doi.org/10.3390/MA8095273

Singh N, Ogunseitan OA, Wong MH, Tang Y (2022) Sustainable materials alternative to petrochemical plastics pollution: a review analysis. Sustain Horiz 2:100016. https://doi.org/10.1016/J.HORIZ.2022.100016

Song JH, Murphy RJ, Narayan R, Davies GBH (2009) Biodegradable and compostable alternatives to conventional plastics. Philos Trans R Soc B Biol Sci 364:2127–2139. https://doi.org/10.1098/RSTB.2008.0289

Susheel K (2011) Handbook of biopolymers and their applications. Wiley

Synani K, Abeliotis K, Velonia K, Maragkaki A, Manios T, Lasaridi K (2024) Environmental impact and sustainability of bioplastic production from food waste. Sustain For 16:5529. https://doi.org/10.3390/SU16135529

Teixeira MO, Marinho E, Silva C, Antunes JC, Felgueiras HP (2023) Pullulan hydrogels as drug release platforms in biomedicine. J Drug Deliv Sci Technol 89:105066. https://doi.org/10.1016/J.JDDST.2023.105066

Thomas AP, Kasa VP, Dubey BK, Sen R, Sarmah AK (2023) Synthesis and commercialization of bioplastics: organic waste as a sustainable feedstock. Sci Total Environ 904:167243. https://doi.org/10.1016/J.SCITOTENV.2023.167243

Thompson RC, Moore CJ, Saal FSV, Swan SH (2009) Plastics, the environment and human health: current consensus and future trends. Philos Trans R Soc B Biol Sci 364:2153–2166. https://doi.org/10.1098/RSTB.2009.0053

Tokiwa Y, Calabia BP, Ugwu CU, Aiba S (2009) Biodegradability of plastics. Int J Mol Sci 10:3722–3742. https://doi.org/10.3390/IJMS10093722

Trinh XT, Van LN, Van Anh LT, Nga PT, Giang NN, Chien PN et al (2022) A comprehensive review of natural compounds for wound healing: targeting bioactivity perspective. Int J Mol Sci 23:9573. https://doi.org/10.3390/IJMS23179573/S1

Trucillo P (2024) Biomaterials for drug delivery and human applications. Materials 17:456. https://doi.org/10.3390/MA17020456

Vepari C, Kaplan DL (2007) Silk as a biomaterial. Prog Polym Sci 32:991–1007. https://doi.org/10.1016/J.PROGPOLYMSCI.2007.05.013

Verma SK, Prasad A, Sonika KV (2024) State of art review on sustainable biodegradable polymers with a market overview for sustainability packaging. Mater Today Sustain 26:100776. https://doi.org/10.1016/J.MTSUST.2024.100776

Virumbrales C, Hernández-Ruiz R, Trigo-López M, Vallejos S, García JM (2024) Sensory polymers: trends, challenges, and prospects ahead. Sensors 24:3852. https://doi.org/10.3390/S24123852

Vrchota J, Pech M, Rolínek L, Bednář J (2020) Sustainability outcomes of green processes in relation to industry 4.0 in manufacturing: systematic review. Sustainability 12:5968. https://doi.org/10.3390/SU12155968

Wang F, Cai X, Shen Y, Meng L (2022) Cell–scaffold interactions in tissue engineering for oral and craniofacial reconstruction. Bioact Mater 23:16–44. https://doi.org/10.1016/J.BIOACTMAT.2022.10.029

Waseem M, Khan MU, Majeed Y, Ntsefong GN, Kirichenko I, Klopova A et al (2023) Seaweed-based films for sustainable food packaging: properties, incorporation of essential oils, applications, and future directions. Potr S J Food Sci 17:899–917. https://doi.org/10.5219/1908

Watkins R, Wu L, Zhang C, Davis RM, Xu B (2015) Natural product-based nanomedicine: recent advances and issues. Int J Nanomedicine 10:6055–6074. https://doi.org/10.2147/IJN.S92162

Wei GK (2011) Green nanotechnology of trends in future energy. Recent Pat Nanotechnol 5:76–88. https://doi.org/10.2174/187221011795909198

Wu RX, Xu XY, Wang J, He XT, Sun HH, Chen FM (2018) Biomaterials for endogenous regenerative medicine: coaxing stem cell homing and beyond. Appl Mater Today 11:144–165. https://doi.org/10.1016/J.APMT.2018.02.004

Xu M, Yang M, Sun H, Gao M, Wang Q, Wu C (2022) Bioconversion of biowaste into renewable energy and resources: a sustainable strategy. Environ Res 214:113929. https://doi.org/10.1016/J.ENVRES.2022.113929

Xu R, Fang Y, Zhang Z, Cao Y, Yan Y, Gan L et al (2023) Recent advances in biodegradable and biocompatible synthetic polymers used in skin wound healing. Materials 16:5459. https://doi.org/10.3390/MA16155459

Yi L-J, Li J-F, Ma M-G, Zhu Y-J (2020) Nanostructured calcium-based biomaterials and their application in drug delivery. Curr Med Chem 27:5189–5212. https://doi.org/10.2174/0929867326666190222193357

Yin L, Yuvienco C, Montclare JK (2017) Protein-based therapeutic delivery agents: contemporary developments and challenges. Biomaterials 134:91–116. https://doi.org/10.1016/J.BIOMATERIALS.2017.04.036

Yu P, Zeng L (2024) The impact of consumer environmental preferences on the green technological innovation of Chinese listed companies. Sustain For 16:2951. https://doi.org/10.3390/SU16072951

Zhang H, Lin X, Cao X, Wang Y, Wang J, Zhao Y (2024) Developing natural polymers for skin wound healing. Bioact Mater 33:355–376. https://doi.org/10.1016/J.BIOACTMAT.2023.11.012

Zia KM, Akram N, Tabasum S, Noreen A, Akbar MU (2021) Processing of bio-based polymers. In: Processing technology for bio-based polymers. Elsevier, pp 151–189. https://doi.org/10.1016/B978-0-323-85772-7.00003-3

Zong J, He Q, Liu Y, Qiu M, Wu J, Hu B (2022) Advances in the development of biodegradable coronary stents: a translational perspective. Mater Today Bio 16:100368. https://doi.org/10.1016/J.MTBIO.2022.100368

Biodegradable Polymers: Advances in Sustainable Biomaterials

2

Satish Mandlik, Deepa Mandlik, and Amarjitsing Rajput

Abstract

Biodegradable polymers play a significant role in various sectors as a novel biomaterial in sustainable development. These biodegradable polymers have gained industry attention due to their characteristics needed in the biomedical sector. They are also helpful in the pharmaceutical, food and agriculture sectors because of their qualities that extend product shelf life and excellent mechanical strength. Natural polymers like starch, cellulose, chitosan, collagen and gelatin are excellent biodegradable polymers. Poly (lactic acid), poly (glycolic acid), polycaprolactone, and poly (butyl succinate) are synthetic polymers that also play important roles as biodegradable polymers. Developing eco-friendly, biodegradable polymers are feasible by bonding the highly efficient functional nanoparticles with these materials. In addition, novel synthesis and processing approaches can be used to get the required barrier level and mechanical characteristics comparable to the strength and deftness of conventional synthetic polymers. This chapter mainly focuses on biodegradable natural and synthetic polymers as novel biomaterials, their role and their features in the pharmaceutical, food, and agriculture sectors. Finally, it covers future trends of biodegradable materials in managing different diseases.

Keywords

Biodegradable polymers · Pharmaceutical · Safety · Challenges

S. Mandlik · A. Rajput (✉)
Department of Pharmaceutics, Bharti Vidyapeeth Deemed to be University, Poona College of Pharmacy, Erandwane, Pune, Maharashtra, India
e-mail: amarjit.rajput@bharatividyapeeth.edu

D. Mandlik
Department of Pharmacology, Bharti Vidyapeeth Deemed to be University, Poona College of Pharmacy, Erandwane, Pune, Maharashtra, India

Md. Faiyazuddin, M. Suryawanshi (eds.), *Design and Processing of Green Materials*, Biomaterials, Bioengineering and Sustainability 4,
https://doi.org/10.1007/978-3-031-91790-5_2

Abbreviations

AFM	Atomic force microscopy
APCs	antigen-presenting cells
ASO	Antisense oligonucleotide
BSA	Bovine serum albumin
CAGR	Compound Annual Growth Rate
CD	Circular dichroism
cp	Centipoise
DDA	Degrees of deacetylation
DES	Drug-eluting stents
DEX-MA	Dextran methacrylate
DMPG	Dimyristoyl glycerophosphoryl glycerol-induced
ECM	Extracellular matrix
EMI	Electromagnetic interference
GPC	Gel permeation chromatography
HA	Hyaluronic acid
hADSCs	Human adipose-derived stem cells
HR TEM	High-resolution transmission electron microscopy
MW	Molecular weights
OVA-PLGA-NP	Ovalbumin-loaded poly (lactic-co-glycolic acid) nanoparticles
PBAT	polybutylene adipate-co-terephthalate
PBMCs	Peripheral blood monocytes
PCL	Polycaprolactone
PCL	Polycaprolactone
PEG	Polyethylene glycol
PEG	Polyethylene glycol
PEG-DMA	Polyethylene glycol dimethacrylate
PEO	Poly (ethylene oxide)
PET	Poly (ethylene terephthalate
PGA	Polyglycolic acid
PGA	Poly(glycolic acid)
PHA	Polyhydroxyalkanoates
PHB	Polyhydroxybutyrate
PHBV	Poly(3-hydroxybutyrate-co-3-hydroxy valerate)
PHEMA	Polyhydroxy ethyl methacrylate
PHH	polyhydroxyalkanoate
PHV	polyhydroxy valerate
PLA	Poly (lactic acid)
PLA	Polylactic acid
PLGA	Poly (lactic-co-glycolic) acid
PLGA	Poly(lactic-co-glycolic Acid)
PLLA	Poly(l-lactide)
PVA	Poly-vinyl alcohol
ROS	Reactive oxygen species

SEM	Scanning electron microscopy
SF	Silk fibroin
TA	Tranexamic acid
TGA	Thermogravimetric analysis
TGF-β1	Transforming Growth Factor-beta 1
TPP	Tripolyphosphate
TPS	Thermoplastic Starch
TPU	Thermoplastic polyurethane
WAXD	Wide-angle X-ray diffraction
X-ray	X-ray photoelectron spectroscopy

2.1 Biodegradable Polymers

2.1.1 Introduction

Materials that can be broken down into smaller molecules by biological processes, like fungi, bacteria, or other microorganisms, are known as biodegradable polymers. Usually, these polymers are derived from synthetic components intended to be readily biodegradable or from natural materials such as proteins, cellulose, or starch (Das and Chaudhary 2021). Biodegradable polymers are made to decompose more quickly than conventional plastics, which can linger in the environment for hundreds of years. Depending on the substance and environmental factors, this process might take several months to several years. They decompose into smaller molecules that microbes can consume and metabolize, eventually returning to the natural environment as carbon dioxide, water, or other substances (Balachandar et al. 2019). There are many possible uses for biodegradable polymers, such as packaging, farming, and medicinal equipment. It's crucial to remember that not all biodegradable polymers degrade well and that some could need particular circumstances to do so.

Furthermore, it is imperative to guarantee the appropriate disposal of biodegradable materials to enable their intended biodegradation (Yoganandam et al. 2020). Non-biodegradable plastics can be replaced with biopolymers to lessen their environmental impact and reliance on fossil fuels. Biopolymers, as substitute materials for biopackaging, enable the packing materials to decompose or decompose fully. The hydrolytic or enzymatic breaking of polymer bonds occurs during the biodegradation of biopolymers. It is important to remember that other processes, including hydrolysis, oxidation, and photodegradation, could potentially affect the polymer chains and structure before or during biodegradation (Zhong et al. 2020).

Through the application of nanotechnology, the science of materials, polymer chemistry, and the engineering of tissues, researchers worldwide are actively exploring new ways to meet the increasing need for organ replacement and healing (Khan et al. 2022b; Singla et al. 2019). To get over these restrictions, tissue regeneration-engineered polymeric biomaterials need to be biodegradable and biocompatible (Ghosh et al. 2022). Biomaterials impersonate the properties of diverse organs and tissues through several types of hybrid, composite, and nanomaterials (Khan et al. 2022b; Liu et al. 2023). They integrate cells into their structure, gradually breaking down to leave only normal tissue in their wake. The ability of polymeric biomaterials to retain an elevated concentration of water, porous structure, and adjust to different conditions for sol-gel has led to their recent surge in popularity (Khan et al. 2022a; Shanmugunantha et al. 2022). Polymeric biomaterials' structural properties facilitate their application as tissue scaffolds within the body by encouraging the inflow of pathways for cell metabolism and removing cell waste through their pores (Al-Arjan et al. 2022). In the previous several decades, polymeric biomaterials have garnered increased attention due to their numerous uses in tissue engineering (Khan et al. 2022a). According to Khan et al. (2020), polymeric biomaterials are either crosslinked or treated materials and may have structures that are soluble in water and covalently bound. Elongated polymeric chains, strong hydrogen bonds or van der Waals contacts, or crystallites that combine more than two macromolecular chains give rise to various physicochemical interactions (Xu et al. 2022). Both manufactured and natural polymers can be used to create polymeric biomaterials. They can also be divided into several groups based on various factors, such as the process of manufacture and the physical-chemical, biological, mechanical, and structural characteristics. Because of their potential responsiveness and characteristics—such as pH level, temperature, ion nature, etc.—these could be utilized in tissue engineering (Nazir et al. 2021; Shah et al. 2016). Amorphous, hydrocolloid, hydrogen-bonded, and semicrystalline structural types are all possible. Applications for polymeric biomaterials in tissue engineering include neural, vascular, cardiac, cutaneous, muscular, oral, and bone (Khan et al. 2022b; Zamri et al. 2021). The scarcity of donors and additional health restrictions on transplantation has increased the demand for organs, so researchers are searching for alternate ways to meet this demand.

To overcome these constraints and other health issues, highly biocompatible but fragile biomaterials for tissue regeneration have been developed (Khan et al. 2021a). Thus, biocompatible and biodegradable biomaterials that can act as a scaffold for treating and repairing many bodily problems are needed (Khan et al. 2021b). Thus, to generate advanced biomaterials such as hydrogels (He et al. 2023), and nanofibers (Samadian et al. 2021). microstructures (Singh and Pramanik 2023), nanostructures (Kumar et al. 2022), and advanced technologies must be used. The most perfect polymeric biomaterials must meet minimum vascularization, host integration, and cell proliferation requirements. The designed polymeric biomaterials ought to break down spontaneously during or after healing. To solve medical issues, integrating these polymeric biomaterials still offers a wide range of applications in biomedical engineering (Khan et al. 2021c). The human body regularly undergoes several biological and pathological changes; hence, polymeric biomaterials' characteristics must mesh well with biological systems (Shafiei et al. 2021). Therefore, finding suitable hydrogel systems is still necessary to promote efficient tissue regeneration for various biomedical applications (Khan et al. 2021c; Kumar et al. 2022).

Polymeric biomaterials have advanced quickly recently in a variety of tissue engineering applications. While many review articles discuss biomaterials based on proteins, carbohydrates, synthetic materials, or ceramics, a thorough review article focused on polymeric biomaterials for tissue engineering is still lacking. The most current advancements and breakthroughs in polymeric biomaterials for tissue engineering are presented in the literature review. The development and evaluation of materials and biomaterials' basic characteristics in tissue regeneration are the main topics of this review paper. A summary of the various biomaterials' manufacturing processes is provided. There has also been a discussion of biomaterials' difficulties and potential applications in tissue engineering. Developing biofabrication techniques and biological settings presents substantial hurdles for repairing and regenerating injured tissue while preserving its characteristics and biological milieu. It also offered potential difficulties and future viewpoints, and by providing important ideas and insight, it would support and direct the reader for further investigation.

2.1.2 Fundamental Characteristics

2.1.2.1 Biodegradability Property

The final goal of tissue engineering is for biomaterials implanted to be replaced by the body's cells. Materials not intended to be longstanding implants are known as biomaterials (Thanigaivel et al. 2022). Therefore, the biomaterials must be biodegradable for cells to produce their extracellular framework. It can be removed from the body without causing damage to the host tissue. An inflammatory response and the carefully regulated infusion of cells, including macrophages, are combined in this procedure (Bao et al. 2022). Medical applications are increasingly using tissue engineering techniques, and during tissue regeneration, deterioration must be allowed to occur. Research on tissue engineering is beginning to emphasize immunology (Bavya Devi et al. 2022).

2.1.2.2 Biocompatibility Property

According to Liu et al. (2020a), tissue engineering requires biomaterials that are both biocompatible and promote cellular adhesion, biofunctionality, proliferation, differentiation, and migration. Tissue regeneration is greatly aided by the interaction of the engineered biomaterials with the cells that reside on the surface of the implanted biomaterials. According to Jin et al. (2023) bioengineered biomaterials ought to produce a bio-friendly environment to maintain injured tissue. To prevent healing or tissue regeneration, the host immunological response must be either nonexistent or very mild, and it shouldn't trigger a significant inflammatory response (Batool et al. 2021).

2.1.2.3 Optical Property

Transparency and fluorescence are two examples of polymeric nanocomposites' optical characteristics. A UV-vis spectrophotometer is commonly used to assess transparency, expressed as the proportion of light transmission via the material in the 200800 nm range. In 2019, Saikia et al. showed that hyperbranched bio-based epoxy thermosetting nanocomposites and cellulose nanofiber exhibit optical transparency (Saikia et al. 2019). The polymer matrix can exhibit notable luminescence when appropriate nanomaterials like Cadmium sulfide (CD), sodium zincate (Na_2ZnO_2), oxides of CD, and TiO_2 are included. By adding Na_2ZnO_2 to polyvinyl alcohol (PVA) using an aqueous solution casting technique, Shivanna et al. developed polymer nanocomposites exhibiting orange-red fluorescence (Shivanna et al. 2019). The created polymer nanocomposites demonstrated strong UVA screening capabilities, excellent thermal stability, and self-cleaning qualities in addition to fluorescence.

2.1.2.4 Thermal Property

The thermogravimetric analyzer (TGA) assesses polymer nanocomposites' thermal durability. Analyzing and identifying the degradation behaviors, initiation, outcome, and peak points of the degradation conditions in controlled atmospheric environments were examined and deemed significant thermal characteristics of the nanocomposites. Factors such as the composition, crosslinking density, dispersion, and interaction of the nanomaterials with the polymer matrix contribute to the thermal resilience of the nanocomposite. Saikia and Karak documented the incorporation of CNFs/PAni/CD nanohybrid to develop bio-based epoxy thermosetting nanocomposites, which significantly enhanced thermal stability (Saikia et al. 2019). Distinctive Scanning Besides TGA thermographs, calorimetry is a well-known technique that offers significant information on the thermal properties of polymers and their nanocomposites. When CD—which has high aspect ratio nanofibers and a variety of surface functionalities—is added, the maximum temperature of the pure epoxy thermoset rises to 343 °C, above the peak breakdown temperature of 313 °C. These elements strengthen the mechanical and chemical characteristics of the polymer matrix by forming physico-chemical connections with it. Subsequent studies yielded additional information about the glass transition temperature (Tg), enthalpy changes with temperature, crystalline polymers' melting peak temperature (Tm), and associated topics. Remis et al.'s study revealed significant changes in enthalpy because of the nanodiamonds' integration into the PVA matrix when added to PVA nanocomposites (Remiš et al. 2021). When the nanodiamond content is increased to 1.9 weight percent, the T_g of the PVA matrix rises to 43.8 °C from 42 °C. This is brought on by PVA's partially crosslinked structure with nanodiamonds, as shown in Fig. 2.1.

2.1.2.5 Mechanical Property

The biomaterial should be engineered to assist the damaged tissue site and provide mechanical support to restore the mechanical qualities. Because it can be challenging to apply mechanically stable biomaterials in bone and cardiac tissue fields, they must be employed in all tissues (Liu et al. 2023). It is still challenging to develop biomaterials for tissue engineering of bone that have the motorized properties and porosity necessary to maintain the host-damaged tissue (Rasouli et al. 2023). The implanted biomaterials should have enough mechanical strength, especially for bone tissues, to enable proper tissue function from implantation to transformation. Another challenge is that the rate at which bone tissue heals and regenerates changes with age. Compared to elderly folks, younger people heal more quickly (Li et al.

Fig. 2.1 Representation of fundamental characteristics of biodegradable polymers

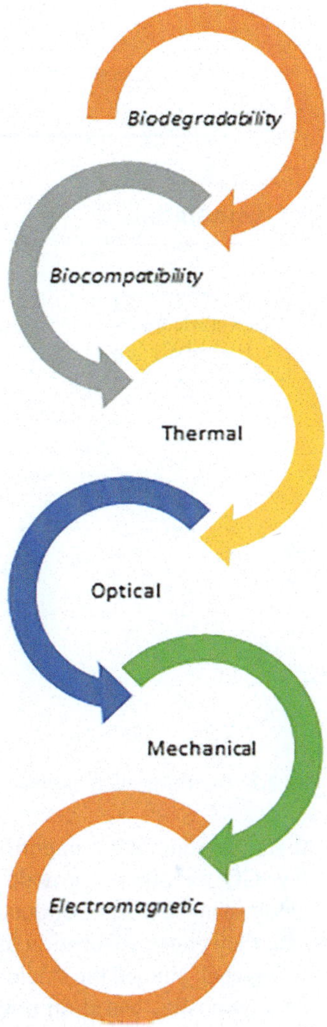

2021b). Consequently, the focus has shifted to creating scaffolds with mechanical characteristics akin to cartilage and bone. Though they lack porosity and pore size, several materials with desired mechanical properties have been created. A porous design that would facilitate cell migration and vascularization and balance the scaffold's mechanical properties would be desirable (Rosa et al. 2023).

2.1.2.6 Electromagnetic Property

Over the past three decades, the electronic industries have grown rapidly, which has increased our dependency on electrical equipment. Though higher frequencies are required for these electrical devices to function, electromagnetic interference (EMI) signals prevent them from doing so. Therefore, it has been reported that a variety of polymer nanocomposites comprising metal oxides like WO3, CdO, NiO, TiO_2, and

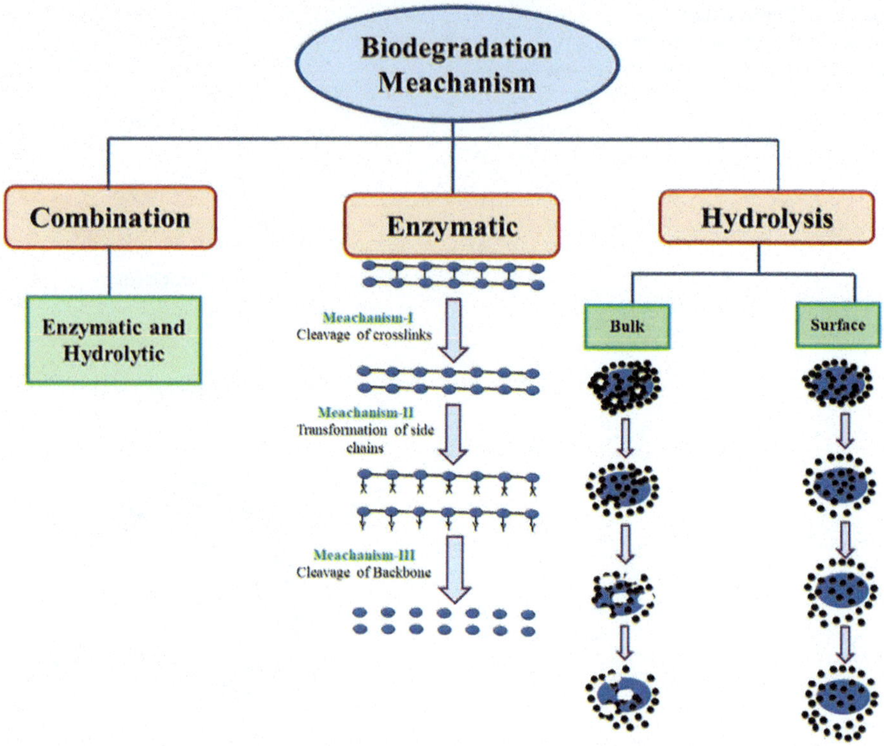

Fig. 2.2 Biodegradation mechanism of polymers

others, or conductive nanofillers like graphene nanoplatelets, carbon black, nanoclay, GO, or others, provide efficient EMI shielding (Harish Kumar et al. 2021; Reddy et al. 2019). Conductive fillers are necessary for effective EMI shielding since they cause absorption and reflection into the polymer matrix.

Das et al. studied the shielding characteristics of thermoplastic polyurethane (TPU) nanocomposites based on poly (lactic acid) (PLA)/ester linkage containing conductive carbon black Vulcan XC-72 (VCB) at varying filler loadings. The results show that, compared to the pristine matrix, the PLA/TPU matrix's EMI shielding capacity in the X band is enhanced by the addition of conductive carbon black filler (Das and Chaudhary 2021). The mechanism of biodegradable polymers is shown in Fig. 2.2.

2.2 Classification of Biodegradable Polymers

2.2.1 Introduction

Biodegradable polymers are materials that can decompose naturally in the environment through the action of microorganisms. Unlike traditional plastics, they break down into harmless by-products such as carbon dioxide, water, and biomass,

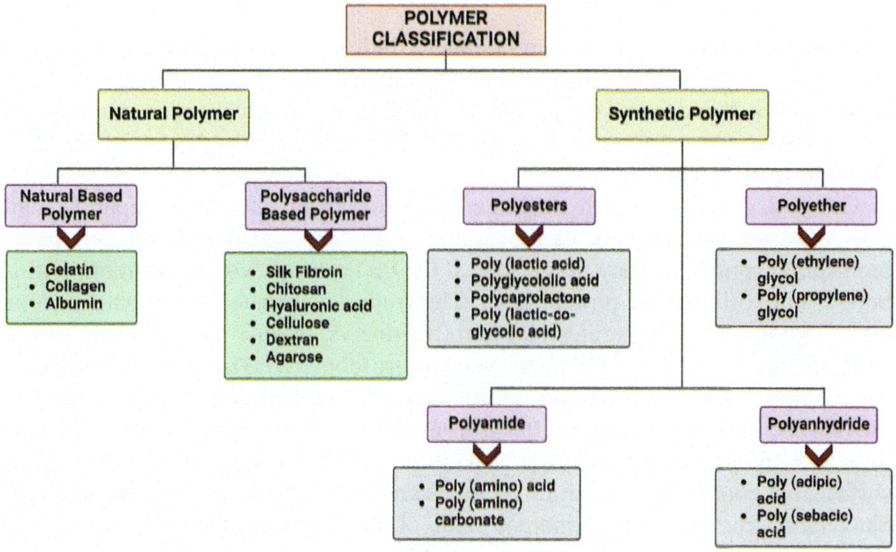

Fig. 2.3 Classification of biodegradable polymers

reducing environmental pollution. This part explores various classifications of biodegradable polymers based on their origin and chemical structure, highlighting their importance in sustainable development and diverse applications across industries (Anjum et al. 2023; Sahani and Sharma 2021). The classification of biodegradable polymers is shown in Fig. 2.3.

2.2.2 Classification of Biodegradable Polymers

This section categorizes biodegradable polymers into natural and synthetic types, each with unique characteristics and applications.

2.2.2.1 Natural Polymers

Natural polymers are biopolymers present in living organisms, such as proteins (like collagen), carbohydrates (such as cellulose), nucleic acids (like DNA), and lipids. They play essential roles in biological processes and are necessary for various industrial applications.

Natural Based Polymers

Natural biodegradable polymers come from renewable biological sources, including plants, animals, and microorganisms. These polymers offer distinct advantages such as biocompatibility, low toxicity, and sustainable sourcing, making them suitable for various biomedical, pharmaceutical, and environmental applications (Mukherjee et al. 2023; Prajapati et al. 2019).

Gelatin

Gelatin, a protein derivative from collagen, finds extensive use in pharmaceutical capsules, wound dressings, and food packaging due to its biocompatibility and biodegradability. It undergoes enzymatic degradation in biological environments, facilitating tissue regeneration and controlled drug release (Mukherjee et al. 2023; Prajapati et al. 2019).

Gio et al. coated poly (ethylene terephthalate) (PET) with a biocompatible, cell-interactive gelatin layer for cardiovascular applications using an aqueous-based, bioinspired approach. They first activated the PET surface with a dopamine coating before immobilizing the protein layer. The study investigated the effects of dopamine deposition and substrate condition (pristine vs. plasma-treated) using various methods, including X-ray photoelectron spectroscopy (XPS), atomic force microscopy (AFM), scanning electron microscopy (SEM), and contact angle measurements. The surface roughness after dopamine coating varied from 96 to 210 nm, depending on the parameters used. Gelatin immobilization generally smoothed the surface, but combining deposition methods and plasma treatment resulted in different gelatin structures. Furthermore, a study of physically changed PET with gelatin (physisorption) and chemically modified PET (dopamine-treated) revealed a different association between the final protein arrangement and blood platelet behavior (Giol et al. 2015).

Collagen

Collagen, the primary structural protein in connective tissues, is a biomaterial in tissue engineering, wound healing, and cosmetic surgery. Its biodegradability and ability to mimic natural extracellular matrices make it valuable for promoting cell adhesion and tissue regeneration (Bhardwaj and Jangde 2023; Chandra and Pandey 2022; Lu et al. 2021).

Banche-Niclot et al. created biodegradable nanocarriers from poly (lactic-co-glycolic) acid (PLGA) intended to deliver Transforming Growth Factor-beta 1 (TGF-β1), a key molecule involved in regulating bone cell activity. They successfully synthesized spherical PLGA nanoparticles, approximately 250 nm in size, loaded with Transforming Growth Factor-beta 1 (TGF-β1), achieving an encapsulation efficiency of around 64%. The hybrid formulation was thoroughly characterized for rheological properties, ensuring suitability for 3D printing. The resulting 10 mm × 10 mm × 1 mm collagen-based scaffolds were validated by immunohistochemical staining to cover approximately 3.4% of the scaffold area, indicating that they successfully reproduced the TGF-β1 environment of the human bone extracellular matrix. Furthermore, the collagen matrix reduced the initial burst release from 38% for the PLGA_TGF-β1 alone to 14.5% within the first 24 hours, demonstrating that incorporating nanocarriers into collagen enables a more sustained release profile (Banche-Niclot et al. 2022).

Albumin

Albumin, a water-soluble protein found abundantly in blood plasma, is employed in drug delivery systems and tissue scaffolds because of its biocompatibility and capacity to carry diverse molecules. Its biodegradability and lack of immunogenicity further increase its suitability for medical uses (Bhardwaj and Jangde 2023; Chandra and Pandey 2022; Lu et al. 2021).

Using various biophysical methods, Hsieh et al. explored how co-polymers influence the structure and thermal stability of bovine serum albumin (BSA). The results revealed that copolymer PGA-1, with a lower PEGMA/AA ratio than copolymer PGA-1.5, significantly altered BSA's conformation, as shown by fluorescence spectroscopy. The copolymers caused BSA's polypeptide chain to unfold, reflected in changes in the negative ellipticity observed in the circular dichroism spectra. Based on differential scanning calorimetry analysis, the transition temperature was dropped by 14.8 °C and 11.5 °C, respectively, with the addition of PGA-1 and PGA-1.5 at 0.05% (w/v).

These findings underscore how variations in synthetic polymer composition can affect protein behavior, which is vital for applications like enzyme immobilization, protein separations, sensor development, and stimuli-responsive systems (Hsieh et al. 2016).

Polysaccharide-Based Polymers

Polysaccharide-based polymers are derived from carbohydrates and exhibit diverse properties that make them valuable in various applications.

Silk Fibroin

Silk fibroin is a protein-based biopolymer derived from silk fibers produced by silkworms. Renowned for its remarkable strength, biocompatibility, and slow degradation, it is ideal for biomedical uses, including tissue engineering scaffolds, wound healing materials, and controlled drug delivery systems. Its distinctive molecular structure enables it to form robust fibers and films that replicate the properties of natural tissues, enhancing cell adhesion and supporting tissue regeneration (Kundu et al. 2013).

Pudkon et al. investigated silk fibroin (SF)-based hydrogels comprising silk fibroin in different concentrations (1–3%). The hydrogels were fabricated using three methods: natural gelation (SF/self-gelled), sodium tetradecyl sulfate-induced (SF/STS), and dimyristoyl glycerophosphoryl glycerol-induced (SF/DMPG). According to the study, 2% SF (2SF) hydrogels showed appropriate storage modulus, shear-thinning behavior, and structural recovery—all advantageous for extrusion. All 2SF-based hydrogel formulations were 3D-printed into a 4-layer box structure without collapsing. The 2SF/STS and 2SF/DMPG hydrogels showed high stability, with structure recovery rates of 70.40% and 53.70%, respectively, compared to 38.90% for the 2SF/self-gelled hydrogel. The 2SF/STS and 2SF/DMPG hydrogels' rheological characteristics, printability, and structural stability make them intriguing candidates for 3D printing applications (Pudkon et al. 2021).

Chitosan

The natural polymer chitin, found in the exoskeletons of crustaceans like shrimp and crab, is the source of chitosan, a polysaccharide. It is known for its biodegradability, biocompatibility, antimicrobial properties, and excellent film-forming capabilities. Chitosan is helpful in wound dressings, surgical sutures, and drug delivery systems. Its antimicrobial properties help prevent infections and enhance tissue regeneration, which is beneficial for wound healing (Jayakumar et al. 2011).

Jesus et al. investigated the immunotoxic effects of chitosan (Chit)-derived nanoparticles (NPs) with varying degrees of deacetylation (DDA) and molecular weights (MW), comparing them to the chitosan polymer itself. Chitosan NPs were produced through ionic gelation with sodium tripolyphosphate (TPP), resulting in two types: Chit 80% NPs (80% DDA) and Chit 93% NPs (93% DDA). The study found that Chit-NPs generally caused more significant reductions in cell viability compared to the chitosan polymers when tested in vitro with human peripheral blood monocytes (PBMCs) and the RAW 264.7 cell line. In particular, Chit 80% NPs produced more reactive oxygen species (ROS) in RAW 264.7 cells (above 156 μg/mL) and were more cytotoxic to PBMCs. They disrupted the intrinsic coagulation pathway (at 1 mg/mL) compared to Chit 93% NPs. However, only Chit 93% NPs led to platelet aggregation (at 2 mg/mL). Chit NPs and chit polymers reduced LPS-induced NO production in RAW 264.7 cells but did not enhance NO production in these cells at any of the tested concentrations. Furthermore, neither Chit polymers nor Chit NPs produced hemolysis or induced PBMCs to release cytokines such as IL-6 and TNF-α. These results suggest that the chitosan polymer's DDA and the chitosan NPs' size significantly impact their *vitro* immunotoxicity profiles (Jesus et al. 2020).

Hyaluronic Acid

Hyaluronic acid (HA) is a naturally found polysaccharide in connective tissues throughout the body, notably in joints and skin. It is known for its high water-binding capacity, viscoelasticity, and lubricating properties. In medicine and cosmetics, HA is used for its moisturizing effects in skin care products and as a visco-supplement in joint therapies to alleviate pain and improve mobility. Its biocompatibility and ability to retain moisture make it a versatile material in tissue engineering and ophthalmic surgeries (Burdick and Prestwich 2011; Highley et al. 2016).

Zheng et al. successfully incorporated 2-methacryloyloxyethyl phosphorylcholine (MPC), a zwitterionic biomaterial known for its excellent hydration lubrication properties, onto hyaluronic acid (HA) of two different molecular weights to create HAMPC. Using an atomic force microscope for lubrication testing, they found that HAMPC significantly reduced friction coefficients compared to HA under various conditions. Additional in vitro experiments demonstrated the biocompatibility of HAMPC and ability to upregulate genes associated with cartilage anabolism while downregulating genes related to cartilage catabolic proteases and pain. Notably, the high molecular weight HAMPC showed better regulatory effects on these genes than the low molecular weight variant. Overall, the high molecular weight HAMPC developed in this study, which enhances lubrication and exhibits anti-inflammatory properties, shows promise as a potential polymer for treating osteoarthritis (Zheng et al. 2019).

Cellulose

Cellulose is a linear polysaccharide composed of glucose units joined by β-1,4-glycosidic linkages. Renowned for its renewability and biodegradability, cellulose also boasts excellent mechanical strength and barrier properties. It is widely utilized in various industries, including papermaking, textiles (e.g., cotton fibers), and packaging materials. Recent advancements in cellulose nanotechnology have expanded its applications to include biomedical uses like wound dressings, scaffolds for tissue regeneration, and drug delivery carriers (Balachandar et al. 2019; Baranwal et al. 2022).

Afolabi et al. studied a novel polymeric surfactant made from sodium cellulose sulfate. The researchers conducted rheological tests on this novel cellulose ester under various external conditions. Their findings revealed that the amphiphilic polymer exhibited associative behavior at a low concentration of 0.15 g/L. Moreover, the biopolymeric surfactant showed resilience to temperature variations (from 35 to 75 °C) and changes in shear rate (from 100 RPM to 250 RPM), preserving its original rheological properties. Its viscosity also increased within a brine salinity range of 10,000 ppm to 60,000 ppm, rising from 45.75 cp to 49.1 cp. These results suggest that this novel cellulose derivative is effective as a mobility control agent, indicating its potential as a cost-effective and environmentally friendly solution for applications in the oilfield industry (Afolabi et al. 2023).

Dextran

Acrylamide units joined by α-1,6-glycosidic linkages and additional α-1,3 branches form the structure of dextran, a water-soluble polysaccharide. Certain bacteria synthesize it during fermentation, and it is renowned for its biocompatibility and non-toxicity. Dextran's ability to form aqueous solutions with varying viscosities makes it useful in pharmaceutical formulations as plasma volume expanders, stabilizers for protein drugs, and carriers for controlled drug release. Its applications extend to biotechnology, where it is used in cell separation, chromatography, and as a medium for microbial growth (Ghimici and Nichifor 2018; Puranen et al. 2023).

Pacelli et al. aimed to develop a biocompatible polymeric cryogel designed as a porous scaffold to improve the delivery of biomolecules. Through radical polymerization at −15 °C, the new formulation combines dextran methacrylate (DEX-MA) with polyethylene glycol dimethacrylate (PEG-DMA). The researchers assessed the improved cryogel's swelling capacity, porosity, mechanical characteristics, and vitamin B12 release profile and also compared it to hydrogels made at normal temperatures. The cryo-gelation technique resulted in scaffolds with better swelling, enhanced interconnected porosity, and greater mechanical strength compared to traditional hydrogels. The cryogels also demonstrated biocompatibility and effectiveness as carriers for water-soluble biomolecules. For biological applications that need a macroporous scaffold for the targeted delivery of bioactive chemicals, this novel cryogel formulation appears promising overall (Pacelli et al. 2021).

Agarose

D-galactose and 3,6-anhydro-L-galactopyranose combine to form a disaccharide called agarobiose, which is repeated in units to form agarose, a linear polymer obtained from red algae. It forms thermally reversible gels in aqueous solutions, which makes it highly useful in biotechnology and molecular biology. Agarose gels are commonly employed for DNA and protein electrophoresis, allowing molecules to migrate through the gel matrix based on their size. Its inert properties, low melting temperature, and biocompatibility make agarose ideal for different uses, including drug delivery systems and tissue engineering scaffolds (Sambrook and Russell 2001; Yazdi et al. 2020).

Bassand et al. investigated how an agarose gel mimics living tissue affects the drug release kinetics of a PLGA (poly (lactic-co-glycolic acid)) implant. The hot melt extrusion technique was used to fabricate Ibuprofen-loaded implants. In phosphate buffer at pH 7.4, drug release was measured in a transwell plate setup and Eppendorf tubes with and without agarose gel. Changes in the implant's mass and dimensions were monitored using gravimetric and optical methods. Implant erosion and polymer degradation were analyzed through scanning electron microscopy (SEM) and gel permeation chromatography (GPC). pH variations in the fluids, gels, and implants during drug release were tracked with pH indicators. The presence of the agarose gel delayed the onset of the second release phase due to restricted implant swelling. After an initial delay, the degrading PLGA matrix became more hydrophilic and mechanically unstable, allowing water penetration and altering drug release conditions, which led to increased mobility and faster release. The mechanical restriction of implant swelling by the surrounding gel contributed to the slower drug release. These results highlight the importance of PLGA swelling in controlling drug release and suggest that this could inform the development of more realistic *in vitro* release models (Bassand et al. 2022).

2.2.2.2 Synthetic Polymers

A wide range of industrial and biological applications can benefit from the unique features of synthetic polymers, which are man-made macromolecules.

Polyesters

Polyesters are a category of synthetic polymers defined by the presence of ester functional groups along their main chain. They are biodegradable, and their characteristics can be modified based on their composition and molecular weight (Dirauf et al. 2022).

Meable et al. combined molecular dynamics simulations with various experimental techniques to investigate ionic transport in polyester-based solid polymer electrolytes (SPEs). The research specifically focused on the physicochemical properties and morphological behaviors of blend SPEs composed of poly (ethylene oxide) (PEO) and poly(caprolactone) (PCL), including their extreme compositions. The findings demonstrated that partial phase segregation between PEO and PCL and the interaction between Li+ ions and ethylene oxide groups significantly impact ionic transport in these blends. This research aims to generate novel approaches for enhancing the efficiency of SPEs and the batteries that utilize them (Meabe et al. 2020).

Poly (Lactic Acid) (PLA)

PLA (polylactic acid) is biodegradable under composting conditions. It is widely used in medical implants, drug delivery systems, and packaging due to its biocompatibility and mechanical strength (Auras et al. 2004).

Carmagnola et al. binary blends of poly(L-lactic) acid (PLLA, molecular weight 201,790 Da) and poly(lactide-co-glycolide) (PLGA, LA:GA = 50:50 mol:mol, molecular weight 32,030 Da) with various compositions were prepared. These PLLA/PLGA blends' physicochemical characteristics were examined, and the results showed that they had a two-phase structure with clear glass transition temperatures similar to the pure polymers. Tensile mechanical properties were evaluated using the Kerner-Uemura-Takayanagi model, which indicated good compatibility for the PLLA/PLGA 75/25 blend. The PLGA phase in the PLLA/PLGA 75/25 blend degraded rapidly (within 2–8 weeks), leading to porous samples that show potential uses. Subcutaneous implantation in Balb/c mice showed that the PLLA/PLGA 75/25 blend induced a minimal inflammatory response (Carmagnola et al. 2015).

Polyglycolic Acid (PGA)

PGA is another biodegradable polyester derived from glycolic acid. Considerable applications include absorbable sutures, scaffolds for tissue engineering, and controlled drug administration systems due to their outstanding strength and rapid degradation rate (Su et al. 2021).

Magazzini et al. conducted a study on combining poly (glycolic acid) (PGA) with other bioplastics to achieve dual benefits: reducing material costs and discovering new properties while preserving eco-friendly attributes. The study involved blending PGA with two commonly used bioplastics, poly(l-lactide) (PLLA) and polycaprolactone (PCL), using the melt blending technique, which is known for its scalability. Field emission scanning electron microscopy (FE-SEM) measurements revealed PGA domains within the blends, with domain size decreasing as the PGA content was reduced. The polymer matrices' thermal characteristics and degradation behavior were significantly affected by PGA despite the restricted compatibility among the components (Magazzini et al. 2021).

Polycaprolactone (PCL)

Polycaprolactone (PCL) is a compound with a low melting point, facilitating its processing into various forms. It degrades slowly *in vivo*, making it useful in long-term implantable medical devices, wound dressings, and drug delivery matrices where sustained release is desired (Middleton and Tipton 2000).

Hou et al. investigated how polycaprolactone's molecular weight (Mn) (PCL) affects various aspects of 3D printed bone scaffolds, including their degradation kinetics, surface characteristics, microstructure, thermal behavior, mechanical strength, and biological properties. Scanning electron microscopy (SEM) examined morphological features and degradation kinetics through accelerated degradation tests. *In vitro* tests with human adipose-derived stem cells (hADSCs) were conducted to evaluate cell proliferation and differentiation using colorimetric assays. The results showed that molecular weight influenced the surface characteristics, mechanical strength, and biological behavior of PCL scaffolds but had no significant impact on degradation rates. These results offer essential information for creating polymeric scaffolds for bone tissue engineering (Hou et al. 2022).

Poly (Lactic-co-Glycolic Acid) (PLGA)

PLGA (poly (lactic-co-glycolic acid)) merges the biocompatibility of PLA (polylactic acid) with the rapid degradation rate of PGA (poly (glycolic acid). It is widely explored in controlled drug delivery systems, tissue engineering scaffolds, and medical implants due to its adjustable degradation kinetics and capacity to encapsulate a broad range of drugs (Makadia and Siegel 2011).

Struzek et al. aimed to implement a systematic approach using quality by-design principles in developing ovalbumin-loaded poly (lactic-co-glycolic acid) nanoparticles (OVA-PLGA-NP) intended for lung delivery. The different process parameters such as OVA, PLGA, and stabilizer concentrations, along with stirring times and speeds for both the inner and double emulsions, to evaluate their effects on critical properties like z-average size, polydispersity index, loading efficiency, and capacity. A central composite response surface design optimized the OVA-PLGA-NP formulation, resulting in an average size of 700 nm with excellent drug-loading capacities. This phase also clarified the effects of stabilizer and PLGA concentrations, including curvature and interaction effects (Struzek and Scherließ 2023).

Polyether

Polyethers are synthetic polymers characterized by ether groups (-O-) in their main chain. They are known for their flexibility, water solubility, and excellent lubricating properties (Flejszar and Chmielarz 2020).

Poly (Ethylene) Glycol (PEG)

PEG (polyethylene glycol) is a polyether compound of repeating ethylene glycol units. It is water-soluble, non-toxic, and biocompatible, making it widely used in pharmaceuticals, cosmetics, and medical uses in drug delivery systems and tissue engineering scaffolds (Gbian and Omri 2022).

For nanocarriers with low protein affinity, Li et al. showed that the density, molecular weight, and shape of polyethylene glycol (PEG) chains on their surface substantially impact cell interaction. By arranging PEG chains in a dense "brush" conformation rather than a collapsed "mushroom" conformation, the researchers effectively minimized the nonspecific uptake of ovalbumin nanocarriers by dendritic cells. The study also showed that adjusting the PEG conformation had a limited impact on dysopsonin adsorption. The brush conformation of PEG imparts stealth properties to the nanocarriers, reducing their uptake by phagocytic cells. This is crucial for the effective in vivo application of nanomedicine, as it ensures prolonged circulation in the bloodstream and targeted delivery. The findings establish a clear link between the PEG brush conformation and reduced phagocytic uptake, highlighting the importance of polymer conformation and surface chemistry in controlling the cellular interactions of nanocarriers (Li et al. 2021a).

Poly (Propylene) Glycol

Poly (propylene) glycol is another polyether compound like PEG but with propylene glycol units. It is also water-soluble and has applications in pharmaceuticals, personal care products, and as a base material for polyurethanes, lubricants, and surfactants (Song et al. 2018).

Pennetta et al. functionalized single-walled and multi-walled carbon nanotubes using a cycloaddition procedure involving carbon nanotubes and a chemical generated from pyrrole. This method created a drug delivery system where doxorubicin (DOX) molecules are efficiently stacked. The resulting supramolecular nanoconveyors were thoroughly characterized using wide-angle X-ray diffraction (WAXD), thermogravimetric analysis (TGA), high-resolution transmission electron microscopy (HR-TEM), and Fourier transform infrared (FT-IR) spectroscopy. Molecular dynamics simulations were conducted to explore the supramolecular interactions while monitoring DOX's emission and absorption spectra offered further insights. Biological studies showed that two of the synthesized nano-vectors effectively delivered DOX into targeted cell lines, significantly increasing cell mortality at lower doses of DOX. This research underscores the effectiveness of the "pyrrole methodology" for functionalizing carbon nanotubes, paving the way for advanced carbon-based drug delivery systems (Pennetta et al. 2020).

Polyamide

Polyamides are synthetic polymers with amide (-CO-NH-) linkages in their main chain. They are known for their toughness, durability, and high melting points.

Poly (Amino) Acid

Poly(amino) acids are biodegradable polyamides made from natural amino acids. They are biocompatible and commonly utilized in biomedical applications like drug delivery systems, tissue engineering scaffolds, and biodegradable sutures, thanks to their ability to break down into non-toxic by-products (Gupta et al. 2021).

Skwarczynski et al. reported an effective method for delivering vaccines. A peptide antigen was linked with poly-hydrophobic amino acid sequences, which act as self-stimulating components, using solid-phase synthesis to create fully characterized single molecular units. These molecules were autonomously assembled in aqueous environments into distinct nanoparticles and chain-like structures. When administered subcutaneously to mice, these particles successfully elicited antibodies specific to opsonic epitopes without requiring external adjuvants. Mice vaccinated with constructs containing 15 leucine residues effectively reduced bacterial presence in targeted organs without causing the release of soluble inflammatory agents. Therefore, the study has developed a precise and efficient self-stimulating delivery system for peptide antigens (Skwarczynski et al. 2020).

Poly (Amino) Carbonate

Poly(amino) carbonates are copolymers formed from cyclic carbonates and amino acids or their derivatives. They blend the biodegradability of polyesters with the mechanical strength of polycarbonates, making them ideal for controlled drug delivery systems, tissue engineering, and other biomedical applications (Numata et al. 2014).

Polyanhydride

Polyanhydrides are synthetic polymers containing anhydride (-CO-O-CO-) linkages in their backbone. They are biodegradable and known for their ability to degrade into non-toxic by-products.

Poly (Adipic) Acid

Poly(adipic) acid is a polyanhydride derived from adipic acid. It degrades via hydrolysis into adipic acid and other metabolizable by-products, making it well-suited for drug delivery systems where controlled release and biocompatibility are essential (Numata et al. 2014). Plastics that have been bio-upcycled offer an innovative way to handle polymer waste that is too polluted for conventional recycling techniques.

Ackermann et al. focused on enhancing the capability of *Pseudomonas putida* KT2440 to metabolize these dicarboxylates efficiently. To achieve this, the team introduced the dcaAKIJP genes from Acinetobacter baylyi, crucial for the initial uptake and activation of dicarboxylates, into *P. putida*. Integrating these genes into the P. putida genome was critical for their expression. Genomic analysis of the evolved strains highlighted the importance of a paa gene cluster, involved in phenylacetate degradation and similar to adipate metabolism. Further modifications included disrupting the psrA gene, which regulates redundant β-oxidation genes, to enhance growth. The engineered strain utilized adipate and other medium-chain-length dicarboxylates like suberate and sebacate. This strain demonstrated a growth rate of $0.35 \pm 0.01\ h^{-1}$ on adipate and a final biomass yield of 0.27 ± 0.00 gCDW/gadipate. The strain demonstrated its ability to transform polymer waste into biodegradable products by producing polyhydroxyalkanoates from adipate, which accounted for 25% of its cell dry weight when grown in nitrogen-limited circumstances (Ackermann et al. 2021).

Poly (Sebacic) Acid

Poly(sebacic) acid is a polyanhydride synthesized from sebacic acid. It degrades into sebacic acid and other metabolizable by-products, making it a promising material for biomedical applications such as tissue engineering scaffolds, implants, and controlled release systems. Its biodegradability and biocompatibility are vital factors that enhance its suitability for these applications (Gupta et al. 2021; Numata et al. 2014).

Varshosaz et al. porous nanofibers were created by blending polyglycerol sebacate (PGS) with polyhydroxy ethyl methacrylate (PHEMA) and incorporating tranexamic acid (TA) through electrospinning. The research aimed to optimize the nanofibers' structure, diameter, porosity, TA content, release kinetics, and mechanical properties. The study involved assessing the cytotoxicity of the nanofibers using the MTT assay on L929 cells and evaluating their hemostatic efficacy in a rat tail-cut model. The optimal formulation included 35% of the total polymer blend, with 20% PGS and 10% TA. The resulting nanofibers exhibited 64% porosity, absorbed 8.59% water, and had a 1.47% weight loss over 28 days without showing cytotoxicity to L929 cells. Thus, the TA-loaded nanofibers effectively managed bleeding and hemorrhagic conditions, improving bleeding time and volume (Varshosaz et al. 2021).

2.3 Preparation Methods of Biodegradable Polymers

Different techniques are utilized to synthesize biodegradable polymers, depending on the type of polymer, its properties and its intended applications. The following are the essential methods of preparation of biodegradable polymers.

2.3.1 Conventional Methods

2.3.1.1 Modified Natural Polymers

Natural biodegradable polymers are materials derived from renewable resources that can readily decompose naturally in the environment through the action of microorganisms or in the body by enzymatic degradation and or metabolism (Ghanbarzadeh and Almasi 2013; Ghuttora 2016). The naturally occurring biodegradable polymers viz polysaccharide (starch, cellulose, chitosan and chitin) have some plasticity, purity and reproducibility issues. Therefore, such biodegradable polymers are often modified chemically. These chemical modifications can improve their mechanical strength, thermal stability, solubility, biodegradability, or biocompatibility, making them more suitable for various industrial and biomedical uses (Okada 2002; Pathak 2017; Zeng et al. 2016).

Chemical and physical modification methods modify natural biodegradable polymers. The natural polymers are chemically modified using grafting- which involves attaching synthetic polymer chains to natural polymer backbones to enhance properties like hydrophilicity, mechanical strength, and thermal stability (Fabbri and Messori 2017; Ojogbo et al. 2020), cross-linking—chemical cross-linking agents are used to create a network structure within the polymer, improving its mechanical strength and stability. In cross-linking method, a network structure is generated by the establishment of covalent bonds between the polymer chains with cross-linkers (Garavand et al. 2017; Şucu and Shaver 2020). In derivatization- introducing functional groups (e.g., acetyl, carboxymethyl) to alter solubility, thermal properties, and biodegradability. Carboxymethyl Cellulose (CMC) is a cellulose modified with carboxymethyl groups to increase its water solubility, used in food, pharmaceuticals, and cosmetics (Rahman et al. 2021).

Natural biodegradable polymers are modified by physical modification methods, namely, blending -mixing natural polymers with other biodegradable polymers or additives to enhance mechanical properties, thermal stability, and processability (Ojogbo et al. 2020; Rai et al. 2021). The co-blending modification technique is used to prepare modified natural polymers, blending two or more natural biopolymers and mixing them with polylactic acid (PLA), polycaprolactone (PCL), thermoplastic starch (TPS) etc. They enhance polymers' mechanical and physical properties (Aderibigbe 2022; Magazzini et al. 2021; Ranakoti et al. 2022). Nanocomposites—incorporating nanoparticles (e.g., clay, graphene) into natural polymers to improve mechanical strength, barrier properties, and thermal stability (Hassan et al. 2021). An intriguing method for producing novel and inventive biocomposite films with the required functional characteristics is using nanoclays (Dharini et al. 2022).

The commonly available examples of modified natural biodegradable polymers are elaborated as follows,

Modified starch is prepared either by cross-linked starch or starch blends; modified cellulose is cellulose acetate, which is produced by acetylating cellulose to improve its solubility and thermal properties, used in textiles and packaging (Amaraweera et al. 2021; Aziz et al. 2022; Chiu and Solarek 2009; Haq et al. 2019).

Other reported modified natural polymers are modified chitosan (N-alkylated Chitosan, Modification with alkyl groups to improve solubility, chitosan grafts, chitosan grafted with synthetic polymers like polyethylene glycol (PEG) to improve biocompatibility and drug delivery properties (Jin and Wang 2022; Suryani et al. 2024; Wang and Zhuang 2022; Wang et al. 2020). The modified alginates are calcium or sodium alginate, alginate derivatives have been chemically modified with various functional groups to tailor their properties for specific applications (Deng et al. 2021; Pawar 2017; Tao et al. 2023).

2.3.1.2 Microbiologically Synthesized Polymers

Microbial polymers, sometimes called biodegradable polymers generated microbiologically, are created by fermentation processes by microorganisms *viz* bacteria, fungi, and algae. These biodegradable polymers have a wide range of applications in the pharmaceutical, agriculture, food and packaging industries (Dalton et al. 2022; Ghanbarzadeh and Almasi 2013; Panchal and Vasava 2020).

Microorganisms are cultivated in a nutrient-rich medium during microbial fermentation, the usual process of producing these polymers. Fermentation, polymer accumulation, and harvesting are the various methods of preparation of microbial polymers. In fermentation, microorganisms are cultivated in bioreactors with nitrogen, other nutrients, and a carbon source (such as glucose or sucrose). To maximize the formation of polymers, factors including pH, temperature, and oxygen concentrations are regulated (Amadu et al. 2021; Behera et al. 2019).

In the polymer accumulation method, microorganisms synthesize and accumulate the polymer as an intracellular storage material under nutrient-limited conditions (e.g., nitrogen or phosphorus limitation) (Amadu et al. 2021; Magazzini et al. 2021). Centrifugation or filtration is used in the harvesting technique to remove the microbial cells from the fermentation broth (More et al. 2021).

Some of the key examples of microbiologically synthesized biodegradable polymers:

(a) Polyhydroxyalkanoates (PHAs), the most common PHA Polyhydroxybutyrate (PHB) are produced by bacteria like *Ralstonia eutropha* and *Bacillus megaterium.* It is used in packaging, biomedical devices, and agricultural films (Fernandez-Bunster and Pavez 2022; Pramanik 2023). Bacteria like *Alcaligenes eutrophus* form polyhydroxyvalerate (PHV) and polyhydroxyhexanoate (PHH) when they thrive on substrates like glucose and propionic acid. These copolymers are employed in applications similar to PHB and have increased flexibility (Agarwal 2020; de Mello et al. 2023). A Poly(3-hydroxybutyrate-co-3-hydroxyvalerate) (PHBV), copolymer of PHB and PHV, offering better mechanical properties and biodegradability, used in packaging and medical applications (Eraslan et al. 2022).
(b) Bacterial cellulose is manufactured by bacterium such as *Acetobacter xylinum*. It is exceedingly pure with unique mechanical qualities and a high water-holding capacity, making it appropriate for wound dressings, tissue engineering, and food applications (Lin et al. 2020; Wahid et al. 2021).
(c) Pullulan is a polysaccharide produced by the fungus *Aureobasidium pullulans*. It is used in food and pharmaceutical industries due to its film-forming ability and biodegradability (Chaari and Smaoui 2024; de Souza et al. 2023).
(d) Xanthan gum is produced by the bacterium *Xanthomonas campestris*. It is used as a thickening agent and stabilizer in the food industry, as well as in cosmetics and pharmaceuticals (Furtado et al. 2022; Kashaudhan et al. 2023).

2.3.1.3 Chemically Synthesized Biopolymers

These synthetic polymers share similar chemical structures with naturally occurring polymers like proteins, polysaccharides, and nucleic acids. These polymeric chains with amide, peptide, and ester linkages quickly biodegrade in the body (Alves et al. 2020; Geyer 2020). The methods of preparation of these polymers include step growth polymerization (PLA), chain growth Polymerization (PCL), controlled polymerization, ring-opening polymerization (PCL from ε-caprolactone), Post-polymerization modifications (chemical functionalization) (Lin et al. 2020; Montané et al. 2020; Narmon et al. 2022; Sommer et al. 2023).

PLA is a linear aliphatic polyester made from lactic acid. It is produced in D- or L- form or as a racemic combination (D, L-PLA). PLA is produced by ring-opening polymerization of lactide (a cyclic dimer of lactic acid) or direct polycondensation of lactic acid (Ramezani Dana and Ebrahimi 2023; Taib et al. 2023).

Poly(lactic-co-glycolic Acid) (PLGA) is a copolymer of variable ratio two monomers: lactic acid and glycolic acid. It can be synthesized by ring-opening copolymerization of lactide and glycolide. PLGA is biodegradable and degrades into lactic acid and glycolic acid, which are naturally metabolized by the body (Little et al. 2021; Nifant'ev et al. 2024; Ramezani Dana and Ebrahimi 2023; Taib et al. 2023).

2.3.2 Enzymatic Synthesis

Enzymatic synthesis is a novel and sustainable process for creating biodegradable polymeric materials. Enzymes are classified as 'green' catalysts because of their predilection for moderate environments and inherent biodegradability. This method is an effective method to manufacture completely biodegradable polymers like polyesters, polysaccharides, and polyamides. Enzymes such as lipases, glycosyltransferases and proteases/peptidases are significantly used to synthesize polyesters, polysaccharides and polyamides, respectively. Compared to conventional chemical methods, polymerization reactions can be facilitated by enzymes and naturally occurring biocatalysts that offer various advantages. Leveraging certain enzymes to accelerate the conversion of monomers into polymers is known as enzymatic polymerization. These enzymes have a high efficiency and specificity (Fernandes et al. 2022; Gkountela and Vouyiouka 2022; Hevilla et al. 2021).

Polylactic acid (PLA) is produced via the ring-opening polymerization process, which an enzyme catalyzes. Lipase or protease enzymes are employed as catalysts (Shalem et al. 2024). Enzymes like transesterases and esterases catalyze the polycondensation of monomers like diols and diacids, which is another technique used to create polyesters like PHA (Koller and Mukherjee 2022; Pospiech et al. 2021).

2.3.3 Chemo-Enzymatic Synthesis

In the polymer industry, chemo-enzymatic polymerization is considered an efficient biopolymer production method due to its moderate synthesis conditions and renewable, non-toxic enzyme catalysts. This approach is particularly effective in producing high molecular weight biodegradable polymers. Enzymes are increasingly synthesizing new materials, either alone or combined with chemical synthesis (chemoenzymatic methods) (Agarwal 2020; Ahmed et al. 2019; Kobayashi et al. 2019).

The chemo-enzymatic synthesis of different glycolipid analogs produced by a lipase from *Candida antarctica* was used as a catalyst to accomplish highly regioselective macrolactonization of disaccharides (maltose, lactose, cellobiose, and melibiose). These glycolipid analogs have potential applications in the cosmetics, formulation, food, and pharmaceutical industries (Bhatt 2006).

Arcens D *et al* synthesized bio-based methacrylated 12-hydroxy stearate glucose and its (co)polymerization with methyl methacrylate by either free- or RAFT radical chemo enzymatic polymerizations. Further improved loading capacity of these polymers was tested on dye-loaded nanoparticles of these polymers (Arcens et al. 2020).

2.4 Applications of Biodegradable Polymers

Biodegradable polymers find diverse applications across various fields due to their ability to degrade into non-toxic by-products, making them suitable for environmentally friendly and biomedical applications.

2.4.1 Tissue Engineering

Biodegradable polymers are widely used in tissue engineering to create scaffolds replicating the extracellular matrix (ECM). These scaffolds offer essential structural support, enabling cells to adhere, proliferate, and differentiate into functional tissues. Commonly used polymers in this field include poly (lactic acid) (PLA), poly (glycolic acid) (PGA), and their copolymer, poly (lactic-co-glycolic acid) (PLGA). These materials are favored for their biocompatibility and ability to gradually degrade, allowing for the seamless formation of new tissue as the scaffold breaks down (Khot et al. 2021).

2.4.2 Implants

Biodegradable polymers are used in medical implants that gradually break down, eliminating the need for extra surgeries to remove them. For instance, orthopedic screws, plates, and pins made from polymers like polycaprolactone (PCL) or PLA help stabilize fractures and support bone healing. These implants offer temporary structural reinforcement and are slowly absorbed and replaced by natural tissue as they degrade (Al-Shalawi et al. 2023; Hennessy et al. 2009).

2.4.3 Wound Dressing

Biodegradable polymers are used in wound dressings to enhance healing and shield wounds from infections. Natural polymers like chitosan and alginate possess inherent antimicrobial qualities. In contrast, synthetic polymers such as PCL and PLGA can be designed to release medications or growth factors in a controlled manner to support healing. These dressings adapt to the shape of the wound, absorb fluids, and break down naturally without requiring removal [31].

2.4.4 Drug Delivery

Biodegradable polymers are essential in drug delivery systems, as they encase medications and release them controlled. Polymers based on PLGA and PEG are commonly employed because of their adjustable degradation rates and compatibility with biological systems. These polymers safeguard drugs from breakdown, direct

Table 2.1 Biodegradable polymers and their application (Prajapati et al. 2019; Rai et al. 2021; Samir et al. 2022)

Sr. No.	Biodegradable Polymer	Application
1.	Polylactic Acid (PLA)	Packaging (food packaging, bottles), textiles, medical implants
2.	Polyhydroxyalkanoates (PHA)	Packaging, disposable tableware, agricultural mulch, biomedical applications
3.	Polybutylene Succinate (PBS)	Packaging, agricultural films, disposable cutlery, mulch films
4.	Polyglycolic Acid (PGA)	Medical sutures, drug delivery systems, tissue engineering
5.	Polycaprolactone (PCL)	Drug delivery systems, orthopaedic implants, packaging films, 3D printing
6.	Starch-based Polymers	Packaging, disposable tableware, agricultural films, compostable bags
7.	Polyhydroxyurethanes (PHUs)	Biomedical applications, tissue engineering, drug delivery
8.	Polyethylene Furanoate (PEF)	Packaging (bottles, films), textile fibers, automotive interiors
9.	Cellulose Derivatives	Packaging, textiles, drug delivery, biomedical applications
10.	Chitosan	Wound dressings, drug delivery, water treatment, agricultural films

them to specific tissues or cells, and can be customized to release drugs over periods ranging from days to months. This approach decreases the need for frequent dosing and reduces potential side effects (Mahalakshmi et al. 2023). Biodegradable polymers and their application are listed in Table 2.1.

2.5 Safety and Efficacy of Biodegradable Polymers

Biodegradable polymers are increasingly recognized for reducing environmental impact while improving safety in various industrial applications, including pharmaceutical, food packaging, agriculture and medical devices. Their safety profile is critical to ensuring that they do not endanger human health or the environment during their lifetime. The biodegradable plastic substance strives to make the planet more sustainable and environmentally friendly. Biodegradable plastics have become increasingly popular in academia and industries as sustainable plastic. Researchers examined biodegradable plastics' sustainability using economic profit, social responsibility, and environmental protection (Amulya et al. 2021; Moshood et al. 2022). Using biodegradable polymers improves the environment by reducing carbon dioxide emissions that cause global warming, biodegradation, and the regeneration of raw materials. Bacteria and fungi may ingest biodegradable polymers and transform them into water, CO_2, and methane (Agarwal 2020; Rai et al. 2021).

Biodegradable polymers hold great promise in addressing the issue of plastic waste. In the environment, these polymers spontaneously break down. They could be categorized as environmentally friendly substitutes for conventional polymers made from petroleum (Agarwal 2020; Kabir et al. 2020; Rai et al. 2021). PHA, PLA, PBS, PCL, PGA, polybutylene adipate-co-terephthalate (PBAT), PLGA, starch and cellulose-based polymers are among the different kinds of biodegradable polymers that show promise. They have a variety of qualities, such as mechanical solid strength, favorable thermal characteristics, and barrier qualities. Applications for biodegradable polymers can be found in consumer goods, packaging, agriculture, and the biomedical industry. They provide more environmentally friendly ways to lessen plastic pollution and encourage sustainability for a cleaner Earth. More research in this field is expected to result in developments and more uses for these eco-friendly materials (Smola-Dmochowska et al. 2023; Westlake et al. 2023). These days, biocomposites are typically created by combining natural or biobased materials with a biobased or biodegradable polymer to create a hybrid product that combines the best qualities of both parent materials. Given that the packaging, construction, automotive, and aerospace industries have shown a great deal of interest in using biocomposites to cut costs while achieving sustainability and environmental safety, it is easy to understand why these materials are becoming more and more popular on the market (Amulya et al. 2021).

Following are the primary safety concerns of biodegradable polymers associated with biomedical applications,

Biocompatibility: Biodegradable polymers are intended to be biocompatible, which means they do not provoke a strong immune reaction when introduced into the body. This makes them appropriate for medicinal applications such as medication delivery systems, sutures, and tissue engineering. For example, poly (lactic acid) (PLA) and poly(glycolic acid) (PGA) are commonly utilized due to their established safety profiles (Cakmak 2024).

Non-toxic Degradation Products: Biodegradable polymers must produce non-toxic degradation products that are quickly metabolized or excreted by the body. For example, PLA degrades into lactic acid, a naturally occurring metabolite in the human body (Liu et al. 2020b).

Regulatory Approval: Regulatory authorities, such as the FDA, have certified many biodegradable polymers for various applications, ensuring they meet strict safety standards. Materials approved for medical devices include polycaprolactone (PCL) and its copolymers (Thaeder et al. 2023).

The efficacy of biodegradable polymers plays a vital role in biomedical applications. Biodegradable polymers are designed to degrade at a controlled rate, which can be tailored to specific applications in drug delivery, namely controlled or sustained drug release from the dosage form (Liu et al. 2020b). Technological developments in polymer synthesis have made it possible to create biodegradable polymers with various mechanical characteristics appropriate for various uses. For instance, medical sutures and implants can benefit from PHB's mechanical strength, flexibility, and copolymers (Cakmak 2024). Biodegradable polymers can be functionalized with various bioactive compounds to increase their effectiveness. Targeted medication administration, enhanced cell adhesion in scaffolds for tissue engineering, and antibacterial qualities in medical equipment are all made possible by this functionalization (Thaeder et al. 2023).

2.6 Current Challenges and Future Perspectives of Biodegradable Polymers

Since the first ever synthetic polymer was synthesized in the 1970s, polymer materials—with their wide range of applications, superior performance, affordable pricing, and other advantages—have almost completely transformed the economy and lifestyle. Due to their widespread usage, the global production of polymer materials is anticipated to cross 500 million metric tons by 2050. Synthetic polymeric materials, like regular plastics, are no longer disposed of; they tend to collect into the environment because of their extreme durability and stability. Dumping in landfills and burning are the main methods of managing solid waste for plastic products (such as bags used for food packaging, plastic containers, and fertilizer). According to reports, only 13% of plastic is currently recycled. 40% of plastic garbage ends up in landfills, 25% is burned, 22% is dumped in unmanaged areas and isn't suitable for recycling. Academic study on environmental concerns has gained increased attention due to society's progressive advancement and growing awareness of the value of environmental protection. Thus, one of the main focus areas in academics should be inventing a wide range of novel and valuable biodegradable polymers using advanced and cost-effective methods (Yin and Yang 2020).

Biodegradable polymers in biomedical applications are primarily categorized as surgical implant devices, wound healing products, and drug delivery systems. Nowadays, there is a lot of interest in developing controlled-release formulations using biopolymeric drug delivery systems. Specifically, new formulations are made with biodegradable polymers, and the buccal mucosa's high permeability makes it a suitable target for drug delivery. In this sense, drug delivery via buccal channels in conjunction with biopolymers is safe, quick, and protective. Likewise, bioresorbable non-wovens that can replace human tissue repair in the case of wound healing, as well as primary sutures, staples, or meshes, are available. Biodegradable polymers are available in different forms and are both affordable and renewable. Notably, because of their bioactive properties that promote cell growth and regeneration potential, provide antimicrobial conditions, and modulate immunological responses, biodegradable polymers are regarded as outstanding candidates for wound healing.

Furthermore, because biodegradable polymers can absorb large amounts of water, they are likely candidates for wound care. These polymers have recently been shown to have the ability to release medications at the site of injury, making them suitable for use in medicinal applications. Due to their film-forming solid properties, several biopolymers can be used in traditional commodity applications. Interestingly, they are utilized in farming film, rubbish bags, wrapping paper, soil retention sheets, and food containers (bottles and jars) (Alaswad et al. 2022).

By 2027, the worldwide plastic packaging industry is projected to grow at a 4.0% compound annual growth rate. Despite the research above efforts, biodegradable food packaging that can please end users while also providing a decent oxygen/water barrier is still a relatively new product on the market. Compostable coffee pod rings and removable foam containers are two examples of packaging made of biodegradable polymers or composites that do not require a gas barrier. However, many obstacles still exist in developing biodegradable packaging for high-barrier end-use, particularly for films made for soft packaging. The cost comes first, the most affordable biodegradable polylactide which is significantly less than the already extensively used petroleum-based polymers, polyethylene and polypropylene. In addition to PBS, PBAT and PHAs, among other cost-effective biodegradable polymers. The sluggish growth in market acceptability and demand contributes to the high cost. The issues of PHAs mass production (with side effects like delayed uptake) serve as a valuable lesson for other industries considering entering this market. Thus, it is crucial to stimulate market demand before increasing manufacturing capacity. Increased market demand is anticipated to drive down the cost of bioplastics as production volume increases. Two crucial factors must be considered to resolve this matter: The initial step is investigating novel approaches for producing low-cost, high-purity monomers from biomass. Most of the costs in the existing technology are related to monomers' production, purification, and fermentation. Second, boosting consumer demand and expanding manufacturing capacity will be crucial. The product's price is anticipated to decrease when production capacity increases. The global bio-plastic market is expected to grow at a consumption annual growth rate of 21.7% over the next 5 years due to shifting consumer preferences toward environmentally friendly plastic products and encouraging government regulations (Wu et al. 2021).

A significant worry regarding synthetic plastics' high toxicity and non-biodegradability has arisen recently due to their mass manufacture and overuse for human comfort and convenience. As a result, single-use plastics will inevitably be prohibited in 2021. As 3D printing becomes increasingly important, the effects of this new technology and its goods won't be felt for some time. The current pollution issue is brought on by (i) improper plastic disposal by industry and consumers as a result of either inadequate trash collection infrastructure or a lack of knowledge about the implications and (ii) the high chemical stability of plastics (Mukherjee et al. 2023).

Several vital issues hamper the 3D-printed biodegradable polymers and their composites. First of all, it can be difficult to provide relevant resources. The requirement for essential mechanical properties like strength and flexibility must be carefully balanced with the biodegradability of these materials. Achieving this equilibrium is a challenging task. Moreover, optimizing the printability of biodegradable materials may provide challenges. Researchers must adjust temperature, speed, and layer height to ensure consistent and trustworthy 3D printing results. The comprehensive evaluation of the environmental impact of 3D printing biodegradable materials presents another significant problem. Researchers must consider using 3D-printed biodegradable materials for tissue engineering applications, medication delivery systems, and medical implants. It is crucial to guarantee the safety and biocompatibility of these materials. Researchers must investigate how 3D-printed biodegradable items might be integrated with sensor technologies and the Internet of Things (IoT) for data collection and monitoring. This integration could have been used in environmental monitoring, agriculture, and other fields (Dananjaya et al. 2024).

Due to their unique qualities, polymers have become the preferred structural material for creating biodegradable polymer vascular stents (BPVSs). These properties include light weight, good machinability, biodegradability, and comparable temporary radial strength to metal stents.

BPVSs have been fabricated using 3D printing technology in recent years. The benefits of 3D printing over traditional stent manufacturing techniques include great customization, low material costs, high fabrication efficiency, and the capacity to produce intricate 3D structures. Furthermore, other subcategories of 3D printing technology enable the production of BPVSs using various build materials and structural designs. Next-generation BPVSs are anticipated to have superior mechanical, biological, and morphological characteristics. Furthermore, in the future, new 3D printing techniques will make it simple to produce personalized BPVSs in any shape (Hua et al. 2022).

Biodegradable polymers will play an important role in personalized medicine, notably in medication delivery systems tuned to individual patient demands. The discovery of new biodegradable scaffolds with improved biocompatibility and mechanical qualities will promote the growth and differentiation of numerous cell types, thereby enhancing regenerative medicine and tissue engineering. Biodegradable polymers are anticipated to supersede traditional plastics in packaging applications, significantly decreasing plastic waste and environmental pollution. Biodegradable polymers will find increased use in agriculture for applications such as mulch films, controlled-release fertilizers, and plant containers, which will degrade naturally, reducing plastic waste in agricultural settings. Biodegradable polymers have a bright future, with enormous potential to influence the environmental, industrial, and biomedical domains. Future sustainability and health consciousness developments will be facilitated by further research and innovation that will solve present constraints and increase the use of these sustainable materials. Marketed formulations containing biodegradable polymers are listed in Table 2.2.

2.7 Conclusion

The amount of plastic recycled globally falls short of what is needed to preserve the environment. Building the infrastructure needed to collect bio-waste and turn it into valuable materials for biodegradable plastics would be a step towards accomplishing the worldwide sustainable development goals. Biopolymers have greatly improved in terms of environmental safety, degradation potential, and applications because of technological advancements. Furthermore, bioplastics now have better mechanical and physical qualities due to natural polymers' modifications.

Table 2.2 Marketed formulations containing biodegradable polymers (Prajapati et al. 2019)

Sr. No.	Trade name	Polymer	Uses	Company	Side effects
1.	Monovisc	Hyaluronic acid	Osteoarthritis	Anika Therapeutics, Inc.	Pain, discomfort, swelling, stiffness, warmth
2.	Juvederm	Hyaluronic acid	Facil wrinkles and folds	Allergan	Itching, tenderness, bruising, redness
3.	Eligard	Poly (DL-Lactide-co-glycolide)	Prostate cancer	Sanofi Aventis	Decreased libido, hot flashes
4.	Abraxane	Albumin	Pancreatic cancer, breast cancer and lung cancer	Abraxis Bioscience	Bone marrow suppression, fatigue, peripheral neuropathy
5.	Opaxio	Poly-γ-glutamic acid	Non-Small Cell Lung Cancer, Ovarian Cancer, Glioblastoma Multiforme, Head and Neck Cancer	Cell Therapeutics, Inc.	Bone marrow suppression, alopecia
6.	Cynviloq	Poly-lactide	Pancreatic cancer, ovarian cancer and bladder cancer	Sorrento Therapeutics, Inc.	Bone marrow suppression, fatigue, alopecia

Bioplastics are becoming increasingly popular as membranes, which have enormous potential for packaging and industrial uses. Because of its significant correlation with environmental safety and its remarkable applications in the medical and agricultural fields, such as in tissue engineering, implants, wound healing, drug delivery, etc., research on biodegradable polymers has been one of the most fascinating areas of study.

Additionally, these refined biodegradable polymers aim to provide an alternative to harmful and unwelcome synthetic single-use plastics. We thoroughly summarized the most current developments in the biological uses of biodegradable polymers here. Biodegradable and biocompatible, environmentally friendly biopolymers are good options for developing emerging applications. However, biopolymers show superiority since they quickly dissolve into tiny particles and are excreted from the human body. In recent years, considerable progress has been made in biological applications of biodegradable polymeric drug delivery systems. More thorough research efforts must be made to develop suitable drug delivery techniques with higher reliability. Using various biodegradable polymeric techniques for therapeutic purposes in the future, including scaling up with planned and targeted activities, may represent a significant step because suitable nanocarrier materials are required to deliver genetic material to the target-specific region. This assessment is well-timed to discuss biodegradable polymers and critically examine their applications and potential uses.

Acknowledgments The authors would like to acknowledge Dr. Atmaram Pawar, Principal, Poona College of Pharmacy, Bharati Vidyapeeth (Deemed to be) University, Pune, for helpful article writing suggestions.

Conflicts of Interest The authors declare that they have no conflict of interest.

References

Ackermann YS, Li W-J, de Hipt LO, Niehoff P-J, Casey W, Polen T et al (2021) Engineering adipic acid metabolism in pseudomonas putida. Metab Eng 67:29–40

Aderibigbe BA (2022) Hybrid-based wound dressings: a combination of synthetic and biopolymers. Polymers 14(18):3806

Afolabi F, Mahmood SM, Dzulkarnain I, Ewere D, Akbari S (2023) A detailed study on the rheological behavior of a novel cellulose-based hydrophobically-modified polymer. Pet Sci Technol 41(13):1313–1327

Agarwal S (2020) Biodegradable polymers: present opportunities and challenges in providing a microplastic-free environment. Macromol Chem Phys 221(6):2000017

Ahmed ST, Leferink NG, Scrutton NS (2019) Chemo-enzymatic routes towards the synthesis of bio-based monomers and polymers. Mol Catal 467:95–110

Al-Arjan WS, Khan MUA, Almutairi HH, Alharbi SM, Razak SIA (2022) pH-responsive PVA/BC-f-GO dressing materials for burn and chronic wound healing with curcumin release kinetics. Polymers 14(10):1949

Alaswad SO, Mahmoud AS, Arunachalam P (2022) Recent advances in biodegradable polymers and their biological applications: a brief review. Polymers 14(22):4924

Al-Shalawi FD, Hanim MA, Ariffin M, Kim CLS, Brabazon D, Calin R, Al-Osaimi MO (2023) Biodegradable synthetic polymer in orthopaedic application: a review. Mater Today Proc 74:540–546

Alves TF, Morsink M, Batain F, Chaud MV, Almeida T, Fernandes DA et al (2020) Applications of natural, semi-synthetic, and synthetic polymers in cosmetic formulations. Cosmetics 7(4):75

Amadu AA, Qiu S, Ge S, Addico GND, Ameka GK, Yu Z et al (2021) A review of biopolymer (Poly-β-hydroxybutyrate) synthesis in microbes cultivated on wastewater. Sci Total Environ 756:143729

Amaraweera SM, Gunathilake C, Gunawardene OH, Fernando NM, Wanninayaka DB, Dassanayake RS et al (2021) Development of starch-based materials using current modification techniques and their applications: a review. Molecules 26(22):6880

Amulya K, Katakojwala R, Ramakrishna S, Mohan SV (2021) Low carbon biodegradable polymer matrices for sustainable future. Compos Part C Open Access 4:100111

Anjum A, Garg R, Kashif M, Eddy NO (2023) Nano-scale innovations in packaging: properties, types, and applications of nanomaterials for the future. Food Chem Adv 3:100560

Arcens D, Le Fer G, Grau E, Grelier S, Cramail H, Peruch F (2020) Chemo-enzymatic synthesis of glycolipids, their polymerization and self-assembly. Polym Chem 11(24):3994–4004

Auras R, Harte B, Selke S (2004) An overview of polylactides as packaging materials. Macromol Biosci 4(9):835–864

Aziz T, Farid A, Haq F, Kiran M, Ullah A, Zhang K et al (2022) A review on the modification of cellulose and its applications. Polymers 14(15):3206

Balachandar M, Ramnath BV, Jagadeeshwar P, Yokesh R (2019) Mechanical behaviour of natural and glass fiber reinforced with polymer matrix composite. Mater Today Proc 16:1297–1303

Banche-Niclot F, Licini C, Montalbano G, Fiorilli S, Mattioli-Belmonte M, Vitale-Brovarone C (2022) 3D printed scaffold based on type I collagen/PLGA_TGF-β1 nanoparticles mimicking the growth factor footprint of human bone tissue. Polymers 14(5):857

Bao L, Dou G, Tian R, Lv Y, Ding F, Liu S et al (2022) Engineered neutrophil apoptotic bodies ameliorate myocardial infarction by promoting macrophage efferocytosis and inflammation resolution. Bioact Mater 9:183–197

Baranwal J, Barse B, Fais A, Delogu GL, Kumar A (2022) Biopolymer: a sustainable material for food and medical applications. Polymers 14(5):983

Bassand C, Verin J, Lamatsch M, Siepmann F, Siepmann J (2022) How agarose gels surrounding PLGA implants limit swelling and slow down drug release. J Control Release 343:255–266

Batool F, Özçelik H, Stutz C, Gegout P-Y, Benkirane-Jessel N, Petit C, Huck O (2021) Modulation of immune-inflammatory responses through surface modifications of biomaterials to promote bone healing and regeneration. J Tissue Eng 12:20417314211041428

Bavya Devi K, Lalzawmliana V, Saidivya M, Kumar V, Roy M, Kumar Nandi S (2022) Magnesium phosphate bioceramics for bone tissue engineering. Chem Rec 22(11):e202200136

Behera SS, Ray RC, Das U, Panda SK, Saranraj P (2019) Microorganisms in fermentation. In: Essentials in fermentation technology. Springer International Publishing, pp 1–39

Bhardwaj H, Jangde RK (2023) Current updated review on preparation of polymeric nanoparticles for drug delivery and biomedical applications. Next Nanotechnol 2:100013

Bhatt S (2006) Chemo-enzymatic route to synthesis of biodegradable polymers and glycolipid analogs. University of South Florida

Burdick JA, Prestwich GD (2011) Hyaluronic acid hydrogels for biomedical applications. Adv Mater 23(12):H41–H56

Cakmak OK (2024) Biodegradable polymers—a review on properties, processing, and degradation mechanism. Circ Econ Sustain 4(1):339–362

Carmagnola I, Nardo T, Gentile P, Tonda-Turo C, Mattu C, Cabodi S et al (2015) Poly (lactic acid)-based blends with tailored physicochemical properties for tissue engineering applications: a case study. Int J Polym Mater Polym Biomater 64(2):90–98

Chaari M, Smaoui S (2024) Pullulan as a biopolymer from microorganisms: role in food packaging. Curr Food Sci Technol Rep 2(2):149–156

Chandra G, Pandey A (2022) Preparation strategies for Mg-alloys for biodegradable orthopaedic implants and other biomedical applications: a review. Irbm 43(3):229–249

Chiu C-W, Solarek D (2009) Modification of starches. In: Starch. Elsevier, pp 629–655

Dalton B, Bhagabati P, De Micco J, Padamati RB, O'Connor K (2022) A review on biological synthesis of the biodegradable polymers polyhydroxyalkanoates and the development of multiple applications. Catalysts 12(3):319

Dananjaya S, Chevali V, Dear J, Potluri P, Abeykoon C (2024) 3D printing of biodegradable polymers and their composites–current state-of-the-art, properties, applications, and machine learning for potential future applications. Prog Mater Sci 101336:101336

Das PP, Chaudhary V (2021) Moving towards the era of bio fibre based polymer composites. Clean Eng Technol 4:100182

de Mello AFM, de Souza Vandenberghe LP, Machado CMB, Brehmer MS, de Oliveira PZ, Binod P et al (2023) Polyhydroxyalkanoates production in biorefineries: a review on current status, challenges and opportunities. Bioresour Technol 393:130078

de Souza CK, Ghosh T, Lukhmana N, Tahiliani S, Priyadarshi R, Hoffmann TG et al (2023) Pullulan as a sustainable biopolymer for versatile applications: a review. Mater Today Commun 36:106477

Deng Y, Shavandi A, Okoro OV, Nie L (2021) Alginate modification via click chemistry for biomedical applications. Carbohydr Polym 270:118360

Dharini V, Selvam SP, Jayaramudu J, Emmanuel RS (2022) Functional properties of clay nanofillers used in the biopolymer-based composite films for active food packaging applications-review. Appl Clay Sci 226:106555

Dirauf M, Muljajew I, Weber C, Schubert US (2022) Recent advances in degradable synthetic polymers for biomedical applications-beyond polyesters. Prog Polym Sci 129:101547

Eraslan K, Aversa C, Nofar M, Barletta M, Gisario A, Salehiyan R, Goksu YA (2022) Poly (3-hydroxybutyrate-co-3-hydroxyhexanoate)(PHBH): synthesis, properties, and applications-a review. Eur Polym J 167:111044

Fabbri P, Messori M (2017) Surface modification of polymers: chemical, physical, and biological routes. In: Modification of polymer properties. Elsevier, pp 109–130
Fernandes CD, Oechsler BF, Sayer C, de Oliveira D, de Araújo PHH (2022) Recent advances and challenges on enzymatic synthesis of biobased polyesters via polycondensation. Eur Polym J 169:111132
Fernandez-Bunster G, Pavez P (2022) Novel production methods of polyhydroxyalkanoates and their innovative uses in biomedicine and industry. Molecules 27(23):8351
Flejszar M, Chmielarz P (2020) Surface modifications of poly (ether ether ketone) via polymerization methods—current status and future prospects. Materials 13(4):999
Furtado IF, Sydney EB, Rodrigues SA, Sydney AC (2022) Xanthan gum: applications, challenges, and advantages of this asset of biotechnological origin. Biotechnol Res Innov J 6(1):e202204
Garavand F, Rouhi M, Razavi SH, Cacciotti I, Mohammadi R (2017) Improving the integrity of natural biopolymer films used in food packaging by crosslinking approach: a review. Int J Biol Macromol 104:687–707
Gbian DL, Omri A (2022) Lipid-based drug delivery systems for diseases managements. Biomedicines 10(9):2137
Geyer R (2020) Production, use, and fate of synthetic polymers. In: Plastic waste and recycling. Elsevier, pp 13–32
Ghanbarzadeh B, Almasi H (2013) Biodegradable polymers. In: Biodegradation-life of science. InTech, pp 141–185
Ghimici L, Nichifor M (2018) Flocculation characteristics of a biodegradable polymer based on dextran. Sep Purif Technol 194:48–55
Ghosh S, Mukherjee S, Patra D, Haldar J (2022) Polymeric biomaterials for prevention and therapeutic intervention of microbial infections. Biomacromolecules 23(3):592–608
Ghuttora N (2016) Increase the usage of biopolymers and biodegradable polymers for sustainable environment
Giol ED, Schaubroeck D, Kersemans K, De Vos F, Van Vlierberghe S, Dubruel P (2015) Bio-inspired surface modification of PET for cardiovascular applications: case study of gelatin. Colloids Surf B: Biointerfaces 134:113–121
Gkountela CI, Vouyiouka SN (2022) Enzymatic polymerization as a green approach to synthesizing bio-based polyesters. Macromolecules 2(1):30–57
Gupta SS, Mishra V, Mukherjee MD, Saini P, Ranjan KR (2021) Amino acid derived biopolymers: recent advances and biomedical applications. Int J Biol Macromol 188:542–567
Haq F, Yu H, Wang L, Teng L, Haroon M, Khan RU et al (2019) Advances in chemical modifications of starches and their applications. Carbohydr Res 476:12–35
Harish Kumar A, Ahamed MB, Deshmukh K, Sirajuddeen MS (2021) Morphology, dielectric and EMI shielding characteristics of graphene nanoplatelets, montmorillonite nanoclay and titanium dioxide nanoparticles reinforced polyvinylidenefluoride nanocomposites. J Inorg Organomet Polym Mater 31:2003–2016
Hassan T, Salam A, Khan A, Khan SU, Khanzada H, Wasim M et al (2021) Functional nanocomposites and their potential applications: a review. J Polym Res 28(2):36
He H, Li H, Pu A, Li W, Ban K, Xu L (2023) Hybrid assembly of polymeric nanofiber network for robust and electronically conductive hydrogels. Nat Commun 14(1):759
Hennessy KM, Pollot BE, Clem WC, Phipps MC, Sawyer AA, Culpepper BK, Bellis SL (2009) The effect of collagen I mimetic peptides on mesenchymal stem cell adhesion and differentiation, and on bone formation at hydroxyapatite surfaces. Biomaterials 30(10):1898–1909
Hevilla V, Sonseca A, Echeverría C, Muñoz-Bonilla A, Fernández-García M (2021) Enzymatic synthesis of polyesters and their bioapplications: recent advances and perspectives. Macromol Biosci 21(10):2100156
Highley CB, Prestwich GD, Burdick JA (2016) Recent advances in hyaluronic acid hydrogels for biomedical applications. Curr Opin Biotechnol 40:35–40
Hou Y, Wang W, Bartolo P (2022) Investigation of polycaprolactone for bone tissue engineering scaffolds: in vitro degradation and biological studies. Mater Des 216:110582

Hsieh S-R, Reddy PM, Chang C-J, Kumar A, Wu W-C, Lin H-Y (2016) Exploring the behavior of bovine serum albumin in response to changes in the chemical composition of responsive polymers: experimental and simulation studies. Polymers 8(6):238

Hua W, Shi W, Mitchell K, Raymond L, Coulter R, Zhao D, Jin Y (2022) 3D printing of biodegradable polymer vascular stents: a review. Chinese J Mech Eng Addit Manufact Front 1(2):100020

Jayakumar R, Prabaharan M, Kumar PS, Nair S, Tamura H (2011) Biomaterials based on chitin and chitosan in wound dressing applications. Biotechnol Adv 29(3):322–337

Jesus S, Marques AP, Duarte A, Soares E, Costa JP, Colaço M et al (2020) Chitosan nanoparticles: shedding light on immunotoxicity and hemocompatibility. Front Bioeng Biotechnol 8:100

Jin H, Wang Z (2022) Advances in alkylated chitosan and its applications for hemostasis. Macromolecules 2(3):346–360

Jin J, Mangal U, Seo J-Y, Kim J-Y, Ryu J-H, Lee Y-H et al (2023) Cerium oxide nanozymes confer a cytoprotective and bio-friendly surface micro-environment to methacrylate based oro-facial prostheses. Biomaterials 296:122063

Kabir E, Kaur R, Lee J, Kim K-H, Kwon EE (2020) Prospects of biopolymer technology as an alternative option for non-degradable plastics and sustainable management of plastic wastes. J Clean Prod 258:120536

Kashaudhan K, Pande PP, Sharma J, Nath A, Shankar R (2023) Production and applications of Xanthan Gum, a polymeric material obtained from Xanthomonas Campestris: a mini review. In: AIJR Proceedings, pp 79–86

Khan MUA, Raza MA, Razak SIA, Abdul Kadir MR, Haider A, Shah SA et al (2020) Novel functional antimicrobial and biocompatible arabinoxylan/guar gum hydrogel for skin wound dressing applications. J Tissue Eng Regen Med 14(10):1488–1501

Khan MUA, Haider S, Haider A, Abd Razak SI, Kadir MRA, Shah SA et al (2021a) Development of porous, antibacterial and biocompatible GO/n-HAp/bacterial cellulose/β-glucan biocomposite scaffold for bone tissue engineering. Arab J Chem 14(2):102924

Khan MUA, Razak SIA, Ansari MNM, Zulkifli RM, Ahmad Zawawi N, Arshad M (2021b) Development of biodegradable bio-based composite for bone tissue engineering: synthesis, characterization and in vitro biocompatible evaluation. Polymers 13(21):3611

Khan MUA, Yaqoob Z, Ansari MNM, Razak SIA, Raza MA, Sajjad A et al (2021c) Chitosan/poly vinyl alcohol/graphene oxide based pH-responsive composite hydrogel films: drug release, anti-microbial and cell viability studies. Polymers 13(18):3124

Khan MUA, Abd Razak SI, Haider S, Mannan HA, Hussain J, Hasan A (2022a) Sodium alginate-f-GO composite hydrogels for tissue regeneration and antitumor applications. Int J Biol Macromol 208:475–485

Khan MUA, Rizwan M, Razak SIA, Hassan A, Rasheed T, Bilal M (2022b) Electroactive polymeric nanocomposite BC-g-(Fe3O4/GO) materials for bone tissue engineering: in vitro evaluations. J Biomater Sci Polym Ed 33(11):1349–1368

Khot S, Rawal SU, Patel MM (2021) Dissolvable-soluble or biodegradable polymers. In: Drug delivery devices and therapeutic systems. Elsevier, pp 367–394

Kobayashi S, Uyama H, Kadokawa J-I (2019) Enzymatic polymerization towards green polymer chemistry. Springer

Koller M, Mukherjee A (2022) Polyhydroxyalkanoates (PHAs)–production, properties, and biodegradation. In: Biodegradable polymers in the circular plastics economy. Wiley, pp 145–204

Kumar A, Sood A, Han SS (2022) Molybdenum disulfide (MoS 2)-based nanostructures for tissue engineering applications: prospects and challenges. J Mater Chem B 10(15):2761–2780

Kundu B, Rajkhowa R, Kundu SC, Wang X (2013) Silk fibroin biomaterials for tissue regenerations. Adv Drug Deliv Rev 65(4):457–470

Li M, Jiang S, Simon J, Paßlick D, Frey M-L, Wagner M et al (2021a) Brush conformation of polyethylene glycol determines the stealth effect of nanocarriers in the low protein adsorption regime. Nano Lett 21(4):1591–1598

Li Z, Du T, Ruan C, Niu X (2021b) Bioinspired mineralized collagen scaffolds for bone tissue engineering. Bioact Mater 6(5):1491–1511

Lin D, Liu Z, Shen R, Chen S, Yang X (2020) Bacterial cellulose in food industry: current research and future prospects. Int J Biol Macromol 158:1007–1019

Little A, Wemyss AM, Haddleton DM, Tan B, Sun Z, Ji Y, Wan C (2021) Synthesis of poly (Lactic acid-co-glycolic acid) copolymers with high glycolide ratio by ring-opening polymerisation. Polymers 13(15):2458

Liu S, Qin S, He M, Zhou D, Qin Q, Wang H (2020a) Current applications of poly (lactic acid) composites in tissue engineering and drug delivery. Compos Part B 199:108238

Liu Y, Song L, Feng N, Jiang W, Jin Y, Li X (2020b) Recent advances in the synthesis of biodegradable polyesters by sustainable polymerization: lipase-catalyzed polymerization. RSC Adv 10(59):36230–36240

Liu S, Yu J-M, Gan Y-C, Qiu X-Z, Gao Z-C, Wang H et al (2023) Biomimetic natural biomaterials for tissue engineering and regenerative medicine: new biosynthesis methods, recent advances, and emerging applications. Mil Med Res 10(1):16

Lu Y, Deshmukh S, Jones I, Chiu Y-L (2021) Biodegradable magnesium alloys for orthopaedic applications. Biomater Transl 2(3):214–235

Magazzini L, Grilli S, Fenni SE, Donetti A, Cavallo D, Monticelli O (2021) The blending of poly (glycolic acid) with Polycaprolactone and poly (l-lactide): promising combinations. Polymers 13(16):2780

Mahalakshmi P, Reshma G, Arthi C, Másson M, Rangasamy J (2023) Biodegradable polymeric scaffolds and hydrogels in the treatment of chronic and infectious wound healing. Eur Polym J 198:112390

Makadia HK, Siegel SJ (2011) Poly lactic-co-glycolic acid (PLGA) as biodegradable controlled drug delivery carrier. Polymers 3(3):1377–1397

Meabe L, Peña SR, Martinez-Ibañez M, Zhang Y, Lobato E, Manzano H et al (2020) Insight into the ionic transport of solid polymer electrolytes in polyether and polyester blends. J Phys Chem C 124(33):17981–17991

Middleton JC, Tipton AJ (2000) Synthetic biodegradable polymers as orthopedic devices. Biomaterials 21(23):2335–2346

Montané X, Montornes JM, Nogalska A, Olkiewicz M, Giamberini M, Garcia-Valls R et al (2020) Synthesis and synthetic mechanism of Polylactic acid. Phys Sci Rev 5(12):20190102

More VS, Ebinesar A, Prakruthi A, Praveen P, Fasim A, Rao A et al (2021) Isolation and purification of microbial exopolysaccharides and their industrial application. In: Microbial polymers: applications and ecological perspectives. Springer Singapore, pp 69–86

Moshood TD, Nawanir G, Mahmud F, Mohamad F, Ahmad MH, AbdulGhani A (2022) Sustainability of biodegradable plastics: new problem or solution to solve the global plastic pollution? Curr Res Green Sustain Chem 5:100273

Mukherjee C, Varghese D, Krishna J, Boominathan T, Rakeshkumar R, Dineshkumar S et al (2023) Recent advances in biodegradable polymers–properties, applications and future prospects. Eur Polym J 192:112068

Narmon AS, Jenisch LM, Pitet LM, Dusselier M (2022) Ring-opening polymerization strategies for degradable polyesters. In: Biodegradable polymers in the circular plastics economy. Wiley, pp 205–271

Nazir S, Khan MUA, Al-Arjan WS, Abd Razak SI, Javed A, Kadir MRA (2021) Nanocomposite hydrogels for melanoma skin cancer care and treatment: in-vitro drug delivery, drug release kinetics and anti-cancer activities. Arab J Chem 14(5):103120

Nifant'ev IE, Tavtorkin AN, Shlyakhtin AV, Ivchenko PV (2024) Chemical features of the synthesis, degradation, molding and performance of poly (lactic-co-glycolic) acid (PLGA) and PLGA-based articles. Eur Polym J 113250:113250

Numata K, Yamazaki S, Katashima T, Chuah JA, Naga N, Sakai T (2014) Silk-pectin hydrogel with superior mechanical properties, biodegradability, and biocompatibility. Macromol Biosci 14(6):799–806

Ojogbo E, Ogunsona EO, Mekonnen TH (2020) Chemical and physical modifications of starch for renewable polymeric materials. Mater Today Sustain 7:100028

Okada M (2002) Chemical syntheses of biodegradable polymers. Prog Polym Sci 27(1):87–133

Pacelli S, Di Muzio L, Paolicelli P, Fortunati V, Petralito S, Trilli J, Casadei MA (2021) Dextran-polyethylene glycol cryogels as spongy scaffolds for drug delivery. Int J Biol Macromol 166:1292–1300

Panchal SS, Vasava DV (2020) Biodegradable polymeric materials: synthetic approach. ACS Omega 5(9):4370–4379

Pathak VM (2017) Review on the current status of polymer degradation: a microbial approach. Bioresour Bioprocess 4(1):1–31

Pawar SN (2017) Chemical modification of alginate. In: Seaweed polysaccharides. Elsevier, pp 111–155

Pennetta C, Floresta G, Graziano ACE, Cardile V, Rubino L, Galimberti M et al (2020) Functionalization of single and multi-walled carbon nanotubes with polypropylene glycol decorated pyrrole for the development of doxorubicin nano-conveyors for cancer drug delivery. Nano 10(6):1073

Pospiech D, Choińska R, Flugrat D, Sahre K, Jehnichen D, Korwitz A et al (2021) Enzymatic synthesis of poly (alkylene succinate) s: influence of reaction conditions. PRO 9(3):411

Prajapati SK, Jain A, Jain A, Jain S (2019) Biodegradable polymers and constructs: a novel approach in drug delivery. Eur Polym J 120:109191

Pramanik N (2023) A tool for biomedical application: synthesis and modification of polyhydroxyalkanoates. Sustain Chem Pharm 32:101041

Pudkon W, Laomeephol C, Damrongsakkul S, Kanokpanont S, Ratanavaraporn J (2021) Comparative study of silk fibroin-based hydrogels and their potential as material for 3-dimensional (3D) printing. Molecules 26(13):3887

Puranen J, Ranta V-P, Ruponen M, Urtti A, Sadeghi A (2023) Quantitative intravitreal pharmacokinetics in mouse as a step towards inter-species translation. Exp Eye Res 235:109638

Rahman MS, Hasan MS, Nitai AS, Nam S, Karmakar AK, Ahsan MS et al (2021) Recent developments of carboxymethyl cellulose. Polymers 13(8):1345

Rai P, Mehrotra S, Priya S, Gnansounou E, Sharma SK (2021) Recent advances in the sustainable design and applications of biodegradable polymers. Bioresour Technol 325:124739

Ramezani Dana H, Ebrahimi F (2023) Synthesis, properties, and applications of polylactic acid-based polymers. Polym Eng Sci 63(1):22–43

Ranakoti L, Gangil B, Mishra SK, Singh T, Sharma S, Ilyas R, El-Khatib S (2022) Critical review on polylactic acid: properties, structure, processing, biocomposites, and nanocomposites. Materials 15(12):4312

Rasouli M, Soleimani M, Hosseinzadeh S, Ranjbari J (2023) Bacterial cellulose as potential dressing and scaffold material: toward improving the antibacterial and cell adhesion properties. J Polym Environ 31(11):4621–4640

Reddy PL, Deshmukh K, Chidambaram K, Ahamed B, Sadasivuni KK, Ponnamma D et al (2019) Effect of poly ethylene glycol (PEG) on structural, thermal and photoluminescence properties of CdO nanoparticles for optoelectronic applications. Mater Today Proc 9:175–183

Remiš T, Bělský P, Kovářík T, Kadlec J, Ghafouri Azar M, Medlín R et al (2021) Study on structure, thermal behavior, and viscoelastic properties of nanodiamond-reinforced poly (vinyl alcohol) nanocomposites. Polymers 13(9):1426

Rosa N, Pouca MV, Olhero SM, Jorge RN, Parente M (2023) Influence of structural features in the performance of bioceramic-based composite scaffolds for bone engineering applications: a prediction study. J Manuf Process 90:391–405

Sahani S, Sharma YC (2021) Advancements in applications of nanotechnology in global food industry. Food Chem 342:128318

Saikia A, Debbarma N, Karak N (2019) Renewable resource based hyperbranched epoxy thermosetting nanocomposite with cellulose nanofiber as a sustainable material. Cellulose 26:4743–4755

Samadian H, Khastar H, Ehterami A, Salehi M (2021) Bioengineered 3D nanocomposite based on gold nanoparticles and gelatin nanofibers for bone regeneration: in vitro and in vivo study. Sci Rep 11(1):13877

Sambrook J, Russell DW (2001) Molecular cloning: Ch. 8. In: Vitro amplification of DNA by the polymerase chain reaction, vol 2. Cold Spring Harbor Laboratory Press

Samir A, Ashour FH, Hakim AA, Bassyouni M (2022) Recent advances in biodegradable polymers for sustainable applications. Npj Mater Degrad 6(1):68

Shafiei M, Ansari MNM, Razak SIA, Khan MUA (2021) A comprehensive review on the applications of exosomes and liposomes in regenerative medicine and tissue engineering. Polymers 13(15):2529

Shah SA, Khan MA, Arshad M, Awan S, Hashmi M, Ahmad N (2016) Doxorubicin-loaded photosensitive magnetic liposomes for multi-modal cancer therapy. Colloids Surf B: Biointerfaces 148:157–164

Shalem A, Yehezkeli O, Fishman A (2024) Enzymatic degradation of polylactic acid (PLA). Appl Microbiol Biotechnol 108(1):413

Shanmuganantha L, Khan MUA, Sulong AB, Ramli MI, Baharudin A, Ariffin HM et al (2022) Characterization of titanium ceramic composite for bone implants applications. Ceram Int 48(16):22808–22819

Shivanna S, Subramani NK, Swamy K, Nagaraj SK, Muthuraj JRB, Siddaramaiah H (2019) Orange-red fluorescent polymer nanocomposite films with large stokes shift: an opto-electronic exercise. J Lumin 208:488–494

Singh AK, Pramanik K (2023) Fabrication and investigation of physicochemical and biological properties of 3D printed sodium alginate-chitosan blend polyelectrolyte complex scaffold for bone tissue engineering application. J Appl Polym Sci 140(12):e53642

Singla R, Abidi SM, Dar AI, Acharya A (2019) Nanomaterials as potential and versatile platform for next generation tissue engineering applications. J Biomed Mater Res B Appl Biomater 107(7):2433–2449

Skwarczynski M, Zhao G, Boer JC, Ozberk V, Azuar A, Cruz JG et al (2020) Poly (amino acids) as a potent self-adjuvanting delivery system for peptide-based nanovaccines. Sci Adv 6(5):eaax2285

Smola-Dmochowska A, Lewicka K, Macyk A, Rychter P, Pamuła E, Dobrzyński P (2023) Biodegradable polymers and polymer composites with antibacterial properties. Int J Mol Sci 24(8):7473

Sommer K, Van Opdenbosch D, Zollfrank C (2023) Synthesis and characterization of functional cellulose–ether-based PCL-and PLA-grafts-copolymers. Polymers 15(2):455

Song R, Murphy M, Li C, Ting K, Soo C, Zheng Z (2018) Current development of biodegradable polymeric materials for biomedical applications. Drug Des Devel Ther Volume 12:3117–3145

Struzek A-M, Scherließ R (2023) Quality by design as a tool in the optimisation of nanoparticle preparation—a case study of PLGA nanoparticles. Pharmaceutics 15(2):617

Su Y, Zhang B, Sun R, Liu W, Zhu Q, Zhang X et al (2021) PLGA-based biodegradable microspheres in drug delivery: recent advances in research and application. Drug Deliv 28(1):1397–1418

Şucu T, Shaver MP (2020) Inherently degradable cross-linked polyesters and polycarbonates: resins to be cheerful. Polym Chem 11(40):6397–6412

Suryani S, Chaerunisaa AY, Joni IM, Ruslin R, Aspadiah V, Anton A et al (2024) The chemical modification to improve solubility of chitosan and its derivatives application, preparation method, toxicity as a nanoparticles. Nanotechnol Sci Appl Volume 17:41–57

Taib N-AAB, Rahman MR, Huda D, Kuok KK, Hamdan S, Bakri MKB et al (2023) A review on poly lactic acid (PLA) as a biodegradable polymer. Polym Bull 80(2):1179–1213

Tao L, Shi C, Zi Y, Zhang H, Wang X, Zhong J (2023) A review on the chemical modification of alginates for food research: chemical nature, modification methods, product types, and application. Food Hydrocoll 147:109338

Thaeder C, Stanek J, Couvreur J, Borrego C, Brunissen F, Allais F et al (2023) Chemo-enzymatic synthesis and biological assessment of p-Coumarate fatty esters: new antifungal agents for potential plant protection. Molecules 28(15):5803

Thanigaivel S, Priya A, Balakrishnan D, Dutta K, Rajendran S, Soto-Moscoso M (2022) Insight on recent development in metallic biomaterials: strategies involving synthesis, types and surface modification for advanced therapeutic and biomedical applications. Biochem Eng J 187:108522

Varshosaz J, Choopannejad Z, Minaiyan M, Kharazi AZ (2021) Rapid hemostasis by nanofibers of polyhydroxyethyl methacrylate/polyglycerol sebacic acid: an in vitro/in vivo study. J Appl Polym Sci 138(5):49785

Wahid F, Huang L-H, Zhao X-Q, Li W-C, Wang Y-Y, Jia S-R, Zhong C (2021) Bacterial cellulose and its potential for biomedical applications. Biotechnol Adv 53:107856

Wang J, Zhuang S (2022) Chitosan-based materials: preparation, modification and application. J Clean Prod 355:131825

Wang W, Meng Q, Li Q, Liu J, Zhou M, Jin Z, Zhao K (2020) Chitosan derivatives and their application in biomedicine. Int J Mol Sci 21(2):487

Westlake JR, Tran MW, Jiang Y, Zhang X, Burrows AD, Xie M (2023) Biodegradable biopolymers for active packaging: demand, development and directions. Sustain Food Technol 1(1):50–72

Wu F, Misra M, Mohanty AK (2021) Challenges and new opportunities on barrier performance of biodegradable polymers for sustainable packaging. Prog Polym Sci 117:101395

Xu H, Zhang S, Yu W (2022) Revealing the mechanism beneath the effects of starch-amino acids interactions on starch physicochemical properties by molecular dynamic simulations. Food Hydrocoll 124:107359

Yazdi MK, Taghizadeh A, Taghizadeh M, Stadler FJ, Farokhi M, Mottaghitalab F et al (2020) Agarose-based biomaterials for advanced drug delivery. J Control Release 326:523–543

Yin G-Z, Yang X-M (2020) Biodegradable polymers: a cure for the planet, but a long way to go. J Polym Res 27(2):38

Yoganandam K, Ramshankar P, Ganeshan P, Raja K (2020) Mechanical properties of alkali-treated Madar and Gongura fibre-reinforced polymer composites. Int J Ambient Energy 41(8):849–850

Zamri MFMA, Bahru R, Amin R, Khan MUA, Abd Razak SI, Hassan SA et al (2021) Waste to health: a review of waste derived materials for tissue engineering. J Clean Prod 290:125792

Zeng S, Duan P, Shen M, Xue Y, Wang Z (2016) Preparation and degradation mechanisms of biodegradable polymer: a review, vol 137. Paper presented at the IOP Conference Series: Materials Science and Engineering, p 012003

Zheng Y, Yang J, Liang J, Xu X, Cui W, Deng L, Zhang H (2019) Bioinspired hyaluronic acid/phosphorylcholine polymer with enhanced lubrication and anti-inflammation. Biomacromolecules 20(11):4135–4142

Zhong Y, Godwin P, Jin Y, Xiao H (2020) Biodegradable polymers and green-based antimicrobial packaging materials: a mini-review. Adv Indust Eng Polym Res 3(1):27–35

Green Chemistry Approaches in Biomaterials Synthesis

3

Priya Jain, Juhi Bhadoriya, Sheetal Mane, Kuldeep Vinchurkar, and Meghraj Vivekanand Suryawanshi

Abstract

The synthesis of biomaterials through green chemistry principles offers a sustainable approach to addressing environmental challenges while advancing biomedical applications. Green chemistry aims to reduce the environmental impact of chemical processes by minimizing hazardous materials, maximizing resource utilization, and enhancing energy efficiency. This review highlights the role of renewable resources, such as plant-derived polymers (cellulose, chitosan) and bio-based monomers (lactic acid for polylactic acid), in the synthesis of environmentally friendly biomaterials. These materials, which are biodegradable, reduce reliance on petroleum-based products, lowering carbon footprints. Non-toxic reagents and solvents, such as e-solvents, are emphasized for their ability to reduce environmental pollutants and improve safety profiles in chemical operations. Energy-efficient techniques like enzymatic catalysis and microwave-assisted synthesis are explored for their potential to decrease reaction times, energy consumption, and waste generation. Examples such as the synthesis of polyhydroxyalkanoates and bioactive ceramics underscore the benefits of these

P. Jain (✉)
Oxford International College, Opposite Pitra Parvat, Indore, Madhya Pradesh, India

University of Galway, Galway, Ireland
e-mail: drpriyajain14@gmai.com

J. Bhadoriya
Oxford International College, Opposite Pitra Parvat, Indore, Madhya Pradesh, India

S. Mane
School of Pharmacy, Devi Ahilya Vishwavidyalaya, Indore, Madhya Pradesh, India

K. Vinchurkar (✉) · M. V. Suryawanshi
Department of Pharmaceutics, Sandip Institute of Pharmaceutical Sciences (SIPS), Affiliated To Savitribai Phule Pune University (SPPU, Pune), Nashik, Maharashtra, India

Md. Faiyazuddin, M. Suryawanshi (eds.), *Design and Processing of Green Materials*, Biomaterials, Bioengineering and Sustainability 4,
https://doi.org/10.1007/978-3-031-91790-5_3

processes in the production of biocompatible and functional biomaterials. Green chemistry also focuses on atom economy and waste minimization, employing strategies like click chemistry and recyclable catalysts to improve yield and reduce by-products. Several case studies illustrate the successful integration of these green methodologies in the production of biodegradable polymers for drug delivery and tissue engineering scaffolds. Overall, green chemistry in biomaterial synthesis enhances environmental sustainability, material functionality, and safety.

Keywords

Green chemistry · Biomaterials synthesis · Renewable resources · Non-toxic reagents · Energy efficiency · Waste minimization sustainable development

3.1 Introduction

In recent years, there has been a growing global emphasis on sustainable development and environmentally responsible practices across all sectors. Among these, the field of biomaterials has seen significant advancements with a focus on eco-friendly and sustainable approaches in material synthesis, including the integration of green chemistry principles. The synthesis of biomaterials using green chemistry approaches is a response to the environmental, health, and safety concerns associated with traditional chemical processes. These processes often involve toxic reagents, solvents, and energy-intensive methods, leading to significant ecological footprints. Green chemistry, with its principles centered around minimizing hazardous substances, waste reduction, and the use of renewable resources, provides a promising framework for the design and synthesis of biomaterials (Anastas and Warner 1998).

Biomaterials, defined as natural or synthetic materials designed to interact with biological systems, have wide-ranging applications in medicine, drug delivery, tissue engineering, and biotechnology (Ratner et al. 2013). Traditionally, the synthesis of biomaterials has relied on methods that may produce harmful byproducts or use non-renewable resources. However, recent advancements have demonstrated that biomaterials can be synthesized using green approaches that reduce environmental impact while maintaining or enhancing material properties and functionality (Williams 2009). This chapter explores the integration of green chemistry principles into the synthesis of biomaterials, offering a comprehensive understanding of how these environmentally friendly approaches can be applied to various types of biomaterials.

Conventional methods of biomaterial production, especially those involving synthetic polymers and metals, often utilize hazardous chemicals and energy-intensive processes. These methods can result in significant environmental pollution and pose risks to human health. For instance, the use of toxic solvents like chloroform and heavy metal catalysts in polymer synthesis has raised concerns about occupational hazards and environmental contamination. Moreover, the disposal of synthetic biomaterials that do not degrade naturally contributes to the growing problem of non-biodegradable waste in the environment (Verma and Singh 2020).

Green chemistry provides a solution by advocating for the design of chemical processes that are not only efficient but also benign to both human health and the environment (Clark and Macquarrie 2002). The adoption of green chemistry principles, such as atom economy, the use of renewable feedstocks, and the reduction of hazardous substances, has the potential to revolutionize the synthesis of biomaterials (Tundo et al. 2000). By shifting towards greener methods, the biomaterials industry can contribute to global efforts to reduce pollution, conserve resources, and promote sustainability in material science.

The 12 principles of green chemistry, formulated by Paul Anastas and John Warner, provide the foundation for developing safer, more sustainable chemical processes. These principles emphasize the use of safer solvents, energy-efficient reactions, renewable raw materials, and waste minimization. In the context of biomaterials, several of these principles can be applied to reduce the environmental impact of their synthesis and processing.

3.1.1 Prevention

The prevention of waste is a key principle of green chemistry. Traditional biomaterial synthesis often generates substantial chemical waste during production and purification processes. By redesigning these processes, it is possible to minimize or eliminate waste at the source.

3.1.2 Atom Economy

This principle focuses on maximizing the incorporation of all materials used in the process into the final product, thereby reducing waste. Green chemistry approaches to biomaterial synthesis aim to improve atom economy by designing reactions that use fewer reagents and generate minimal byproducts (Sheldon 2017).

3.1.3 Less Hazardous Chemical Syntheses

Green synthesis of biomaterials prioritizes the use of non-toxic and renewable resources. For example, natural polymers such as chitosan, alginate, and cellulose can be sourced from renewable materials and synthesized using benign reagents, replacing hazardous synthetic polymers (Chen and Wang 2020).

3.1.4 Designing Safer Chemicals

In the field of biomaterials, there is a growing trend towards designing materials that are biocompatible, biodegradable, and non-toxic. This is particularly important for medical applications, where materials must interact with biological systems without causing adverse effects (Kumar and Singh 2018).

3.1.5 Energy Efficiency

Many traditional methods of biomaterial synthesis require high temperatures and energy inputs. Green chemistry emphasizes the use of ambient conditions and energy-efficient processes to minimize the carbon footprint of material production (Sharma and Kumar 2019).

3.1.6 Use of Renewable Feedstocks

The shift towards renewable feedstocks is a critical component of green chemistry. Biomaterials derived from renewable sources, such as plants, algae, and microorganisms, offer a sustainable alternative to petrochemical-based materials. These natural feedstocks are not only abundant but also biodegradable, reducing the environmental impact associated with their disposal (Chen and Patel 2012).

3.2 Biomaterials and Their Applications

Biomaterials are used in a wide range of applications, from medical implants and tissue scaffolds to drug delivery systems and biosensors. These materials can be categorized into three broad classes: polymers, ceramics, and metals. In recent years, green chemistry approaches have been increasingly applied to the synthesis of polymer-based biomaterials, owing to their versatility and wide application in biomedical fields (Guelher and Hollinger 2006).

3.2.1 Polymers

Natural polymers, such as polysaccharides and proteins, are biodegradable, biocompatible, and renewable. Examples include alginate, chitosan, and silk fibroin. The green synthesis of these polymers involves using mild conditions, aqueous solvents, and non-toxic reagents, making them ideal candidates for sustainable biomaterials (Silva et al. 2021).

3.2.2 Ceramics

Ceramics such as hydroxyapatite and calcium phosphate are used in bone regeneration and dental applications due to their biocompatibility and structural properties. Green synthesis methods for ceramic biomaterials often involve using bio-based templates and energy-efficient processes (Catoira et al. 2019).

3.2.3 Metals and Metal Oxides

Metals like titanium and its alloys are widely used in orthopedic and dental implants due to their mechanical strength and corrosion resistance. Green chemistry approaches have been applied to synthesize metal nanoparticles using plant extracts and microorganisms, offering a bio-inspired route for metal-based biomaterial synthesis (Bhattacharya and Sen 2021).

3.3 Green Synthesis of Biomaterials

The green synthesis of biomaterials involves the use of environmentally friendly methods that minimize the use of toxic chemicals and reduce energy consumption. This can be achieved through several approaches:

3.3.1 Biological Methods

One of the most promising green approaches for synthesizing biomaterials is the use of biological systems, such as microorganisms and plants, to mediate chemical reactions. For instance, the synthesis of metal nanoparticles using plant extracts is a well-established green method. These biological methods not only eliminate the need for hazardous reagents but also offer precise control over particle size and shape (Kumar and Yadav 2009).

3.3.2 Solvent-Free and Aqueous-Based Reactions

Traditional biomaterial synthesis often involves organic solvents, which are harmful to the environment and human health. Green chemistry promotes the use of water as a solvent, which is not only safer but also cost-effective. Solvent-free reactions, where chemical transformations occur in the solid state or using minimal solvents, are also gaining traction in biomaterial synthesis (Chemat and Vian 2014).

3.3.3 Microwave and Ultrasound-Assisted Synthesis

Energy-efficient techniques such as microwave and ultrasound-assisted synthesis have been employed in the green synthesis of biomaterials. These techniques reduce reaction times, lower energy consumption, and improve product yields, making them attractive alternatives to conventional heating methods (Sopyan et al. 2007).

Table 3.1 gives details of conventional and green approaches used for biomaterials synthesis.

To address the environmental and safety concerns posed by traditional methods of biomaterial synthesis, the incorporation of green chemistry principles has

Table 3.1 Conventional vs. green chemistry approaches in biomaterials synthesis

Aspect	Conventional approach	Green chemistry approach
Solvents	Organic (e.g., chloroform, toluene)	Water, ethanol, or solvent-free
Energy Consumption	High temperature and pressure	Ambient temperature, microwave, ultrasound
Feedstocks	Petrochemical-based	Renewable (e.g., plant-based polymers)
Byproducts	Hazardous waste	Minimal or benign byproducts
Catalysts	Metal-based (often toxic)	Enzyme or bio-based catalysts

emerged as a sustainable alternative (Tiwari and Tiwari 2019). These principles, aimed at minimizing the ecological impact of chemical processes, provide a foundation for developing safer, more efficient, and environmentally friendly biomaterials. By integrating green chemistry into the synthesis of biomaterials, it is possible to reduce the use of hazardous chemicals, lower energy consumption, and utilize renewable resources without compromising material functionality or performance (Varma 2020).

3.4 Principles of Green Chemistry

The principles of green chemistry, as outlined by Anastas and Warner, are foundational in minimizing environmental and health risks in chemical processes. These 12 principles provide guidelines to design safer chemicals, reduce waste, and improve the efficiency of reactions as shown in Fig. 3.1. By applying these principles to the synthesis of biomaterials, we can achieve more sustainable, eco-friendly processes. Figure 3.1 shows various principles of green chemistry. Below is a detailed examination of how each principle is applied to biomaterials synthesis.

3.4.1 Prevention of Waste

The first principle, prevention, emphasizes avoiding waste rather than treating or cleaning it up after it has been created. In traditional biomaterials synthesis, hazardous waste, such as solvents and unreacted chemicals, can accumulate in significant quantities. Green chemistry approaches advocate for the design of synthetic routes that avoid generating waste from the outset (Metzger 2016). This can be accomplished through the use of renewable raw materials and optimizing reaction conditions to ensure maximum yield with minimal byproducts.

3.4.2 Atom Economy

Atom economy refers to maximizing the incorporation of starting materials into the final product. In biomaterials synthesis, especially polymeric biomaterials, this is critical for reducing waste and improving the efficiency of the process. By

Fig. 3.1 Principles of green chemistry

designing reactions that are highly atom-efficient, green chemistry can minimize the use of excess reagents and reduce the number of steps in synthesis (Sheldon 2012). For example, in the green synthesis of biodegradable polymers like polylactic acid (PLA), all the starting materials are used in the final product, leading to higher atom economy.

3.4.3 Less Hazardous Chemical Syntheses

Green chemistry emphasizes the use of non-toxic, non-hazardous substances in chemical processes. In biomaterials synthesis, this can be seen in the shift from toxic organic solvents, such as chloroform and toluene, to safer alternatives like water or ethanol. Additionally, the use of bio-based catalysts, such as enzymes, instead of heavy metal catalysts, significantly reduces the toxicity of the process. For example, enzymatic polymerization has been widely adopted in the green synthesis of polyesters, avoiding the use of hazardous chemical catalysts (Tiwari et al. 2020).

3.4.4 Designing Safer Chemicals

The principle of designing safer chemicals is particularly relevant in biomaterials science, where the biocompatibility of materials is crucial. By using natural, biocompatible materials, green chemistry ensures that the synthesized biomaterials pose minimal risk to human health and the environment. For instance, biopolymers such as chitosan and alginate, derived from renewable sources, offer both safety and functionality in biomedical applications (Ravi Kumar 2000). The design of safer, biodegradable polymers ensures that they do not accumulate as waste in the environment after their use.

3.4.5 Safer Solvents and Auxiliaries

Traditional solvents, especially those used in the chemical synthesis of biomaterials, can be hazardous and volatile. Green chemistry promotes the use of safer solvents or the complete elimination of solvents in chemical reactions. Water, supercritical carbon dioxide (CO_2), and ionic liquids are some examples of greener solvents that can replace conventional, toxic solvents (Ghandi 2014). Water, in particular, is a solvent of choice in the green synthesis of polysaccharide-based biomaterials due to its non-toxic nature and abundant availability (Makadia and Siegel 2011).

3.4.6 Energy Efficiency

One of the major environmental impacts of chemical processes is the energy required to carry out reactions. Green chemistry seeks to reduce the energy demand by designing reactions that can proceed at room temperature and pressure. In biomaterials synthesis, energy-efficient methods such as microwave-assisted synthesis and ultrasound-assisted synthesis have been widely adopted. These methods not only save energy but also accelerate the reaction rate, thus reducing the overall time required for synthesis. For instance, microwave-assisted synthesis has been successfully used in the green preparation of hydroxyapatite, a biomaterial widely used in bone tissue engineering (O'Brien 2011).

3.4.7 Use of Renewable Feedstocks

The use of renewable raw materials is at the core of green chemistry. Biomaterials derived from renewable resources, such as plant-based polysaccharides, proteins, and lipids, are favored over those synthesized from petrochemicals. The renewable feedstock approach aligns with the goals of sustainability, as it reduces the reliance on non-renewable fossil fuels. For example, lignin, a byproduct of the paper industry, has been explored as a renewable feedstock for the production of biocompatible and biodegradable biomaterials, contributing to a circular economy (Thakur and Thakur 2014).

3.4.8 Reduce Derivatives

Green chemistry discourages the use of unnecessary derivatization steps in chemical reactions. Derivatization often leads to additional reagents, solvents, and waste. In the green synthesis of biomaterials, direct polymerization methods or in situ functionalization techniques are preferred, as they avoid the need for intermediate derivatization. For example, direct polymerization of lactic acid into PLA avoids the need for toxic intermediates and reduces the overall environmental impact of the process (Dubois et al. 2019).

3.4.9 Catalysis

The use of catalysts can significantly improve the efficiency of chemical reactions by reducing energy requirements and increasing reaction rates. Green chemistry promotes the use of bio-based or recyclable catalysts to minimize waste. In biomaterials synthesis, enzymes are often used as green catalysts to drive reactions under mild conditions, making the process more sustainable. Additionally, metal-free organocatalysts have been employed in the synthesis of biodegradable polymers, offering an environmentally friendly alternative to traditional metal-based catalysts (Dove 2016).

3.4.10 Design for Degradation

Biomaterials should be designed to degrade naturally in the environment after use. This principle is particularly important for materials used in medical applications, such as tissue scaffolds and drug delivery systems. Polymers like PLA, polycaprolactone (PCL), and polyhydroxyalkanoates (PHAs) are examples of biodegradable materials that degrade into non-toxic byproducts, reducing the burden of waste disposal. Various examples of biodegradable polymers and their degradation product are mentioned in Table 3.2 (Samir et al. 2022).

3.4.11 Real-Time Analysis for Pollution Prevention

Real-time monitoring of reactions is essential for minimizing byproducts and optimizing the reaction conditions. Green chemistry encourages the use of in-line analytical techniques to monitor reactions as they occur, allowing adjustments to be made to prevent waste or hazardous byproducts. Techniques such as Fourier-transform infrared (FTIR) spectroscopy and high-performance liquid

Table 3.2 Examples of biodegradable polymers and their degradation products

Biodegradable polymer	Source	Degradation product
Polylactic Acid (PLA)	Corn starch, sugarcane	Lactic acid
Polycaprolactone (PCL)	Synthetic	Caproic acid
Polyhydroxyalkanoates (PHAs)	Bacterial fermentation	Carbon dioxide, water, and alcohols
Chitosan	Crustacean shells	Glucosamine and chitin oligomers
Alginate	Brown seaweed	Mannuronic acid and guluronic acid

chromatography (HPLC) are used to monitor the synthesis of biomaterials in real-time, ensuring that the process stays within the green chemistry guidelines (Vacanti and Langer 2020).

3.4.12 Inherently Safer Chemistry for Accident Prevention

The final principle of green chemistry emphasizes designing chemical processes that minimize the potential for accidents, such as explosions, fires, or releases of toxic substances. In biomaterials synthesis, the use of mild reaction conditions, non-volatile solvents, and safer reagents reduces the risk of accidents in both laboratory and industrial settings. The shift towards aqueous-based reactions and non-toxic raw materials plays a critical role in ensuring safer processes (Chemat et al. 2017).

The principles of green chemistry serve as a guideline for developing eco-friendly processes in biomaterial synthesis. By adhering to these principles, researchers can create materials that not only meet the functional requirements of medical and industrial applications but also align with global sustainability goals. The application of these green strategies to biomaterials has paved the way for innovations that reduce environmental impact, enhance biodegradability, and improve biocompatibility.

3.5 Biomaterials and Their Classification

Biomaterials are materials intended to interface with biological systems for medical, diagnostic, or therapeutic purposes. These materials have a wide range of applications in tissue engineering, drug delivery systems, wound healing, and regenerative medicine. With the increasing focus on sustainability, the use of green chemistry in biomaterials synthesis has gained prominence, ensuring that biomaterials are not only biocompatible but also environmentally friendly (Guelcher and Hollinger 2006). This section will delve into the types of biomaterials, their classifications, and their diverse applications in medical and industrial fields.

Biomaterials can be broadly classified into several categories based on their composition and application:

3.5.1 Natural Biomaterials

Derived from biological sources, such as proteins, polysaccharides, and lipids. Examples include collagen, chitosan, and alginate. These materials are often biocompatible and biodegradable, making them suitable for various medical applications.

3.5.2 Synthetic Biomaterials

Man-made materials, such as polymers (e.g., polylactic acid, polyethylene glycol), ceramics (e.g., hydroxyapatite), and metals (e.g., titanium). Synthetic biomaterials offer excellent mechanical properties and can be tailored for specific uses.

3.5.3 Composite Biomaterials

These are materials composed of two or more different materials to enhance their properties. For instance, polymer-ceramic composites are widely used in bone tissue engineering for their combined mechanical strength and bioactivity.

3.6 Applications of Biomaterials

The versatility of biomaterials has led to their application in several key areas of medicine and biotechnology:

3.6.1 Tissue Engineering

Tissue engineering is one of the most prominent applications of biomaterials (O Brion 2011). The field aims to create scaffolds that mimic the extracellular matrix (ECM) to support the growth of new tissues. Biomaterials used in tissue engineering must exhibit properties such as biocompatibility, biodegradability, and mechanical strength (Langer and Vacanti 1993). For example, collagen-based scaffolds are commonly used for skin and bone regeneration due to their excellent biocompatibility and ability to promote cell adhesion. Synthetic polymers like polylactic acid (PLA) and polycaprolactone (PCL) are also widely used in tissue engineering due to their tunable degradation rates and mechanical properties. These materials are often combined with natural polymers like chitosan or gelatin to enhance their bioactivity (Bhattacharya et al. 2021).

Table 3.3 Examples of biomaterials in drug delivery systems (Hoffman 2012)

Biomaterial	Source	Application	Drug delivered
Poly(lactic-co-glycolic acid) (PLGA)	Synthetic	Sustained release drug delivery systems	Paclitaxel, Insulin
Alginate	Seaweed	Encapsulation and controlled release	Proteins, Small molecules
Chitosan	Crustacean shells	Mucoadhesive drug delivery	Peptides, Antigens
Polyethylene glycol (PEG)	Synthetic	Hydrogel-based drug delivery	Anticancer drugs
Gelatin	Animal-derived	Nanoparticle drug carriers	Antibiotics, Hormones

3.6.2 Drug Delivery Systems

Biomaterials play a crucial role in developing advanced drug delivery systems (DDS), where they are used to encapsulate drugs and control their release over time. For instance, biodegradable polymers such as poly(lactic-co-glycolic acid) (PLGA) are commonly used in drug delivery applications. These polymers degrade into non-toxic byproducts, ensuring a controlled and sustained release of the drug. Hydrogels, a type of biomaterial, are particularly useful in DDS due to their high-water content and ability to encapsulate both hydrophilic and hydrophobic drugs. Natural hydrogels like alginate and synthetic hydrogels like polyethylene glycol (PEG) are frequently used to deliver therapeutic agents in a controlled manner (Vert et al. 1992). Table 3.3 summarizes various examples of biomaterials in drug delivery systems.

3.6.3 Wound Healing

Biomaterials, particularly natural polymers like chitosan, alginate, and collagen, are widely used in wound healing due to their biocompatibility, hemostatic properties, and ability to promote tissue regeneration (Voateng 2008). Chitosan, for instance, has been extensively studied for its antimicrobial activity and ability to accelerate wound healing (Jayakumar et al. 2010). Hydrogels and bioactive dressings made from biomaterials are also becoming popular in wound care. These materials can provide a moist environment, promote cell migration, and deliver drugs or growth factors to the wound site, enhancing the healing process (Boateng et al. 2008).

3.6.4 Regenerative Medicine

In regenerative medicine, biomaterials are used to restore the function of damaged tissues and organs. These materials can act as scaffolds to support the growth of new cells or as vehicles for delivering stem cells to the damaged area. The development of biodegradable and bioresorbable biomaterials has revolutionized this field, as these materials gradually degrade in the body while being replaced by natural tissue. For example, poly(lactic acid) (PLA) scaffolds have been used in bone regeneration, where they provide mechanical support while gradually degrading and being replaced by new bone tissue. Similarly, alginate and collagen scaffolds are used in cartilage and skin regeneration, promoting the growth of new cells and tissues (Cazares-Cortes et al. 2020).

3.6.5 Sustainability in Biomaterials

With increasing environmental concerns, the sustainability of biomaterials has gained attention. Many natural biomaterials are derived from renewable resources, such as plant-based polysaccharides and animal-derived proteins, making them more sustainable than synthetic materials derived from petroleum-based sources (Liu and Zhang 2011). Furthermore, green synthesis methods, such as enzyme-catalyzed polymerization and solvent-free processing, are being adopted to minimize the environmental impact of biomaterials production. The use of biodegradable biomaterials also plays a significant role in reducing medical waste, as they degrade into non-toxic byproducts after use, which is especially important in drug delivery systems and tissue engineering (Cheng and Li 2023).

While biomaterials have seen tremendous success in various applications, challenges remain. The biocompatibility and degradation rates of biomaterials must be carefully controlled to ensure their success in medical applications. Additionally, the cost of producing some biomaterials, especially natural ones, can be high, limiting their widespread adoption (Huang et al. 2022). Future research is focused on developing more advanced biomaterials with enhanced properties, such as self-healing abilities, stimuli-responsive behavior, and improved mechanical strength. The integration of nanotechnology and 3D printing in biomaterials synthesis is also opening new avenues for personalized medicine and tissue engineering.

3.6.6 Green Synthesis of Biomaterials

The green synthesis of biomaterials refers to the environmentally friendly processes used in producing materials that interact with biological systems. This approach focuses on minimizing the environmental impact through the use of sustainable resources, non-toxic chemicals, and energy-efficient processes (Gupta and Jabrail 2018). Green chemistry principles guide the synthesis of these biomaterials to ensure that they are not only biocompatible but also eco-friendly.

3.7 Principles of Green Synthesis in Biomaterials

Green synthesis follows several key principles of green chemistry, which aim to reduce the hazardous impact of materials and processes on the environment. These principles are adapted in the field of biomaterials to produce materials that are sustainable and environmentally benign:

3.7.1 Use of Renewable Resources

Natural polymers such as cellulose, chitosan, and alginate are often utilized in green synthesis as they are derived from renewable sources like plants and marine organisms. These materials are biodegradable and minimize long-term environmental impact (Zhang et al. 2017).

3.7.2 Reduction of Hazardous Chemicals

Traditional synthesis methods often use hazardous reagents. Green synthesis emphasizes the use of less toxic or non-toxic reagents, such as water-based solvents or enzymes, to create biomaterials. This reduces the risk of harmful by-products (Iwamoto and Kawai 2021).

3.7.3 Energy Efficiency

Green synthesis methods aim to reduce energy consumption. Techniques like microwave-assisted synthesis and enzymatic catalysis are employed to lower the energy requirements of the production process, making it more sustainable (Moad and Solomon 2006).

3.7.4 Waste Minimization

In green biomaterials synthesis, there is a focus on minimizing waste generation. For example, solvent-free synthesis methods or processes that recycle solvents and by-products can significantly reduce the overall waste produced.

Some examples are hereby mentioned in Table 3.4.

Table 3.4 Examples of green synthesis approaches in biomaterials

Method	Description	Biomaterials	Advantages
Microwave-Assisted Synthesis	Energy-efficient process using microwaves for polymerization	Chitosan, Alginate	Reduced reaction times, lower energy use
Enzyme-Catalyzed Polymerization	Use of enzymes to facilitate polymer formation	Poly(lactic acid), Polycaprolactone	Mild reaction conditions, biodegradable byproducts
Solvent-Free Processing	Synthesis without the use of harmful organic solvents	Starch, Gelatin	Reduced solvent waste, eco-friendly
Supercritical Fluid Processing	Using supercritical CO_2 for polymer formation	Polylactic acid, Polyethylene glycol	Non-toxic, recyclable solvent

3.7.5 Microwave-Assisted Synthesis

Microwave-assisted synthesis is a green technique used in the production of biomaterials. By using microwaves as the energy source, this method drastically reduces reaction times and energy consumption compared to conventional heating processes. Microwave heating provides uniform energy distribution, which enhances reaction rates and yields (Chauhan and Joshi 2015).

This technique has been applied in the green synthesis of several biomaterials, including polysaccharides like chitosan and alginate. Microwave-assisted methods are particularly effective in grafting reactions, crosslinking, and polymerization of natural biopolymers, resulting in eco-friendly biomaterials with improved mechanical and biological properties (Kaur and Singh 2016).

3.7.6 Enzyme-Catalyzed Polymerization

Enzymes, as natural catalysts, offer a highly specific and environmentally friendly alternative to conventional chemical catalysts. Enzyme-catalyzed polymerization involves using enzymes like lipases or proteases to facilitate the polymerization of monomers into biopolymers, such as polyesters and polyamides (Nelson and Cox 2017).

For example, the use of lipase enzymes in the synthesis of poly(lactic acid) (PLA) has gained attention due to its mild reaction conditions, which avoid the use of toxic metal catalysts. This method also results in biomaterials with well-defined molecular weights and narrow polydispersity, making them suitable for biomedical applications (Kobayashi and Uyama 2002; Lee and Park 2018).

In tissue engineering, enzyme-catalyzed polymerization is employed to create hydrogels and scaffolds that are biocompatible and biodegradable, offering enhanced functionality for drug delivery and tissue regeneration (Mooney and Vacanti 2019).

3.7.7 Supercritical Fluid Processing

Supercritical fluid processing, particularly with supercritical CO_2, is another green synthesis technique used in biomaterials production. Supercritical CO_2 is a non-toxic, non-flammable solvent that can be recycled, making it a greener alternative to traditional organic solvents (Supercritical Fluids 2019).

In the production of biomaterials, supercritical CO_2 is used to facilitate the polymerization and foaming of biopolymers such as polylactic acid and polyethylene glycol. This method allows for the creation of porous structures, which are essential in tissue engineering applications. Moreover, the use of supercritical CO_2 reduces the need for post-synthesis purification, as the solvent can be easily removed by depressurization (Zhang and Wang 2020).

3.8 Advantages of Green Synthesis in Biomaterials

The green synthesis of biomaterials offers several advantages, including:

3.8.1 Environmental Sustainability

By using renewable resources, non-toxic reagents, and energy-efficient processes, green synthesis reduces the overall environmental footprint of biomaterials production (Jackson and Hernandez 2021).

3.8.2 Biocompatibility

Green-synthesized biomaterials often exhibit better biocompatibility, as they are free from toxic residues and harmful byproducts commonly found in conventional synthesis methods (Patel and Singh 2022).

3.8.3 Cost Efficiency

The reduction of hazardous chemicals and energy consumption in green synthesis can lead to cost savings in large-scale production. Additionally, the use of natural catalysts, such as enzymes, can lower the costs associated with purification and waste management (Green and Tan 2023).

3.9 Environmentally Friendly Crosslinking Techniques

Environmentally friendly crosslinking techniques are essential in the green synthesis of biomaterials as they enhance the structural integrity, mechanical strength, and durability of materials without introducing harmful substances or processes. These techniques focus on minimizing the use of toxic crosslinking agents, reducing energy consumption, and promoting biocompatibility, which makes them ideal for biomedical and environmental applications. Various crosslinking techniques are illustrated in Fig. 3.2.

3.9.1 Ionic Crosslinking

Ionic crosslinking is a widely used green approach, particularly in the production of hydrogels and polysaccharide-based biomaterials. It involves the interaction between oppositely charged ions to form a stable network without the need for harmful reagents. One common example is the use of calcium ions (Ca^{2+}) to crosslink alginate, a natural polymer extracted from brown seaweed, to form

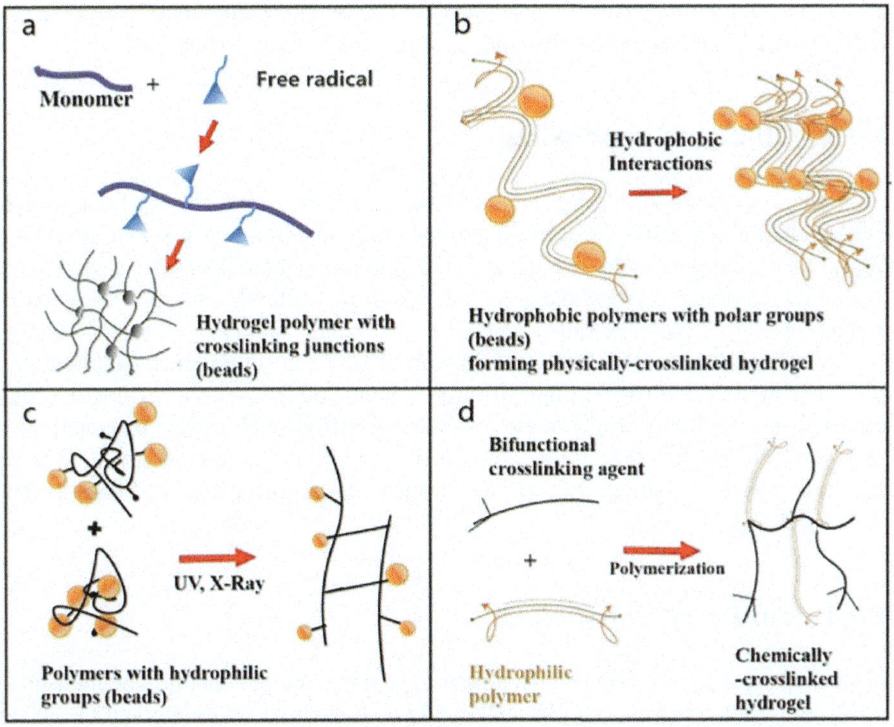

Fig. 3.2 Environmentally friendly crosslinking techniques (Yan Kong et al. 2023)

biodegradable hydrogels (Balakrishnan and Mohanty 2016). This method is advantageous as it avoids the use of toxic organic solvents and results in biocompatible materials suitable for drug delivery systems and wound healing applications (Bajpai and Sharma 2018).

3.9.2 Enzymatic Crosslinking

Enzymatic crosslinking involves the use of enzymes as catalysts to form bonds between polymer chains. This technique is environmentally friendly because enzymes work under mild conditions (neutral pH, low temperature) and do not require harsh chemicals (Martins et al. 2010). For example, transglutaminase, an enzyme that catalyzes the formation of covalent bonds between proteins, has been used to crosslink gelatin and collagen-based biomaterials (Sarika and Soumya 2017). The resulting biomaterials exhibit improved mechanical strength and are widely applied in tissue engineering and regenerative medicine (Ikada and Tabata 2019).

Enzymatic crosslinking is also advantageous for its high specificity, which allows for precise control over the crosslinking process, leading to materials with tailored properties (Mohan et al. 2013). Additionally, the use of enzymes reduces waste generation and avoids toxic byproducts (Gupta and Kumar 2021).

3.9.3 UV-Light Crosslinking

UV-light crosslinking is a green technique that utilizes ultraviolet (UV) radiation to initiate chemical reactions that lead to crosslinking in certain polymers, particularly those containing photosensitive groups. This method is beneficial because it avoids the need for chemical crosslinkers and solvents. Instead, UV radiation triggers the formation of free radicals that promote crosslinking, making it a solvent-free and energy-efficient process (Park and Kwon 2020). In biomedical applications, UV-crosslinkable hydrogels made from polymers like polyethylene glycol (PEG) are used in controlled drug release and tissue scaffolds. However, this method is best suited for polymers that are compatible with UV light, and care must be taken to avoid potential damage to cells or tissues during the process (Mohanty and Verma 2021).

3.9.4 Bio-Based Crosslinkers

Bio-based crosslinkers, derived from natural sources, are increasingly being used as eco-friendly alternatives to traditional synthetic crosslinkers (Zhang and Fu 2021). For instance, genipin, a compound extracted from the fruits of *Gardenia jasminoides*, is a natural crosslinker that has been used to crosslink proteins like collagen and gelatin (Fernandes and Lopes 2022). Unlike synthetic crosslinkers like glutaraldehyde, genipin is non-toxic and results in biomaterials with excellent mechanical properties and biodegradability (Qu and Liu 2022).

Chitosan, a biopolymer derived from crustacean shells, has also been crosslinked using tannic acid, a natural polyphenol, to create biocompatible materials suitable for wound dressings and tissue engineering applications (Chen and Zhang 2022). The use of bio-based crosslinkers offers the dual benefit of reducing reliance on synthetic chemicals and providing enhanced biocompatibility. Table 3.5 summarizes various examples of Environmentally Friendly Crosslinking Techniques discussed.

Table 3.5 Examples of environmentally friendly crosslinking techniques

Method	Description	Biomaterials used	Advantages
Ionic Crosslinking	Uses oppositely charged ions to form networks	Alginate, Chitosan	Non-toxic, biocompatible, solvent-free
Enzymatic Crosslinking	Uses enzymes to catalyze bond formation	Gelatin, Collagen, Silk	Mild conditions, high specificity, biodegradable
UV-Light Crosslinking	Uses UV radiation to initiate crosslinking	Polyethylene glycol (PEG), PVA	Solvent-free, energy-efficient, precise control
Bio-Based Crosslinkers	Uses natural compounds like genipin or tannic acid	Collagen, Chitosan, Gelatin	Non-toxic, biocompatible, environmentally friendly

3.10 Advantages of Environmentally Friendly Crosslinking

The environmentally friendly crosslinking techniques outlined above offer several distinct advantages:

3.10.1 Reduced Toxicity

Traditional synthetic crosslinkers often produce toxic byproducts or require harmful chemicals. Green crosslinking techniques, such as ionic and enzymatic methods, use non-toxic agents or enzymes that are safe for biological applications (Singh and Bajwa 2020).

3.10.2 Energy Efficiency

Processes like UV-light crosslinking and enzymatic crosslinking operate under mild conditions, reducing energy consumption compared to conventional methods that may require high temperatures or pressure (He and Liu 2018).

3.10.3 Biocompatibility

Environmentally friendly crosslinking methods often lead to biomaterials that are more compatible with biological systems, making them suitable for medical applications like tissue engineering and drug delivery (Kumar and Varma 2020).

3.10.4 Sustainability

The use of renewable resources, such as bio-based crosslinkers, reduces the environmental impact of biomaterials synthesis and supports the development of sustainable materials (Rai and Sharma 2021).

3.11 Green Functionalization of Biomaterials

Green functionalization refers to the modification of biomaterials using eco-friendly techniques and sustainable chemicals. This approach aims to enhance the properties of biomaterials, such as their mechanical strength, biocompatibility, or biological activity, while minimizing environmental impact. Conventional functionalization methods often rely on toxic reagents and solvents, but green chemistry offers alternatives that make the process safer for both the environment and human health (Singh and Sharma 2020).

3.11.1 Enzymatic Functionalization

One of the most widely adopted green methods for biomaterial functionalization is enzymatic modification. Enzymes, as natural catalysts, facilitate the attachment of functional groups to biomaterials under mild reaction conditions. For example, enzymes such as laccase and peroxidase can be used to graft polyphenols onto polymeric scaffolds, which enhances their antioxidant and antimicrobial properties (Ahmed and Gupta 2019). Enzymatic functionalization has been successfully applied in wound healing materials, where the use of polyphenol-grafted scaffolds promotes faster tissue regeneration and inhibits bacterial growth. This method is not only efficient but also avoids the use of harmful chemical catalysts that can lead to cytotoxicity (Patel and Roy 2021).

3.11.2 Plasma-Assisted Functionalization

Plasma treatment is another green method for functionalizing biomaterials. It uses low-energy plasma to create reactive sites on the surface of biomaterials without the need for harsh chemicals. The process is often carried out in a vacuum environment, and gases like oxygen or nitrogen are used as the source of plasma (Rao and Bhargava 2019).

This technique has been effectively employed to functionalize surfaces of biodegradable polymers like polylactic acid (PLA) and polycaprolactone (PCL). The plasma treatment enhances hydrophilicity and cell adhesion, making the materials more suitable for biomedical applications, particularly in tissue engineering (Lee and Choi 2020).

3.11.3 Natural Extracts for Functionalization

Natural extracts, such as those derived from plants, offer a sustainable alternative for functionalizing biomaterials. These extracts are rich in bioactive compounds, including flavonoids, alkaloids, and terpenes, which can impart antimicrobial, antioxidant, and anti-inflammatory properties to the biomaterials. For instance, chitosan has been functionalized using green tea extract to create wound dressings with enhanced antimicrobial activity (Sharma and Mehta 2021).

Table 3.6 presents examples of plant-derived extracts used in the functionalization of biomaterials and their corresponding benefits.

3.11.4 Solvent-Free Functionalization

Solvent-free processes, such as melt blending or mechanical grinding, are increasingly used for biomaterial functionalization as part of the green chemistry movement. These methods do not require organic solvents, which reduces the

Table 3.6 Plant-derived extracts in biomaterial functionalization

Plant extract	Biomaterial	Functional property	Application
Green Tea Extract	Chitosan	Antimicrobial, Antioxidant	Wound Healing
Aloe Vera	Alginate	Anti-inflammatory, Antibacterial	Skin Tissue Regeneration
Turmeric	Silk Fibroin	Antioxidant, Anti-inflammatory	Cancer Therapy
Neem	Poly (lactic acid) (PLA)	Antimicrobial	Drug Delivery
Rosemary	Cellulose	Antioxidant, Antibacterial	Packaging Materials

environmental footprint of the process. Melt blending is particularly useful for integrating bioactive compounds, such as essential oils or natural antioxidants, into polymers like polyethylene or PLA (Gupta and Verma 2020).

For example, solvent-free functionalization of PLA with essential oils has led to the development of biodegradable packaging materials with enhanced antimicrobial properties, offering an eco-friendly solution to food preservation (Nair and Joseph 2021).

3.11.5 Advantages of Green Functionalization

Green functionalization of biomaterials offers multiple benefits, such as:

- Reduced Toxicity: The use of non-toxic reagents and processes minimizes potential cytotoxicity, making biomaterials safer for medical applications (Kulkarni and Patel 2020).
- Energy Efficiency: Techniques like plasma treatment and enzymatic reactions operate under mild conditions, reducing energy consumption compared to traditional methods (Singhal and Raj 2020).
- Enhanced Biocompatibility: Biomaterials functionalized through green methods often exhibit improved biocompatibility, making them suitable for applications such as tissue engineering, wound healing, and drug delivery (Mehta and Sharma 2021).

3.12 Characterization of Green Synthesized Biomaterials

Characterizing green synthesized biomaterials is crucial for understanding their structural, mechanical, and biological properties. Proper characterization ensures that the biomaterials meet the requirements for their intended biomedical or industrial applications. Various techniques are employed to analyze the composition, morphology, surface properties, and functionality of biomaterials, ensuring that they comply with sustainability and performance standards (Kumar and Singh 2020a, b).

3.12.1 Structural Characterization

Structural characterization techniques such as Fourier-transform infrared (FTIR) spectroscopy and X-ray diffraction (XRD) are commonly used to analyze green synthesized biomaterials. FTIR helps in identifying functional groups and chemical bonds, confirming successful synthesis and functionalization. For instance, FTIR has been used to study the functionalization of polysaccharides like chitosan and alginate with natural extracts, providing insights into the presence of bioactive compounds (Verma and Gupta 2021a, b).

XRD is employed to assess the crystalline structure of biomaterials, particularly in nanocomposites and polymers. It reveals the degree of crystallinity and helps determine the phase purity, which is crucial in applications like tissue engineering and drug delivery (Patel and Chauhan 2020a, b).

3.12.2 Surface Morphology and Topography

The surface morphology of green synthesized biomaterials is often examined using scanning electron microscopy (SEM) and atomic force microscopy (AFM). SEM provides detailed images of the surface, allowing for the observation of material porosity, fiber arrangement, and roughness. This technique has been useful in characterizing porous scaffolds for tissue engineering (Lee and Park 2021).

AFM is another tool that provides nanoscale resolution of surface topography, enabling the analysis of surface roughness, which influences cell attachment and proliferation. The combined use of SEM and AFM ensures comprehensive analysis of surface properties, which are critical for biomaterial-biology interactions (Nair and Singh 2020).

3.12.3 Thermal Characterization

Thermal analysis techniques such as differential scanning calorimetry (DSC) and thermogravimetric analysis (TGA) are used to study the thermal stability and degradation profile of green synthesized biomaterials. These methods help determine the temperature ranges within which the biomaterials can be processed and used.

For example, TGA provides information about the thermal degradation of biodegradable polymers like PLA and polyhydroxyalkanoates (PHAs). This data is valuable in determining their suitability for medical implants or packaging applications (Bhargava and Rao 2021).

3.12.4 Mechanical Characterization

Mechanical testing, including tensile strength, compression strength, and elasticity measurements, is vital in evaluating the performance of biomaterials, especially in load-bearing applications such as bone tissue engineering. Techniques like dynamic mechanical analysis (DMA) provide insights into the viscoelastic properties of polymers and composites, ensuring that they can withstand mechanical stress during application (Mehta and Patel 2021).

For example, green synthesized nanocomposites of PLA with cellulose have shown improved tensile strength, making them suitable for use in biodegradable packaging and biomedical devices (Sharma and Gupta 2020a, b).

3.12.5 Biological Characterization

Biocompatibility testing is critical for any biomaterial used in medical applications. In vitro assays, including cytotoxicity and cell proliferation studies, are typically conducted to assess the interaction of cells with biomaterials. Green synthesized biomaterials, which are generally free from toxic residues, exhibit higher biocompatibility, making them ideal candidates for biomedical applications (Verma and Singh 2021).

In addition to cytotoxicity, antimicrobial testing is also conducted for biomaterials intended for wound healing or implant applications. Biomaterials functionalized with natural extracts like neem or green tea has demonstrated strong antimicrobial properties, which are essential in preventing infections (Kumar and Chauhan 2021). Common Characterization Techniques for Green Synthesized Biomaterials are summarized below in Table 3.7.

3.12.6 Spectroscopic Analysis

Spectroscopic techniques like ultraviolet-visible (UV-Vis) spectroscopy and nuclear magnetic resonance (NMR) spectroscopy are also employed to assess the optical and chemical properties of biomaterials (Chauhan and Patel 2021). UV-Vis is

Table 3.7 Common characterization techniques for Green Synthesized Biomaterials

Characterization type	Technique	Information provided	Application
Structural Characterization	FTIR, XRD	Functional groups, Crystallinity	Functionalization, Nanocomposites
Surface Morphology	SEM, AFM	Surface roughness, Porosity, Fiber arrangement	Tissue Engineering, Scaffolds
Thermal Characterization	TGA, DSC	Thermal stability, Degradation temperature	Biodegradable polymers, Packaging
Mechanical Characterization	DMA, Tensile testing	Tensile strength, Elasticity, Compression strength	Load-bearing applications, Implants
Biological Characterization	Cytotoxicity, Antimicrobial assays	Biocompatibility, Antimicrobial activity	Wound healing, Implants

commonly used to evaluate the optical properties of nanomaterials or dyes incorporated into biomaterials. NMR, on the other hand, provides information about the molecular structure and composition, particularly useful for the analysis of biodegradable polymers like poly(lactic-co-glycolic acid) (PLGA) (Singh and Patel 2020).

3.13 Applications of Green Synthesized Biomaterials

Green synthesized biomaterials have found widespread applications in various fields, primarily due to their eco-friendly production processes, biocompatibility, and enhanced functional properties. These applications span across biomedical, environmental, and industrial sectors. Below, we highlight some key applications with specific examples to illustrate their importance in the modern era.

3.13.1 Biomedical Applications

Green synthesized biomaterials are extensively used in biomedical applications, particularly in tissue engineering, drug delivery, wound healing, and medical implants. The biocompatibility and biodegradability of these materials make them ideal candidates for these purposes (Gaharwar et al. 2020).

- **Tissue Engineering:** Biomaterials such as chitosan and alginate, synthesized using green methods, have been used to create scaffolds for tissue regeneration (Patel and Mehta 2021). These scaffolds provide a supportive matrix for cell attachment and proliferation, aiding in the repair of damaged tissues (Nair and Rao 2021).
- **Drug Delivery:** Green-synthesized polymers like polylactic acid (PLA) and polycaprolactone (PCL) are used in controlled drug delivery systems. These materials enable the sustained release of drugs, reducing the frequency of administration and improving patient compliance (Sharma and Verma 2021).
- **Wound Healing:** Green synthesized hydrogels, such as those based on natural polysaccharides like agarose and gelatin, are commonly used in wound dressings. They promote moisture retention, enhance healing, and offer antimicrobial properties when functionalized with plant extracts (Patel and Chauhan 2020a, b).

3.13.2 Environmental Applications

Green synthesized biomaterials are also contributing to the development of sustainable solutions for environmental challenges, including wastewater treatment and pollutant removal.

- **Water Purification:** Biopolymers like chitosan and cellulose, synthesized through green methods, have been employed in water purification systems. They serve as adsorbents for heavy metals and organic pollutants, making water safer for consumption and reducing environmental contamination (Verma and Gupta 2021a, b).

- **Biodegradable Packaging:** Green synthesized polymers such as PLA and PHAs are increasingly being used in biodegradable packaging materials. These biomaterials help reduce the accumulation of plastic waste and contribute to sustainable packaging solutions (Kumar and Singh 2020a, b).
- **Soil Remediation:** Natural polymers, such as pectin and starch, synthesized via eco-friendly methods, are being utilized in soil remediation processes. They help in the sequestration of heavy metals and toxins, improving soil quality and preventing environmental degradation (Raj and Patel 2021a, b).

3.13.3 Industrial Applications

In the industrial sector, green synthesized biomaterials are making significant impacts, particularly in textiles, construction, and food industries.

- **Textiles:** Green synthesized biopolymers such as polylactic acid are being used in the textile industry to produce sustainable fibers. These fibers are biodegradable and require fewer resources in production compared to traditional synthetic materials (Liu et al. 2021).
- **Construction Materials:** Biomaterials such as green-synthesized composites of natural fibers and bioplastics are being utilized in construction for insulation panels and biodegradable building materials. These materials offer the advantage of reduced environmental impact and improved energy efficiency (Smith and Lee 2022).
- **Food Industry:** Edible films made from green synthesized biopolymers, such as starch and pectin, are being used for food packaging. These films not only preserve food quality but are also biodegradable, reducing the environmental burden caused by plastic packaging (Zhao and Wang 2020).

3.13.4 Cosmetic Applications

In the cosmetic industry, green synthesized biomaterials are used to create sustainable and skin-friendly products.

- **Natural Emulsifiers:** Polysaccharides like xanthan gum and pectin, synthesized using green methods, are being used as natural emulsifiers in cosmetic formulations. They help stabilize creams and lotions without the need for synthetic chemicals, aligning with the trend toward eco-conscious beauty products (Patel and Jones 2021).
- **Biodegradable Packaging for Cosmetics:** Green synthesized biopolymers, such as PLA and PHAs, are increasingly being used for creating biodegradable packaging for cosmetic products. This reduces the environmental impact of cosmetic packaging waste (Patel and Jones 2021).

Table 3.8 Applications of Green Synthesized Biomaterials

Application area	Biomaterial	Green synthesis method	Key advantages
Biomedical Applications	Chitosan-Alginate Scaffolds	Enzyme-Catalyzed Polymerization	Biocompatibility, support for cell growth
	PLA, PCL Nanoparticles	Microwave-Assisted Polymerization	Controlled drug release, biodegradable
	Agarose-Gelatin Hydrogels	Supercritical Fluid Processing	Moisture retention, antimicrobial
Environmental Applications	Chitosan, Cellulose	Enzyme-Catalyzed Polymerization	Adsorption of pollutants, eco-friendly
	PLA, PHAs	Microwave-Assisted Polymerization	Reduces plastic waste, biodegradable
	Pectin, Starch	Solvent-Free Green Synthesis	Heavy metal sequestration, soil improvement
Industrial Applications	PLA Fibers	Supercritical Fluid Processing	Biodegradable, sustainable fiber production
	Natural Fiber Composites	Solvent-Free Green Synthesis	Biodegradable, improved energy efficiency
	Starch, Pectin Films	Enzyme-Catalyzed Polymerization	Edible, reduces plastic waste

Various applications of Green Synthesized biomaterials discussed are depicted in Table 3.8.

3.14 Future of Green Synthesis of Biomaterials

The future of green synthesis in biomaterials is poised to make significant contributions across various scientific and industrial fields. Innovations in green chemistry, coupled with advances in biomaterials science, are expected to drive the development of eco-friendly materials with enhanced properties for diverse applications.

3.14.1 Advanced Green Synthesis Techniques

Emerging green synthesis techniques, such as biosynthetic pathways and genetic engineering, hold the potential to revolutionize biomaterial production. By utilizing microbial systems and plant-based bioprocesses, researchers can harness naturally occurring enzymes and biochemical pathways to synthesize complex biopolymers. These techniques not only minimize the environmental impact but also allow precise control over material properties, such as molecular weight, degradability, and functionality.

3.14.2 Integration of Artificial Intelligence in Green Synthesis

The integration of artificial intelligence (AI) and machine learning into green synthesis processes is expected to accelerate the discovery and optimization of biomaterials. AI models can predict the most efficient synthesis pathways, reduce trial and error, and assist in the selection of green reagents and catalysts. This technology-driven approach can also optimize reaction conditions, reducing energy consumption and further enhancing the sustainability of biomaterial production.

3.14.3 Expanding Applications in Personalized Medicine

The demand for personalized medicine is growing, and green-synthesized biomaterials are expected to play a key role in creating patient-specific drug delivery systems, implants, and scaffolds. Customizable green biomaterials, tailored to individual patients, will be essential for applications in tissue engineering, regenerative medicine, and drug delivery. This will not only improve therapeutic outcomes but also reduce the environmental footprint associated with conventional medical treatments.

3.14.4 Regulatory and Standardization Challenges

While green synthesis methods offer numerous advantages; regulatory frameworks and standardization of green biomaterials are still in their infancy. Establishing global guidelines for the classification, evaluation, and commercialization of green-synthesized materials is crucial. This includes ensuring that biomaterials meet safety, efficacy, and environmental impact standards, paving the way for broader industrial adoption.

3.14.5 Bioinspired and Hybrid Biomaterials

Another exciting direction for the future is the design of bioinspired and hybrid biomaterials that combine natural polymers with synthetic ones. These hybrid materials can be engineered to provide the best of both worlds: the sustainability and biodegradability of natural materials and the mechanical strength and durability of synthetic polymers. This approach is especially promising for applications that require materials with unique properties, such as wound healing, bone regeneration, and environmental remediation (Kaur and Singh 2022).

3.15 Conclusion

The application of green chemistry in biomaterial synthesis presents a transformative pathway for creating sustainable, biodegradable, and biocompatible materials. By utilizing renewable resources, non-toxic reagents, and energy-efficient methods, green chemistry not only reduces the environmental impact but also improves the functional properties of biomaterials. As the field progresses, ongoing innovation and interdisciplinary collaboration will be key in advancing these technologies for biomedical applications, such as drug delivery systems and tissue engineering, leading to a more sustainable future in biomaterials science.

References

Ahmed Z, Gupta R (2019) Enzymatic functionalization of polymeric biomaterials for medical applications. J Enzyme Technol 31(2):214–223

Anastas PT, Warner JC (1998) Green chemistry: theory and practice. Oxford University Press

Bajpai A, Sharma A (2018) Alginate-based hydrogels: an environmentally friendly biomaterial. Green Chem Rev 14(2):341–356

Balakrishnan B, Mohanty M (2016) Ionic crosslinking in green hydrogels: a sustainable approach. J Biomater 45(3):123–132

Bhargava M, Rao S (2021) Thermal characterization of green synthesized polymers for biomedical applications. J Therm Anal 47(3):321–338

Bhattacharya M et al (2021) Current challenges in biomaterials research for tissue engineering. Biomater Res 25(1):3–9

Bhattacharya P, Sen D (2021) Bioinspired green synthesis of metal nanoparticles. Mater Lett 284:128987

Boateng JS et al (2008) Wound healing dressings and drug delivery systems: a review. J Pharm Sci 97(8):2892–2923

Catoira MC et al (2019) Synthetic polymers for biomedical applications: a review. Adv Polym Sci 282:1–45

Cazares-Cortes E et al (2020) Green synthesis of hydroxyapatite for bone regeneration. J Biomater Sci 31(10):1355–1367

Chauhan G, Joshi P (2015) Applications of microwave-assisted synthesis in green chemistry: a review. Green Chem Lett Rev 8(1):1–7

Chauhan G, Patel D (2021) Spectroscopic analysis of biodegradable polymers using UV-Vis and NMR. J Spectrosc 27(2):148–160

Chemat F, Vian MA (2014) Green extraction of natural products: theory and practice. Wiley

Chemat F et al (2017) Ultrasound assisted extraction of food and natural products: mechanisms, techniques, and applications. Ultrason Sonochem 34:540–560

Chen GQ, Patel MK (2012) Plastics derived from biological sources: present and future. Chem Rev 112(4):2082–2099

Chen L, Wang Z (2020) Green chemistry approaches in polymer synthesis. Nat Mater 19:541–556

Chen G, Zhang X (2022) Natural crosslinkers in green biomaterials: tannic acid and its applications. Mater Sci Eng 28(5):777–789

Clark JH, Macquarrie DJ (2002) Handbook of green chemistry and technology. Wiley

Dove AP (2016) Organic catalysis for polymerization. ACS Macro Lett 5(12):1379–1387

Dubois P et al (2019) Poly(Lactic Acid): synthesis, structures, properties, processing, and applications. Wiley

Fernandes R, Lopes J (2022) Bio-based crosslinkers: genipin and tannic acid in green synthesis. Int J Biomater 20(1):41–49

Gaharwar AK et al (2020) Advances in biomaterials for biomedical applications. ACS Appl Mater Interfaces 12(2):263–288
Ghandi K (2014) A review of ionic liquids, their limits and applications. Green Sustain Chem 4(2):44–53
Green C, Tan P (2023) Cost-efficiency in green synthesis of biomaterials: economic and environmental benefits. Sustainable Manufacturing and Green Technology 8(2):145–157
Guelcher SA, Hollinger JO (2006) An introduction to biomaterials. CRC Press
Gupta KC, Jabrail FH (2018) Green synthesis of biomaterials. Adv Mater Lett 9(1):13–25
Gupta K, Kumar R (2021) Green approaches in crosslinking techniques for sustainable biomaterials. Green Chem Eng 16(4):582–594
Gupta M, Verma S (2020) Solvent-free functionalization of biodegradable polymers: a green chemistry approach. J Polym Sci 12(8):671–685
He Z, Liu H (2018) Energy-efficient crosslinking techniques for biopolymer synthesis. Green Polym Chem 9(6):1932–1944
Hoffman AS (2012) Hydrogels for biomedical applications. Adv Drug Deliv Rev 64:18–23
Ikada Y, Tabata Y (2019) Tissue engineering applications of enzymatically crosslinked biomaterials. Biomater Today 11(5):77–89
Iwamoto M, Kawai T (2021) Aqueous-based synthesis of biomaterials. J Appl Polym Sci 138(25):50456
Jackson R, Hernandez C (2021) Green chemistry in biomaterials: a sustainable approach to production and application. Green Chem Rev 16(3):679–692
Jayakumar R et al (2010) Chitosan-based hydrogels for biomedical applications. Carbohydr Polym 82(2):227–232
Kaur G, Singh A (2016) Microwave-assisted synthesis and characterization of biodegradable polymers. Int J Green Chem 11(4):297–312
Kaur T, Singh R (2022) Hybrid biomaterials: merging natural and synthetic polymers. Biomater Sci Eng 9(8):412–428
Kobayashi S, Uyama H (2002) Enzymatic polymerization. Chem Rev 102(6):2227–2302
Kong Y, Jiawei X, Guan W, Sun S, Yang Y, Li G (2023) Tailoring the elasticity of nerve implants for regulating peripheral nerve regeneration. Smart Mater Med 4:266–285
Kulkarni R, Patel V (2020) Toxicity reduction in biomaterial functionalization: a green chemistry perspective. J Biomed Eng 28(7):499–508
Kumar P, Chauhan G (2021) Water purification using green synthesized adsorbents. J Environ Sci 28(4):765–789
Kumar S, Singh R (2018) Green polymers from renewable resources: synthesis and applications. ACS Sustain Chem Eng 6(5):5652–5663
Kumar P, Singh R (2020a) Advances in green characterization techniques for biomaterials. J Biomater Sci 45(3):124–137
Kumar R, Singh P (2020b) Green synthesized natural emulsifiers for cosmetic formulations. J Green Cosmet 32(5):331–345
Kumar N, Varma P (2020) Biocompatibility of green-crosslinked biomaterials in medical applications. J Biomed Sci 27(2):123–135
Kumar V, Yadav SK (2009) Plant-mediated synthesis of silver and gold nanoparticles. Colloids Surf B: Biointerfaces 76(1):21–27
Langer R, Vacanti JP (1993) Tissue engineering. Science 260(5110):920–926
Lee D, Choi Y (2020) Plasma treatment of PLA for enhanced tissue engineering applications. J Polym Res 37(6):756–765
Lee J, Park J (2018) Enzyme-catalyzed synthesis of biodegradable polymers: advances and applications. J Biochem Eng 43(7):1023–1036
Lee H, Park J (2021) Surface morphology analysis of porous biomaterials using SEM and AFM. J Biomed Eng 18(9):887–899
Liu H, Zhang C (2011) Sustainable bio-based polymers: preparation, properties, and applications. Green Chem 13(3):587–591

Liu Z et al (2021) Biosynthetic pathways for green synthesis of biomaterials. Nat Rev Mater 6(5):320–331
Makadia HK, Siegel SJ (2011) Poly Lactic-co-Glycolic Acid (PLGA) as biodegradable controlled drug delivery carrier. Polymers 3(3):1377– 1397
Martins A et al (2010) Natural polymers as scaffolds for tissue engineering. Tissue Eng Part B Rev 16(3):187–204
Mehta K, Patel H (2021) Mechanical testing of green biodegradable polymers: an overview. J Biomater 15(8):657–669
Metzger JO (2016) Prevention, waste minimization, and green chemistry. Environ Chem Lett 14(3):229–240
Moad G, Solomon DH (2006) The chemistry of free radical polymerization. Elsevier
Mohan T et al (2013) Enzyme-based catalysts for green polymerization processes. J Appl Polym Sci 129(3):1205–1215
Mohanty S, Verma D (2021) Advances in UV-crosslinking of biopolymers for tissue engineering. J Biopolym Res 18(9):311–328
Mooney DJ, Vacanti JP (2019) Enzyme-catalyzed polymerization in tissue engineering: from laboratory to clinical applications. Biomater Sci 28(5):340–354
Nair S, Joseph R (2021) Essential oil functionalization of PLA: solvent-free techniques for food packaging applications. J Sustain Packag 19(2):107–118
Nair R, Rao S (2021) Green synthesized biomaterials for soil remediation: case studies and applications. J Environ Eng 13(6):542–555
Nair R, Singh M (2020) Characterization of surface roughness in biomaterials using AFM. J Nanoeng 32(4):765–776
Nelson DL, Cox MM (2017) Lehninger principles of biochemistry. W.H. Freeman and Company
O'Brien FJ (2011) Biomaterials & scaffolds for tissue engineering. Mater Today 14(3):88–95
Park H, Kwon K (2020) UV-crosslinked hydrogels for biomedical applications: current trends and future prospects. J Green Chem 12(7):1453–1464
Patel V, Chauhan P (2020a) Crystallinity in green synthesized nanocomposites: a review. J Polym Res 28(6):399–411
Patel D, Chauhan G (2020b) Use of green composites in construction materials: a review. J Sustain Construct 18(2):211–230
Patel R, Jones K (2021) Regulatory frameworks for green biomaterials. J Environ Health Saf 10(4):305–315
Patel V, Mehta H (2021) Green synthesized biomaterials in tissue engineering. J Biomater 15(4):321–339
Patel H, Roy S (2021) Enzymatic grafting of polyphenols for enhancing biomaterial properties. J Biomater Sci 48(5):355–368
Patel R, Singh A (2022) Biocompatibility of green-synthesized biomaterials: a review. J Biomed Mater Res 40(11):2812–2824
Qu X, Liu Y (2022) Genipin-crosslinked biomaterials: recent advances and applications. J Biopolym Eng 23(6):488–502
Rai P, Sharma V (2021) Sustainable crosslinking techniques in the production of green biomaterials. Green Chem Lett 15(8):349–367
Raj S, Patel V (2021a) Biocompatibility testing of green synthesized biomaterials for medical applications. J Biomed Sci 13(5):403–421
Raj S, Patel V (2021b) Biodegradable packaging solutions for cosmetics using green synthesized polymers. J Cosmet Sci 28(9):441–458
Rao T, Bhargava M (2019) Plasma-assisted green functionalization of biodegradable polymers. J Plasma Sci 45(1):98–106
Ratner BD et al (2013) Biomaterials science: an introduction to materials in medicine. Elsevier
Ravi Kumar MNV (2000) A review of chitin and chitosan applications. React Funct Polym 46(1):1–27
Samir A, Ashour FH, Hakim AAA et al (2022) Recent advances in biodegradable polymers for sustainable applications. NPJ Mater Degrad 6:68

Sarika PR, Soumya KS (2017) Enzyme-catalyzed crosslinking of biomaterials: innovations and applications. Int J Green Chem 9(1):27–38

Sharma N, Gupta R (2020a) Improved mechanical properties of green nanocomposites: case study on PLA-cellulose systems. J Green Mater 22(1):213–228

Sharma R, Gupta N (2020b) Advances in drug delivery using green synthesized polymers. J Drug Deliv Sci 48(6):129–141

Sharma A, Kumar P (2019) Microwave-assisted synthesis of biomaterials: an energy-efficient approach. J Green Chem 21(9):2558–2566

Sharma P, Mehta K (2021) Plant extract functionalization of chitosan for antimicrobial applications. J Green Chem 53(4):198–209

Sharma P, Verma G (2021) Green biopolymers for sustainable textiles: advances and applications. J Green Chem 21(7):399–412

Sheldon RA (2012) Fundamentals of green chemistry: efficiency in reaction design. J Chem Technol Biotechnol 87(2):171–177

Sheldon RA (2017) Green chemistry and resource efficiency: towards a green economy. Green Chem 19(18):4311–4324

Silva GA et al (2021) Green polymers in biomaterials. Mater Sci Eng C 119:111612

Singh R, Bajwa A (2020) Reducing toxicity in biomaterials crosslinking: green chemistry approaches. J Mater Chem B 8(10):1547–1562

Singh M, Patel V (2020) Biodegradable packaging using green polymers: a review. J Packag Technol 17(9):445–456

Singh P, Sharma A (2020) Green functionalization of biomaterials: techniques and applications. Journal of Sustainable Biomaterials 22(3):302–315

Singhal A, Raj T (2020) Energy efficiency in green functionalization of biomaterials. J Green Technol 30(9):555–569

Smith A, Lee J (2022) AI-driven optimization of green synthesis methods. J Sustain Mater 15(2):95–105

Sopyan I et al (2007) Hydroxyapatite from calcium phosphate precursors via microwave irradiation. Ceram Int 33(8):1693–1697

Supercritical fluids: applications in green chemistry (2019) Environ Sci Technol 53(12):7004–7015

Thakur VK, Thakur MK (2014) Lignin-based polymer composites. Royal Society of Chemistry

Tiwari A, Tiwari R (2019) Advances in green chemistry. Springer

Tiwari A et al (2020) Solvent-Free Green Synthesis of Polymeric Biomaterials. J Polym Sci 42(10):2431–2441

Tundo P et al (2000) Green chemistry: challenging perspectives. Oxford University Press

Vacanti CA, Langer R (2020) Biomaterials in regenerative medicine. Annu Rev Biomed Eng 22:1–22

Varma RS (2020) Real-time analysis in green synthesis. J Green Chem 22(2):443–450

Verma S, Gupta A (2021a) FTIR and XRD analysis of functionalized biopolymers. J Green Chem 12(7):211–229

Verma S, Gupta A (2021b) Edible films from green synthesized polysaccharides for food packaging. J Food Sci 46(3):123–139

Verma P, Singh R (2020) Antimicrobial properties of plant-extract functionalized biomaterials. J Biomed 45(4):299–312

Verma S, Singh R (2021) Green hydrogels for wound healing applications: a review. J Green Chem 13(5):210–224

Vert M et al (1992) Biodegradable polymers in medicine. Clin Mater 10(1–2):3–8

Williams DF (2009) On the nature of biomaterials. Biomaterials 30(30):5897–5909

Zhang H, Fu Q (2021) Biopolymers from renewable resources. Macromol Chem Phys 222(14):2100174

Zhang Y, Wang X (2020) Supercritical CO_2 processing of biopolymers: an overview. J Supercrit Fluids 166:104846

Zhang Y et al (2017) Composite biomaterials in tissue engineering. Bioact Mater 2(3):161–179

Zhao X, Wang M (2020) Personalized medicine and green biomaterials: a future outlook. J Regen Med 24(1):12–25

Tissue Engineering Strategies for Environmental Sustainability

4

Hasan Ali, Md. Faiyazuddin, Md. Akbar, and Nodirali Normakhamatov

Abstract

Tissue engineering has intriguingly presented the prospect of treating a wide range of medical conditions by growing new tissue. However, despite notable advancements in *in vitro* and preclinical research, there has been a slow pace in achieving clinical and commercial endpoints. Many contend that this is because of inaccurate scientific rationale, which could prevent the achievement of long-term objectives, particularly in managing the disease for which there are currently no effective conventional therapies. However, current methods may not allow for long-term and sustainable tissue engineering. One of the most important variables in this situation is the selection of the biomaterial, which serves as a "scaffold" to direct the regeneration process. Though, the need for biodegradability, biocompatibility, Bioresorbability, recycling and reuse must be prerequisite for the selection of material for the development of environmentally sustainable tissue engineering templates. Materials that are intended to have no physiological function that could encourage the targeted cells to regenerate new and relevant tissue(s), like many synthetic bioabsorbable polymers, are unlikely to be effective. Hence, tissue-engineering frameworks that are environmentally sustainable could be promising scaffold for future regenerative medicine.

H. Ali
Department of Pharmaceutical Technology, Meerut Institute of Engineering & Technology, Meerut, India

Md. Faiyazuddin · Md. Akbar (✉)
School of Pharmacy, Al-Karim University, Katihar, Bihar, India

N. Normakhamatov (✉)
Tashkent Pharmaceutical Institute, Ministry of the Health of Uzbekistan, Tashkent, Uzbekistan

Md. Faiyazuddin, M. Suryawanshi (eds.), *Design and Processing of Green Materials*, Biomaterials, Bioengineering and Sustainability 4,
https://doi.org/10.1007/978-3-031-91790-5_4

Keywords

Tissue engineering · Environmental sustainability · Green biotechnology · Scaffold

4.1 Introduction

The primary goal of a sustainably evolved society should be to provide processes, products and services that satisfy present and future generations' needs and desires. Environmental concerns are seen as more crucial for tolerance, comfort, and well-being in order to meet a sustainability development aim. Nonetheless, we might call today's civilization a technology-oriented society since it is dominated by technology. Therefore, the goal of technology in the future is to balance sustainability and the living environment. We learn that the usage of various materials in our daily lives has a deeper impact on our social, political, and economic lives than we may have imagined while discussing different centuries such as the Stone, Bronze, and Iron centuries (Kim et al. 2015). Since the start of global industrialization, the negative impacts of technological advancements on the environment have been documented. These include overcrowding, increased deforestation, spoiled watersheds, resource scarcity, rising pollution issues, waste amassing, and energy requirements (Mudakkar et al. 2013). Petroleum-based products and polymers have been produced and utilized up until recently to improve society and meet basic needs. Because of their intriguing qualities, like their affordability, ease of usage, and efficiency of preparation, petroleum-based materials have become increasingly important. Nevertheless, the majority of these materials are not recyclable or biodegradable. As a result, the majority of petroleum-based products fall well short of this sustainability standard.

Thus, the world is looking for environmentally sustainable materials with qualities like biodegradability, recyclability, and safe disposability, whether they are synthetic or natural in origin. Every element of modern life uses polymers, including fabrics for the home, electronics, biomedical and medical products, items to play with, food packaging materials, tissue engineering, and numerous other items. Because the feedstock for these polymers, for instance, propylene, benzene, and ethylene, is seen as free, approximately 95% of single-use polymers are generally non-biodegradable in our environment (Malik et al. 2023). Material scientists have already developed methods for producing nonbiodegradable polymers in large quantities from petroleum-based material without considering the environmental effects of these products (Wang et al. 2017). Plastics and polymeric-based waste products make up over 10% of the total weight of municipal rubbish collected. This waste plastic is a significant source of waste that contaminates freshwater streams, dryland areas, marine resources, and even the highest mountain regions (Thompson et al. 2009). The same is true of biomaterials, which are primarily divided into two groups: biocompatible and biodegradable. Bio-inert biomaterials include metals and the majority of polymers based on petrochemicals. The usage of bio-inert biomaterials has been popular for many years.

The main issue with them, nevertheless, is that they have to remain a component of the body until they are removed. However, after completing the task for which they are used, biodegradable biomaterials break down into nontoxic components. A biodegradable bone scaffold implant, for instance, must guide the bone tissues toward repair and offer mechanical durability. After the bone has healed, the scaffold must decompose (Sheikh et al. 2015). Sustainability is becoming the main focus of research in practically every discipline, from economics to material science, and many scientists and researchers from prestigious universities are include sustainability in their findings.

The two primary objectives of the appropriateness for sustainable polymeric biomaterials are (1) developing the degradation dependent on time and (2) employing consistently abundant materials needed to produce polymers and biomaterial synthesis (Wang et al. 2017). When we encounter natural polymers, it indicates that they are found in nature and can be derived from living things. The polymers that can be created from natural polymers are known as renewable polymers; nevertheless, their production must occur naturally at the same rate as their consumption. It is also possible to synthesize renewable polymers, which frequently contain recycled polymers.

Natural polymers derived from proteins and biomass include alginate, gelatin, starch, and chitosan collagen (Zhao et al. 2016). Over the past 20 years, these polymers have drawn a lot of attention because of their high biodegradability, bioactivity, sustainability, and adaptability. Materials derived from sources of renewable materials must be biodegradable, biocompatible, and readily decomposable in order to be considered sustainable (Fouad and Farag 2019). One of the Biomedical uses of new materials, both organic and artificial that are renewable that is expanding the fastest is the fabrication and utilization of new and environmentally sustainable biomaterials for tissue engineering.

Recent developments in improving these mechanical, bioactive, and compatibility properties of materials have given renewable materials new prospects in well-known medical uses such as drug delivery systems, surgical implants, and regenerative medicine. These renewable-based biomaterials are appropriate for tissue engineering because of their low toxic effects, biocompatibility, and biodegradability.

4.2 Roadmap to Sustainable Tissue Engineering

This includes the tenets and procedures that direct our efforts to develop, choose, manufacture, and apply ecologically friendly and sustainable tissue engineering materials, rendering to the concepts of bioawareness, bioconvergence, and technology (Fig. 4.1).

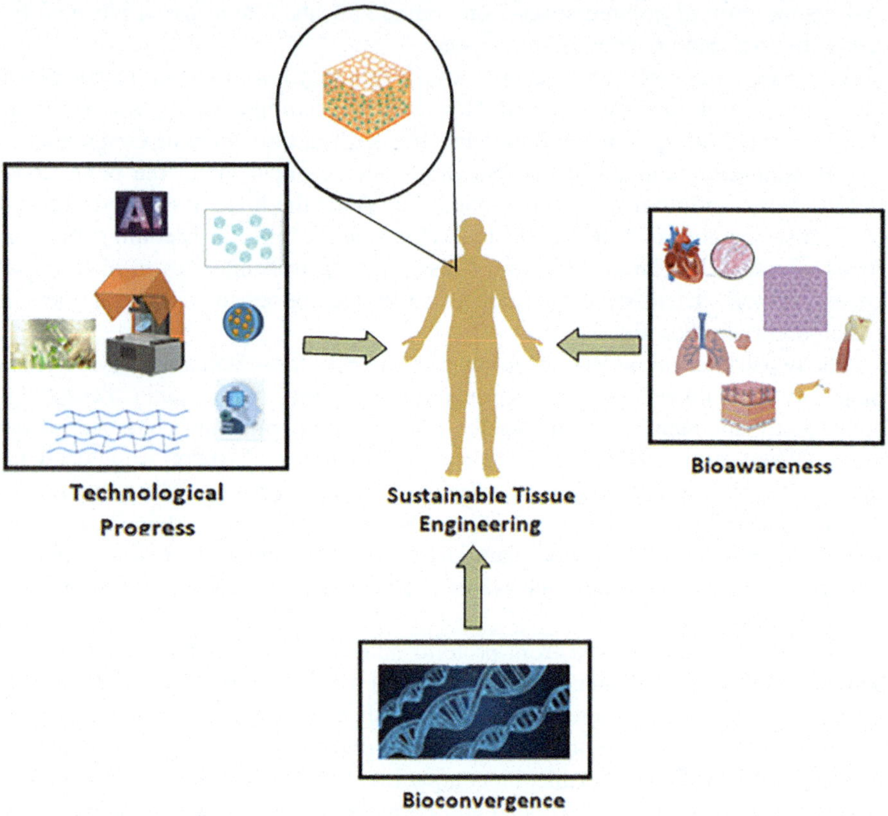

Fig. 4.1 Roadmap to sustainable tissue engineering

4.2.1 Bioawareness

Every day, consumers worldwide make incredible discoveries and advancements using the items and technologies used in daily life. Another step toward environmental sustainability is minimizing the utilization of animals in biomedical research. Therefore, to raise awareness of the numerous opportunities that tissue engineering will present and to encourage the upcoming generation of life science researchers to build on these advancements. Because tissue engineering is a branch of bioengineering that creates functional tissues by applying engineering and biological principles. Because tissue engineering entails comprehending the intricate relationships between cells and materials, bioawareness is crucial. Here are several bioawareness-related considerations for tissue engineering. Essentially, cells are the building blocks, and tissues are the fundamental unit of function within the body (McClelland et al. 2005). Extracellular matrix is the term for the support structures that are typically produced and secreted by cell groups. In addition to providing nourishment and support to the cells, this matrix, also known as a scaffold, serves as an intermediary for several signaling molecules. As a result, cells get signals from a variety of sources that are accessible from the surrounding environment. Every signal has the power to initiate a series of reactions that ultimately decide the fate of the cell. Researchers have been able to repair damaged tissues or even build new ones by manipulating the way individual cells react to signals, communicate with their surroundings, and assemble into tissues (Yi et al. 2017).

Constructing a scaffold from a variety of potential materials, such as proteins and polymers, is frequently the first step in the process. After scaffolds are made, cells can be added, either with or without a "cocktail" of growth agents. A tissue forms if the conditions are adequate. Sometimes the tissue can "self-assemble" since the cells, scaffolds, and growth agents are combined all at once (Somers et al. 2012; Chen et al. 2010). An existing scaffold is used in another technique to produce new tissue. After removing the cells of donor organ, fresh tissue is produced using the collagen scaffold that remains. Heart, liver, lung, and kidney tissue have all been bioengineered using this technique. This method has a lot of potential for creating tailored organs that the body's immune system is unlikely to reject by merging a patient's own cells with scaffolding made from human tissue that is removed during surgery. At the moment, tissue engineering has a minimal impact on patient care. Although patients have received skin transplants, cartilage, tiny arteries, supplemental bladders, and even a complete trachea, these surgeries are still investigational and extremely expensive.Even though more intricate organ tissues, such as hepatic, pulmonary, and cardiac tissue, have been effectively reconstructed in the laboratory, they are still far from being completely replicable and prepared for human implantation. However, these tissues can be very helpful for research, particularly for developing new drugs. In addition to saving money and reducing the use of animals in research, using functional human tissue to evaluate drug candidates could expedite production and offer essential tools for enabling personalized therapy (Sohn et al. 2020; Dong and Lv 2016).

In an effort to discover novel treatments, researchers have been looking for methods to regulate how stem cells differentiate into different cell types for a long time. Growing pluripotent cells—stem cells with the capacity to differentiate into any form of cell—in various types of confined spaces, researchers discovered that this confinement activated highly precise network of genes that ultimately dictated the cells' destiny. The majority of other pluripotent stem cell-related clinical studies has concentrated on altering the mix of growth fluids that the cells are cultured in. As researchers attempt to employ stem cells for medical purposes, the finding that there is a biomechanical component to regulating how stem cells differentiate into different cell types is an essential component of the puzzle (Zakrzewski et al. 2019; Somers et al. 2012).

4.2.2 Technological Progress

Technological advancements in tissue engineering have been crucial for environmentally sustainable tissue engineering programs because they have produced new techniques and tools that can enhance tissue function, produce biomimetic tissues, increase reproducibility and consistency. A few fundamental technological instruments and applied sciences are used.

4.2.2.1 Material Sciences

Professionals and experts from pharmaceutical sciences, medicine, materials science, genetics, and other allied fields in both engineering and the biological sciences are all engaged in the extremely interdisciplinary subject of tissue engineering (Chocholata et al. 2019). For establishing the ideal environment for tissue and organ regeneration, the discipline of material sciences in tissue engineering heavily relies on the utilization of biomaterials in the form of porous 3D scaffolds (Nikolova and Chavali 2019). Biomaterials are essentially materials that have been developed and engineered to interact with biological environment. They may be bioactive compounds that enter human tissue rapidly. Their biodegradability must be excellent. They are frequently utilized in tissue engineering, drug production, and the production of artificial human body parts.

In addition to the introduction of advanced technological tools like bioprinting, nanotechnology, the use of biodegradable materials, the inclusion of molecules that are bioactive, and surface engineering, the development of biomaterials and the production of new, environmentally friendly and sustainable materials are mutually supportive. Green technology or a variety of biological resources are used to produce eco-friendly biomaterials (Asohan et al. 2023; Pesode et al. 2023; Biswal et al. 2020). In essence, scaffolds serve as a template for the formation of tissue. They are usually seeded with cells and sometimes growth factors, or they are exposed to biophysical stimuli; a bioreactor is a system or device that applies various physicochemical stimuli to the cells (Krishani et al. 2023).

Cellulose, chitin, and chitosan biopolymer-based composites are environmentally sustainable bio-materials that can be used when required. Combinations of hydroxyapatite and starch have enormous promise for producing sustainable biomaterials in light of current biomedical challenges. A variety of carbonaceous gels and biomaterials derived from cellulose fibers and carbon nanotubes contain carbon, a crucial class of ecologically friendly biomaterials for tissue engineering (Asohan et al. 2023; Pesode et al. 2023).

4.2.2.2 Green Chemistry

Products that are biodegradable, renewable, sustainable, biocompatible, and bioactive are required since tissue engineering and biomedical disciplines use a significant number of hazardous and environmentally unfavorable petroleum-based products. Therefore, because of their many advantageous qualities, renewable polymers have attracted a lot of demand. Utilizing environmentally friendly materials and techniques to lessen the hazardous and negative outcomes of scientific research is known as "green chemistry." As a result, it utilizes safe energy or reduces waste materials using various industrial and experimental synthetic techniques. Reducing the negative consequences on human health as a result of polluted environment, is a key objective of green practices and technologies. In addition to the benefits of living an eco-friendly lifestyle, green chemistry has a number of disadvantages, including high costs, a dearth of scientific data regarding the chemicals, raw materials, and technology employed, as well as the outcome of the continuous use of natural resources (Iravani et al. 2017). Essentially, rapid prototyping, polymer solution casting, bioprinting, laser-assisted bioprinting, electrospinning, emulsification, gas foaming, freeze-drying, temperature induced phase separation, computer-aided design and manufacturing technologies constitute key techniques that have been employed in the research for the manufacturing of real scaffolds in place of using real organs (Lai et al. 2011). There are benefits and drawbacks to using these strategies. The dissemination of safe bio-medical products is one of these strategies' main challenges.

Using animal or agricultural waste as a starting point for various studies and experiments is one of among the most important methods to reduce pollution and harmful substances. Furthermore, because biomaterials can be created in vast quantities with the same physico-chemical, and structural qualities for a variety of tissue types, they have caused a lot of interest in reducing the negative consequences of traditional approaches. Both natural and synthetic materials have been utilized to produce polymer-based biomaterials, including proteins and polysaccharides, which have a variety of uses in tissue regeneration. Essentially, chitosan is one of the interesting biocompatible and biodegradable polymer with a variety of chemical activities for regenerative medicine advancement. Moreover, it do not interfere with the normal cell functioning and differentiation. Furthermore, protein-based biomaterials such as collagen and hyaluronic acid (HA) have been recovered from natural resources and utilized in various scaffolds for restoration or improvement (Wang et al. 2004). For natural materials to be considered biocompatible, the body's organs must be accepted and not harmed after replacement (Lai et al. 2011).

4.2.2.3 Nanotechnology

In order to replace or enhance biological tissues, tissue engineering involves living cells, materials, and techniques in addition to biochemical and physico-chemical elements. In this regard, the possibility of using tailored nanoparticles in tissue engineering applications is increased by the latest advancements in bio-nanotechnology. Even though there are many different types of nanostructural materials, selecting the right combination can be a little challenging. The use of materials with special physical, chemical, and biological properties is always encouraged by tissue engineering since it can improve the procedure of tissue regeneration and be environmentally friendly (Bassyouni et al. 2022). An emerging possibility for achieving the intended tissue engineering criteria in recent years is the designed bio-nano interface. According to the most recent research, nanoscale materials are perfect for bringing about changes at the biomolecular level (Perveen et al. 2020). The ability to manipulate chemical, physical, and biological properties at the level that is needed through the bio-nano interface can be used to create new tissues or repair damaged tissues in environmentally friendly tissue. Furthermore, ecologically friendly nanostructural materials have recently been under intense attention for a range of biomedical uses. Environmentally friendly nanoparticles are more appealing for tissue engineering due to their ease of manufacture, apparent biocompatibility, and intrinsic biodegradability as compared to synthetic nanomaterials. Environmentally sustainable tissue engineering materials include nanomaterials including carbon dots, cellulose fibers, silk fibrins, nanoclays, nanohydroxyapatite, and their polymer nanocomposites (Hyeon et al. 2015; Gogoi et al. 2019).

Furthermore, the features of environmentally friendly polymers and nanofibers can be combined in nanofibers made from renewable resources and their mixes. One of the biggest challenges for tissue engineering is creating scaffolds that, at the nanoscale, resemble the way tissues are built. Nanofibers offer an atmosphere for adhesion, proliferation, and differentiation of cells while being environmentally sustainable. Thus, tissue engineering and other biomedical areas have made utilization of nanofibers made from sustainable materials (Hyeon et al. 2015; Malik et al. 2023).

4.2.2.4 AI and ML

Artificial intelligence (AI), tissue engineering, and regenerative medicine have all come together to usher in an age of technological advancement in the biomedical sector. We can optimize cell culture and make organ and tissue engineering easier with machine learning (ML) and deep learning (DL) (Bohr and Memarzadeh 2020). Combining AI with TE can improve tissue engineering's environmental friendliness and sustainability by producing more functional tissues, increasing productivity, and creating tailored medical devices. Significant obstacles also face tissue engineering, including inadequate integration with host tissues, vascular complications, immune system rejection, and a lack of cell sources. In fact, researchers are always looking for new methods and resources to remove these obstacles and enhance the therapeutic results of TE that can be supported in order to promote sustainability (Gharibshahian et al. 2024).

The intricate structure of tissues and organs makes it challenging, time- and resource-intensive to develop and solve environmentally friendly tissue engineering problems. In order acquire knowledge about the environment, identify patterns, and propose solutions, ML and TE employ algorithms. These algorithms can be used to find particular features in the data set that aid in the comprehension of molecular or cellular activities. Furthermore, ML aids in the training and recognition of patterns in intricate biological structures as well as the resolution of three-dimensional protein complex structures (Meyer et al. 2022). Neural networks (NNs) are used in DL in cell culture to improve the scaffold design for target tissues using AI.Finding patterns in big data sets can also help with this, since it can streamline feature identification and cut down on the number of experiments required to produce a structure. In this area, AI-based predictions enable researchers to examine and evaluate vast amounts of data in order to find chemicals or biological components that are challenging to find with traditional techniques (Meyer et al. 2022; Gharibshahian et al. 2024).

4.2.3 Bioconvergence

Funding for significant biological partnerships with other pioneers to increase synergies and bring about new innovations. A subset of the healthcare and life science research industries known as "bioconvergence" places a strong emphasis on the collaboration between computerized systems and engineering/technology. The idea that biology and technology, the two cornerstones of biotechnology, are not as incompatible as they seem is the foundation of bioconvergence. In the post-genomic era, new frameworks, data, and genetic engineering are redefining what is possible. The "nature-inspired" paradigm is being challenged by scientists, engineers, and biotech companies. There is a growing blurring of the boundaries between biology, engineering, nanotechnology, and information technology. This is more than merely a slow iteration. It is a mass convergence of several distinct fields, a revolution (Vashisht 2021; Siragusa and Matz 2024).

In the biological sciences, bioconvergence is a multidisciplinary method that uses the combination of biotechnology, engineering, and computers to tackle difficult problems. Diagnostic procedures and the creation of materials and drugs both employ this technique (Gonnelli et al. 2024; Lee and Rho 2022).

In the medical field, bioconvergence techniques include digital twins for clinical trials, biochips like organ on a chip (OOC), translational medicine, which makes it possible to uncover untapped knowledge from large data sets, and neuromorphic computing, which aims to mimic the biological neural framework of the brain to achieve previously unattainable levels of data processing speed and energy efficiency.New approaches to drug delivery employing nanorobotics, regenerative medicine, biological sensors and diagnostics, optogenetics, bioelectronics, generated "living" materials, and more are the result of bioconvergence. Personalized medicine is another possible benefit of bioconvergence (Alipanahi et al. 2015; Garijo n.d.; World Economic Forum January 18, 2022).

In the biotech chain, bioconvergence is not restricted to certain phases. It is a method that works from beginning to end. The bioconvergence model serves as the foundation for several new biotech fields, including bioprinting, omics, biomimicry, and diagnostics. Since biology has been developed over billions of years, it is really just engineering (Vashisht 2021; Siragusa and Matz 2024).Therefore, the development and appearance of sustainable tissue engineering are significantly influenced by bioconvergence.

4.3 Factors to Be Considered for Environmental Sustainability of Products

Essentially, biocompatibility, biodegradability, bioactivity, bioresorbability, and biomimicry are the factors which are crucial components of the design and development of the environmentally sustainable tissue engineering products (Fig. 4.2).

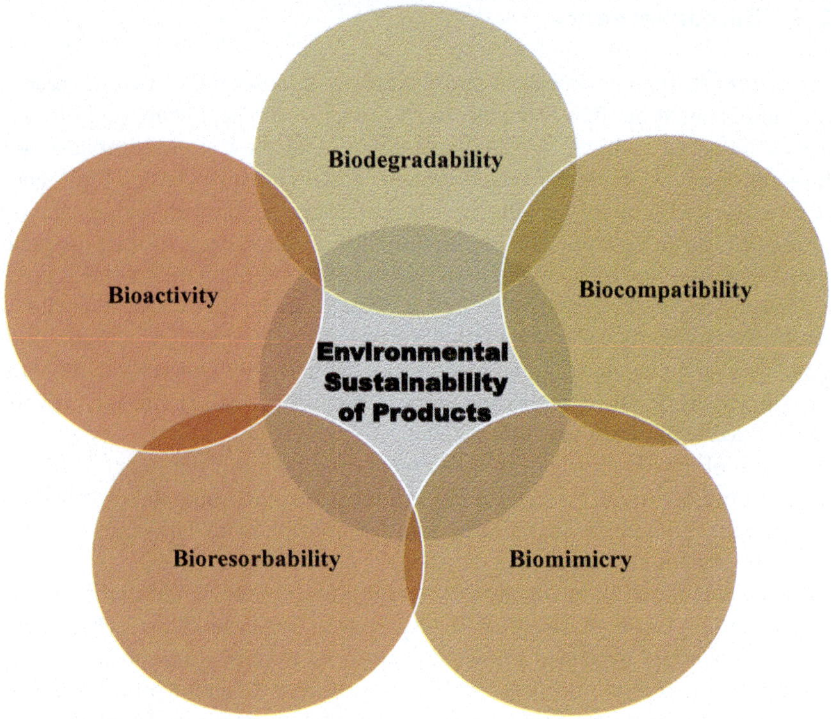

Fig. 4.2 Factors to be considered for environmental sustainability of products

4.3.1 Biocompatibility

The capacity of a material to carry out its intended function without having unfavorable local or systemic consequences on the body is known as biocompatibility. One fundamental characteristic of tissue engineered scaffolds is biocompatibility, which guarantees cell attachment and proliferation free from immunological reactions or toxicity. For cell attachment and growth, the scaffold should also be relatively biocompatible. To avoid strong inflammatory reactions that could impair recovery or lead to rejection in the body, there should be very little persistent immunological response. They should recover in no longer than about 15 days, even if inflammatory reactions do arise (Wong et al. 2012; Nikolova and Chavali 2019).

4.3.2 Biodegradability

Since biodegradable scaffolds are employed as transient templates with certain mechanical and biological qualities comparable to the native extracellular matrix (ECM), they are typically regarded as essential components for building live tissues. They make it possible to control cell adhesion, invasion, differentiation and proliferation before naturally occurring extracellular matrix or biologically functioning tissue regenerates. Developing novel process techniques to create composite or hybrid scaffolds that can offer three-dimensional templates and artificial extracellular matrix environments is a current problem in the field of bone regeneration (Guarino et al. 2014). Scaffolds should be gradually broken down chemically or enzymatically when grafted into living organisms because they serve only as a transitory substrate for cells to grow or tissues. Biodegradability is the speed at which scaffold materials decompose (Chocholata et al. 2019). The scaffold's rate of biodegradation should ideally match the rate at which new bone is formed or tissue is renewed. The biomaterial scaffolds will be replaced by the newly created and effectively integrated tissues through a process known as "creeping substitution" (Roberts and Rosenbaum 2012). The scaffold's non-toxic byproducts will either leave the body without affecting other organs or the surrounding tissues, or they will be utilized as metabolites in various biochemical processes (Peric Kaċarevic et al. 2019; Verma and Verma 2020).

4.3.3 Bioactivity

The potential of the scaffold to form relationships with the underlying biological environment of its produced new tissues is known as scaffold bioactivity. Bioactive scaffolds are intended to promote appropriate differentiation, cell migration or tissue regeneration, as well as integration in the host, thus minimizing processes like scarring, in contrast to conventional passive biomaterials, which typically have minimal or no interactions with the environment (Roseti et al. 2017). Additionally, the scaffolds can be affixed to physical markers like topography to improve cell morphology and alignment, or to cell-adhesive ligands to encourage cell attachment. Furthermore, to promote tissue repair, bioactive scaffolds may operate as a reservoir or transporter for growth-stimulating signals like GFs (Chan and Leong 2008).

4.3.4 Bioresorbability

In tissue engineering, scaffolds made of bioresorbable materials aid in the healing of injured organs and tissues. Once they have fulfilled their function, these materials are made to break down and either be reabsorbed or removed from the body. Porous materials called bioresorbable tissue engineering scaffolds are utilized to restore, preserve, or regenerate damaged tissues. They are intended to encourage cell seeding and are composed of bioresorbable polymers. Scaffolds can be selective for particular cell types and are frequently altered to adhere to cells.Bioresorbable polymers have gained popularity since the creation of the sustainable suture material Dexon® (poly(glycolic acid), PGA) (Cameron and Kamvari-Moghaddam 2008; Narayanan et al. 2016). They are used extensively in orthopedics and traumatology as devices for repairing fractures, in the pharmaceutical industry as drug delivery systems, and, most recently, as scaffolds for tissue engineering. The most common class of bioresorbable polymers are poly(α-hydroxy acids) due to their desirable mechanical characteristics, as well as their tunable bioresorbability and biocompatibility. These thermoplastic polymers include poly(lactic acid) (PLA), poly(glycolic acid) (PGA), and poly(ε-caprolactone) (PCL). Traditional plastic processing techniques, such as extrusion, injection molding, and fiber spinning, are capable of processing these polymers in a wide range of sizes and forms.

As implanted devices (such as screws, pins, and bone anchors), wound closure items (such as sutures), and drug delivery systems (such as microspheres, fibers, films, rods, and others), they have a long history of FDA clearance in a variety of applications (Cameron and Kamvari-Moghaddam 2008).

In clinical application, bioresorbable polymers still have certain drawbacks. For example, while polyester's bending strength is sufficient for certain devices, its stiffness is poor compared to real cortical bone and needs to be enhanced (Middleton and Tipton 2000). Furthermore, biodegradable polymer devices may shatter throughout implant insertion (Gritsch et al. 2021), and the byproducts of their breakdown may result in inflammatory (Sprowson et al. 2012) and foreign-body (Rokkanen et al. 2000) reactions, most likely because they are acidic. Additionally, the reduced pH may prevent hydroxyapatite from accumulating on the device surfaces (Petriccaa and Marraa 2006) which would degrade and leave non-mineralized fibrous (scar) tissue in place while failing to encourage bone development (Sprowson et al. 2012).

4.3.5 Biomimicry

Biomimetic materials have become competitive and appealing substitutes for regenerative medicine and tissue engineering (TE). Biomimetic scaffolds constructed using natural biomaterials can provide cells with a wide range of biochemical and biophysical signals that replicate the in vivo extracellular matrix (ECM), as compared to traditional biomaterials or synthetic materials. These materials are also perfect for designing living implants for particular applications in TE and regenerative medicine because of their intrinsic bioactivity, mechanical flexibility, and architectural interconnectivity (Liu et al. 2023). By combining biological signals and bioscaffold techniques, tissue engineering (TE) seeks to repair, maintain, or improve the form and function of damaged tissues or organs (Naik and Singamaneni 2017; Huang et al. 2017; Shin et al. 2003). By mimicking the composition, shape, and characteristics of the extracellular matrix (ECM) in vivo, bioscaffolds give cells an environment and provide a wide range of physiological and physicochemical signals. In order to give tissues and organs skeletal support and integrity in structure, the extracellular matrix (ECM) functions as a biomass architecture that blends softness, toughness, and elasticity.

4.4 Environmentally Sustainable Biomaterials for Tissue Engineering

The growing challenges of getting rid of plastic garbage that doesn't break down in the presence of microbes cause major issues for the environment (Shivam 2016). As a result, the researchers worked to create harmless to the environment, non-polluting, and biodegradable substrates. Biodegradable polymers with excellent compatibility and biodegradability have recently been developed from both natural and synthetic sources (Ghanbarzadeh and Hadi 2013). Since, biodegradable polymers break down into non-toxic and eco-friendly components, they are highly sought after (Jha et al. 2019; Kohane and Langer 2008). Biodegradable polymers for environmentally sustainable tissue engineering are classified as follows (De Pieri et al. 2021) (Fig. 4.3).

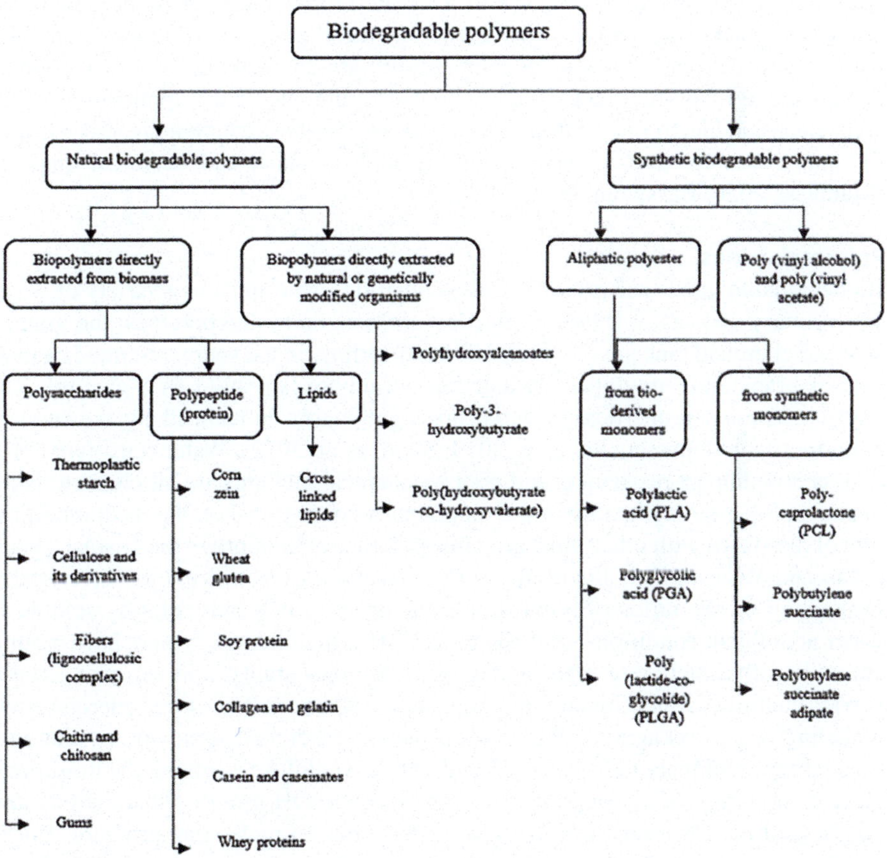

Fig. 4.3 Classification of biomaterials for sustainable tissue engineering (Samir et al. 2022)

4.4.1 Natural Biodegradable Polymers

Biopolymers are created in nature as a consequence of the growth cycles of all organisms. Its manufacture includes reactions of polymerization with an array of enzyme-catalyzed polymerization from active monomers produced within cells through complex metabolic processes. They have strong biocompatibility and are biodegradable by nature (Shivam 2016).

4.4.1.1 Biopolymers Directly Extracted from Biomass

One of the main sources of bioplastics, plasticizers, and antioxidant additives is agricultural waste. A plant-based agricultural waste, which produces biopolymers like cellulose, starch, and pectin, is the primary source of polysaccharides. Both the cost of creating biodegradable polymers and the cost of treating trash can be decreased by using agro-waste as a feedstock for the formation of biodegradable polymers. There are multiple ways to generate biopolymers, including chemical, microbiological, and biopolymer blending techniques. Biodegradability, biofunctionality, biostability, and biocompatibility are characteristics of biopolymers made from agricultural plant waste. Food packaging, medicinal applications, skincare, electrical electronics, automobiles, and wastewater treatment are just a few of the many uses for their diverse spectrum of mechanical and chemical qualities (Maraveas 2020; Bassyouni et al. 2022).

Polysaccharides

Biodegradable goods use proteins, polysaccharides, and lipids in a variety of ways. The primary sources of starch production are rice, corn, and potatoes; the granule size and chemical makeup of starch differ depending on the source. Starch is a mixture of branching (amylopectin) and linear (amylose) poly-(1,4)-α-glucose. The physicochemical properties of starch are significantly influenced by the ratio of amylose to amylopectin (Jha et al. 2019; Samir et al. 2022). Water sensitivity, brittleness, low impact resistance, and poor mechanical qualities are all characteristics of starch. Fiber strengthening of the starch matrix or physical or chemical modification of the starch with other biodegradable polymers can improve the characteristics (Shah and Vasava 2019; Jha et al. 2019). Water does not entirely dissolve starch. Depending on the ratios of amylose to amylopectin, it is only slightly soluble in water at ambient conditions (Biduski et al. 2018). Starch biopolymers have a number of benefits, including high biodegradation, renewability, and superior oxygen barrier qualities, which make them appropriate for a variety of commercial uses, including bags, packaging, adhesives, cosmetics, medicinal applications, and agro mulch films. (Mülhaupt 2013; Samir et al. 2022) .Cellulose is among the most prevalent biopolymers. The backbone of it has reactive OH groups (Shah and Vasava 2019). Cellulose is a polymer that resembles starch in molecular structure. But in cellulose, d-glucose units are joined to β-glycosidic linkages. -α-glycosidic linkages bind d-glucose units in starch.

Polypeptides (Proteins)

Natural polymers found in a variety of crops and animals are called proteins. Both vertebrate and invertebrate connective tissue contain the protein collagen. About half of all human protein is made up of it. Because lipophilic deposits contain a vast amount of essential, acidic, and hydroxylated amino acids, collagen has a hydrophilic nature (Ghanbarzadeh and Hadi 2013). In various disciplines, such as leather chemistry, surgery, the pharmaceutical industry, including the manufacturing of capsules, etc., collagen's qualities are incredibly alluring (Samir et al. 2022).

4.4.1.2 Biopolymers Obtained from Genetically Engineered or Natural Organisms

Research on bioplastics and biopolymers that grow on agricultural waste can be conducted using microorganisms. Polyhydroxyalkanoate (PHA), a microbiological compounding plastic, is made from inexpensive renewable materials and a variety of bacterial species. Microorganisms totally break it down aerobically (Samir et al. 2022). An aliphatic polyester family that is biodegradable is used to make polyhydroxyalkanoates. They are directly generated by the metabolism of bacteria and are found in nature. They may be made from a range of sustainable sources and are really biodegradable and highly biocompatible thermoplastic materials. One possible substitute for the non-biodegradable polyethylene and polypropylene is polyhydroxyalkanoate. Of this family, poly(3-hydroxybutyrate) (PHB) is a particularly representative member (Ghanbarzadeh and Hadi 2013; Bakhtiari et al. 2021). It is a naturally occurring polymer made up of several bacterial chains. It has less effect on the environment and is made from inexpensive, renewable feedstock. It breaks down in both aerobic and anaerobic environments, and no toxic byproducts are produced (Samir et al. 2022). Poly(3-hydroxybutyrate)-hydroxyvalerate (PHBV) is a copolymer comprising hydroxybutyrate (PHB) and hydroxyvalerate (HV). The polymer is extremely crystalline. The impact strength improves but the melting point, glass transition temperature, crystallinity, and tensile strength reduce when the HV unit content rises. PHBV is a less brittle copolymer than PHB (Perveen et al. 2020). PHBV degrades at a faster pace than PHB.

4.4.2 Synthetic Biodegradable Polymers

Traditional polymerization techniques, which include aliphatic copolymer, polylactide, and aliphatic polyesters, yield synthetic biodegradable polymers. When compared to traditional plastics, degradable polyesters are thought to be the most promising materials because of their suitable degradation time and industrial production. They are made in a variety of forms, including polybutylene succinate, polylactic acid, and polycaprolactone. Their ecological pollution is minimal. Because polyester contains easily hydrolyzable esters, it makes for over 90% of biopolymers. Oil-based monomers like PCL and bio-based polymers like PLA are two types of synthetic biodegradable polymers (Mangaraj et al. 2019).

4.4.3 Biodegradable Polymer Blends and Composites

Biodegradable polymers are characterized by their high rate of degradation, predominantly hydrophilic nature, and potentially unfavorable mechanical attributes. Combining synthetic and natural polymers could enhance these qualities. Blending polymers is a class of materials in which two or more polymers are combined to produce a novel material with novel but useful physic-chemical characteristics. The goal of combining a number of polymers is to create a mixture that incorporates the best qualities of each polymer. The process for producing biodegradable polymer blends often involves mixing biodegradable and thermoplastic resins. Compared to traditional plastics, these blends should be more biodegradable. Following the biodegradable elements' breakdown, the remaining components are more eco-friendly and generally do not cause environmental pollution (Ibrahim et al. 2021; Ghanbarzadeh and Hadi 2013).

4.4.4 Natural Fibers

Composites made of polymers (thermoplastics, thermosets, and elastomers) benefit greatly from the reinforcement characteristics of natural fibers. Hence, natural fiber-reinforced polymer composites are becoming more popular because of their superior mechanical qualities, significant manufacturing advantages, and ability to address environmental contaminants. Natural fiber-based composites have a greater influence on the business because of their low density and high specific characteristics. Since they are non-toxic, they pose no health risks. Natural fibers are inexpensive, lightweight, renewable, and have a comparatively high tensile strength. Due to environmental factors and their characteristics, natural fibers have replaced costly and non-renewable synthetic fibers, for instance, carbon, glass, and kevlar fibers, in a variety of applications. High specific strength and modulus, flexibility after treatment, and superior corrosion resistance are some of the desirable qualities of natural fibers (Gholampour and Ozbakkaloglu 2020). In contrast to glass and carbon fibers, they exhibit several drawbacks, including high moisture absorption, increased anisotropy, limited compatibility with traditional resins, and poor homogeneity (Rong et al. 2001). Natural fibers have been utilized for generations in a variety of products, including apparel, basketry, ropes, and numerous auto parts (Samir et al. 2022).

4.5 Market Status

With a compound annual growth rate (CAGR) of 8.4%, the production of fibers and polymers climbed from 2 metric tons in 1950 to 380 metric tons in 2015. Only metric tons of bio-based plastics and/or biodegradable polymers are produced worldwide (Geyer et al. 2017). The volume of the tissue engineering market across the globe was projected at United States Dollar (USD) 3.75 billion by the end of year

2022 and is anticipated to augment at a CAGR of 15.23% from 2024 to 2031, from USD 4.42 billion in 2023 to USD 13.74 billion by 2031 (SkyQuest Technology Consulting Pvt. Ltd n.d.). The financial consequences of regenerative medicine are still too great for both public and commercial healthcare systems to handle, especially when you factor in the additional expenses of costly hospitalization, rehabilitation and operations.

The absence of comparative effectiveness evidence is another problem that restricts tissue engineering medicines from obtaining reimbursement after market authorization (De Pieri et al. 2021; Bubela et al. 2015). In order to ensure long-term safety and efficacy, manufacturers must show that their product offers additional benefits over the current standard of treatment, both economically and—most importantly—clinically (Bubela et al. 2015). Due to the rising incidence of musculoskeletal issues, the orthopedics, musculoskeletal, and spine sectors dominated the tissue engineering market. Treatments including bone replacements, bone transplants, and spine surgery were important in the orthopedic regeneration industry. Due to the over 15 million fracture cases that occur each year, bone replacement or repair treatments accounted for around 900,000 surgical interventions in the United States alone, which significantly increased healthcare expenses by $59.9 billion (SkyQuest Technology Consulting Pvt. Ltd n.d.).

Due to the increasing prevalence of cardiovascular disorders globally, the cardiology and vascular tissue engineering sector is anticipated to develop at the fastest rate over the analysis period. Leading companies in this field are working hard to create stem cell treatments that will repair, regenerate, and revascularize damaged heart tissue. To encourage the regeneration of damaged cardiac cells, research is also being done on gene therapy, sophisticated biologics, and small chemicals. Research on the effects of cardiovascular illnesses is increasingly using autologous and allogeneic stem cells, either separately or in combination. Furthermore, there is encouraging potential for future market expansion due to developments in tissue engineering technology, such as the creation of 3D heart muscles and Engineered Heart Tissue (EHT) (SkyQuest Technology Consulting Pvt. Ltd n.d.). Some scaffold-based marketed cell system products (Table 4.1).

4.6 Future Scope

There are advantages and disadvantages for the environment associated with tissue engineering, a subject that focuses on developing functional tissues and organs for biomedical purposes. Despite significant advancements in tissue engineering and nanotechnology, there are still a number of obstacles to overcome before more complex tissues may be produced using nanotechnology. In order to incorporate newly produced scaffolds into clinical trials, a systematic investigation of their biodegradability and biocompatibility is required. Continued contacts and cooperation between academics from many disciplines and corporations with financing and regulatory businesses are necessary for further growth in this sector. Furthermore, tissue engineering and regenerative medicine development and research costs are

Table 4.1 Scaffold-based commercially available cell systems (De Pieri et al. 2021)

Indicated site	Item	Technology details	Purpose for treatment
Skin	Apligraf®	Dermal equivalent made up of bovine collagen matrix obtained from neonatal keratinocytes and dermal fibroblasts.	In therapy of diabetic foot and venous
	Dermagraft®	Skin replacement obtained from polyglactin cultured with neonatal dermal fibroblasts.	In therapy of diabetic foot ulcers
	DenovoDerm™	Skin replacement obtained from bovine collagen cultured with dermal fibroblasts.	In therapy of of burns
	DenovoSkin™	Dermal substitute obtained from bovine collagen cultured with keratinocytes and dermal fibroblasts	For the therapy of burns
	OrCel®	Dermal substitute obtained from bovine collagen cultured with keratinocytes and dermal fibroblasts	For burns therapy
	TransCyte®	Dermal substitute obtained from bioresorbable polyglycolic acid cultured with neonatal dermal fibroblasts	For burns therapy
Cartilage	Biocart™ II	Hyaluronic acid and fibrin-based platform decked with chondrocytes.	For the treatment of cartilage injuries
	BioSeed®-C	Polylactic acid/polydioxanone/ polyglycolic, fibrin-based platform decked.	For the treatment of joint cartilage damages
	CaReS®	Collagen decked with chondrocytes.	In the treatment of articular cartilage injury
	MACI®	Porcine peritoneum derived collagen decked with chondrocytes.	In therapy of cartilage injuries of the joints
	NeoCart®	Bovine derived porous collagen platform decked with chondrocytes.	Therapy of articular cartilage lesions
	NOVOCART®	Bilayered collagen scaffold having chondroitin sulfate decked with chondrocytes.	Therapy of articular cartilage lesions
Bone	BIO4®	Bone matrix decked with mesenchymal stem cells, osteoblasts and osteoprogenitor cells.	In treatment of bone allograft
	Osteocel®	Mesenchymal stem cells decked with bone matrix.	Therapy of bone grafting after lateral inter-body fusion
	Trinity elite®	Mesenchymal stem cells and osteoprogenitor cells decked with bone matrix.	As bone allograft
	ViBone®	Mesenchymal stem cells and osteoprogenitor cells decked with bone matrix.	As bone allograft
	ViviGen®	Mesenchymal stem cells and osteoprogenitor cells decked with bone matrix.	In therapy ofbone allograft

(continued)

Table 4.1 (continued)

Indicated site	Item	Technology details	Purpose for treatment
Blood vessels	VascuGel™	Gelfoam® gelatin platform decked with human aortic endothelial cells.	For the treatment of peripheral artery disease
Oesophagus	Cellspan®	Biostage™ polyurethane platform decked with adipose-derived stem cells.	In treatment of oesophageal injuries
Cornea	NT 501	Porous polyethersulfone fibers decked with genetically engineered retinal pigment epithelial cells.	The treatment of retinal degenerative diseases

excessively high, and these systems must be made affordable.Therefore, the cutting-edge use of nanotechnology in CRISPR and regenerative medicine will undoubtedly transform the foundation for the diagnosis, treatment, and prevention of a number of illnesses. The sustainable development of tissue engineering is becoming more and more important to researchers and practitioners as the field develops. More sustainable procedures are being developed, including the use of environmentally friendly biomaterials, energy optimization, and the investigation of waste elimination and recycling techniques. All things considered, tissue engineering holds promise for major medical advancements, however like any scientific and technical undertaking, it needs constant attention to reduce its negative effects on the environment and improve sustainability. In conclusion, any future developments in the area of sustainable biomaterials require a thorough understanding of the composition and structure of the building block for a number of applications.

While there are some challenges associated with sustainable biomaterials, they also provide promising answers for a variety of pharmacological and biological uses. One of the biggest challenges is developing sustainable biomaterials that are as good as or better than traditional materials without sacrificing functionality or safety. Obstacles still include guaranteeing the scalability and cost-effectiveness of sustainable biomaterials, as well as regulatory requirements and standards for their usage in pharmaceutical and biomedical applications. Furthermore, for sustainable biomaterials to be widely accepted and adopted within the pharmaceutical industry and biomedical community, barriers related to perception, education, and awareness must be removed. Furthermore, as some materials may cause harmful reactions in live cells, the biocompatibility of environmentally friendly biomaterials is essential. To avoid negative impacts on cell viability and function, biocompatibility and material functioning must be balanced. One of the field's continuous challenges is comprehending the processes behind the cytotoxicity of environmentally friendly biomaterials and creating practical solutions to lessen these effects. Furthermore, to guarantee precise and repeatable results across investigations, standardized techniques for evaluating the cell toxicity of sustainable biomaterials must be developed. To fully realize the potential of sustainable biomaterials in biomedicine and pharmaceuticals, ongoing research efforts are concentrated on resolving these problems.

4.7 Conclusion

In order to advance environmentally friendly biopolymer technologies into the mainstream, cooperation between researchers, doctors, and industry stakeholders will be essential as the field develops. Public awareness campaigns, financing programs, and government backing will all be crucial in promoting the use of sustainable biomaterials in the medical field. We can hasten the transition of biopolymer-based innovations from the lab to clinical practice by leveraging interdisciplinary expertise and adopting creative strategies. In conclusion, this review emphasizes the sustainable nature, biocompatibility, and adaptability of environmentally friendly biopolymers, highlighting their enormous promise in regenerative medicine. The future of regenerative medicine and patient care will be shaped by the advancements in biopolymer research, which have opened the door for a more ethical and environmentally friendly approach to biomedical applications. Eco-friendly biopolymers have the potential to completely transform the biomedical field and usher in a new age of successful and ecologically friendly medicinal solutions with sustained commitment, research, and teamwork.

References

Alipanahi B, Delong A, Weirauch MT, Frey BJ (2015) Predicting the sequence specificities of DNA- and RNA-binding proteins by deep learning. Nat Biotechnol 33(8):831–838. https://doi.org/10.1038/nbt.3300

Asohan AW, Jasme N, Kahar NNFNMN, Bustami Y, Hamid ZAA (2023) 3D bioprinted scaffolds from sustainable materials for tissue engineering: evolution and current challenges. In: Wan Kamarul Zaman WS, Abdullah NA (eds) Sustainable material for biomedical engineering application. Springer, Singapore. https://doi.org/10.1007/978-981-99-2267-3_13

Bakhtiari SE, Karbasi S, Toloue EB (2021) Modified poly(3-hydroxybutyrate)-based scaffolds in tissue engineering applications: a review. Int J Biol Macromol 166:986–998

Bassyouni M, Zoromba MS, Abdel-Aziz MH, Mosly I (2022) Extraction of nanocellulose for eco-friendly biocomposite adsorbent for wastewater treatment. Polymers (Basel) 14:1852

Biduski B et al (2018) Starch hydrogels: the influence of the amylose content and gelatinization method. Int J Biol Macromol 113:443–449

Biswal T, BadJena SK, Pradhan D (2020) Sustainable biomaterials and their applications: a short review. Mater Today: Proc 30:274–282

Bohr A, Memarzadeh K (2020) The rise of artificial intelligence in healthcare applications. In: Artificial Intelligence in Healthcare. Elsevier, pp 25–60

Bubela T et al (2015) Bringing regenerative medicines to the clinic: the future for regulation and reimbursement. Regen Med 10:897–911

Cameron RE, Kamvari-Moghaddam A (2008) Synthetic bioresorbable polymers. In: Degrad. Rate Bioresorbable Mater, pp 46–66. https://doi.org/10.1016/B978-1-84569-929-1.50005-2

Chan BP, Leong KW (2008) Scaffolding in tissue engineering: general approaches and tissue-specific considerations. Eur Spine J 17:467–479

Chen FM, Zhang M, Wu ZF (2010) Toward delivery of multiple growth factors in tissue engineering. Biomaterials 31(24):6279–6308. https://doi.org/10.1016/j.biomaterials.2010.04.053

Chocholata P, Kulda V, Babuska V (2019) Fabrication of scaffolds for bone-tissue regeneration. Materials 12:568. [Google Scholar] [CrossRef] [Green Version]

De Pieri A, Rochev Y, Zeugolis DI (2021) Scaffold-free cell-based tissue engineering therapies: advances, shortfalls and forecast. npj Regen Med 6:18. https://doi.org/10.1038/s41536-021-00133-3

Dong C, Lv Y (2016) Application of collagen scaffold in tissue engineering: recent advances and new perspectives. Polymers 8(2):42. https://doi.org/10.3390/polym8020042

Fouad D, Farag M (2019) Design for sustainability with biodegradable composites. In: Design Engineering and Manufacturing [Working Title]. IntechOpen, London. https://doi.org/10.5772/intechopen.88425

Garijo B (n.d.) Bioconvergence: a multidisciplinary approach to advance human health. Ernst Young

Geyer R, Jambeck JR, Law KL (2017) Production, use, and fate of all plastics ever made. Sci Adv 3:e1700782

Ghanbarzadeh B, Hadi A (2013) Biodegradable polymers. In: Biodegradation-life of science, pp 141–185. https://doi.org/10.5772/56230

Gharibshahian M, Torkashvand M, Bavisi M, Aldaghi N, Alizadeh A (2024) Recent advances in artificial intelligent strategies for tissue engineering and regenerative medicine. Skin Res Technol 30(9):e70016. https://doi.org/10.1111/srt.70016. PMID: 39189880; PMCID: PMC11348508

Gholampour A, Ozbakkaloglu T (2020) A review of natural fiber composites: properties, modification and processing techniques, characterization, applications. J Mater Sci 55:829. Springer US

Gogoi S, Das B, Chattopadhyay D, Khan R (2019) Sustainable nanostructural materials for tissue engineering. In: Dynamics of advanced sustainable nanomaterials and their related nanocomposites at the bio-nano interface, pp 75–100

Gonnelli A, Sarogni P, Giannini N, Linsalata S, Di Martino F, Zamborlin A et al (2024) A bioconvergence study on platinum-free concurrent chemoradiotherapy for the treatment of HPV-negative head and neck carcinoma. Artif Cells Nanomed Biotechnol 52(1):122–129. https://doi.org/10.1080/21691401.2024.2309233

Gritsch L, Perrin E, Chenal J-M, Fredholm Y, Maçon ALB et al (2021) Combining bioresorbable polyesters and bioactive glasses: orthopedic applications of composite implants and bone tissue engineering scaffolds. Appl Mater Today 22:100923. https://doi.org/10.1016/j.apmt.2020.100923

Guarino V, Raucci MG, Ronca A, Cirillo V, Ambrosio L (2014) Multifunctional scaffolds for bone regeneration. In: Bone substitute biomaterials. Elsevier, pp 95–117

Huang G, Li F, Zhao X, Ma Y, Li Y, Lin M et al (2017) Functional and biomimetic materials for engineering of the three-dimensional cell microenvironment. Chem Rev 117(20):12764–12850

Hyeon T, Manna L, Wong SS (2015) Sustainable nanotechnology. Chem Soc Rev 44:5755–5757

Ibrahim NI et al (2021) Overview of bioplastic introduction and its applications in product packaging. CoatingsTech 11:1423

Iravani A, Akbari MH, Zohoori M (2017) Advantages and disadvantages of green technology; goals, challenges and strengths. Int J Sci Eng Appl 6(9):272–284

Jha K, Kataria R, Verma J, Pradhan S (2019) Potential biodegradable matrices and fiber treatment for green composites: a review. AIMS Mater Sci 6:119–138

Kim JH, Shim BS, Kim HS et al (2015) Review of nanocellulose for sustainable future materials. Int J Precis Eng Manuf-Green Technol 2(2):197–213

Kohane D, Langer R (2008) Polymeric biomaterials in tissue engineering. Pediatr Res 63:487–491. https://doi.org/10.1203/01.pdr.0000305937.26105.e7

Krishani M, Shin WY, Suhaimi H, Sambudi NS (2023) Development of scaffolds from bio-based natural materials for tissue regeneration applications: a review. Gels 9:100. https://doi.org/10.3390/gels9020100

Lai Y, Asthana A, Kisaalita WS (2011) Biomarkers for simplifying HTS 3D cell culture platforms for drug discovery: the case for cytokines. Drug Discov Today 16(7–8):293–297

Lee SJ, Rho M (2022) Multimodal deep learning applied to classify healthy and disease states of human microbiome. Sci Rep 12(1):824. https://doi.org/10.1038/s41598-022-04773-3

Liu S, Yu JM, Gan YC et al (2023) Biomimetic natural biomaterials for tissue engineering and regenerative medicine: new biosynthesis methods, recent advances, and emerging applications. Military Med Res 10:16. https://doi.org/10.1186/s40779-023-00448-w

Malik S, Sundarrajan S, Hussain T, Nazir A, Ayyoob M, Berto F, Ramakrishna S (2023) Sustainable nanofibers in tissue engineering and biomedical applications. Mat Design Process Commun

Mangaraj S, Yadav A, Bal LM, Dash SK, Mahanti NK (2019) Application of biodegradable polymers in food packaging industry: a comprehensive. Rev J Packag Technol Res 3:77–96

Maraveas C (2020) Production of sustainable and biodegradable polymers from agricultural waste. Polymers (Basel) 12:1127

McClelland R, Dennis R, Reid LM, Palsson B, Macdonald JM (2005) Tissue Engineering. In: Introduction to biomedical engineering (second edition) biomedical engineering, pp 313–402

Meyer TA, Ramirez C, Tamasi MJ, Gormley AJ (2022) A user's guide to machine learning for polymeric biomaterials. ACS Polym Au 3(2):141–157

Middleton JC, Tipton AJ (2000) Synthetic biodegradable polymers as orthopedic devices. Biomaterials 21:2335–2346. https://doi.org/10.1016/S0142-9612(00)00101-0

Mudakkar SR, Zaman K, Khan MM, Ahmad M (2013) Energy for economic growth, industrialization, environment and natural resources: living with just enough. Renew Sust Energ Rev 25:580–595

Mülhaupt R (2013) Green polymer chemistry and bio-based plastics: dreams and reality. Macromol Chem Phys 214:159–174

Naik RR, Singamaneni S (2017) Introduction: bioinspired and biomimetic materials. Chem Rev 117(20):12581–12583

Narayanan G, Vernekar VN, Kuyinu EL, Laurencin CT (2016) Poly (lactic acid)-based biomaterials for orthopaedic regenerative engineering. Adv Drug Deliv Rev 107:247–276. https://doi.org/10.1016/j.addr.2016.04.015

Nikolova MP, Chavali MS (2019) Recent advances in biomaterials for 3D scaffolds: a review. Bioact Mater 4:271–292

Perić Kačarević Ž, Rider P, Alkildani S, Retnasingh S, Pejakic M, Schnettler R, Gosau M, Smeets R, Jung O, Barbeck M (2019) An introduction to bone tissue engineering. Int J Artif Organs 43:69–86

Perveen K, Masood F, Hameed A (2020) Preparation, characterization and evaluation of antibacterial properties of epirubicin loaded PHB and PHBV nanoparticles. Int J Biol Macromol 144:259–266

Pesode P, Barve SB, Wankhede S, Ahmad A (2023) Sustainable materials and technologies for biomedical applications. Adv Mater Sci Eng 2:1–22. https://doi.org/10.1155/2023/6682892

Petriccaa PKS, Marraa K (2006) Chemical synthesis of poly(lactic-co-glycolic acid)/hydroxyapatite composites for orthopaedic applications. Acta Biomater 2:277–286

Roberts TT, Rosenbaum AJ (2012) Bone grafts, bone substitutes and orthobiologics: bone grafts, bone substitutes and orthobiologics the bridge between basic science and clinical advancements in fracture healing. Organogenesis 8:114–124

Rokkanen PU, Bo O, Hirvensalo E, Ma EA, Partio EK, Vainionpa S (2000) Bioabsorbable fixation in orthopaedic surgery and traumatology, 23. Biomaterials 21:2607–2613

Rong MZ, Zhang MQ, Liu Y, Cheng Yang G, Min Zeng H (2001) Effect of fiber treatment on the mechanical properties of unidirectional sisal-reinforced epoxy composites. Compos Sci Technol 61:1437–1447

Roseti L, Parisi V, Petretta M, Cavallo C, Desando G, Bartolotti I, Grigolo B (2017) Scaffolds for bone tissue engineering: state of the art and new perspectives. Mater Sci Eng C 78:1246–1262

Samir A, Ashour FH, Hakim AAA et al (2022) Recent advances in biodegradable polymers for sustainable applications. npj Mater Degrad 6:68. https://doi.org/10.1038/s41529-022-00277-7

Shah TV, Vasava DV (2019) AL-Self-dual Leonard pairs A glimpse of biodegradable polymers and biomedical applications. E-Polymers 19:385–410

Sheikh Z, Najeeb S, Khurshid Z, Verma V, Rashid H, Glogauer M (2015) Biodegradable materials for bone repair and tissue engineering applications. Materials (Basel) 8:5744–5794

Shin H, Jo S, Mikos AG (2003) Biomimetic materials for tissue engineering. Biomaterials 24(24):4353–4364

Shivam P (2016) Recent Developments on biodegradable polymers and their future trends. Int Res J Sci Eng 4(1):17–26

Siragusa N, Matz L (2024) Bioconvergence: bridging science and nature to shape tomorrow. In: Siragusa N, Nemat AT (eds) Bioconvergence revolution. Springer, Cham. https://doi.org/10.1007/978-3-031-61593-1_1

SkyQuest Technology Consulting Pvt. Ltd. (n.d.). https://www.skyquestt.com/report/tissue-engineering-market#:~:text=What%20is%20the%20global%20market,period%20(2024%2D2031)

Sohn S, Buskirk MV, Buckenmeyer MJ, Londono R, Faulk D (2020) Whole organ engineering: approaches, challenges, and future directions. Appl Sci 10(12):4277. https://doi.org/10.3390/app10124277

Somers P, Cornelissen R, Thierens H, Van Nooten G (2012) An optimized growth factor cocktail for ovine mesenchymal stem cells. Growth Factors 30(1):37–48. https://doi.org/10.3109/08977194.2011.634411

Sprowson AP, Aldridge SE, Noakes J, Read JW, Wood DG (2012) Bio-interference screw cyst formation in anterior cruciate ligament 31 reconstruction-10-year follow up. Knee:644–647. https://doi.org/10.1016/j.knee.2012.01.004

Thompson RC, Moore CJ, Saal FSV, Swan SH (2009) Plastics, the environment and human health: current consensus and future trends. Philos Trans R Soc B Biol Sci 364:2153–2166

Vashisht M (2021) BioConvergence: assembling the forces of nature and technology for a remarkable future. Mouse Electronics

Verma P, Verma V (2020) Concepts of tissue engineering. In: Animal biotechnology, 2nd edn. Academic Press, Cambridge, MA, pp 295–307

Wang YX, Robertson JL, Spillman WB, Claus RO (2004) Effects of the chemical structure and the surface properties of polymeric biomaterials on their biocompatibility. Pharm Res 21(8):1362–1373

Wang Z, Yuan L, Tang C (2017) Sustainable elastomers from renewable biomass. Acc Chem Res 50:1762–1773

Wong YS, Tay CY, Wen F, Venkatraman SS, Tan LP (2012) Engineered polymeric biomaterials for tissue engineering. Curr Tissue Eng 1:41–53

World Economic Forum (18 January 2022) "What are the biggest healthcare shifts we'll see in 2022? Here's what health experts think". europeansting.com

Yi S, Ding F, Gong L, Gu X (2017) Extracellular matrix scaffolds for tissue engineering and regenerative medicine. Curr Stem Cell Res Ther 12(3):233–246. https://doi.org/10.2174/1574888X11666160905092513

Zakrzewski W, Dobrzyński M, Szymonowicz M et al (2019) Stem cells: past, present, and future. Stem Cell Res Ther 10:68. https://doi.org/10.1186/s13287-019-1165-5

Zhao Y, Qiu Y, Wang H, Chen Y, Jin S, Chen S (2016) Preparation of nanofibers with renewable polymers and their application in wound dressing. Int J Polym Sci 2016:4672839–4672856. https://doi.org/10.1155/2016/4672839

Bio-Fabrication Techniques for Eco-Friendly Biomedical Devices

5

Moumita Das Kirtania, Radheshyam Pal, Sumel Ashique, Ankit Kumar, and Sanjay Dey

Abstract

Biomedical devices have been an integral part of healthcare system since last few decades for various categories of patients. Some of such devices operate on power like cochlear implants, cardiac pacemakers, etc. while others do not like contact lenses or knee implants, respiratory masks, catheters etc. They offer specific advantages for the consumers. However, their production and use are challenging. Further the disposal of such devices poses huge amount of dump of medical waste which are majorly made of non-biodegradable plastics, polyvinyl chloride or phthalates out of which only a small percentage can be recycled, posing a threat to the environment. Thus, the need of the hour is to fabricate eco-friendly biomedical devices using bio-fabrication techniques to reduce the environmental load and toxic reactions in the environment. Recently the use of bio-plastics obtained from sugarcane and cornstarch have been studied upon.

M. D. Kirtania
School of Pharmacy, Techno India University, Kolkata, West Bengal, India

R. Pal
School of Pharmaceutical Sciences, Lovely Professional University, Phagwara, Punjab, India

S. Ashique
School of Pharmaceutical Sciences, Lovely Professional University, Phagwara, Punjab, India

Department of Pharmaceutical Technology, Bharat Technology, Uluberia, West Bengal, India

A. Kumar
Department of Pharmaceutical Technology, Meerut Institute of Engineering and Technology, Meerut, UP, India

S. Dey (✉)
Department of Pharmaceutical Technology, School of Health and Medical Sciences, Adamas University, Kolkata, India

Md. Faiyazuddin, M. Suryawanshi (eds.), *Design and Processing of Green Materials*, Biomaterials, Bioengineering and Sustainability 4,
https://doi.org/10.1007/978-3-031-91790-5_5

Further using batteries which can be recharged can reduce the load of disposable batteries. Paper based fuel cells can be used in batteries for energy generation for eco-friendly devices. The following review would illustrate the bio-fabrication techniques for generation of eco-friendly biomedical devices.

Keywords

Eco friendly · Biomedical devices · Bio-fabrication techniques · Bio-plastics

5.1 Introduction

Diabetic patients should monitor their blood glucose levels continuously; especially the diabetic patient should check their daily blood glucose levels before & after the intake of meals & after any workout. Two types of glucose sensor are currently available to monitor the blood glucose level, blood glucose monitor (BGMs) and continuous glucose monitor (CGMs). BGMs are self-monitored, whereas, CGMs don't need any kind of patient attention also its check the blood glucose level independently. The measurement of the blood glucose level of fresh blood samples is based on the principles of (i) amperometry, (ii) colorimetry, or (iii) fluorescence. The reagents like glucose oxidase or dehydrogenase are used in amperometry-based BGMs. In this method, the reagents react electrochemically with glucose, and the alteration of current is observed through sense of electrode (Wang 2008). The current passes throughout the solution according to where glucose is present. In the colorimetric method, sampling was checked due to their absorbance level. Lastly, the fluorescence-based BGMs utilize light for the checking of fluorescent in a particular glucose sample also the reaction in reversible in nature. The glucose concentration is proportional to change in fluorescence. CGMs continuously monitor the glucose level of subcutaneous tissues in the systemic circulation by differentiate in solution concentration. The biocompatibility issue restricts the use of CGMs for prolonged period. CGMs more often show some error that is why it get calibrated by different types of measurement of CGM and by the external sensor, generally the sensor is BGM. Diabetic patients require self-administration of insulin on daily basis using syringes, pens, or patches to control their blood glucose level. Sometimes, multiple administration are required to maintain normoglycemia throughout the whole day, which leads to discomfort and poor compliance. Nowadays, to deliver insulin daily, the insulin pen has been most widely acceptable by diabetic patients. Insulin pen can reach up to 900 U of insulin per pen and able to reduce pain, hospital visit, and adverse social and psychological stigma; making them more user-friendly in contrast to traditional syringes. Newly the research has been implanted to get non-invasive glucose sensor thus resulting more biocompatibility (Lee et al. 2018; Kim et al. 2018a). The new devices are attached to the skin or implanted to monitor the glucose level in body fluids. At low concentrations, Glucose Oxidase (GOx) immobilized on CNTs is capable of detecting glucose level electrochemically (Yan et al. 2008; Tsai et al. 2009; Kang et al. 2019). The glucose reacts with oxygen and gets converted into gluconic acid upon getting in touch with hybrid material, resulting in changes of the electrochemical properties of the sensor. Kang et al. fabricated a wearable sensor from CNTs and found the fabricated sensor was capable of detecting glucose at a low level (50 μM) within 5 s (Feng et al. 2019). The detection principle is based on measuring the faradic current generated by the redox process. The in vivo monitoring of blood glucose level efficacy of CNT fiber-based electrochemical sensor was investigated by Wang et al. (Das et al. 2022). The sensor was able to determine the glucose level in the range of 2.5–5 mM. The amount of hydrogen peroxide release from the systematic circulation during glucose oxidation thus sensing the level of peroxide. However, this technology-driven solution is not commercially implemented yet, its need further study for the ultimate progress to the Carbon nanotechnology.

5.2 Biomedical Devices

5.2.1 Glucose Monitoring Device

Diabetes is nowadays a daily challenge for worldwide patients especially monitoring the sugar level of the patients is more complicated and time-consuming. Currently, diabetic patients go through two main systems for the diagnostic measurement of their blood sugar level first (Zhang et al. 2021). The Blood Glucose Monitoring (BGM) system is most useful for measuring the blood sugar level between the capillaries. The other technique is Continuous Glucose Monitoring CGM. The CGMs have new daiabetes management techniques due to the real time data management (Krakauer et al. 2021). Unlike the traditional finger-picking methods, new treatments deal with the new types of sensors attached to the patients and give results on the glucose level every minute. In the current scenario the CGM values are determined by its real-time value measurement (rtCGM) or intermittently scanned CGM also this kind of technique is called "flash glucose monitoring"(Hina and Saadeh 2022). The glucose level starts by measuring the glucose level in the interstitial fluid. Three basic components are needed to measure fluid: sensor, transmitter, and receiver. If the results are showing inaccurate in the sensors, so check the sensor repeatedly and calibrate with the fingerstick method. There are a few limitations are there like the cost of the machine. Environmental effects and the operator influence in the meter also the some inaccuracy in detecting hypoglycemia. Similarly, the other medications like icodextrin presence can potentially have an impact on the performance of the glucose meter (Blachowicz et al. 2021).

5.2.2 Electrocardiography (ECG)

Electrophysiological signals, such as electrocardiogram are the measurement of potential differences in the cell membrane due to polarization of ion channels during cell polarization. The non-invasive techniques have been adopted to obtain necessary information by keeping on the skin. Electro cardiogram data shows the electric transmittance to the heart during cardiac arrest and ischemia (Nigusse et al. 2021; Yeh et al. 2021; Huang et al. 2021; Takamatsu et al. 2015). The investigation of intensity of peaks (i.e., P-QRS-T complex) in the data sheet of electrocardiogram has very influential to identify different heart diseases. ECG signals shows an absence in P waves in cardiac arrest and in atrial fibrillation. However, the R-R intervals slightly change in the sleep arena and the peak is observed. Thereby, the continuous monitoring of ECG signals can able to improve survivals rate and treatment efficacy of patients. The PEDOT: PSS (Ankhili et al. 2018a, b, 2019; Lidón-Roger et al. 2018; Yapici and Alkhidir 2017), graphene (Taylor et al. 2021), carbon nanotube (Jin et al. 2019), silver flakes (Kazani et al. 2012), silver nanoparticles (Nigusse et al. 2020; Arquilla et al. 2020; Arquilla et al. 2021; Dore et al. 2021; Piuzzi et al. 2020; Yao et al. 2019), and Ag NWs (Taylor et al. 2021) based e-textiles have been employed for gathering of ECG signals. New wearable sensors are highly capable of detecting electrophysiological signals because of the new mechanical instrumentation. Additionally, these sensors are highly compatible with new manufacturing textiles, leads to production of cost-effective and scalable sensors. The stretchable and sewable conductive thread-based ECG-sensing e-textile was fabricated by Taylor et al. through a wet-spinning of CNT fibres (Chi et al. 2019). The average diameter, length, and aspect ratio of the prepared CNT fiber were exhibited ~1.8 mm, ~7.4 μm, and ~4100, respectively. The required tensile strength of 1.7 ± 0.2 GPa was ensured suitability for applying to standard sewing machine and laundry machine. The high SNR as compare to commercial 3M electrodes indicates the appropriateness of thread stitched in a shirt for recording of ECG signals.

The application of CNTs for ECG acquisition has been reported in various articles (Jung et al. 2012; Wang and Loh 2017; Liu et al. 2016; Zhao et al. 2018; Chu and Naguib 2021; Hwang et al. 2018). To ensure appropriate adhesion to the skin and to make it user friendly, the nanocarbon component is incorporated within the flexible polymer matrix, such as polyurethane or polydimethylsiloxane (Kim et al. 2018b; Kolanowska et al. 2021). After repeating wash only 6% decrease of electrical resistance is observed, which is encouraging in practical point of view. Chi et al. developed a CNT-based sensor are very capable of detecting signals on regular basis, outclassing the traditional Ag/AgCl electrodes (Acar et al. 2019).

5.2.3 Electroencephalography (EEG)

EEG signals denote the electrical activity of the brain through placing of electrode on the scalp (Chan et al. 2018). This is very useful to study the brain disorders such as epileptic event, coma state, unresponsive wakefulness state, and minimally conscious state. The alpha and beta waves in EEG signals represents blinking and active thinking, respectively. The hair on the head ultimately affects the signal quality during recording of EEG signals by using many prototypes. Thereby, the recording of EEG signals on forehead is generally preferred (Fiedler et al. 2015). Silver-based materials (Löfhede et al. 2010; La et al. 2018; Shu et al. 2019; Muthukumar et al. 2015), conductive polymers (Golparvar et al. 2021) and graphene derivatives (Awara et al. 2014) based electrode embedded within e-textiles have been useful for collection of EEG signals. Awara et al. reported the use of thin, metal free, CNT electrode for of different signals (Malfa et al. 2020). This electrode can able to overcome the problem associated with new use of Ag/AgCl electrodes. Generally, Ag/AgCl restrict X-rays, resulting in creation of artifacts on Computed Topography (CT) during surgery and complication of procedure. Kim et al. fabricated simple, fast, and cost-effective sensor using CNT in association with polydimethylsiloxane elastomer for EEG monitoring (Ruffini et al. 2008). They investigated the alteration of alpha band after placing on the forehead or earlobe. However, the fabricated CNT/PDMS sensor was found very rapid as its properties altered minutely after subjecting it to new cycle`. Finally, Ruffini et al. applied CNTs for recording of EEG and reported that the fabricated CNTs act similarly as compare to state of electrodes have no side effect on the skin were exhibited after six month of their application directly to the skin (Lee et al. 2014). Thereby, it can be postulated that the application of CNTs for monitoring of EEG signals in the form of smart textile will become an interesting topic got cutting edge research.

5.2.4 Electromyography (EMG)

The wellbeing of muscles and nerve cells is monitored by EMG. Detection of the signals can be captured through intramuscularly by non-inversely. Lee et al. approach a new combination of CNTs for the detection of electrical signals in the part of skin through wrinkling (Ullah et al. 2022). In this approach, the application of gel on the skin need not to apply for adhesion, resulting in comfortable to use and long-term monitoring can be possible. In another study, the accurate measurement of EMG is possible electrodes for the new wearable textile base (Qi et al. 2020). The signal-to-noise ratio is increased tremendously upon increase in activity from walking to running. The CNT sensors showed similar performance to Ag/AgCl electrodes. The notable advantage of CNT sensor is that the performance is not affected even at a high bend angle because of high flexibility of the components. Qi et al. fabricated piezoresistive sensor made up of CNT-coated polyurethane nanofibers to monitor swallowing, respiration, bulging, or wrist bending (Liu et al. 2010). The textile-based EEG sensor can be successfully used even physically challenged to athletic persons and also able to monitor the state of muscle during exercising to avoid injury. For integration of textile-based EMG sensors into the fabrics, Ag flakes consisting of different printing with highly conductor ink and fluorine rubber/surfactant is utilized (Zhang et al. 2022a). Dias et al. fabricated textile electrode as a wearable sensor and reported that the fabricated device can be able to successfully capturing of EMG signal and subsequent transmission to the sensor electronics. This can be attributed due to high elastic conductive inks and textile substrates (Lee and Kim 2023; Baik et al. 2019). In the other side, Linz et al. demonstrated the use of contactless EMG sensors inflated onto fabrics provide similar EMG signals and shapes as like commercial EMG electrodes (Root et al. 2020).

5.2.5 Body Temperature Monitoring Devices

The human body temperature is altered in several health disorders and thereby requires special attention to monitor the same. Sometimes, exhaustion and strokes may result in accumulation of excessive heat in the body. It is very essential for athletes and persons working at high temperature as such condition may lead to fetal consequence. The devices have been used to monitor the body temperature is generally based on the principles of a thermocouple or alteration of electric resistance with temperature (Blasdel and Monty 2015). In former case, there is generation of voltage due to temperature difference between the dissimilar conductors placed at different temperatures. In future studies the alteration of material may be depend upon the increase in the temperature . The temperature is also monitored by optically, where angular deformation of waves due to reflection of sensor from human body. However, this approach is not widely used, as CNTs based monitoring of electrical properties is more user-friendly. Blasdel and Monty reported the use of multi-walled CNTs impregnated on Nylon 6 for determination of temperature between 25 °C and 45 °C (Rosace et al. 2017). The use of Polypyrrole is beneficial for the protection from external environment and improves the electrical percolation. The negative TCR of −0.228 ± 0.03% was achieved. Sensitivity of the sensor depends upon the bending and exposing new heating cooling cycle. Even though, the parameters were affected by the humidity. Rosace et al. fabricated wearable sensor using in combination of multi-walled CNTs and cotton to investigate the effect of humidity and temperature on surface resistance (Tai et al. 2020). The prepared sensor was possessed good mechanical strength, electric conductivity, appropriate adherence to the substrate well, suitable sensitivity and reproducibility.

5.2.6 Respiration Monitoring Devices

Last few decades, the new high-tech devices for detection have been received enormous interest in both professional and scientific world for respiration monitoring and breadth analysis due to its compact size, prolong durability, and light weight (Kano et al. 2022; Güntner et al. 2019). During COVID-19 pandemic, the analysis of breadth and monitoring of respiration were become a crucial indicator to investigate the health status. In addition, respiration monitoring can be a well targeted device for the detection of different disease several diseases, such as asthma, sleep apnea, and cardiac arrest (Mitchell et al. 2010). The breathing pattern is also monitored in sports or rehabilitation by breathing sensors (Guo et al. 2011; Ramos-Garcia et al. 2016; Rovira et al. 2011; Zhang and Liu 2016). The schematic presentation of respiration monitoring system is given in Fig. 5.1.

Wang et al. discovered new trielctric powdered weabable new sensor for investigating respiration (Hughes-Riley et al. 2018). Similarly, the mounting of acoustic sensor on the neck can be successfully able to monitor the respiration (Koncar 2016). Folke et al. and Al-Khalidi et al. postulated the monitoring of breathing by placing the sensor near to nose (Ehrmann and Ehrmann 2020, 2021). However,

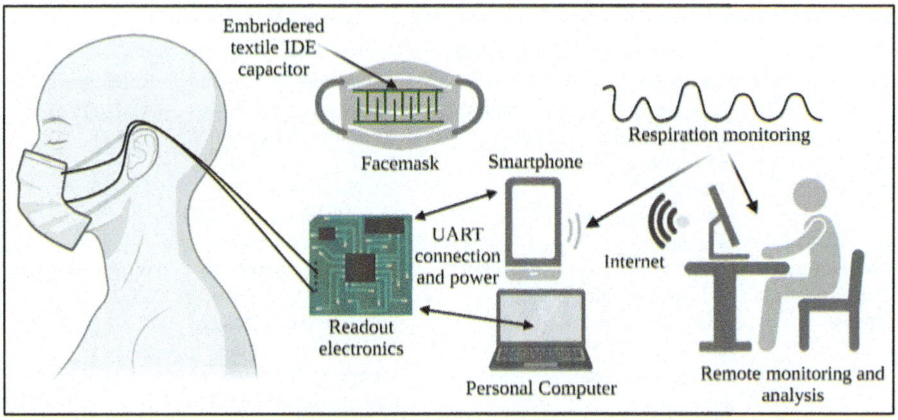

Fig. 5.1 Block schematic of the system for respiration monitoring (Omura 2014). (Copyright: © 2022 by the authors. Licensee MDPI, Basel, Switzerland. This article is an open access article distributed under the terms and conditions of the Creative Commons Attribution (CC BY) license (https://creativecommons.org/licenses/by/4.0/)

wearing of sensor for long period of time on the nose and neck are not acceptable by the public (Paiva et al. 2018). In addition, the conductive textile technology has been extensively adopted for fabrication of sensors for breadth monitoring (Khundaqji et al. 2020a; Scataglini et al. 2020; Li et al. 2023; Khundaqji et al. 2020b) Huang et al. fabricated a wearable yarn-based piezo-resistive sensors for tracking respiration rate (Wu et al. 2024). The usage of displacement transducers and fiber-meshed strain were proposed by Scataglini et al. (Güder et al. 2016) The use of ordinary face mask for monitoring of respiration rate was explored by Güder et al. and Zhong et al., as well as customized facemasks by Tipparaju et al. (Zhong et al. 2021; Tipparaju et al. 2020; Huang et al. 2008). The use of various types of respiration monitoring sensors and their limitations are summarized in Table 5.1.

5.2.7 Sweat Monitoring Devices

In post-COVID-19 era, the application of wearable bioelectronics becomes an emerging technology to continuously and seamlessly monitor the various human body physiological parameters (Amjadi et al. 2016; Nagamine and Tokito 2021; Kohagura et al. 2013). Among these various physiological parameters, sweat has been paid much more attention as its obtained non-invasively and safely (He et al. 2019). The qualitative and quantitative estimation of various physiological parameters of sweat such as glucose, ascorbic acid, lactate, pH value and rate of sweating during exercise can be monitored from sweat (Coyle et al. 2009; Chen et al. 2013). Hyponatremia and chronic nephritis can be recognized by uric acid and sodium ion concentrations in sweat (Morgan et al. 2004; Feig et al. 2008; Potts et al. 2002). In early 2001, Potts et al. reported using a wearable sensor named GlucoWatch for

Table 5.1 Different types of respiration monitoring devices and their limitations

Types of sensor	Description	Mounting Details	limitations	References
Piezo Resistive sensor	They worked by changes in the measurement in electrical resistance due to pressure.	Respiratory belt	The readout circuit is somewhat complicated; it is huge; the sensor response varies with changes in the skin and electrode conditions	Atalay et al. (2014)
Well knitted resistive strain sensor	Changes in resistance occur when the knitted fabric stretched during chest and abdominal movement.	Respiratory belt	over time; the sensor placement is extremely constrained because of its sensitivity to the breathing of the chest and abdomen; and the sampling rate is limited.	Wijesiriwardana (2006)
An inductive fiber meshed strain sensor	Measures the changes in inductance caused by variations in the geometry of meshed coiled strains.	Respiratory belt		Min et al. (2014)
Textile capactive force sensor	Changes in capacitance occqur due to variations in the separation or overlap of conductive when force is applied.	Respiratory belt		Guo et al. (2013)
Piezoresistive sensor	The sensors resistance changes proportionally.	Respiratory belt		Kundu et al. (2013)
Textile wearable capacitive sensor	Capacitance changes as the geometry (distance or area) between conductive layers varies due to external stimuli like strain or pressure.	Two electrodes are fixed on the inner anterior and posterior sides of a T-shirt.	Due to the sensor's sensitivity to breathing in the chest and abdomen and its low sampling rate, its location is quite constrained.	Kim et al. (2020)

(continued)

Table 5.1 (continued)

Types of sensor	Description	Mounting Details	limitations	References
Paper-based moisture sensor	Measures changes in electrical properties (resistance, capacitance, or conductivity) when paper absorbs or loses moisture.	Sensor embedded in the facemask.	Non-toxicity, rapid reaction and recovery speed, and linear response in a very small humidity range are yet unmet.	Zhang et al. (2022b)
Flexible pressure sensor	Measures changes in electrical resistance under applied pressure.	Sensor embedded in the facemask.	Inability to get air ahead of the sensor and a challenging manufacturing procedure.	Yang et al. (2020)

monitoring human blood sugar (Gao et al. 2016). Recently, Gao et al. described the application of a flexible and fully integrated new plastic base patched for the detection of glucose, lactate, Na^+, and K^+ ions in sweat (Chen et al. 2023). Yang et al. described the efficient sampling of microfluidic by applying fully laser-engraved patch (Yan et al. 2020). However such 2D planner wearable bioelectronics, which are often made using photolithography or printing innovations, have a little surface area that comes into contact with the body, which means they can only sample a small number of biomarkers.. Additionally, there is the significant reduction of comfort as its need to be attached directly to the skin.

Last one decade, fiber-based electronics have demonstrated significant promise in wearable applications due to unique geometrical shape, aspect ratio, wearable qualities, and mechanical characteristics (Nishimoto et al. 2022). However, fiber-based electrochemical probes obtained by thermal drawing suffered from shortcomings such as weak signal, limited scalability, small surface area, and low sensitivity (Booth et al. 2021; Richard et al. 2021; Possanzini et al. 2020). Possanzini et al. described the application of functionalized conductive PEDOT:PPS with bromothymol blue dye for the measurement of pH value of sweat (Promphet et al. 2019). In another study, Promphet et al. used different colour dye (Chen et al. 2020).

5.2.8 Moisture Detection Devices

Measuring the impedance or direct current resistance between two conductive fibers or conduction lines is the basic method used to detect moisture. In order to use the embroidered sensor without a battery, Chem et al. paired it with a passive RFID (radio-frequency identification) tag (Chen et al. 2019; Grethe et al. 2018). Grethe et al. adopted spinning, printing and coating techniques to produce flexible and highly textile-integrated moisture sensors (Gaubert et al. 2020). High sensitivity in the 25–80% relative humidity region has been demonstrated by their manufactured sensors, indicating their promising application potential in a variety of domains, including personal safety systems, medical textiles, environmental control, and work clothing. To prevent nocturnal enuresis of children, Gaubert et al. designed underwear embedded with moisture sensors which are able to detect wake up children even one drop of urine is detected (Tanaka et al. 2016). The two layer underpants system is consisting of textile leakage sensor in the inner side and electronics on the outside and the two layers are connected by conductive connections. Likewise, Tanaka et al. produced a wireless transmitter and a urine sensor in the shape of a diaper (Devaux et al. 2011). The battery is wrapped in absorbent material and situated between electrodes made of carbon sponge and metal. The device was able to convert 0.6 V to 2 V and also transmit a signal over 5 m if a defined amount of urine was spotted. Devaux et al. created a textile-based humidity sensor that is consistent and reversible by utilizing carbon nanotubes and multifilament yarn composed of poly(lactic acid) (PLA) (Mecnika et al. 2015). Mecnika et al. used conductive yarns to fabricate embroidered humidity sensors (Wendler et al. 2020). Multi-layer braiding is used to construct a capacitive moisture sensor with high sensitivity and able to work from 22 to 94% relative humidity (Wang et al. 2020). It is decided that the textile-based moisture sensor is less problematic over the other textile sensors. To get the right results, sensors must always be calibrated after cleaning and even after wearing, as aging may be caused by the oxidation of metallic components.

5.2.9 Strain and Pressure Monitoring Devices

Discovering the core biomechanics of muscular response requires constant monitoring of strain and pressure during physical activity (Wu et al. 2019). Instruments have been fabricated using resistive, capacitive, or triboelectric methods. The resistive sensing comprises of Ag nanowires (Li et al. 2018), Ag nanoparticles (Hao et al. 2018), polypyrrole (Hu et al. 2017; Jiyong et al. 2016; Eom et al. 2017), poly(3,4-ethylenedioxythiophene) (Li et al. 2017), CNTs (Zhang et al. 2019; Park et al. 2019; Yang et al. 2021), carbon black (Du et al. 2016), and / or rGO (Yang et al. 2018; Lee et al. 2015). Whereas, polydimethylsiloxane (PDMS) (Keum et al. 2021) [42/63], ion-gel (Yu et al. 2019), Ecoflex (Uzun et al. 2019) and/or nitrile rubber (Meng et al. 2020) is incorporated in capacitive type of sensor. Triboelectric sensing is based on the contact separation between two different frictional materials with opposing triboelectric polarities (Fan et al. 2020; Fang et al. 2020). Because of its comfort and special qualities, textile-based wearable activity sensors have recently been created as strain and pressure sensors. In order to detect changes in a wide range of deformations, a sensor with excellent mechanical characteristics and sensitivity is essential. Liu et al. constructed multi-layer stacked structured flexible rGO pressure sensor on silk fiber by thermal reduction process (Cao et al. 2023). The manufactured sensors were able to detect at different pressure and subtle body pressure.

5.2.10 Neurotransmitter Monitoring Devices

Neurotransmitter monitoring devices are advanced tools used to measure and track the levels of neurotransmitters in the brain and other parts of the nervous system. These devices are essential in clinical diagnostics, neuroscience research, and the treatments of neurological disorders. In the past, neurotransmitter levels were tracked using carbon nanoelectrodes made from carbon fibers and 3D printing based on direct laser writing. Qun et al. fabricated CNSs are grown onto etched metal wires to create carbon nanoelectrodes. The produced electrodes demonstrated appropriate electrochemical characteristics and a high surface to volume ratio. The promising potential of fabricated nanoelectrode sensors for determination of neurotransmitter level is reported (Ronkainen et al. 2010).

5.2.11 Electrochemical Sensor Devices

Electrochemical sensor devices can transform biological and chemical stimuli into a quantifiable electrical output signal (Cordeiro et al. 2018). A transducer (the electrode), an identifying element (proteins, enzymes, or receptors), an element that is biological (the analyte), and a signal processor make up the majority of electrochemical sensor devices. The transducer produces signal in proportion to the analyte-bioreceptor interaction, finally, signal processor collects, amplify and display the signal (Cordeiro et al. 2018). The amperometry, potentiometry, and conductometry approaches are based on oxidation or reduction of an electroactive species, accumulation of charge, and modification of conductive properties of the solution (Malfa et al. 2020; Rovira et al. 2011). Electrochemical sensors have been utilized most frequently lately in the pharmacy and clinical fields (such as diabetic and cardiac self-monitoring) (Cordeiro et al. 2018). Cordeiro et al. investigated *in vivo* brain monitoring using enzyme-based glucose biosensor based on Au-coated W needle microelectrodes (Accessed 2019). The fabricated biosensor could detect a very small amount of H_2O_2 concentration, with high sensitivity and poses quick response. In another study, an amperometric glucose biosensor containing various concentrations of enzyme (0.2, 0.4, and 0.6 U/μL) was developed and investigated against glucose concentrations up to 64 mM. The nonlinear increase the oxidative response to glucose. Biological sensors are helpful for the identification of glucose moiety in the physiological bound between 0.5 and 2.5 Mm WITH limited range between 96.32 and 222.84 microeter.

5.2.12 Allergy Monitoring Devices

In the field of allergology, medical devices are vital in monitoring the allergic reactions. Several medical devices have been used in the field of allergology by healthcare professionals to manage allergic reactions to improve the quality of life for individuals with allergies. The devices generally used in allergology can be broadly classified into two groups i.e. (1) general devices and (2) organ-specific devices. Skin prick test (SPT) is a primary diagnostic tool and is most widely used as a first-line approach for diagnosis of allergic reactions (Ansotegui et al. 2020; Wu et al. 2020). This test is very essential for detecting sensitization to various allergens, including airborne and food allergens. SPT can vary in terms of materials, shape and number. SPT elicits visible IgE-medicated wheal and flare reaction that is specific for the allergic reaction. Extraction of allergen should attached with the mast cell by different adaptable device. The accuracy of result in SPT is totally depend on operator's skill. Otherwise, false negative results may obtain because of insufficient puncture depth or interference from certain medications. Patch tests (PT) plays a crucial role for diagnosis of contact dermatitis. Theses test helps to identify delayed (type IV) allergic reaction of the skin in presence of potential allergens. PTs include allergen-containing patches and generally applied to the upper back of patients.

5.3 Bio-Fabrication Techniques

5.3.1 Stereolithography

Charles Hull created stereolithography (SLA), the first vat polymerization method, in 1986 (Hull 1984; Oliaei and Nasseri 2020). A new ray source (UV light) is used to initiate polymerization interaction in photosensitive materials. It is founded on the photopolymerization concept (Sadasivuni et al. 2020; Gibson et al. 2021). Precursors (monomers or oligomers), binders, absorbers, photoinitiators (PI), additives, and chain transfer agents make up the photopolymer resin (Zhu et al. 2023). The primary component of the resin hardens as a result of cross-linking is the precursor (Heller et al. 2009). When the device in contact with the light, and with different photoinitiators thus changes into radicals that cross-link with the monomers and oligomers to form a chain reaction (Wang et al. 2017; Schmidleithner and Kalaskar 2018). Additionally, to enhance the printed object's qualities or personalize it, additives in the form of pigments and nanoparticles are added (Manapat et al. 2017; Yilmaz et al. 2021). Layer-by-layer structure development is the outcome of the process, which takes place on the resin's surface because it is a single photon response (Al Rashid et al. 2021). Galvano meters are employed to maintain the ray accuracy because the end product is greatly impacted by the light source's precision and accuracy (Gao et al. 2015). Beyond its early applications in modeling and prototyping, SLA has undergone substantial development. It is increasing its capacity to produce elaborate and highly complex designs (Robles Martinez et al. 2018; Leigh et al. 2011). From conventional polymers, the variety of materials has expanded to include metals, ceramics, and composites. Product development for industries such as aerospace, robotics, sensors (Wazeer et al. 2023), medical implants (Sinha 2020; Tikhonov et al. 2020; Fu et al. 2017), energy storage components (Egorov et al. 2020; Tian et al. 2017; He et al. 2016), microfluidic devices (Huang et al. 2020), and more (Leijten et al. 2017) has been made possible by the

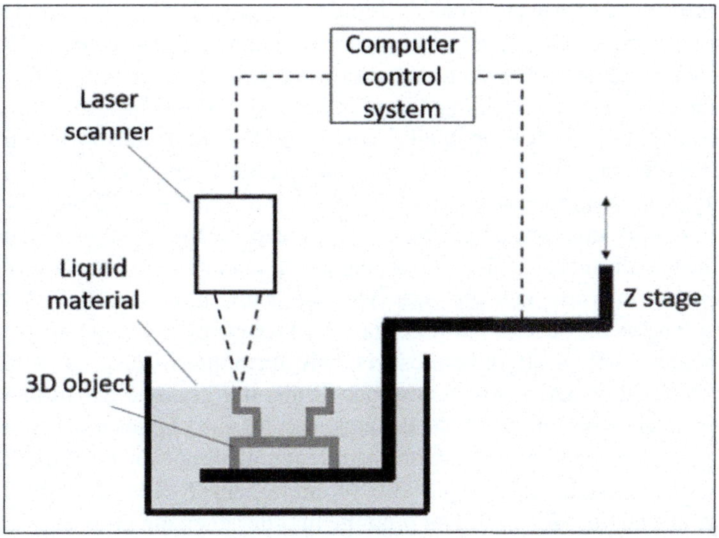

Fig. 5.2 Stereolithography system of Charles Hull for fabrication of biomedical device

increased material adaptability. A schematic presentation has been given in Fig. 5.2 (Du et al. 2016).

5.3.2 Cell Alignment

Cell alignment is essential for the organization of different organelles like adhesion complexes and the cytoskeleton, as well as for a variety of other processes like cellular differentiation, proliferation, and matrix deposition. Aligning cells within designed tissues for in vitro 3D cell culture and in vivo therapeutic application is therefore significant (Li et al. 2014; Mirbagheri et al. 2019; Vo et al. 2021; Prendergast et al. 2021; Lu et al. 2021). The preservation of particular functions is significantly influenced by the ordered alignment of cells in different tissues. Ideal scaffolds should mimic the properties and morphologies of real tissues in order to promote such an alignment. Tissue regeneration and repair are aided by aligned components that direct cell direction. Fabrication of several aligned structures, such as aligned electrospun nanofibers, aligned porous or channeled structures, micropatterns, and combinations of these, as well as their use in the regeneration of nerves, skeletal muscles, tendons, and tubular dentin can be done (Hernandez et al. 2008).

5.3.3 Liquid Exfoliation

Liquid-phase exfoliation (LPE) is a solution-processing technique that was first shown in 2008 (Coleman et al. 2011) and is used to produce huge amounts of two-dimensional nanosheets from multilayer crystals (Ferrari et al. 2015). At the moment, it is one of the main techniques for creating 2D nanosheets (Paton et al. 2014). This process entails combining powdered layered crystals, like graphite, with suitable solvents and adding energy, usually by ultrasonication, though high-shear mixing is frequently employed (Li et al. 2020). Small nanosheets are removed from the layered crystals as a result of a combination of fragmentation and exfoliation brought on by the input of energy (Daniel et al. 2022). Graphite can be transformed into vast amounts of graphene nanosheets in this manner (Claudia et al. 2019). These nanosheets typically have lateral diameters between tens of nanometres and many microns, and they are typically a few monolayers thick (Hu et al. 2021). As long as the solvents utilised have surface energies comparable to those of the nanosheets, these scattered nanosheets will form quasi-stable suspensions. It is possible to reach dispersed concentrations of about 1 gramme per litre. Molecular stabilisers, such as polymers or surfactants, can be used in addition to solvents to coat the nanosheets and prevent them from reaggregating (Torrisi et al. 2012). The benefit of this is that it makes it possible for nanosheets to float in water. This technique was initially used to exfoliate graphite in order to create graphene nanosheets, but it has subsequently been applied to the production of a variety of 2D compounds, such as black phosphorus, SnP_3, nickel (II) hydroxide, tungsten diselenide, boron nitride, molybdenum disulphide, and germanium monosulfide. A variety of functional structures can be made from the liquid suspensions that liquid phase exfoliation produces. For instance, they can be produced using common methods like inkjet printing to create networks and thin films (Papadimitriou et al. 2020).

5.3.4 Cell-Electrospinning and Bio-Electrospraying

The process of electrospinning uses electrostatic forces to extract thin fibres from a viscous fluid. The high concentration of the solution and the irregularly distributed charges cause the charged jet to whip or bend as it is expelled towards the collector during this operation. Consequently, a solid, thin fibre is deposited on the grounded collector following the jet's elongation and the solvent's quick evaporation. Usually, the finished product is a non-woven mat that is orientated randomly. Several specially constructed collectors, such spinning drums, metal frames, or two conductive surfaces divided by an insulating gap (Park et al. 2007; Xue et al. 2017), have been employed in recent decades to produce aligned electrospun fibres. The set up for electrospraying and electrospinning is given in Fig. 5.3 (Park et al. 2007). The final nanofiber structure is also influenced by the collector's form. Copper, aluminium, gold, wood, and other materials can be used to create the deposited designs, which can be in the form of circles, triangles, squares, crosses, rectangles, and more (Park et al. 2007; Xue et al. 2017). Additionally, to create a continuous yarn from electrospun fibres, a liquid system with a coagulation bath can be employed in place of a collector or in conjunction with a spinning mandrel. Generally speaking, mechanical, magnetic, or electrostatic methods can be used to achieve the required fibre alignment (Shi et al. 2015; Taheri and Jafari 2019).

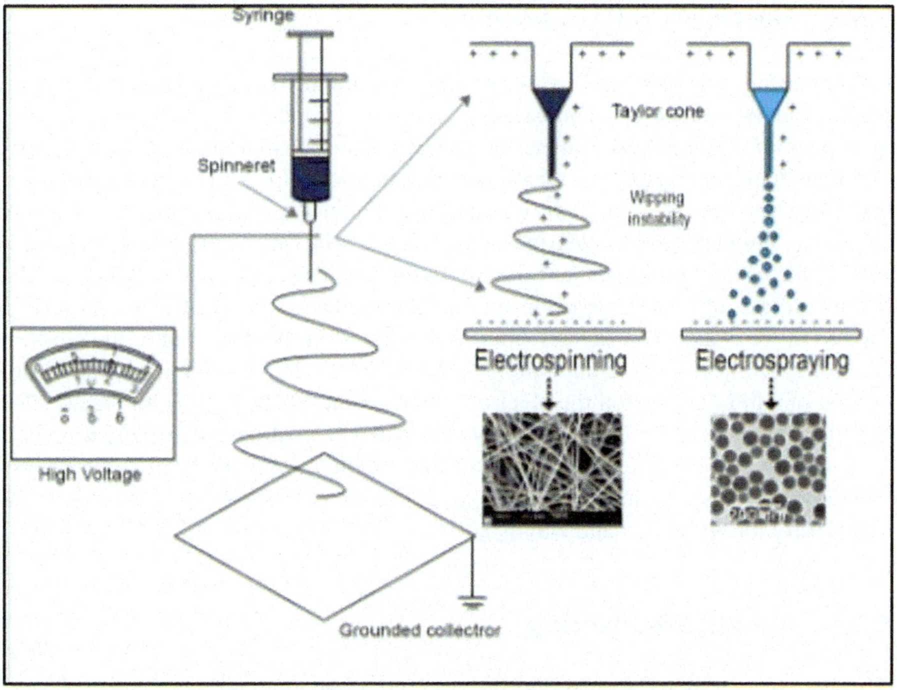

Fig. 5.3 Fabrication of biomedical device by electrospraying and electrospinning

Electro-spraying is the method for atomizing different liquids using electrical forces. As previously, the different concentration of the solution is main construct core also the electrospinning. Due to varicose instability, the low concentration of the solution in electrospraying causes the jet to become unstable, resulting in the formation of small droplets. Coagulation of the particles, as well as the aggregation of droplets, are avoided due to disperse in space. Additionally, as time passes, the +evaporation of different solvents encourages the droplets to compress and solidify, depositing solid particles on the grounded collector. Since liquid atomization requires no additional mechanical energy, the jet deforms and breaks up into droplets during the electrospraying process mostly due to electrical force. By altering the flow velocity and voltage applied to the nozzle, it is simple to regulate the droplet's size and charge to a certain degree. Therefore, compared to traditional mechanical spraying systems, where droplets are charged by induction, electrospraying offers the following benefits: (i) it generates smaller droplets; (ii) its size distribution is typically narrow with low standard deviation; (iii) the charged droplets self-disperse and do not agglomerate or coagulate; (iv) electric fields, such as deflection or focusing, can readily control the motion of the charged droplets; and (v) compared to uncharged droplets, the charged spray has a substantially better deposition efficiency (Park et al. 2007; Boccaccini et al. 2010).

5.3.5 Electrophoretic Deposition

An increasingly popular method for processing biomaterials, particularly bioactive coatings and biomedical nanostructures, is electrophoretic deposition (EPD). Starting with well-dispersed suspensions of biomaterials in particulate form (micro-sized and nanoscale particles, nanotubes, and nanoplatelets), the well-known benefits of EPD for the production of a broad range of microstructures and nanostructures as well as unique and complex material combinations are being utilized (Heavens 1990). The EPD of biological entities, including bacteria, cells, and enzymes, is also being studied. EPD is often performed in a two-electrode cell. Electrophoresis is used to move charged particles scattered in a liquid in the direction of the working electrode, while particle coagulation is the main mechanism for the creation and growth of solid deposits on the electrode (Sarkar and Nicholson 1996; Gani 1994). EPD is applicable to a wide range of materials that are available as colloidal suspensions or tiny powders, often with particle sizes smaller than 30 μm. EPD can deposit metals, polymers, ceramics, glasses, and their composites (van der Biest and Vandeperre 1999; Robert and Baylis 2008).

5.3.6 Genetic Engineering

A variety of methods for purposefully modifying genetic material—mostly deoxyribonucleic acid, or DNA—to change, fixes, or improve shape or function are referred to as genetic engineering. It is a collection of technologies that alter cells' genetic composition, including transferring genes both inside and between species to create new or improved organisms. Recombinant DNA technologies were created in the second half of the twentieth century and involve the chemical splicing (recombination) of different DNA strands, usually by direct microinjection or the use of bacteria (like Escherichia coli) or bacteriophages (viruses that infect bacteria, like λ phage). New methods for creating and constructing—literally, engineering—new living forms, commonly known as synthetic biology, have recently been added to these conventional tools (Genetic Engineering 2022; Terms and Acronyms 2015). Either artificial DNA synthesis or recombinant DNA techniques are used to isolate and replicate the desired genetic material to create new DNA. This DNA is often inserted into the host organism via a construct. In addition to introducing genes, the procedure can be used to "knock out" or eliminate genes. The new DNA might be targeted to a particular region of the genome or inserted at random. A genetically modified organism (GMO) is the result of genetic engineering, and an organism created by this process is referred to as genetically modified (GM). In 1973, Stanley Cohen and Herbert Boyer created the first genetically modified organism (GMO) (Terms and Acronyms 2015; Mendes et al. 2013).

5.3.7 Self-Assembly

The automatic grouping of individual molecules into distinct and repeatable higher-ordered structures by the use of non-covalent interactions such van der Walls, hydrogen, hydrophobic, and electrostatic forces is known as self-assembly. Self-assembly is nature's main method of fabrication, transforming small molecules into hierarchical and functional structures, from the precise folding of individual proteins or DNA molecules to higher-order assemblies of phospholipids into cell membranes or proteins into the tobacco mosaic virus (Smith et al. 2011). There is an unrealized potential for creating innovative techniques that combine self-assembly and biofabrication. The benefits of one strategy typically outweigh the drawbacks of the other in many respects. The capacity to bioprint using several biomolecule types, for instance, that instantly come together to form a milieu of distinct nanostructures that selectively activate cells while arranging them into exact anatomical geometries with hierarchical order. Self-assembling bioinks (SABs) are made up of smaller parts, including proteins, peptides, polymers, or DNA, that can be reliably assembled into distinct, higher-ordered structures (Moroni et al. 2018).

5.3.8 Bioprinting

There is an unrealized potential for creating innovative techniques that combine self-assembly and bio-fabrication. The benefits of one strategy typically outweigh the drawbacks of the other in many respects. The capacity to bio-print using several biomolecule types, for instance, that instantly come together to form a milieu of distinct nanostructures that selectively activate cells while arranging them into exact anatomical geometries with hierarchical order. Self-assembling bioinks (SABs) are made up of smaller parts, including proteins, peptides, polymers, or DNA, that can be reliably assembled into distinct, higher-ordered structures (Nicola et al. 2020). Technologies for 3D bioprinting (3DBP) create a lot of opportunities for creating extremely intricate cellular structures. Depending on their inherent qualities and how they have been presented in the biologic environment, nano-biomaterials have been widely used in tissue engineering and regenerative medicine (TERM) for a variety of goals and activities. By strengthening these manufacturing processes' fundamental weaknesses and increasing their efficiency by strategically placing nano-features, the combination of bioprinting and nano-biomaterials opens up new possibilities in the biofabrication scenario. For native tissues to be successfully designed and matured, 3D cell organization is essential (Gong et al. 2022). 3D printing techniques are highlighted in Fig. 5.4 (Gong et al. 2022).

5.3.9 Screen Printing

Flexible sensors are thought to be best manufactured using screen printing technique. Through the use of conventional printing techniques, screen printing

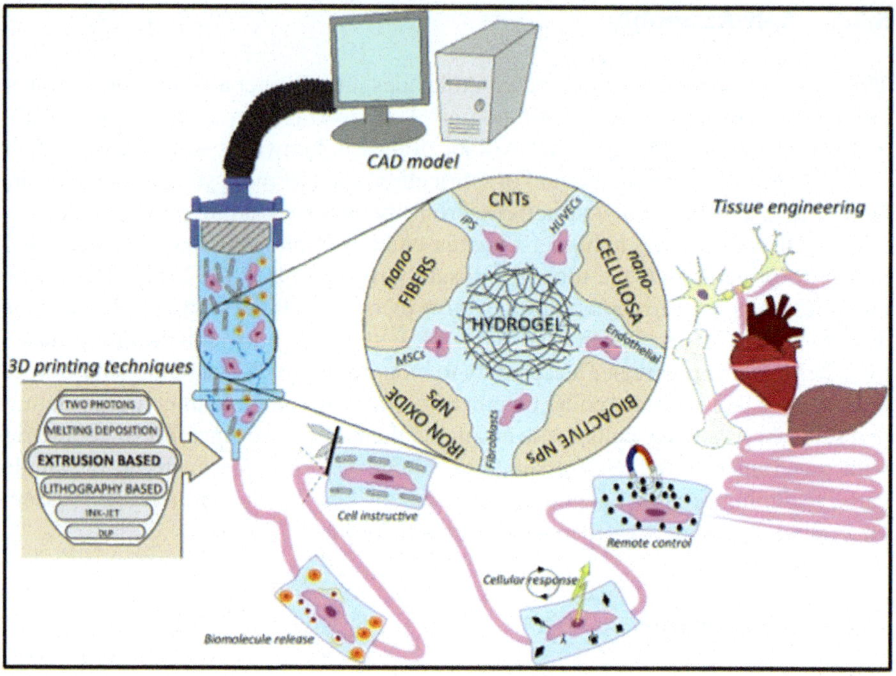

Fig. 5.4 Exploring 3D printing methods and the use of nanocomposite bio-inks in the biofabrication process for tissue engineering

prepares electronic devices by creating inks that are appropriate for printing procedures and contain particular functional elements. In addition to inheriting the benefits of mass production, low initial investment, and low R&D cost of traditional printing technology, screen printing is a very straightforward, quick, and efficient printing technique that can produce many identical patterns in a single printing process at a low cost under normal environmental conditions. It can also be applied to a variety of substrates, including fibers, fabrics, and plastics. Micro/nano electronic devices have prospered due to the quick development of modern manufacturing technology, and in recent years, flexible sensors based on screen printing have gained a lot of attention to satisfy the demands of wearable electronics and human health monitoring. Flexible sensors have also been successfully demonstrated to provide very special benefits and a wide range of mass production opportunities when paired with screen printing technology. Three types of screen-printed flexible sensors are used in the field of human health monitoring: sensing arrays (gait, sitting posture, electronic skin, body temperature monitoring), non-strain sensors (human blood, sweat, and electrocardiogram monitoring), and mechanical strain sensors (human physiological signal monitoring, human joint motion monitoring). Ink controllability, high resolution printing, and thickness controllability are the three areas where screen printing technology presents difficulties (Butcher et al. 2014).

5.3.10 Crosslinking

The adoption of biomaterials in biomedical applications is limited by a number of factors, despite their well-established benefits and broad range of uses (Jiang et al. 2010). First and foremost, biopolymeric materials frequently lack the durability in aqueous and physical environments needed for medical applications, as well as sufficient mechanical characteristics (Jiang et al. 2013). For example, protein-based films and electrospun structures break down in aqueous solutions or at high humidity levels (Hennink and Nostrum 2012; Martinez et al. 2014). The most popular method for getting over biomaterials' drawbacks is crosslinking (Koide and Daito 1997; Reddy et al. 2015). In addition to increasing molecular weight and connecting molecules, crosslinkers typically offer better stability and mechanical qualities. Crosslinking, however, also results in altered rheology, reduced availability of functional groups in the cross-linked polymer, and decreased degradability, which might enhance cytotoxicity and cause further processing issues (Aarthi et al. 2024). Depending on the kind of biopolymer to be crosslinked and the desired level of property improvement, several crosslinkers and crosslinking techniques are employed. Glutaraldehyde is the most widely employed of the several chemical crosslinkers because it reacts with the functional groups in proteins and carbohydrates to produce materials with significantly better tensile qualities. Despite the fact that glutaraldehyde improves mechanical qualities, there is conflicting data about the cytotoxicity of materials crosslinked with glutaraldehyde (Gharib et al. 2022).

5.3.11 Microfluidics

Integrated structures made up of multiple micro- and nano-sized integrated devices are known as biomedical microdevices. These devices perform a variety of functions, such as particle manipulation and sensing. Passive microfluidic systems are primarily used for particle manipulation (Özbey et al. 2016; Shen et al. 2020) and liquid mixing (Özbey et al. 2016; Zhao et al. 2021), while active types contribute more to particle trapping (Zhao et al. 2021; Mandal and Banerjee 2022) and sensing (Kocheril et al. 2022; Niculescu et al. 2021), even though different types of microfluidic devices can accomplish similar tasks in biomedical applications. While active microfluidic devices create streams relying on external energy to disturb particles or fluids inside microfluidic devices, passive devices are controlled by diffusion, inertial forces, secondary flows, geometry-induced turbulence, and particle manipulation. In microfluidic channels, separation of particles with varying sizes and densities under the influence of internal forces can achieve high-efficiency values, and fluid mixing ratios may be reasonably high depending on the geometric design. In biomedical applications, external forces such as acoustic pressure fields, electric fields, magnetic fields, thermal fields, pressure fields, and optical fields could mix fluids and control chemical or biological particles. Microfluidics devices have a number of biomedical uses. The diagnosis of illnesses, such as infectious and

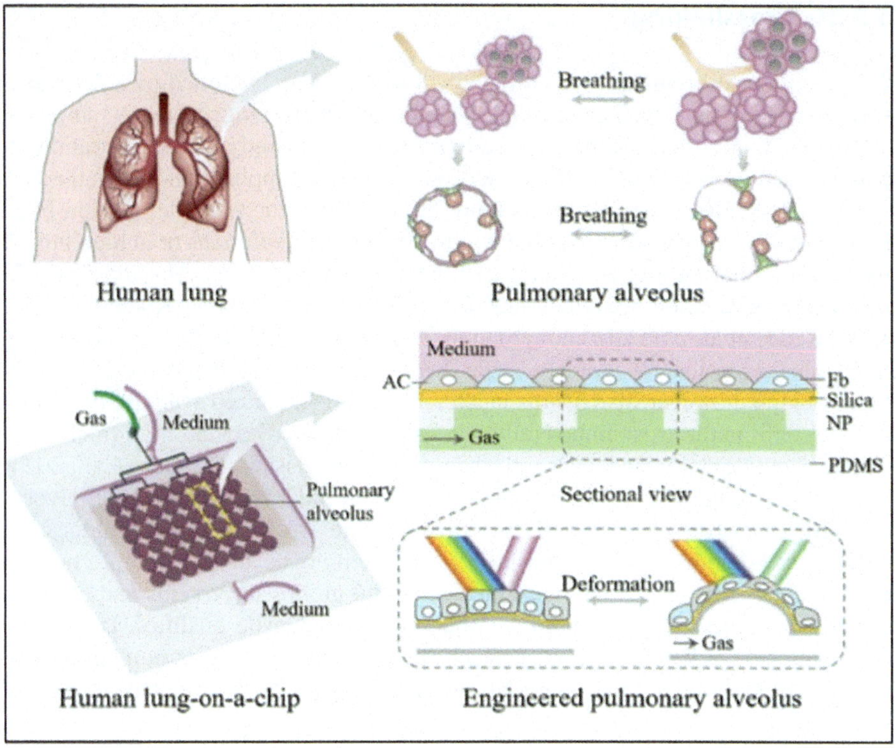

Fig. 5.5 Lung-on-a-chip platform

cancerous conditions, is one of the biomedical sciences' most promising uses of microfluidics. However, as microfabrication advances, microfluidics devices can be used in organ-on-a-chip, tissue engineering, and disease modeling (Sharmila et al. 2024; Hassan et al. 2024). Furthermore, microfluidics–biosensing technology has gained popularity for uses like cell manipulation, biosensors, and point-of-care diagnostics (Assadi et al. 2016; Özbey et al. 2016). Figure 5.5 shows a representation of a lung-on-a-chip platform (Özbey et al. 2016).

5.3.12 Cold Spray

Cold-spray is a solid-state material deposition technique in which a carrier gas (air, nitrogen (N_2), or helium (He) propels micron-sized powder particles onto a substrate at an exceptionally high velocity (often supersonic). Because of metallurgical bonding at the particle-substrate contact or mechanical interlocking, these particles become bound to the substrate upon impact. High velocity is achieved by expanding the hot carrier gas through a convergent-divergent nozzle (Champagne 2007; Sobon 2018). The cold spray (CS) system can be classified as high pressure or low pressure based on its operating pressure capabilities. The fundamental distinction between

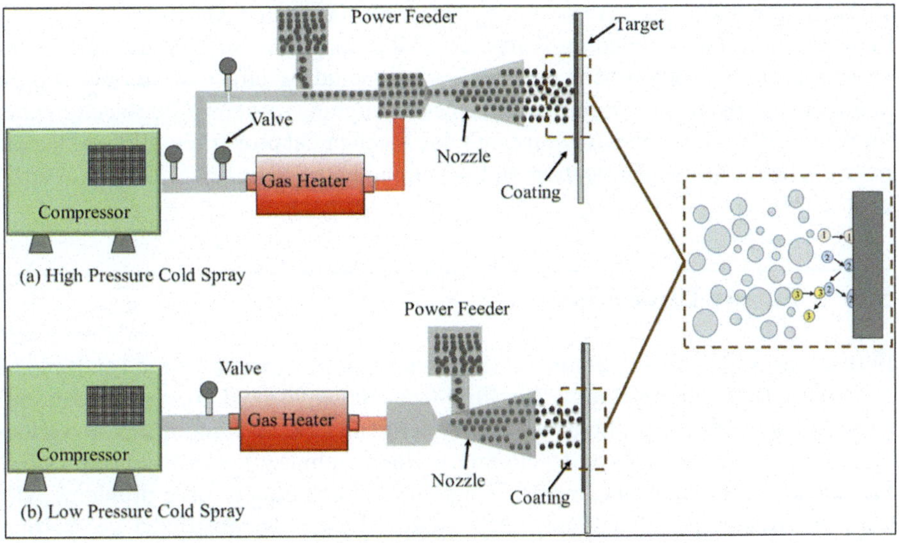

Fig. 5.6 Cold spray technology for fabrication of biomedical device

the CS and other thermal spray techniques is that, during the whole process, the feedstock particles maintain their solid state, which is what gives the CS its name, even if the carrier gas is heated to provide acceleration and thermal softening. Therefore, high kinetic energy is the driving force in CS, which results in severe plastic deformation during deposition rather than the full or partial melting of the feedstock as is the case with other thermal spray systems. We may deposit a variety of materials, including metals, alloys, polymers, and composites, using CS. Numerous industries, including automotive (Liao et al. 2022), biomedical (Boruah et al. 2019), aerospace (Kumar et al. 2022), medical (Widener et al. 2016), energy, and marine (Yeom and Sridharan 2021; Gholap et al. 2024, 2025), employ CS coating technology extensively (Saminathan et al. 2024). CS coatings offer defense against chemicals, oxidation, corrosion, and erosion. In contrast to other coating techniques, the coatings created by CS are dense, free from thermal stresses, and contain less oxide. The schematic diagram for cold spray technology is given in Fig. 5.6 (Gholap et al. 2025).

5.4 Challenges

Bio-fabrication and biomedical devices hold a great promise for revolutionizing the healthcare industry. However, challenges pertain to the development of new technologies, their development and maintenance. Development of in-vitro models in 3D structure for their study and evaluation has been challenging (Gholap et al. 2023). Effective model development for regenerative medicine or tissue engineering applications also requires great skill and expertise. Simultaneous bio-fabrication of vascular networks capable of perfusing bigger bio-fabricated constructions is one of the most important continuing issues for a wide range of tissue types. Further the various processes involved in bio-fabrication including effective nutrient supply, maintaining effective environment, chances of immune response and rejection after implantation in body or maintaining suitable mechanical properties, all these factors pose challenge to bio-fabrication and development of biomedical devices (Aarthi et al. 2024).

5.5 Future Directions

Further developments in volumetric (3rd-order unit processes), hybrid (multimodal and assisted), and bio-assembly (scaffold-free) strategies hold promise for the strategic arrangement of multiple materials, cells, and ECM components at various length scales in the quickly developing multidisciplinary field of bio-fabrication. Particularly, recent advancements in volumetric techniques solve problems with popular layer-by-layer methods, such as low throughput, restricted geometrical freedom, and higher post-processing needs. Functional tubular structures have recently been demonstrated by coupling high cellular density spheroid organizations with magnetic and acoustic bio-assembly (Gholap et al. 2023). The development of novel bioinks will also affect the fine-tuning of target tissues' structural and biomimetic components. The multiple chemotactic and haptotactic cues present in natural tissue maturation may be replicated by stimuli-responsive materials that can imitate time-dependent morphological tissue changes (by temperature, pH, and light), as in 4D printing processes. Furthermore, bio-fabrication techniques are a key part of advancements in customized medical care; examples include corneal replacements, meniscal scaffolds, and cardiac patches tailored to each patient. The success of personalized bio-fabrication strategies depends on the therapeutic-scale production of stem cells, cell aggregates, and organoids. Scalability can be attained through methods such as aggregate suspension bioreactors, optimized use of expensive growth factors, and a better understanding of the in vivo mechanisms of autonomous tissue development.

5.6 Conclusions

The present book chapter describes the various biomedical devices like glucose monitoring device, ECG/EEG device, allergy diagnostic devices, wearable sensors, body temperature monitoring devices, etc. Further bio-fabrication techniques including liquid exfoliation, electrospinning and electrospraying, 3D printing, crosslinking methods have been described with their suitable applications and methods. Finally, the challenges faced for the development and use of such biomedical devices have been discussed, along with the future directions in the technology for optimized development and application of biomedical devices.

References

Aarthi J, Kaviya SS, Kirubavathi K, Selvaraju K, Shakeel F, Gowri S, Faiyazuddin M (2024) Green synthesis of nickel oxide nanoparticles using leaf extract of Mimosa pudica for killing triple-negative breast cancer MDA-MB-231 cells and bacteria. J Mol Struct 1321:140178. https://doi.org/10.1016/j.molstruc.2024.140178

Acar G, Ozturk O, Golparvar AJ, Elboshra TA, Böhringer K, Yapici MK (2019) Wearable and flexible textile electrodes for biopotential signal monitoring: a review. Electronics 8:479. https://doi.org/10.3390/electronics8050479

Accessed 18 Dec 2019. https://www.ema.europa.eu/en/human-regulatory/overview/medical-devices

Al Rashid A, Ahmed W, Khalid MY, Koç M (2021) Vat photopolymerization of polymers and polymer composites: processes and applications. Addit Manuf 47:Article 102279. https://doi.org/10.1016/j.addma.2021.102279

Amjadi M, Kyung KU, Park I, Sitti M (2016) Stretchable, skin-mount-able, and wearable strain sensors and their potential applications: a review. Adv Funct Mater 26:1678–1698. https://doi.org/10.1002/adfm.201504755

Ankhili A, Tao X, Cochrane C, Coulon D, Koncar V (2018a) Washable and reliable textile electrodes embedded into underwear fabric for electrocardiography (ECG) monitoring. Materials 11:256. https://doi.org/10.3390/ma11020256

Ankhili A, Tao X, Cochrane C, Koncar V, Coulon D, Tarlet JM (2018b) Comparative study on conductive knitted fabric electrodes for long-term electrocardiography monitoring: silver-plated and PEDOT:PSS coated fabrics. Sensors 18:3890. https://doi.org/10.3390/s18113890

Ankhili A, Tao X, Cochrane C, Koncar V, Coulon D, Tarlet JM (2019) Ambulatory evaluation of ECG signals obtained using washable textile-based electrodes made with chemically modified PEDOT:PSS. Sensors 19:416. https://doi.org/10.3390/s19020416

Ansotegui IJ, Melioli G, Canonica GW et al (2020) IgE allergy diagnostics and other relevant tests in allergy, a World Allergy Organization position paper. World Allergy Organ J 13(2):100080. https://doi.org/10.1016/j.waojou.2019.100080. Erratum in: World Allergy Organ J. 2021 Jun 17;14(7):100557. doi: 10.1016/j.waojou.2021.100557. PMID: 32128023; PMCID: PMC7044795

Arquilla K, Webb AK, Anderson AP (2020) Textile electrocardiogram (ECG) electrodes for wearable health monitoring. Sensors 20:1013. https://doi.org/10.3390/s20041013

Arquilla K, Devendorf L, Webb AK, Anderson AP (2021) Detection of the complete ECG waveform with woven textile electrodes. Biosensors 11:331. https://doi.org/10.3390/bios11090331

Assadi H, Kreye H, Gartner F, Klassen T (2016) Cold spraying—a materials perspective. Acta Mater 116:382–407. https://doi.org/10.1016/j.actamat.2016.06.034

Atalay O, Kennon WR, Demirok E (2014) Weft-knitted strain sensor for monitoring respiratory rate and its electro-mechanical modelling. IEEE Sensors J 15(1):110–122. https://doi.org/10.1109/JSEN.2014.2339739

Awara K, Kitai R, Isozaki M, Neishi H, Kikuta K, Fushisato N, Kawamoto A (2014) Thin-film electroencephalographic electrodes using multi-walled carbon nanotubes are effective for neurosurgery. Biomed Eng Online 13:166. https://doi.org/10.1186/1475-925x-13-166

Baik S, Lee HJ, Kim DW, Kim JW, Lee Y, Pang C (2019) Bioinspired adhesive architectures: from skin patch to integrated bioelectronics. Adv Mater 31(34):e1803309. https://doi.org/10.1002/adma.201803309

Blachowicz T, Ehrmann G, Ehrmann A (2021) Textile-based sensors for biosignal detection and monitoring. Sensors 21:6042. https://doi.org/10.3390/s21186042

Blasdel NJ, Monty CN (2015) Temperature sensitive fabric for monitoring dermal temperature variations. In: Mukhopadhyay SC (ed) Wearable electronics sensors: for safe and healthy living. Springer International Publishing, Cham, pp 193–220. ISBN 978-3-319-18191-2. https://doi.org/10.1007/978-3-319-18191-2_8

Boccaccini AR, Keim S, Ma R, Li Y, Zhitomirsky I (2010) Electrophoretic deposition of biomaterials. JR Soc Interface 7:S581–S613. https://doi.org/10.1098/rsif.2010.0156.focus

Booth MA, Gowers SAN, Hersey M et al (2021) Fiber-based electrochemical biosensors for monitoring pH and transient neurometabolic lactate. Anal Chem 93:6646–6655. https://doi.org/10.1021/acs.analchem.0c05108

Boruah D, Ahmad B, Lee TL, Kabra S et al (2019) Evaluation of residual stresses induced by cold spraying of Ti6Al-4V on Ti-6Al-4V substrates. Surf Coat Technol 374:591–602. https://doi.org/10.1016/j.surfcoat.2019.06.028

Butcher AL, Offeddu GS, Oyen ML (2014) Nanofibrous hydrogel composites as mechanically robust tissue engineering scaffolds. Trends Biotechnol 32:564–570. https://doi.org/10.1016/j.tibtech.2014.09.001

Cao Q, Shao Z, Hensley D, Venton BJ (2023) Carbon nanospike coated nanoelectrodes for measurements of neurotransmitters. Faraday Discuss 233:303–314. https://doi.org/10.1039/D1FD00053E

Champagne VK (2007) The cold spray materials deposition process: fundamentals and applications. Woodhead Publishing

Chan HL, Kuo PC, Cheng CY, Chen YS (2018) Challenges and future perspectives on electroencephalogram-based biometrics in person recognition. Front Neuroinform 12:66. https://doi.org/10.3389/fninf.2018.00066

Chen SC, Chang JM, Yeh SM, Su HM, Chen HC (2013) Association of uric acid and left ventricular mass index with renal outcomes in chronic kidney disease. Am J Hypertens 26(2):243–249. https://doi.org/10.1093/ajh/hps020

Chen XC, He H, Khan Z, Sydänheimo L, Ukkonen L, Virkki J (2019) Design, fabrication and wireless evaluation of a passive 3D-printed moisture sensor on a textile substrate. In: Proceedings of the 2019 photonics & electromagnetics research symposium–spring (PIERS-spring), Rome, Italy, pp 1027–1030. https://doi.org/10.1109/PIERS-Spring46901.2019.9017301

Chen XC, He H, Khan Z, Sydanheimo L, Ukkonen L, Virkki J (2020) Textile-based batteryless moisture sensor. IEEE Antennas Wirel Propag Lett 19:198–202. https://doi.org/10.1109/LAWP.2019.2957879

Chen S, Cao Z, Zhou K, Li S, Li H, Xu K, Tang H, Deng H, Zhou Q, Pan J, Xia F (2023) Screen printing and laser-induced flexible sensors for the simultaneous sensitive detection of uric acid, tyrosine, and ascorbic acid in sweat. Analyst 148(13):2965–2974. https://doi.org/10.1039/d3an00591g

Chi M, Zhao J, Dong Y, Wang X (2019) Flexible carbon nanotube-based polymer electrode for long-term electrocardiographic recording. Materials 12:971. https://doi.org/10.3390/ma12060971

Chu M, Naguib HE (2021) Soft flexible conductive CNT nanocomposites for ECG monitoring. Smart Mater Struct 30:065003. https://doi.org/10.1088/1361-665X/abefb6

Claudia B, Campi D, Szydlowska B, Synnatschke K et al (2019) Equipartition of energy defines the size–thickness relationship in liquid-exfoliated nanosheets. ACS Nano 13(6):7050–7061. https://doi.org/10.1021/acsnano.9b02234

Coleman JN, Lotya M, O'Neill A et al (2011) Two-dimensional nanosheets produced by liquid exfoliation of layered materials. Science 331(6017):568–571. https://doi.org/10.1126/science.1194975

Cordeiro CA, Sias A, Koster T, Westerink BHC, Cremers TIFH (2018) *In vivo* "real-time" monitoring of glucose in the brain with an Amperometric enzyme-based biosensor based on gold coated tungsten (W-Au) microelectrodes. Sens Actuators B Chem 263:605–613. https://doi.org/10.1016/j.snb.2018.02.116

Coyle S, Morris D, Lau KT, Diamond D, Taccini N, Costanzo D, Salvo P, di Francesco F, Trivella MG, Porchet JA (2009) Textile sensors to measure sweat pH and sweat-rate during exercise. In: Proceedings of the 2009 3rd international conference on pervasive computing technologies for healthcare, London, UK, pp 1–6. https://doi.org/10.4108/ICST.PERVASIVEHEALTH2009.5957

Daniel GUC, Patrizia C, Omar SF (2022) Revisión sobre la síntesis de grafeno por exfoliación en fase líquida: Mecanismos, factores y técnicas. Uniciencia 36(1):1–14. https://doi.org/10.15359/ru.36-1.35

Das SK, Nayak KK, Krishnaswamy PR, Kumar V, Bhat N (2022) Electrochemistry and other emerging technologies for continuous glucose monitoring devices. ECS Sensors Plus 1(3):031601

Devaux E, Aubry C, Campagne C, Rochery M (2011) PLA/carbon nanotubes multifilament yarns for relative humidity textile sensor. J Eng Fibers Fabr 6:13–24. https://doi.org/10.1177/155892501100600302

Dore H, Aviles-Espinosa R, Luo Z, Anton O, Rabe H, Rendon-Morales E (2021) Characterization of textile embedded electrodes for use in a neonatal smart mattress electrocardiography system. Sensors 21:999. https://doi.org/10.3390/s21030999

Du D, Li P, Ouyang J (2016) Graphene coated nonwoven fabrics as wearable sensors. J Mater Chem C 4:3224–3230. https://doi.org/10.1039/C6TC00350H

Egorov V, Gulzar U, Zhang Y, Breen S, O'Dwyer C (2020) Evolution of 3D printing methods and materials for electrochemical energy storage. Adv Mater 32(29):Article 2000556. https://doi.org/10.1002/adma.202000556

Ehrmann G, Ehrmann A (2020) Suitability of common single circuit boards for sensing and actuating in smart textiles. Commun Dev Assem Text Prod 1:170–179. https://doi.org/10.25367/cdatp.2020.1.p170-179

Ehrmann G, Ehrmann A (2021) Electronic textiles. Encyclopedia 1:115–130. https://doi.org/10.3390/encyclopedia1010013

Eom J, Jaisutti R, Lee H, Lee W, Heo JS, Lee JY, Park SK, Kim YH (2017) Highly sensitive textile strain sensors and wireless user-interface devices using all-polymeric conducting fibers. ACS Appl Mater Interfaces 9:10190–10197. https://doi.org/10.1021/acsami.7b01771

Fan W, He Q, Meng K, Tan X, Zhou Z, Zhang G, Yang J, Wang ZL (2020) Machine-knitted washable sensor array textile for precise epidermal physiological signal monitoring. Sci Adv 6:eaay2840. https://doi.org/10.1126/sciadv.aay2840

Fang C, He B, Wang Y, Cao J, Gao S (2020) EMG-centered multisensory based technologies for pattern recognition in rehabilitation: state of the art and challenges. Biosensors 10:85. https://doi.org/10.3390/bios10080085

Feig DI, Kang DH, Johnson RJ (2008) Uric acid and cardiovascular risk. N Engl J Med 359:1811–1821. https://doi.org/10.1056/nejmra0800885

Feng J, Chen C, Sun X, Peng H (2019) Implantable fiber biosensors based on carbon nanotubes. Acc Mater Res 2:138–146. https://doi.org/10.1021/accountsmr.0c00109

Ferrari AC, Bonaccorso F, Fal'ko V et al (2015) Science and technology roadmap for graphene, related two-dimensional crystals, and hybrid systems. Nanoscale 7(11):4598–4810. https://doi.org/10.1039/C4NR01600A

Fiedler P, Pedrosa P, Griebel S, Fonseca C, Vaz F, Supriyanto E, Zanow F, Haueisen J (2015) Novel multipin electrode cap system for dry electroencephalography. Brain Topogr 28:647–656. https://doi.org/10.1007/s10548-015-0435-5

Fu K, Yao Y, Dai J, Hu L (2017) Progress in 3D printing of carbon materials for energy-related applications. Adv Mater 29(9):Article 1603486. https://doi.org/10.1002/adma.201603486

Gani MSJ (1994) Electrophoretic deposition. A review. Ind Ceram 14:163–174. https://doi.org/10.1002/9781118176665.ch15

Gao W, Zhang Y, Ramanujan D, Ramani K et al (2015) The status, challenges, and future of additive manufacturing in engineering. Comput Aided Des 69:65–89. https://doi.org/10.1016/j.cad.2015.04.001

Gao W, Emaminejad S, Nyein HYY et al (2016) Fully integrated wearable sensor arrays for multiplexed in situ perspiration analysis. Nature 529:509–514. https://doi.org/10.1038/nature16521

Gaubert V, Gidik H, Koncar V (2020) Boxer underwear incorporating textile moisture sensor to prevent nocturnal enuresis. Sensors 20:3546. https://doi.org/10.3390/s20123546

Genetic Engineering. Genome.gov. Retrieved 20 February 2022

Gharib G, Bütün I, Muganli Z, Kozalak G et al (2022) Biomedical applications of microfluidic devices: a review. Biosensors 12(11):1023. https://doi.org/10.3390/bios12111023
Gholap AD, Gupta J, Kamandar P, Bhowmik DD, Rojekar S, Faiyazuddin M, Hatvate NT, Mohanto S, Ahmed MG, Subramaniyan V, Kumarasamy V (2023) Harnessing nanovaccines for effective immunization— a special concern on COVID-19: facts, fidelity, and future prospective. ACS Biomat Sci Eng 10(1):271–297. https://doi.org/10.1021/acsbiomaterials.3c01247
Gholap AD, Pardeshi SR, Hatvate NT, Dhorkule N, Sayyad SF, Faiyazuddin M, Khalid M (2024) Environmental implications and nanotechnological advances in Octocrylene-enriched sunscreen formulations: a comprehensive review. Chemosphere 142235. https://doi.org/10.1016/j.chemosphere.2024.142235
Gholap AD, Sawant SD, Pardeshi SR, Khuspe PR, Sayyad SF, Chavan M, Hatvate NT, Faiyazuddin M, Uddin MJ (2025) Omics tools for determining microbial composition: a systematic review. Microb Eng:35–50. https://doi.org/10.1201/9781003394662-3
Gibson I, Rosen D, Stucker B (2021) Development of additive manufacturing technology, additive manufacturing technologies. Springer International Publishing, Cham, pp 23–51
Golparvar A, Ozturk O, Yapici MK (2021) Gel-free wearable electroencephalography (EEG) with soft graphene textiles. In: Proceedings of the 2021 IEEE sensors, virtual format, pp 1–4. https://doi.org/10.1109/SENSORS47087.2021.9639711
Gong X, Huang K, Wu YH, Zhang XS (2022) Recent progress on screen-printed flexible sensors for human health monitoring. Sensors Actuators A Phys 345:113821. https://doi.org/10.1016/j.sna.2022.113821
Grethe T, Borczyk S, Plenkmann K, Normann M, Rabe M, Schwarz-Pfeiffer A (2018) Textile humidity sensor. In: Proceedings of the 2018 symposium on design, test, integration & packaging of MEMS and MOEMS (DTIP), Rome, Italy, pp 1–3
Güder F, Ainla A, Redston J, Mosadegh B, Glavan A, Martin TJ, Whitesides GM (2016) Paper-based electrical respiration sensor. Angew Chem Int Ed 55:5727–5732. https://doi.org/10.1002/anie.201511805
Güntner AT, Abegg S, Königstein K, Gerber PA, Schmidt-Trucksäss A, Pratsinis SE (2019) Breath sensors for health monitoring. ACS Sens 4:268–280. https://doi.org/10.1021/acssensors.8b00937
Guo L, Berglin L, Li YJ, Mattila H, Kalantar Mehrjerdi A, Skrifvars M (2011) 'Disappearing sensor'-textile based sensor for monitoring breathing. In: Proceedings of the 2011 international conference on control, automation and systems engineering (CASE), Singapore, pp 1–4. https://doi.org/10.1109/ICCASE.2011.5997723
Guo L, Berglin L, Wiklund U, Mattila H (2013) Design of a garment-based sensing system for breathing monitoring. Text Res J 83(5):499–509. https://doi.org/10.1177/0040517512444336
Hao D, Xu B, Cai Z (2018) Polypyrrole coated knitted fabric for robust wearable sensor and heater. J Mater Sci Mater Electron 29:9218–9226. https://link.springer.com/article/10.1007/s10854-018-8950-2
Hassan J, Faiyazuddin M, Uddin MJ (2024) Marine Drug Delivery Systems. In: Handbook of research in marine pharmaceutics. Apple Academic Press, pp 49–135
He Y, Wu Y, Fu J, Gao Q, Qiu J (2016) Developments of 3D printing microfluidics and applications in chemistry and biology: a review. Electroanalysis 28(8):1658–1678. https://doi.org/10.1002/elan.201600043
He W, Wang C, Wang HM, Jian MQ, Lu WD, Liang XP, Zhang X, Yang FC, Zhang YY (2019) Integrated textile sensor patch for real-time and multiplex sweat analysis. Sci Adv 5:eaax0649. https://doi.org/10.1126/sciadv.aax0649
Heavens SN (1990) Electrophoretic deposition as a processing route for ceramics. In: Advanced ceramic processing and technology, pp 255–283
Heller C, Schwentenwein M, Russmueller G, Varga F, Stampfl J, Liska R (2009) Vinyl esters: low cytotoxicity monomers for the fabrication of biocompatible 3D scaffolds by lithography based additive manufacturing. J Polym Sci Polym Chem 47(24):6941–6695. https://doi.org/10.1002/pola.23734

Hennink WE, Nostrum CF (2012) Novel crosslinking methods to design hydrogels. Adv Drug Deliv Rev 64:223–236. https://doi.org/10.1016/s0169-409x(01)00240-x

Hernandez Y, Nicolosi V, Lotya M et al (2008) High-yield production of graphene by liquid-phase exfoliation of graphite. Nat Nanotechnol 3(9):563–568. https://doi.org/10.1038/nnano.2008.215

Hina A, Saadeh W (2022) Noninvasive blood glucose monitoring systems using near-infrared technology—a review. Sensors 22(13):4855. https://doi.org/10.3390/s22134855

Hu J, Zhou S, Shi J, Zhang H, Zhu F, Yang X (2017) Determinants of electrical resistance change of in situ PPy-polymerized stretch plain woven fabric under uniaxial tensile strain. J Text Inst 108:1545–1551. https://doi.org/10.1080/00405000.2016.1261603

Hu CX, Yuyoung S, Oliver R, Cinzia C (2021) Dispersant-assisted liquid-phase exfoliation of 2D materials beyond graphene. Nanoscale 13(2):460–484. https://doi.org/10.1039/D0NR05514J

Huang CT, Shen CL, Tang CF, Chang S, H. (2008) A wearable yarn-based piezo-resistive sensor. Sensors Actuators A Phys 141(2):396–403. https://doi.org/10.1016/j.sna.2007.10.069

Huang J, Qin Q, Wang J (2020) A review of stereolithography: processes and systems. PRO 8(9):1138. https://doi.org/10.3390/pr8091138

Huang Y, Song Y, Gou L, Zou Y (2021) A novel wearable flexible dry electrode based on cowhide for ECG measurement. Biosensors 11:101. https://doi.org/10.3390/bios11040101

Hughes-Riley T, Dias T, Cork C (2018) A historical review of the development of electronic textiles. Fibers 6:34. https://doi.org/10.3390/fib6020034

Hull CW (1984) Apparatus for production of three dimensional objects by stereolithography. US4575330A

Hwang I, Kim HN, Seong M, Lee SH, Kang M, Yi H, Bae WG, Kwak MK, Jeong HE (2018) Multifunctional smart skin adhesive patches for advanced health care. Adv Healthc Mater 7:1800275. https://doi.org/10.1002/adhm.201800275

Jiang Q, Reddy N, Yang Y (2010) Cytocompatible cross-linking of electrospunzein fibers for the development of water-stable tissue engineering scaffolds. Acta Biomater 6:4042–4051. https://doi.org/10.1016/j.actbio.2010.04.024

Jiang Q, Reddy N, Zhang S, Roscioli N, Yang Y (2013) Water stable electrospun collagen fibers from a non-toxic solvent and crosslinking system. J Biomed Mater Res Part A 101A:1237–1247. https://doi.org/10.1002/jbm.a.34422

Jin H, Nayeem MOG, Lee S, Matsuhisa N, Inoue D, Yokota T, Hashizume D, Someya T (2019) Highly durable nanofiber-reinforced elastic conductors for skin-tight electronic textiles article. ACS Nano 13:7905–7912. https://doi.org/10.1021/acsnano.9b02297

Jiyong H, Xiaofeng Z, Guohao L, Xudong Y, Xin D (2016) Electrical properties of Ppy-coated conductive fabrics for human joint motion monitoring. Autex Res J 16:7–12. https://doi.org/10.1515/aut-2015-0048

Jung HC, Moon JH, Baek DH, Lee JH, Choi YY, Hong JS, Lee SH (2012) CNT/PDMS composite flexible dry electrodes for long-term ECG monitoring. IEEE Trans Biomed Eng 59:1472–1479. https://doi.org/10.1109/tbme.2012.2190288

Kang BC, Park BS, Ha TJ (2019) Highly sensitive wearable glucose sensor systems based on functionalized single-wall carbon nanotubes with glucose oxidase-Nafion composites. Appl Surf Sci 470:13–18. https://doi.org/10.1016/j.apsusc.2018.11.101

Kano S, Jarulertwathana N, Mohd-Noor S, Hyun JK, Asahara R, Mekaru H (2022) Respiratory monitoring by ultrafast humidity sensors with nanomaterials: a review. Sensors 22:1251. https://doi.org/10.3390/s22031251

Kazani I, Hertleer C, De Mey G, Schwarz A, Guxho G, Van Langenhove L (2012) Electrical conductive textiles obtained by screen printing. Fibres Text East Eur 20:57–63

Keum K, Eom J, Lee JH, Heo JS, Park SK, Kim YH (2021) Fully-integrated wearable pressure sensor array enabled by highly sensitive textile-based capacitive Ionotronic devices. Nano Energy 79:105479. https://doi.org/10.1016/j.nanoen.2020.105479

Khundaqji H, Hing W, Furness J, Climstein M (2020a) Smart shirts for monitoring physiological parameters: scoping review. JMIR Mhealth Uhealth 8:e18092. https://doi.org/10.2196/18092

Khundaqji H, Hing W, Furness J, Climstein M (2020b) Smart shirts for monitoring physiological parameters: scoping review. JMIR Mhealth Uhealth 8(5):e18092. https://doi.org/10.2196/18092
Kim J, Campbell AS, Wang J (2018a) Wearable non-invasive epidermal glucose sensors: a review. Talanta 177:163–170. https://doi.org/10.1016/j.talanta.2017.08.077
Kim JH, Hwang JY, Hwang HR, Kim HS, Lee JH, Seo JW, Shin US, Lee SH (2018b) Simple and cost-effective method of highly conductive and elastic carbon nanotube/polydimethylsiloxane composite for wearable electronics. Sci Rep 8:1375. https://doi.org/10.1038/s41598-017-18209-w
Kim J, Ahn D, Sun J, Park S et al (2020) Vertically and horizontally drawing formation of graphite pencil electrodes on paper by frictional sliding for a disposable and foldable electronic device. ACS Omega 6(3):1960–1970. https://doi.org/10.1021/acsomega.0c04792
Kocheril PA, Lenz K, Mascareñas DDL, Morales-Garcia JE et al (2022) Portable waveguide-based optical biosensor. Biosensors 12:195. https://doi.org/10.3390/bios12040195
Kohagura K, Kochi M, Miyagi T, Kinjyo T, Maehara Y, Naga-hama K, Sakima A, Iseki K, Ohya Y (2013) An association between uric acid levels and renal arteriolopathy in chronic kidney disease: a biopsy-based study. Hypertens Res 36:43–49. https://doi.org/10.1038/hr.2012.135
Koide T, Daito M (1997) Effects of various collagen crosslinking techniques on mechanical properties of collagen film. Dent Mater J 16:1–9. https://doi.org/10.4012/dmj.16.1
Kolanowska A, Herman AP, Jędrysiak RG, Boncel S (2021) Carbon nanotube materials for electrocardiography. RSC Adv 11(5):3020–3042. https://doi.org/10.1039/d0ra08679g. PMID: 35424207
Koncar V (2016) Introduction to smart textile and their applications. In: Smart textiles and their applications. Woodhead Publishing Series in Textiles, Cambridge, UK, pp 1–8. https://doi.org/10.1016/B978-0-08-100574-3.00001-1
Krakauer M, Botero JF, Lavalle-González FJ, Proietti A, Barbieri DE (2021) A review of flash glucose monitoring in type 2 diabetes. Diabetol Metab Syndr 13:10. https://doi.org/10.1186/s13098-021-00654-3
Kumar A, Kant R, Singh H (2022) Microstructural and tribological properties of laser - treated cold - sprayed titanium/baghdadite deposits. J Mater Res 37(16):2698–2709. https://ui.adsabs.harvard.edu/link_gateway/2022JMatR..37.2698K/doi:10.1557/s43578-022-00675-2
Kundu SK, Kumagai S, Sasaki M (2013) A wearable capacitive sensor for monitoring human respiratory rate. Jpn J Appl Phys 52(4S):04CL05. https://doi.org/10.7567/JJAP.52.04CL05
La TG, Qiu S, Scott DK, Bakhtiari R, Kuziek JWP, Mathewson KE, Rieger J, Chung HJ (2018) Two-layered and stretchable e-textile patches for wearable healthcare electronics. Adv Healthc Mater 7:1801033. https://doi.org/10.1002/adhm.201801033
Lee S, Kim C (2023) Highly flexible, stretchable, durable conductive electrode for human-body-attachable wearable sensor application. Polym Test 122:108018. https://doi.org/10.1016/j.polymertesting.2023.108018
Lee SM, Byeon HJ, Lee JH, Baek DH, Lee KH, Hong JS, Lee SH (2014) Self-adhesive epidermal carbon nanotube electronics for tether-free long-term continuous recording of biosignals. Sci Rep 4:6074. https://doi.org/10.1038/srep06074
Lee J, Kwon H, Seo J, Shin S, Koo JH, Pang C, Son S, Kim JH, Jang YH, Kim DE, Lee T (2015) Conductive fiber-based ultrasensitive textile pressure sensor for wearable electronics. Adv Mater 27:2433–2439. https://doi.org/10.1002/adma.201500009
Lee H, Hong YJ, Baik S, Hyeon T, Kim DH (2018) Enzyme-based glucose sensor: from invasive to wearable device. Adv Healthc Mater 7:1701150. https://doi.org/10.1002/adhm.201701150
Leigh SJ, Purssell CP, Bowen J, Hutchins DA, Covington JA, Billson DR (2011) A miniature flow sensor fabricated by micro-stereolithography employing a magnetite/acrylic nanocomposite resin. Sens Actuators A Phys 168(1):66–71. https://doi.org/10.1016/j.sna.2011.03.058
Leijten J, Seo J, Yue K, Santiago GT, Tamayol A, Ruiz-Esparza GU, Shin SR, Sharifi R, Noshadi I, Álvarez MM, Zhang YS, Khademhosseini A (2017) Spatially and temporally controlled hydrogels for tissue engineering. Mater Sci Eng R Rep 119:1–35. https://doi.org/10.1016/j.mser.2017.07.001

Li Y, Huang G, Zhang X, Wang L, Du Y, Lu TJ, Xu F (2014) Engineering cell alignment in vitro biotechnology. Advances 32(2):347–365. https://doi.org/10.1016/j.biotechadv.2013.11.007

Li L, Bai Y, Li L, Wang S, Zhang T (2017) A Superhydrophobic smart coating for flexible and wearable sensing electronics. Adv Mater 29:1702517. https://doi.org/10.1002/adma.201702517

Li X, Hu H, Hua T, Xu B, Jiang S (2018) Wearable strain sensing textile based on one-dimensional stretchable and weavable yarn sensors. Nano Res 11:5799–5811. https://doi.org/10.1007/s12274-018-2043-7

Li Z, Young RJ, Backes C et al (2020) Mechanisms of liquid-phase exfoliation for the production of graphene. ACS Nano 14(9):10976–10985. https://doi.org/10.1021/acsnano.0c03916

Li S, Li H, Lu Y, Zhou M, Jiang S, Du X, Guo C (2023) Advanced textile-based wearable biosensors for healthcare monitoring. Biosensors (Basel) 13(10):909. https://doi.org/10.3390/bios13100909

Liao TY, Biesiekierski A, Berndt CC, King PC et al (2022) Multifunctional cold spray coatings for biological and biomedical applications: a review. Prog Surf Sci 97(2):100654. https://doi.org/10.1016/j.progsurf.2022.100654

Lidón-Roger JV, Prats-Boluda G, Ye-Lin Y, Garcia-Casado J, Garcia-Breijo E (2018) Textile concentric ring electrodes for ECG recording based on screen-printing technology. Sensors 18:300. https://doi.org/10.3390/s18010300

Liu X, Chang H, Li Y, Huck WTS, Zheng Z (2010) Polyelectrolyte-bridged metal/cotton hierarchical structures for highly durable conductive yarns. ACS Appl Mater Interfaces 2:529–535. https://doi.org/10.1021/am900744n

Liu B, Luo Z, Zhang W, Tu Q, Jin X (2016) Carbon nanotube-based self-adhesive polymer electrodes for wireless long-term recording of electrocardiogram signals. J Biomater Sci Polym Ed 27:1899–1908. https://doi.org/10.1080/09205063.2016.1239951

Löfhede J, Seoane F, Thordstein M (2010) Soft textile electrodes for EEG monitoring. In: Proceedings of the 10th IEEE international conference on information technology and applications in biomedicine (ITAB), Corfu, Greece, pp 1–4. https://doi.org/10.1109/ITAB.2010.5687755

Lu K, Qian Y, Gong J, Zhu Z, Yin J, Ma L, Yu M, Wang H (2021) Biofabrication of aligned structures that guide cell orientation and applications in tissue engineering. Bio-Des Manuf 4:258–277. https://doi.org/10.1007/s42242-020-00104-5

Malfa FL, Puce S, Rizzi F, Vittorio MD (2020) A flexible carbon nanotubes-based Auxetic sponge electrode for strain sensors. Nano 10(12). https://doi.org/10.3390/nano10122365

Manapat JZ, Chen Q, Ye P, Advincula RC (2017) 3D printing of polymer nanocomposites via stereolithography. Macromol Mater Eng 302(9):Article 1600553. https://doi.org/10.1002/mame.201600553

Mandal D, Banerjee S (2022) Surface Acoustic Wave (SAW) sensors: physics, materials, and applications. Sensors 22:820. https://doi.org/10.3390/s22030820

Martinez AW, Caves JM, Ravi S, Li W, Chaikof EL (2014) Effects of crosslinking on the mechanical properties, drug release and cytocompatibility of protein polymers. Acta Biomater 10:26–33. https://doi.org/10.1016/j.actbio.2013.08.029

Mecnika V, Hoerr M, Jockenhoefel S, Gries T, Krievins I, Schwarz-Pfeiffer A (2015) Preliminary study on textile humidity sensors. In: Proceedings of the smart SysTech 2015–European conference on smart objects, systems and technologies, Aachen, Germany, pp 1–9

Mendes AC, Baran ET, Reis RL, Azevedo HS (2013) Self-assembly in nature: using the principles of nature to create complex nanobiomaterials. Wiley Interdiscip Rev Nanomed Nanobiotechnol 5:582–612. https://doi.org/10.1002/wnan.1238

Meng K, Zhao S, Zhou Y, Wu Y, Zhang S, He Q, Wang X, Zhou Z, Fan W, Tan X, Yang J, Chen J (2020) A wireless textile-based sensor system for self-powered personalized health care. Matter 2:896–907. https://doi.org/10.2139/ssrn.3451450

Min SD, Yun Y, Shin H (2014) Simplified structural textile respiration sensor based on capacitive pressure sensing method. IEEE Sensors J 14(9):3245–3251. https://doi.org/10.1109/JSEN.2014.2327991

Mirbagheri M, Adibnia V, Hughes BR, Waldman SD, Banquy X, Hwang DK (2019) Advanced cell culture platforms: a growing quest for emulating natural tissues. Mater Horiz 6:45–71. https://doi.org/10.1039/C8MH00803E

Mitchell E, Coyle S, O'Connor N, Diamond D, Ward T (2010) Breathing feedback system with wearable textile sensors. In: Proceedings of the 2010 IEEE international conference on body sensor networks, Singapore, pp 56–61. https://doi.org/10.1109/BSN.2010.31

Morgan RM, Patterson MJ, Nimmo MA (2004) Acute effects of dehydration on sweat composition in men during prolonged exercise in the heat. Acta Physiol Scand 182:37–43. https://doi.org/10.1111/j.1365-201x.2004.01305.x

Moroni L, Boland T, Burdick JA, Maria CD et al (2018) Biofabrication: a guide to technology and terminology. Trends Biotechnol 36:384–402. https://doi.org/10.1016/j.tibtech.2017.10.015

Muthukumar N, Thilagavathi G, Kannaian T (2015) Polyaniline-coated foam electrodes for electroencephalography (EEG) measurement. J Text Inst 107:283–290. https://doi.org/10.1080/00405000.2015.1028248

Nagamine K, Tokito S (2021) Organic-transistor-based biosensors interfaced with human skin for non-invasive perspiration analysis. Sens Actuators B Chem 349:130778. https://doi.org/10.1016/j.snb.2021.130778

Nicola DM, David E, Tiziano S, Lorenzo M (2020) Bio-fabrication: convergence of 3D bioprinting and nano-biomaterials in tissue engineering and regenerative medicine. Front Bioeng Biotechnol 8. https://doi.org/10.3389/fbioe.2020.00326

Niculescu AG, Chircov C, Bîrcă AC, Grumezescu AM (2021) Fabrication and applications of microfluidic devices: a review. Int J Mol Sci 22:2011. https://doi.org/10.3390/ijms22042011

Nigusse AB, Malengier B, Mengistie DA, Tseghai GB, Langenhove L (2020) Van development of washable silver printed textile electrodes for long-term ECG monitoring. Sensors 20:6233. https://doi.org/10.3390/s20216233

Nigusse AB, Mengistie DA, Malengier B, Tseghai GB, Langenhove LV (2021) Wearable smart textiles for long-term electrocardiography monitoring—a review. Sensors 21:4174. https://doi.org/10.3390/s21124174

Nishimoto R, Sato Y, Wu J et al (2022) Thermally drawn CNT-based hybrid nanocomposite fiber for electrochemical sensing. Biosensors (Basel) 12(8):559. https://doi.org/10.3390/bios12080559

Oliaei SNB, Nasseri B (2020) Stereolithography and its applications. In: Paulo Davim J (ed) Additive and subtractive manufacturing: emergent technologies. De Gruyter, Berlin, Boston, pp 229–250

Omura Y (2014) Sensor technology for monitoring of health-related conditions. In: Hashmi S (ed) Comprehensive materials processing; sensor materials, technologies and applications, vol 13. Elsevier, Amsterdam. https://doi.org/10.3390/s21186042

Özbey A, Karimzade U, Akgönül S, Gozuacik D, Koşar A (2016) Inertial focusing of microparticles in curvilinear microchannels. Sci Rep 6:38809. https://doi.org/10.1038/srep38809

Paiva A, Ferreira F, Catarino A, Carvalho M, Carvalho H (2018) Design of smart garments for sports and rehabilitation. IOP Conf Ser Mater Sci Eng 459:012083. https://doi.org/10.1088/1757-899X/459/1/012083

Papadimitriou L, Manganas P, Ranella A, Stratakis E (2020) Biofabrication for neural tissue engineering applications. Mater Today Bio 6:100043. https://doi.org/10.1016/j.mtbio.2020.100043

Park S, Park K, Yoon D et al (2007) Apparatus for preparing electrospun nanofibers: designing an electrospinning process for nanofiber fabrication. Polym Int 56:1361–1366. https://doi.org/10.1002/pi.2345

Park H, Kim JW, Hong SY, Lee G, Lee H, Song C, Keum K, Jeong YR, Jin SW, Kim DS, Ha JS (2019) Dynamically stretchable supercapacitor for powering an integrated biosensor in an all-in-one textile system. ACS Nano 13:10469–10480. https://doi.org/10.1021/acsnano.9b04340

Paton KR, Varrla E, Backes C et al (2014) Scalable production of large quantities of defect-free few-layer graphene by shear exfoliation in liquids. Nat Mater 13(6):624–630. https://doi.org/10.1038/nmat3944

Piuzzi E, Pisa S, Pittella E, Podestà L, Sangiovanni S (2020) Wearable Belt with built-in textile electrodes for cardio—respiratory monitoring. Sensors 20:4500. https://doi.org/10.3390/s20164500

Possanzini L, Decataldo F, Mariani F, Gualandi I, Tessarolo M, Scavetta E, Fraboni B (2020) Textile sensors platform for the selective and simultaneous detection of chloride ion and pH in sweat. Sci Rep 10:17180. https://doi.org/10.1038/s41598-020-74337-w

Potts RO, Tamada JA, Tierney MJ (2002) Glucose monitoring by reverse iontophoresis. Diabetes Metab Res Rev 18:49–53. https://doi.org/10.1002/dmrr.210

Prendergast ME, Davidson MD, Burdick JA (2021) A biofabrication method to align cells within bioprinted photocrosslinkable and cell-degradable hydrogel constructs via embedded fibers. Biofabrication 13(4):10. https://doi.org/10.1088/1758-5090/ac25cc

Promphet N, Rattanawaleedirojn P, Siralertmukul K et al (2019) Non-invasive textile based colorimetric sensor for the simultaneous detection of sweat pH and lactate. Talanta 192:424–430. https://doi.org/10.1016/j.talanta.2018.09.086

Qi K, Zhou Y, Ou K, Dai Y, You X, Wang H, He J, Qin X, Wang R (2020) Weavable and stretchable Piezoresistive carbon nanotubes-embedded nanofiber sensing yarns for highly sensitive and multimodal wearable textile sensor. Carbon 170:464–476. https://ui.adsabs.harvard.edu/link_gateway/2020Carbo.170..464Q/doi:10.1016/j.carbon.2020.07.042

Ramos-Garcia RI, da Silva F, Kondi Y, Sazonov E, Dunne LE (2016) Analysis of a coverstitched stretch sensor for monitoring of breathing. In: Proceedings of the 2016 10th international conference on sensing technology (ICST), Nanjing, China, pp 1–6

Reddy N, Reddy R, Jiang Q (2015) Crosslinking biopolymers for biomedical applications. Trends Biotechnol 33(6):362–369. https://doi.org/10.1016/j.tibtech.2015.03.008

Richard I, Schyrr B, Aiassa S, Carrara S, Sorin F (2021) All-in-fiber electrochemical sensing. ACS Appl Mater Interfaces 13:43356–43363. https://doi.org/10.1021/acsami.1c11593

Robert JS, Baylis F (2008) Genetic engineering. In: Harald Kristian (Kris) Heggenhougen (ed) International Encyclopedia of public health. Academic Press, pp 35–39

Robles Martinez P, Basit A, Gaisford S (2018) The history, developments and opportunities of stereolithography. In: Basit AW, Gaisford S (eds) 3D printing of pharmaceuticals, vol 31, 1st edn. Springer, Cham, pp 55–79

Ronkainen NJ, Halsall HB, Heineman WR (2010) Electrochemical biosensors. Chem Soc Rev 39:1747–1763. https://doi.org/10.1039/B714449K

Root W, Bechtold T, Pham T (2020) Textile-integrated thermocouples for temperature measurement. Materials 13:626. https://doi.org/10.3390/ma13030626

Rosace G, Trovato V, Colleoni C et al (2017) Structural and morphological characterizations of MWCNTs hybrid coating onto cotton fabric as potential humidity and temperature wearable sensor. Sens Actuators B Chem 252:428–439. https://ui.adsabs.harvard.edu/link_gateway/2017SeAcB.252..428R/doi:10.1016/j.snb.2017.05.175

Rovira C, Coyle S, Corcoran B, Diamond D, Stroiescu F, Daly K (2011) Integration of textile-based sensors and shimmer for breathing rate and volume measurement. In: Proceedings of the 2011 5th international conference on pervasive computing technologies for healthcare (PervasiveHealth) and workshops, Dublin, Ireland, pp 238–241. https://doi.org/10.4108/icst.pervasivehealth.2011.246104

Ruffini G, Dunne S, Fuentemilla L, Grau C, Farrés E, Marco-Pallarés J, Watts PCP, Silva SRP (2008) First human trials of a dry electrophysiology sensor using a carbon nanotube array interface. Sens Actuators A Phys 144:275–279. https://doi.org/10.48550/arXiv.physics/0701159

Sadasivuni KK, Deshmukh K, Almaadeed MA (eds) (2020) 3D and 4D printing of polymer nanocomposite materials. Elsevier, pp 1–24

Saminathan S, Sundaram G, Jayapal A, Shakeel F, Rajalingam S, Murugan S, Kalaichelvan K, Faiyazuddin M (2024) Green engineering of NiO nanoparticles decorated with Arachis hypogaea shell extract for biomedical applications. Z Phys Chem. https://doi.org/10.1515/zpch-2024-0856

Sarkar P, Nicholson PS (1996) Electrophoretic deposition (EPD): mechanism, kinetics and application to ceramics. J Am Ceram Soc 79:1987–2002. https://doi.org/10.1111/j.1151-2916.1996.tb08929.x
Scataglini S, Moorhead AP, Feletti F (2020) A systematic review of smart clothing in sports: possible applications to extreme sports. Muscles Ligaments Tendons J 10:333–342. https://doi.org/10.32098/mltj.02.2020.19
Schmidleithner C, Kalaskar DM (2018) 3D printing. In: Cvetković D (ed) Stereolithography. Intech Open. https://doi.org/10.5772/intechopen:78147
Sharmila S, Saranya A, Arulprakasajothi M, Saranya R, Srimanickam B, Abel SK, Shakeel F, Faiyazuddin M (2024) Expediting the bioactivity of zinc sulfide nanoparticles with copper oxide as a nanocomposite. BMC Chem 18(1):204. https://doi.org/10.1186/s13065-024-01320-1
Shen C, Jiang Z, Li L, Gilchrist JF, Ou-Yang HD (2020) Frequency response of induced-charge electrophoretic metallic Janus particles. Micromachines 11:334. https://doi.org/10.3390/mi11030334
Shi XM, Zhou W, Ma D, Ma Q et al (2015) Electrospinning of nanofibers and their applications for energy devices. J Nanomater 2015:Article ID 140716, 20. https://doi.org/10.1155/2015/140716
Shu L, Xu T, Xu X (2019) Multilayer sweat-absorbable textile electrode for EEG measurement in forehead site. IEEE Sensors J 19:5995–6005. https://doi.org/10.1109/JSEN.2019.2912667
Sinha SK (2020) Chapter 5 - additive manufacturing (AM) of medical devices and scaffolds for tissue engineering based on 3D and 4D printing. In: Sadasivuni KK, Deshmukh K, Almaadeed MA (eds) 3D and 4D printing of polymer nanocomposite materials. Elsevier, pp 119–160
Smith KH, Tejeda-Montes E, Poch M, Mata A (2011) Integrating top-down and self-assembly in the fabrication of peptide and protein-based biomedical materials. Chem Soc Rev 40:4563–4577. https://doi.org/10.1039/C1CS15064B
Sobon D (2018) Application of cold sprayed coatings in aviation and automotive, 11th Int. Sci Tech Conf Automot Saf, Automot Saf:1–4. https://doi.org/10.1109/AUTOSAFE.2018.8373311
Taheri A, Jafari SM (2019) Nanostructures of gums for encapsulation of food ingredients. In: Biopolymer nanostructures for food encapsulation purposes. Elsevier, pp 521–578. https://doi.org/10.1016/B978-0-12-815663-6.00018-5
Tai H, Wang S, Duan Z, Jiang Y (2020) Evolution of breath analysis based on humidity and gas sensors: potential and challenges. Sens Actuators B Chem 318:128104. https://doi.org/10.1016/j.snb.2020.128104
Takamatsu S, Lonjaret T, Crisp D, Badier JM, Malliaras GG, Ismailova E (2015) Direct patterning of organic conductors on knitted textiles for long-term electrocardiography. Sci Rep 5:15003. https://doi.org/10.1038/srep15003
Tanaka M, Utsunomiya F, Douseki T (2016) Wearable self-powered diaper-shaped urinary-incontinence sensor suppressing response-time variation with 0.3 V start-up converter. IEEE Sensors J 16:3472–3479. https://doi.org/10.1109/JSEN.2015.2483900
Taylor LW, Williams SM, Yan JS, Dewey OS, Vitale F, Pasquali M (2021) Washable, sewable, all-carbon electrodes and signal wires for electronic clothing. Nano Lett 21:7093–7099. https://doi.org/10.1021/acs.nanolett.1c01039
Terms and Acronyms U.S. Environmental Protection Agency online. Retrieved 16 July 2015
Tian X, Jin J, Yuan S, Chua CK, Tor SB, Zhou K (2017) Emerging 3D-printed electrochemical energy storage devices: a critical review. Adv Energy Mater 7(17):Article 1700127. https://doi.org/10.1002/aenm.201700127
Tikhonov A, Evdokimov P, Klimashina E et al (2020) Stereolithographic fabrication of three-dimensional permeable scaffolds from CaP/PEGDA hydrogel biocomposites for use as bone grafts. J Mech Behav Biomed Mater 110:Article 103922. https://doi.org/10.1016/j.jmbbm.2020.103922
Tipparaju VV, Wang D, Yu J, Chen F, Tsow F, Forzani E, Tao N, Xian X (2020) Respiration pattern recognition by wearable mask device. Biosens Bioelectron 169:112590. https://doi.org/10.1016/j.bios.2020.112590
Torrisi F, Hasan T, Wu W, Sun Z et al (2012) Inkjet-Printed Graphene Electronics. ACS Nano 6(4):2992–3006. https://doi.org/10.1021/nn2044609

Tsai TW, Heckert G, Neves LF, Tan Y, Kao DY, Harrison RG, Resasco DE, Schmidtke DW (2009) Adsorption of glucose oxidase onto single-walled carbon nanotubes and its application in layer-by-layer biosensors. Anal Chem 81:7917–7925. https://doi.org/10.1021/ac900650r

Ullah H, Wahab MA, Will G, Karim MR, Pan T, Gao M, Lai D, Lin Y, Miraz MH (2022) Recent advances in stretchable and wearable capacitive electrophysiological sensors for long-term health monitoring. Biosensors 12(8):630. https://doi.org/10.3390/bios12080630

Uzun S, Seyedin S, Stoltzfus AL, Levitt AS, Alhabeb M, Anayee M, Strobel CJ, Razal JM, Dion G, Gogotsi Y (2019) Knittable and washable multifunctional MXene-coated cellulose yarns. Adv Funct Mater 29:1905015. https://doi.org/10.1002/adfm.201905015

van der Biest OO, Vandeperre LJ (1999) Electrophoretic deposition of materials. Ann Rev Mater Sci 29:327–352. https://doi.org/10.1146/annurev.matsci.29.1.327

Vo J, Mastoor Y, Mathieu PS, Clyne AM (2021) A simple method to align cells on 3D hydrogels using 3D printed molds. Biomed Eng Adv 1:100001–100025. https://doi.org/10.1016/j.bea.2021.100001

Wang J (2008) Electrochemical glucose biosensors. Chem Rev 108:814–825. https://doi.org/10.1021/cr068123a

Wang L, Loh KJ (2017) Wearable carbon nanotube-based fabric sensors for monitoring human physiological performance. Smart Mater Struct 26:055018. https://doi.org/10.1088/1361-665X/aa6849

Wang X, Jiang M, Zhou Z, Gou J, Hui D (2017) 3D printing of polymer matrix composites: a review and prospective. Compos Part B Eng 110:442–458. https://doi.org/10.1016/j.compositesb.2016.11.034

Wang J, Lu C, Zhang K (2020) Textile-based strain sensor for human motion detection. Energy Environ Mater 3:80–100. https://doi.org/10.1002/eem2.12041

Wazeer A, Das A, Sinha A, Inaba K, Ziyi S, Karmakar A (2023) Additive manufacturing in biomedical field: a critical review on fabrication method, materials used, applications, challenges, and future prospects. Prog Addit Manufact 8(5):857–889. https://doi.org/10.1007/s40964-022-00362-y

Wendler J, Maraite D, Nocke A, Cherif C (2020) Novel textile moisture sensors based on multi-layered braiding constructions. Text Res J 90:469–477. https://doi.org/10.1177/0040517519868172

Widener CA, Carter M, Ozdemir OC, Hrabe RH et al (2016) Application of high-pressure cold spray for an internal bore repair of a navy valve actuator. J Therm Spray Technol 25:193–201. https://doi.org/10.1007/s11666-015-0366-4

Wijesiriwardana R (2006) Inductive fiber-meshed strain and displacement transducers for respiratory measuring systems and motion capturing systems. IEEE Sensors J 6(3):571–579. https://doi.org/10.1109/JSEN.2006.874488

Wu R, Ma L, Hou C, Meng Z, Guo W, Yu W, Yu R, Hu F, Liu XY (2019) Silk composite electronic textile sensor for high space precision 2D combo temperature-pressure sensing. Small 15:1901558. https://doi.org/10.1002/smll.201901558

Wu H, Fahy WP, Kim SD, Kim H, Zhao N, Pilato L, Kafi A, Bateman S, Koo JH (2020) Recent developments in polymers/polymer nanocomposites for additive manufacturing. Prog Mater Sci 111:Article 100638. https://doi.org/10.1016/j.pmatsci.2020.100638

Wu P, Gu J, Liu X, Ren Y, Mi X, Zhan W, Zhang X, Wang H, Ji X, Yue Z, Liang J (2024) A robust core-shell nanofabric with personal protection, health monitoring and physical comfort for smart sportswear. Adv Mater e2411131:e2411131. https://doi.org/10.1002/adma.202411131

Xue J, Xie J, Liu W, Xia Y (2017) Electrospun nanofibers: new concepts, materials, and applications. Acc Chem Res 50:1976–1987. https://doi.org/10.1021/acs.accounts.7b00218

Yan YM, Baravik I, Yehezkeli O, Willner I (2008) Integrated electrically contacted glucose oxidase/carbon nanotube electrodes for the bioelectrocatalyzed detection of glucose. J Phys Chem C 112:17883–17888. https://doi.org/10.1021/jp805637e

Yan W, Dong C, Xiang Y et al (2020) Thermally drawn advanced functional fibers: new frontier of flexible electronics. Mater Today 35:168–194. https://doi.org/10.1016/j.mattod.2019.11.006

Yang Z, Pang Y, Han XL, Yang Y, Ling J, Jian M, Zhang Y, Yang Y, Ren TL (2018) Graphene textile strain sensor with negative resistance variation for human motion detection. ACS Nano 12:9134–9141. https://doi.org/10.1021/acsnano.8b03391

Yang Y, Song Y, Bo X et al (2020) A laser-engraved wearable sensor for sensitive detection of uric acid and tyrosine in sweat. Nat Biotechnol 38:217–224. https://doi.org/10.1038/s41587-019-0321-x

Yang S, Li C, Wen N, Xu S, Huang H, Cong T, Zhao Y, Fan Z, Liu K, Pan L (2021) All-fabric-based multifunctional textile sensor for detection and discrimination of humidity, temperature, and strain stimuli. J Mater Chem C 9:13789–13798. https://doi.org/10.1039/D1TC02755G

Yao S, Yang J, Poblete FR, Hu X, Zhu Y (2019) Multifunctional electronic textiles using silver nanowire composites. ACS Appl Mater Interfaces 11:31028–31037. https://doi.org/10.1021/acsami.9b07520

Yapici MK, Alkhidir TE (2017) Intelligent medical garments with graphene-functionalized smart-cloth ECG sensors. Sensors 17:875. https://doi.org/10.3390/s17040875

Yeh LR, Chen WC, Chan HY, Lu NH, Wang CY, Twan WH, Du WC, Huang YH, Hsu SY, Chen TB (2021) Integrating ECG monitoring and classification via IoT and deep neural networks. Biosensors 11:188. https://doi.org/10.3390/bios11060188

Yeom H, Sridharan K (2021) Cold spray technology in nuclear energy applications: a review of recent advances. Ann Nucl Energy 150:107835. https://doi.org/10.1016/j.anucene.2020.107835

Yilmaz B, Al Rashid A, Mou YA, Evis Z, Koç M (2021) Bioprinting: a review of processes, materials and applications. Bioprinting 23:Article e00148. https://doi.org/10.1016/j.bprint.2021.e00148

Yu L, Feng Y, Selven DSMT, Yao L, Soon RH, Yeo JC, Lim CT (2019) Dual-core capacitive micro-fiber sensor for smart textile applications. ACS Appl Mater Interfaces 11:33347–33355. https://doi.org/10.1021/acsami.9b10937

Zhang D, Liu Q (2016) Biosensors and bioelectronics on smartphone for portable biochemical detection. Biosens Bioelectron 75:273–284. https://doi.org/10.1016/j.bios.2015.08.037

Zhang L, He J, Liao Y, Zeng X, Qiu N, Liang Y, Xiao P, Chen T (2019) A self-protective, reproducible textile sensor with high performance towards human–machine interactions. J Mater Chem A 7:26631–26640. https://doi.org/10.1039/C9TA10744D

Zhang J, Xu J, Lim J, Nolan JK, Lee H, Lee CH (2021) Wearable glucose monitoring and implantable drug delivery systems for diabetes management. Adv Healthc Mater 10(17):2100194

Zhang X, Ke L, Zhang X, Xu F, Hu Y, Lin H, Zhu J (2022a) Breathable and wearable strain sensors based on synergistic conductive carbon nanotubes/cotton fabrics for multi-directional motion detection. ACS Appl Mater Interfaces 14(22):25753–25762. https://doi.org/10.1021/acsami.2c04790

Zhang K, Li Z, Zhang J, Zhao D, Pi Y, Shi Y, Wang R, Chen P, Li C, Chen G, Lei IM, Zhong J (2022b) Biodegradable smart face masks for machine learning-assisted chronic respiratory disease diagnosis. ACS Sens 7(10):3135–3143. https://doi.org/10.1021/acssensors.2c01628

Zhao Y, Cao Y, Liu J, Zhan Z, Li X, Li WJ (2018) Single-wall carbon nanotube-coated cotton yarn for electrocardiography transmission. Micromachines 9:132. https://doi.org/10.3390/mi9030132

Zhao H, Chin LK, Shi Y, Liu PY et al (2021) Continuous optical sorting of nanoscale biomolecules in integrated microfluidic-nanophotonic chips. Sens Actuators B Chem 331:129428. https://doi.org/10.1016/j.snb.2020.129428

Zhong J, Li Z, Takakuwa M, Inoue D, Hashizume D, Jiang Z, Shi Y, Ou L, Nayeem OG, Umezu S (2021) Smart face mask based on an ultrathin pressure sensor for wireless monitoring of breath conditions. Adv Mater 34:2107758. https://doi.org/10.1002/adma.202107758

Zhu N, Zhang L, Hou Y, Xu M, Zhong C, Chen Y, Wen G (2023) Influence of number average molecular weight on the properties of 3D printed precursor ceramics. Ceram Int 49(4):5604–5612. https://doi.org/10.1016/j.ceramint.2022.10.148

Regenerative Medicine-A Promising Approach to Address the Global Healthcare Challenges and Promoting Sustainability

6

Shahid Nazir Wani, Ishrat Zahoor, Junaid Tantray,
Goli Venkateshwarlu, Pradyumn Tiwari, Shahbaz Eqbal,
Md. Faiyazuddin, and Amer Ahmed Syed

Abstract

Organ regeneration through regenerative medicine is very important in the future sustainable health care systems because it can offer new treatment solutions for the damaged tissues and organs rather than relying on organ transplantation. It has the ability of addressing issues in the long-term and reduce on the frequency of the interventions. This interdisciplinary field concerns the use of concepts derived from stem cell biology, tissue engineering and material science in order to produce therapies that offer the functionality and architecture of the natural tissue. Stem cell research is at the forefront of regenerative medicine because they provide information on optimizing tissue and organ regeneration as well as developing new treatment approaches for numerous diseases. Tissue engineering

S. N. Wani (✉)
Chitkara College of Pharmacy, Chitkara University, Rajpura, Punjab, India

Aman Pharmacy College, Jhunjhunu, Rajasthan, India

I. Zahoor
M.M. College of Pharmacy, Maharishi Markandeshwar (Deemed to be University), Mullana, Ambala, Haryana, India

J. Tantray
Department of Pharmacology, NIMS Institute of Pharmacy, NIMS University, Jaipur, Rajasthan, India

G. Venkateshwarlu · P. Tiwari · S. Eqbal
Aman Pharmacy College, Jhunjhunu, Rajasthan, India

Md. Faiyazuddin
School of Pharmacy, Al – Karim University, Katihar, India

A. A. Syed (✉)
School of Professional Studies, Northwestern University, Chicago, IL, USA

Md. Faiyazuddin, M. Suryawanshi (eds.), *Design and Processing of Green Materials*, Biomaterials, Bioengineering and Sustainability 4,
https://doi.org/10.1007/978-3-031-91790-5_6

makes the development of biological substitutes for helping damaged tissues and organs, where scaffold material design plays a vital role. Most regenerative treatments are outpatient procedures, making it easy for doctors to tailor the treatment to help boost the efficiency of the intervention, shorten the healing period, and reduce susceptibility to post-surgical complications. There are innovative biomaterials, new cell origin, and tissue engineering technologies. Regenerative medicine can be aligned with sustainable healthcare as it provides long term effective solutions since people do not have to visit the hospital frequently over and over again, enhance patients' quality of life and decrease the prevalence of chronic diseases. The thoughtful transition to applied regenerative medicine defines the new efficient and sustainable trend in the healthcare system.

Keywords

Regenerative medicine · Stem cell · Tissue engineering · Biomaterials · Sustainable healthcare

6.1 Introduction to Regenerative Medicine

This recent scientific specialty is based in the management of science and engineering that applies biology and medicine to create sophisticated new therapies designed to repair, replace, or engineer tissues and organs (Deng et al. 2024). This academic field is expanding at a fast pace and includes stem cell therapy, tissue engineering, and gene therapy, with the general objective of regeneration of the affected tissues and organs (Kumbhar et al. 2024). Regenerative medicine is arguably one of the exciting areas in current medical innovation promising new treatments for current incurable diseases and passage to new forms of health care delivery. The field of regenerative medicine can be described as the use of cells and tissues engineering to restore, treat, or replace tired, damaged, or failed tissue/organ function caused by various factors including aging, diseases, injuries, or congenital abnormalities (Wei et al. 2024). Regenerative medicine is not limited to the application of a particular medical specialty because its goal is to help people affected by diseases that may have limited treatment options or are chronic in nature. It seeks to work on the causes of diseases and injuries owing to the body's self-healing tissues and enhancing pel by new and more sophisticated biotechnological approaches. Though defined in a broad sense, regenerative medicine comprises several innovative techniques of reconstructing the function of damaged tissues and organs. Cellular therapies involve use of stem cells and progenitor cells to replace lost structures while tissue engineering involves use of scaffolds, cells, bioactive molecules to produce functional tissues (Augustine et al. 2024). The same as the gene therapy represents the intervention to treat or prevent diseases through changing or adding genes into cells and tissues, the biomaterials and artificial organs refer to the application of synthetic or biological material to enhance or replace organic functions (Dong et al. 2024). Furthermore, small molecule drugs and biologic products work to modulate the body's own natural healing response, using signaling molecules to call in the body's own regenerative machinery (Suhag 2024). Altogether, these diverse strategies present a contemporary vision of regenerative remedy and might act as more sparing approaches to a great deal of diseases due to which patients' outcomes can be improved.

6.1.1 Historical Development and Milestones

The concept of regenerative medicine originated from scratch, the history of the epoch when the civilization observed and noted the tissue regeneration. Nevertheless, the contemporary program of regenerative medicine can be launched from the middle of the twentieth century with several significant discoveries and advances. The', the regenerative medicine history can be described as the transit from a myth story toward the advanced scientific accomplishments. This brings out the major developments that have marked the field and transformed its principles and uses. It is with this context that this paper provides an overview of RM including the basic tenets, key landmark findings, as well as the state of research in the field as we discover it today. The idea of tissue regeneration can be dated back to Greek mythology when Prometheus teach human to replace their body parts (Bizios and Wechsler 2024). Regeneration has its roots it fourth century BCE Aristotle's biological theories, it impacted in future scientific thinking (Kumari 2024). The nineteenth century saw the formulation of the cell theory by Schwann and Schleiden, emphasizing that all cells arise from pre-existing cells, a principle crucial to RM. Howard Green's innovations in epithelial stem cell culture during the twentieth century revolutionized skin regenerative therapies, leading to significant clinical applications. Current RM practices integrate stem cell therapies, biomaterials, and tissue engineering to restore or replace damaged tissues. Despite the promise of engineered tissues, fully functional products for clinical use remain elusive, highlighting ongoing challenges in the field. While regenerative medicine has made remarkable strides, some critics argue that the field has yet to fulfill its initial promises, emphasizing the need for continued research and innovation to realize its full potential (Fig. 6.1).

6.1.2 Key Principles and Concepts of Regenerative Medicine

The field of regenerative medicine is underpinned by several fundamental principles and concepts, including stem cell plasticity, extracellular matrix interactions, biomimicry, immunomodulation, scalability and manufacturing, personalized medicine, tissue homeostasis and repair, biocompatibility and integration, developmental biology, and epigenetic regulation (Velikic et al. 2024). Stem cell plasticity allows for the generation of various cell types for therapeutic applications, while understanding ECM interactions informs tissue organization and regeneration (Fatma and Siddique 2024). Biomimicry means development of counterfeit tissue and organs resembling the regular ones and immunomodulation is vital for rejection of transplanted cells or tissues (Dhamdar et al. 2024). Cost-efficient production methods

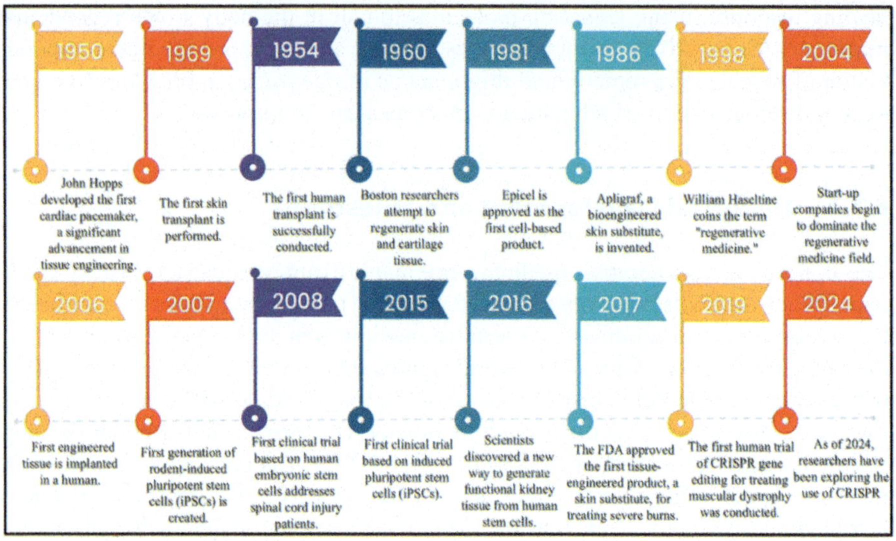

Fig. 6.1 Historical development and milestones of regenerative medicine

and manufacturing infrastructure are key factors in large scale regenerative therapy production, on the other hand, individualized medicine considers the patient's needs and objectives. Even though the study of this book is from the perspective of tissue homeostasis and repair, knowledge of the homeostasis and repair mechanisms helps in providing regenerative strategies (Dzobo et al. 2018). Regeneration and formation of tissues and organs in adults are influenced by insights generated by developmental biology, and epigenetics is examining how control of change in gene and protein expression affects cell behaviors and tissue reparative capabilities.

6.2 Principles of Sustainable Healthcare

Sustainable healthcare needs to deliver different healthcare solutions that have to be of good quality and cost-efficient, without negatively impacting the environment. This idea is vital, now more than ever since the world is putting itself back together after experiencing a harsh pandemic, which means that the healthcare structures should be able to organize themselves well and ensure sustainable healthcare without taking up too much on the health resources or end up harming the natural environment (Rahat et al. 2024). Sustainable healthcare is necessary since it offers a chance for optimization of patient results, cost reduction in healthcare provision, and a decrease in the adverse impacts of healthcare-related activity on the environment (Zurynski et al. 2022). The increased quality of and access to care for patients; the maximization of societal, economic and public health benefits; preparing future health care leaders, influencing policy and practicing effective and efficient patient care; all function to increase the strength and durability of health care delivery systems within society and communities (Braithwaite et al. 2019). Sustainability in healthcare encompasses three primary dimensions: each of them, environmental, economical and social. The environmental dimensin concerns itself with the lowering of the carbon dioxide emission and waste of the health care system through policies on environmental supply chain management as well as rational use of resources (Mostepaniuk et al. 2023; Rattan et al. 2022). Economic domain is to finance sustainable healthcare systems by working on diseases of health systems, avoiding leverage on futile costs, and implementing cost effective practices (Zurynski et al. 2022; Braithwaite et al. 2019). The social perspective of the triple aim presupposes equal access to quality health care, enhancement of patient-centric care, the involvement of communities and patients in health care management (Berniak-Wozny and Rataj 2023; Hoop et al. 2022). All these dimensions are related and form a synergy that defines overall sustainability of health care systems (Wieringa et al. 2024). There are various factors which threaten the sustainability of healthcare systems. These include the increasing complexity of patient populations, the rising costs of healthcare, and the inefficiencies and wasteful practices prevalent in the sector (Zurynski et al. 2022; Braithwaite et al. 2019). Additionally, the healthcare sector is a significant emitter of environmental pollutants, which paradoxically harm public health (Berniak-Wozny and Rataj 2023; Sherman et al. 2020). Fragmented healthcare systems and siloed professional groups further complicate efforts to integrate sustainable practices. There is also a lack of consensus on defining and measuring sustainability in healthcare, which hampers the development and implementation of effective sustainability strategies. Addressing these challenges requires robust, flexible, and context-relevant frameworks and measures to support the long-term sustainability and performance of healthcare systems.

6.3 Regenerative Medicine Technologies and Techniques

Regenerative medicine is a new field that focuses on restoring specific tissues and organs in patients with serious injuries or chronic diseases when the body's natural regenerative responses are insufficient. In the current situation, donated tissues and organs are unable to fulfill the transplantation requirements of ailing and elderly individuals, which has resulted in the need for the search for alternatives. In recent years, stem cells have been recognized as a leading source of regenerative medicine for the purpose of repairing tissues and organs that have been damaged as a result of congenital defects, diseases, and age-related effects (Mason and Dunnill 2008) (Fig. 6.2).

6.3.1 Stem Cell Therapies and Types

Stem cells are defined by their indefinite division, ability to assume characteristics of other cell types, and capacity for transdifferentiation. They form the foundation for tissues and organs and are crucial in disease progression, development, and tissue repair. Stem cells are categorized based on their transdifferentiation potential

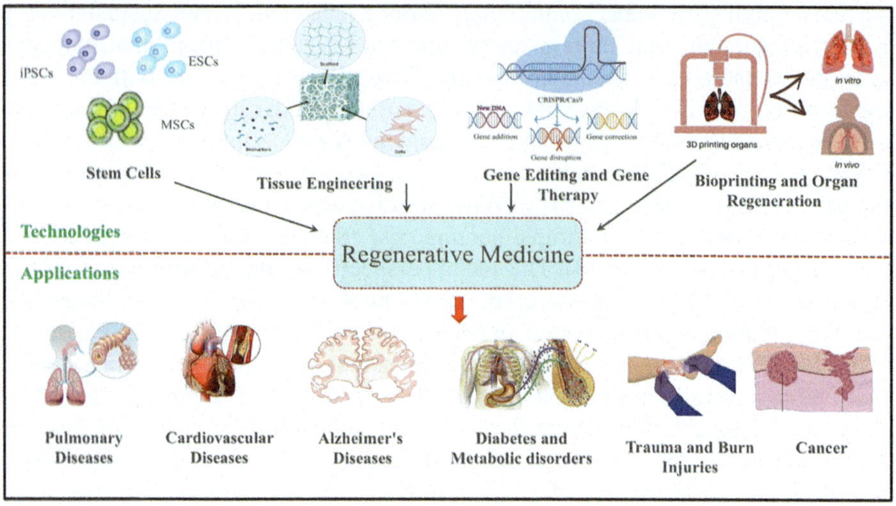

Fig. 6.2 Technologies and applications of regenerative medicine for sustainable healthcare

into four types: unipotent, multipotent, pluripotent, and totipotent (Fortier 2005). The zygote, the only totipotent stem cell, can develop into a complete organism, while inner cell mass (ICM) cells of the embryo are pluripotent and can differentiate into the three germ layers but not extraembryonic tissue (Fortier 2005).

Stem cells are categorized by regenerative applications into embryonic stem cells (ESCs), tissue-specific progenitor stem cells (TSPSCs), mesenchymal stem cells (MSCs), umbilical cord stem cells (UCSCs), bone marrow stem cells (BMSCs), and induced pluripotent stem cells (iPSCs). Transplants can be autologous, allogenic, or syngeneic to aid tissue regeneration and immunolysis of pathogens or malignant cells. To reduce host-versus-graft rejection, tissue typing of human leukocyte antigens (HLA) and immunosuppressants are recommended (Petersdorf et al. 2007). Stem cells have low major histocompatibility complex (MHC) expression and secrete chemokines, recruiting endothelial and immune cells to promote graft site tolerance (Petersdorf et al. 2007). Current regenerative techniques use tissue engineering, combining cell transplantation, material science, and microengineering to develop organoids for restoring damaged tissues and organs. This method creates new tissue on biodegradable 3D scaffolds (Leferink et al. 2015; Gubareva et al. 2016). Ideal scaffolds support cell adhesion and growth, mimic target tissue mechanics, encourage angiogenesis and neovascularization, and are non-immunogenic, removing the need for systemic immune suppression (Garzón et al. 2012). The number of stem cells in transplants affects regeneration, requiring ex vivo growth of transplantable cells. For optimal results, transplanted stem cells must survive, proliferate, differentiate site-specifically, and integrate into the host circulatory system.

6.3.1.1 Embryonic Stem Cells

Embryonic stem cells (ESCs) are pluripotent, capable of differentiating into over 200 cell types, and hold potential for treating various diseases (Park et al. 2004).

Their pluripotency is regulated by transcription factors OCT4, SOX2, and NANOG, known as pluripotency factors. OCT4 alleles are separated in the pluripotent stage; their pairing during embryogenesis and transdifferentiation is crucial for lineage commitment (Hogan et al. 2015). ESCs' diverse lineage commitment capacity makes them ideal for regenerative therapies targeting diseases and tissue abnormalities. They can transdifferentiate into cell types from the three germ layers, making them promising for tissue regeneration and disease treatment. Ethical considerations limit ESC use, necessitating adherence to guidelines; thus, TSPSCs, MSCs, UCSCs, BMSCs, and iPSCs are explored as alternatives (Mahla 2016).

6.3.1.2 Tissue-Specific Progenitor Stem Cells

TSPSCs uphold tissue homeostasis through continuous cell division but, unlike ESCs, they maintain stem cell plasticity and differentiation specific to tissues, forming a limited range of cell types. The proportion of TSPSCs in the total cell population is minimal, posing significant challenges for their collection and modification in vitro for therapeutic applications. Numerous TSPSC types exist in the human body, making it impractical to explore all for therapeutic use. TSPSCs aid tissue regeneration, with advancements in organoid culture technology underscoring the importance of niche effects in tissue regeneration and the therapeutic potential of ex vivo expanded stem cells (Rohban and Pieber 2017).

6.3.1.3 MSCs/Stromal Cells in Regenerative Medicine

Multilineage stem cells (MSCs) develop into mesodermal tissues like muscles, tendons, bone, cartilage, ligaments, and neurons (Dominici et al. 2006). MSCs are characterized by markers such as HLA-DR, CD73+, CD90+, CD105+, CD11b−, CD14−, CD19−, CD34−, CD45−, and CD79a−. Their potential in regenerative medicine is evident from various clinical trials (Squillaro et al. 2016). Donor and recipient physiological and anatomical traits significantly affect therapeutic outcomes. Baboon bone marrow-derived MSCs, similar in morphology and phenotype to bladder stem cells, can regenerate bladder tissue. MSCs also show promise in regenerating liver tissue and treating cirrhosis (Sharma et al. 2011). MSCs can be used for direct transplantation or transdifferentiation followed by transplantation. Ex vivo transdifferentiation via retroviral delivery can potentially cause malignancy. MSCs' multipotency makes them suitable for durable tissue constructions through induced three-dimensional organoid culture. However, their heterogeneous dispersion slows cell growth, affecting therapeutic applications (Dameshghi et al. 2016). MSCs' ability to differentiate into multiple lineages and their use in next-generation organoid culture make them a promising prospect for regenerative medicine (Delorme et al. 2006).

6.3.1.4 UCSCs in Regenerative Medicine

The umbilical cord, typically discarded at childbirth, is a significant source of stem cells. It can be obtained non-invasively and with fewer ethical concerns than ESCs, containing hematopoietic stem cells (HSCs) and MSCs with high regenerative capacity (Rosemann 2014). HSCs in cord blood continually regenerate blood and immune cells. Musashi-2 protein mediates the attenuation of Aryl hydrocarbon receptor (AHR) signaling, regulating HSC proliferation (Rentas et al. 2016). The FDA has approved five HSCs for treating hematological and immunological disorders (Rosemann 2014). UCSCs are ideal for patients lacking fully compatible HLA-type donors for peripheral blood, PBMCs, and bone marrow (Gluckman et al. 2005). Transplanted UCSCs are critical for neurodegenerative disease therapies, such as Alzheimer's, ALS, traumatic brain injuries, Parkinson's, spinal cord injuries, and stroke. In animal models, UCSCs migrate to the central nervous system and peripheral organs post-injection (Ehrhart et al. 2016). UCSC transplantation in severe congenital neutropenia (SCN) patients effectively restores neutrophil counts without adverse effects, making UCSCs a viable SCN treatment option in the absence of HLA-matched bone marrow donors (Osone et al. 2016). Myocardial infarction therapy efficacy depends on age, systemic inflammation, and cell preparation. Human hyaluronan hydrogel infusion in pig UCSCs promotes angiogenesis, reduces scarring, and improves cardiac performance, suggesting potential human application (Chang et al. 2016).

6.3.1.5 iPSCs in Regenerative Medicine

iPSC technology and research emerged in 2006 when Takahashi and Yamanaka created embryonic stem cell-like cells by incorporating Sox2, Oct3/4, Klf4, and c-Myc into skin fibroblasts (Takahashi and Yamanaka 2006). iPSCs, generated through nuclear reprogramming, are similar to ESCs in transcriptome, epigenetic marks, and functionality (Hirschi et al. 2014). The use of retrovirus in transdifferentiation raises concerns about iPSC technology. Recent advances allow iPSCs to be generated from adult cells using ESCs or direct transdifferentiation. This technique transforms terminally differentiated skin cells into kidney organoids resembling functional kidney tissue in vivo (Amabile and Meissner 2009). While kidneys can repair minor damage, they struggle with severe damage due to limited regenerative ability. During recovery, progenitor stem cells differentiate into 20 cell types responsible for waste elimination, pH control, and ion restoration. Human ex vivo kidney organoid production, including functional nephrons, has been established and these organoids resemble first-trimester fetal kidneys. They can be used for nephrotoxicity testing of drugs, disease modeling, and organ transplantation. However, creating fully functional kidneys with current scientific methods is highly unlikely (Osafune 2021; Takasato et al. 2015).

6.3.1.6 BMSCs in Regenerative Medicine

Bone marrow found in soft spongy bones is responsible forformation of all peripheral blood and comprises hematopoietic stem cells (producing blood cells) and stromal cells (producing fat, cartilage, and bones) (Burdon et al. 2011). Visually bone marrow has two types, red marrow (myeloid tissue; producing RBC, platelets, and most of WBC) and yellow marrow (producing fat cells and some WBC). Imbalance in marrow composition can culminate to the diseased condition (Travlos 2006).

6.3.2 Tissue Engineering and Biomaterials

Tissue engineering combines cells with porous scaffolds to regenerate damaged tissues, utilizing advanced biofabrication techniques like 3D bioprinting and textile methods; these scaffolds are applied in tissue repair and replacement for organs such as skin, bone, and neural tissues, and recent advances in biomaterials focus on sustainability through the use of biocompatible, biodegradable, and bioactive materials. Tissue engineering (TE) leverages scaffolds—biocompatible and biodegradable structures that support cell attachment, growth, and differentiation—to regenerate damaged tissues. These scaffolds can be fabricated using various techniques such as three-dimensional (3D) bioprinting, electrospinning, and freeze-drying, which allow for the creation of highly porous and structurally complex constructs (O'brien 2011; Nikolova and Chavali 2019; Collins et al. 2021). The development of biofabrication in the past years has shifted to more accurate reproduction of the native tissue structure and the use of such techniques as textile engineering and the integration of different kinds of biomaterials in order to improve the mechanical characteristics and cell responses (Nikolova and Chavali 2019; Elkhoury et al. 2021). Scaffolds are very important in TE applications for tissue repair and replacement in such tissues as bone, skin, neural as well as cardiac tissues. For example, the scaffolds designed from chitosan and collagen in the bone tissue engineering support bone formation and regeneration (Soundarya et al. 2018; Collins et al. 2021). Likewise, natural hydrogels to reconstructed cardiac and neural tissues in cardiac and neural tissue engineering processes because of their biocompatibility and close similarity to the ECM. Using such applications, authors outlined the prospects of TE in solving the problem of the shortage of donor organs and the treatment of patients with extensive tissue injury (Pedde et al. 2017; Soundarya et al. 2018). As part of the newly emerging scientific field, the development of sustainable biomaterials has become more of concern within TE. Scientists are investigating biodegradable polymers and composites for scaffolds in tissue engineering without harm to the environment (Ambekar and Kandasubramanian 2019; Reddy et al. 2021). The advances in the bioactive molecules and nanoparticles that improve the functionality and antibacterial activity of scaffolds. Organized into nine headings, the advancements described here by Xia et al. seek to develop 'innovative and environmentally friendly technologies that could provide better solutions for clinical services to facilitate greater TE application in medical practice.'

6.3.3 Role of Gene Editing and Gene Therapies in Regenerative Medicine

The gene editing as well as gene therapies are instrumental in serving as regenerative medicine by allowing adjustments at genomic level toward the realisation of tissue repair and regeneration. These technologies have advanced and are used in many fields such as bone growth promotion; induced pluripotent stem cells to be used to help regenerate cardiac tissue. CRISPR and other Gene editing tools like Zinc finger nucleases and TALE nucleases, enable the cutting and pasting of DNA sequences to correct a defective gene, add a therapeutic gene, or remove an undesirable gene (Doudna 2020; Maeder and Gersbach 2016; Kues et al. 2022). These advancements have opened a wide variety of therapeutic possibilities and they are used in treatment of monogenic diseases, cancer and tissue engineering directly

(Memi et al. 2018; Uddin et al. 2020; Hsu et al. 2019). For instance, in clinical trials, CRISPR was applied in treating diseases as sickle cell anemia, muscular dystrophy, which makes this technology capable to cure previously unmanageable diseases (Hosseinkhani et al. 2023). However, the advancement of these technologies is fast bringing to force ethical, legal and social questions. Some of these areas are as follows: concerns with possible further effects of the therapy on the beneficiary's genes, something that Robert wants to consider, fairness in provision of the treatments, as well as requirements in legal requirements concerning the delivery of such therapies (Subica 2023). Ensuring responsible use of gene editing in regenerative medicine requires balancing innovation with ethical considerations to maximize benefits while minimizing risks.

6.3.4 Bioprinting and Organ Regeneration

Bioprinting and organ regeneration is a revolutionary branch of regenerative medicine use cutting edge methods of 3D bioprinting to form living tissues and organs (Xu et al. 2014; Ramadan and Zourob 2021). This strategy requires the use of cells and biomaterials to build multilayered and organized structures that are similar to vascularized tissues. Some advancements have been achieved in regenerating several body part and tissues including skin, cartilage, and even heart muscle to support patient specific solution (Matai et al. 2020). Researchers have had successes in organ and tissue engineering development and latest efforts include development of bioprinted skin, cardiac muscle, bone, cartilage, liver, and neural tissues. These developments are not only promising for transplantation but also in drug testing and disease modeling (Matai et al. 2020; Cui et al. 2017; Murphy and Atala 2014).

6.4 Applications and Impact of Regenerative Medicine on Sustainable Healthcare

6.4.1 Regenerative Medicine in Chronic Disease Management

Regenerative medicine holds immense potential in the management of chronic diseases, offering innovative solutions that go beyond traditional treatments. By leveraging the capabilities of stem cells, growth factors, and advanced technologies, regenerative medicine aims to restore function and improve outcomes for patients with cardiovascular diseases, diabetes, metabolic disorders, and neurodegenerative diseases. Continued research and clinical trials are essential to overcome current challenges and fully realize the potential of these therapies.

6.4.1.1 Cardiovascular Diseases

New approaches are available through regenerative medicine for cardiovascular diseases CVDs which are on the leading cause of death in the world. Recent discoveries serve on cellular reprogramming, stem cell treatments and on tissue engineering and all of these aspects are engaged in the possibility of cardiac remodeling. The second treatment option is induced pluripotent stem cells because of the differentiation of cardiomyocytes, which can form micro arteries and redirect blood flow from areas affected by myocardial infarctions to avoid fatality and recidivism (Garg 2022). Moreover, mesenchymal stem cells (MSCs) have been mentioned for their availability, proliferation, and purification to be the most suitable for cardiac regeneration without the use of immunosuppressive agents (Arjmand et al. 2021). MSCs possess characteristics that make them suitable for cardiac regeneration namely, cardiomyogenic potential and anti-inflammation. Another candidate for regenerative medicine which targets the cardiovascular system has been shown to induce cell migration, proliferation, differentiation and angiogenesis by incorporating Insulin Growth Factor-1 (IGF-1) into the therapy despite the concerns of its link to cancer growth, and site-specific migration (Gan et al. 2023). Tissue engineering combines biomaterials and 3D bioprinting to create vascular grafts, addressing the shortage of autologous grafts (Sassi et al. 2023).

6.4.1.2 Diabetes and Metabolic Disorders

The evidence of nanomedicine in diabetes and its related metabolic complications is based only on possibilities of regenerative medicine and is committed to the development of innovations that will establish the pancreas functionality. Modern studies focus on stem cell therapies, islet transplant, and the new peptide different ways to overcome the drawbacks of standard diabetic management (Poojari et al. 2023). Stem cell-based therapies, including the use of autologous and allogeneic stem cells, have shown promise in treating these conditions by regenerating damaged tissues and organs (Al-Ghadban et al. 2022). Islet transplantation is still performed today to reclaim insulin secreting capacity in type 1 diabetic patients, however at the cost of immunosuppression (Lakey et al. 2024). Current islet transplantation clinical trials have longer treatment duration and investigate the effectiveness and safety of transplantation in patients for better long-term results. Based on the integration of engineering and life sciences regenerative medicine wants to return the functionality of damaged tissues and organs and thus the dependency on traditional treatment methods and also the related patient benefits (Mao and Mooney 2015). However, the application of phage display technology for isolating peptide ligands for any number of targets also provides new prospects for the treatment of metabolic disorders through regenerative methods (Martins et al. 2016). Research on Mito Organelle Peptides in recent work has shown effectiveness in the preservation of pancreatic beta cells from autoimmune attack which provides new form of treatment for type I diabetes (Lakey et al. 2024).

6.4.1.3 Neurodegenerative Diseases

Alzheimer's, Parkinson's, and amyotrophic lateral sclerosis (ALS) remain particularly difficult to treat because they involve the gradual degeneration of certain neuronal types. Regenerative medicine, especially through the application of neurotrophic factors (NTFs), holds potential for therapy to stimulate neuronal repair (El Ouaamari et al. 2023). Studies show that the stem cell environment can be adjusted to promote differentiation as well as tissue regeneration and reparative processes in diseases such as Alzheimer's and Parkinson's (Velikic et al. 2024). Heterologous stem and immune cells, viral vectors, biomaterials have been used to deliver NTFs to CNS, but the route faces unique barriers including BBB and side effects which persist (El Ouaamari et al. 2023). Furthermore, polyphenolic phytochemicals and new generation stem cell therapies are currently being explored to devise sophisticated strategies to dealing the factors of neurodegenerative diseases and neuronal growth (Moradi et al. 2022). The anti-inflammatory/fibrotic effects of the secretome of adipose stem cells (ASCs) as well as their cell-matrix interactions in neurodegenerative diseases add to the mention concept of regenerative medicine in this field (Al-Ghadban et al. 2022).

6.4.2 Role of Regenerative Medicine in Acute and Emergency Care

The application of regenerative medicine to treat trauma and burn injuries that commonly occur in acute and emergency care fields is examined. Regenerative approaches can benefit traumatic injuries, often leading to permanent disability, which acquire biological substitutes for damaged tissues. Engineered skin substitutes are FDA approved and are new avenues of treatment for severe skin injuries (Moradi et al. 2015). In the field of burn care, regenerative strategies such as tissue engineering seeks to address complicated burn wound healing that produces skin barrier function with reduced infection, scarring and contracture, thereby improving patient outcome (Stone et al. 2018). It is used to accelerate the healing of traumatic and burn injuries, such as with platelet rich plasma (PRP) (Etulain 2018). Regenerative medicine reduces healing and recovery time in surgical interventions. For example, gene therapy elicits local production of a growth factor that accelerates tissue regeneration and repair (Giatsidis et al. 2013). This approach holds promise for the preclinical treatment of chronic wounds and burn injuries. Graft fabrication and tissue mimics evolve to make grafting procedure more successful with less reliance on donor tissue available (Mao and Mooney 2015). Mesenchymal and hematopoietic stem cells promote surgical wound treatment, and in addition improve the body's capacity for regeneration (Imran et al. 2022). Regenerative medicine is shown through clinical trials to be effective in acute and emergency care. Significant improvement of wound healing and recovery times (Moradi et al. 2015) can be achieved with engineered skin substitutes. In burn care, both pharmacological agents, biomaterials, and cellular based therapies are promising regenerative treatments that are trail and hold significant promise to restore normal skin physiology, minimize adverse outcomes (Stone et al. 2018). Hypertrophic scars and keloids have been used in PRP studies reporting positive outcomes with less side effects and satisfied patients (Jafarzadeh et al. 2024). Regenerative medicine boasts the potential to transform acute and emergency care with effective, minimally invasive treatments leading to a rapid recovery and better patient quality of life (Adamič and Vengust 2023).

6.4.3 Cancer Therapy and Regenerative Medicine

Tumor microenvironment (TME) is recognized as a key factor that influences the cancer development and treatment. MSCs of regenerative medicine have the ability to switch the Tumor microenvironment to either support or suppress tumor growth. By the secretome containing cytokines and growth factors, MSCs can affect the TME pro-angiogenic, immunomodulatory and tissue remodelling properties. However, this bimodal capability can be problematic as the same contributors to tissue repair can enhance tumor growth and spread (Zimmerlin et al, 2013; Donnenberg et al. 2010). In light of this it is crucial to comprehend the multifaceted interactions of MSCs with cancer cells in the TME for the purpose of improving safe and efficient regenerative therapies (Zimmerlin et al. 2013). Inventive of cancer treatment is another aspect of regenerative medicine where tissue engineering, cell therapy, Gene therapy is used. They include those that seek to restore or regenerate diseased or injured tissues, and enhance the effectiveness of cancer therapeutics. For example, nanotheranostics and smart drug delivery may increase the specificity and efficacy of chemotherapy drugs, while minimizing their harms to patients (Garbayo et al. 2020). Additionally, gene therapy techniques are being explored to modify cancer cells or the surrounding tissue to inhibit tumor growth and promote regeneration (Hosseinkhani et al. 2023). The integration of these regenerative approaches in oncology holds promise for more effective and less invasive cancer treatments. Emerging therapies in regenerative medicine are paving the way for novel cancer treatments. Recent advancements include the use of non-viral gene delivery systems, which offer a safer alternative to viral vectors for gene therapy applications (Hosseinkhani et al. 2023). Furthermore, the role of the immune system, particularly macrophages, in regenerative medicine is gaining attention. Macrophages can be polarized to support tissue regeneration and inhibit tumor growth, representing a shift from traditional views of immune responses as purely detrimental in cancer therapy (Brown et al. 2014). Future prospects in regenerative medicine for cancer therapy involve a deeper understanding of the molecular pathways shared between regeneration and tumorigenesis, such as the role of the TP53 gene in controlling cell proliferation and preventing malignant transformation (Charni et al. 2017). Continued research in these areas is essential for developing therapies that can effectively balance tissue regeneration with cancer inhibition (Oviedo and Beane 2009).

6.4.4 Regenerative Medicine for Aging and Longevity

Cellular senescence is a condition where cells become committed to long-term growth arrest and can be associated with the production of pro-inflammatory cytokines referred to as the senescence-associated secretory phenotype (SASP) (Borghesan et al. 2020; Di Micco et al. 2021). This process is initiated by multiple stressors as DNA damage, oxidative stress, and telomere shortening and is considered inessential to the ageing and the age-related diseases (Khalil et al. 2023;

Amaya-Montoya et al. 2020). Ageing cells start to pile up in the tissues, thus negatively affecting tissue's functionality and its ability to regenerate, contributing to chronic inflammation, and diseases (Turinetto et al. 2016; Childs et al. 2015). Thus, therapeutic approaches in cellular senescence include direct elimination of senescent cells (senolytics) and/or modification of their detrimental outputs (senomorphics) (Khalil et al. 2023; Di Micco et al. 2021). These approaches have been effective for given diseases in pre-clinical models enhancing eventually health span and lifespan through attenuation of senescent cell. Furthermore, elimination of senescent cells or their reprogramming into iPSCs was shown to rejuvenate human cells, which can be preliminarily applied for tissues' regeneration without ethic issues derived from utilizing human embryos (Lapasset et al. 2011). Selective elimination of senescent cells in recepritive tissues, for instance, fat tissue, may also help to reduce age-related pathology and extend healthy life span (Palmer and Kirkland 2016). That is why the development and further usage of regenerative therapies for aging originate some interesting societal and ethically related questions. The massive opportunity to elongate people's existence and enhance their appreciation of life should be taken with consideration to efficacy, fairness, implications to the health sector, and the consequences of having more elderly citizens (Miller and Roubenoff 2019). They also include the consequences of occupying regulatory aspects of body processes and the impacts not foreseen in the process of intervention. It will therefore be necessary for scientific research to continue engaging in discussion with ethicists, policymakers, and the public to ensure that the development and implementation of these therapies is done in a responsible manner.

6.5 Patents and Clinical Status of Regenerative Medicine

Several regenerative medicine therapies have received FDA approval and are commercially available. Over 5000 clinical trials involving stem cell therapies are registered, focusing on various organs such as the lung, kidney, heart, and liver (Petrosyan et al. 2022). Searches for patents and related information were conducted using the official World Intellectual Property Organization website, employing analytics to examine and categorize recent advancements in Regenerative Medicine (Table 6.1). Numerous novel therapeutic approaches classified as regenerative medicine are currently undergoing various stages of clinical trials, primarily focusing on either treating or investigating potential treatments for several life-threatening diseases. A selection of these trials is presented in Table 6.2 (https://clinicaltrials.gov).

6.6 Challenges and Future Perspectives

Regenerative medicine presents a groundbreaking approach to healthcare, offering the potential to mend or substitute impaired tissues and organs. However, the field faces significant challenges that must be addressed to realize its full potential. Key obstacles include the complexity of translating stem cell biology into effective

Table 6.1 Published patent literature about the regenerative medicine

Patent name	Patent Number	Applicant	Publication date	References
Method for assessing the potency of cell populations in regenerative medicine using biomarkers	WO/2024/227139	University of Florida Research Foundation, Incorporated	31.10.2024	Phinney et al. (2024)
Composition for preservation of cells, tissues or organs, comprising isolated mitochondria, and use thereof	US20240341299	Paean Biotechnology Inc.	17.10.2024	Young-Cheol et al. (2024)
Regenerative medicine devices and foam methods of manufacture	CA2604525	Johnson & Johnson Regenerative Therapeutics, LLC.	29.03.2008	Keeley et al. (2008)
Degradable regenerative medical material for promoting tissue in-situ regeneration and preparation method therefor	US20220296785	Beijing Best Life Regenerative Medicine Technology Co. Ltd Best Life Regenerative Medicine Technology Co. Ltd	22.09.2022	Chuangzhouet al. (2022)
Kefiran for use in regenerative medicine and/or tissue engineering	EP3506946	Association For the Advancement of Tissue Engineering and Cell Based Tech & Therapies A4TEC Associac	10.07.2019	Radhouani et al. (2019)
Nanotechnology smart chip regenerative medicine	IN202141011918	Bharath Institute of Higher Education and Research	26.03.2021	Durga (2021)
Polysaccharide for use in regenerative medicine and/or tissue engineering	WO2020075135	Association For The Advancement of Tissue Engineering and Cell Based Technologies & Therapies	16.04.2020	López et al. (2020)
Primed placental tissue and uses in regenerative medicine	WO2022074523	Osiris Therapeutics, Inc	14.04.2022	Li et al. (2022)

Table 6.2 Current status of clinical trials which are in-progress as regenerative medicine

Study Title/type	Condition or Disease	Therapeutic Approach	Sponsor	NCT No.	Phase
Effectiveness of Regenerative Medicine (PRP and PBD-VSEL Stem Cell Therapy) for Parkinson's Disease/ Interventional	Parkinson's Disease	Stem Cell therapy	Fatima Jinnah Medical University	NCT06142981	Not Applicable
Role of Autologous Dermal Micrografts in Regenerative Surgery/ Interventional	Wound healing	Tissue Engineering	Istituti Clinici Scientifici Maugeri SpA	NCT03912675	Not Applicable
Safety of Autologous Adipose-Derived Stroma/Stem Cells to Treat Critical Leg Ischemia/ Interventional	Critical Leg Ischemia	Stem Cell therapy	University Hospital, Toulouse	NCT01211028	Phase 2
Regenerative Medicine of Articular Cartilage: Characterization and Comparison of Chondrogenic Potential and Immunomodulatory Adult Mesenchymal Stem Cells/ Interventional	Knee Osteoarthritis	Stem Cell therapy	Nantes University Hospital	NCT01879046	Not Applicable
Transplantation of Mesenchymal Stem Cells for Prevention of Acute Myocardial Infarction Induced Heart Failure: a Phase III Randomized Clinical Trial/Interventional	Heart Failure	Stem Cell therapy	Shiraz University of Medical Sciences	NCT05043610	Phase 3
Cardiovascular Clinical Project to Evaluate the Regenerative Capacity of CardioCell in Patients With Acute Myocardial Infarction (AMI)/ Interventional	Acute Myocardial Infarction	Stem Cell therapy	John Paul II Hospital, Krakow	NCT03404063	Phase 3

(continued)

Table 6.2 (continued)

Study Title/type	Condition or Disease	Therapeutic Approach	Sponsor	NCT No.	Phase
Randomized Clinical Trial to Evaluate the Regenerative Capacity of CardioCell in Patients with Chronic Ischaemic Heart Failure (CIHF)/ Interventional	Heart Failure	Stem Cell therapy	John Paul II Hospital, Krakow	NCT03418233	Phase 3
GCSF Therapy in Decompensated Cirrhosis—A Double Blinded RCT/Interventional	Decompensated Liver Cirrhosis	Stem Cell therapy	Post Graduate Institute of Medical Education and Research, Chandigarh	NCT03911037	Phase 3
SkinTE® for the Treatment of Wagner 1 Diabetic Foot Ulcers (COVER DFUS II)/ Interventional	Diabetic Foot	Tissue Engineering	PolarityTE	NCT06140303	Phase 3

therapies, the high costs and logistical difficulties of developing and manufacturing treatments, and the need for robust regulatory frameworks to ensure safety and efficacy. Advancements in stem cell research suggest significant potential for regenerative treatments. Over the next decade, it is expected that a variety of tissues, organoids, and organs will be produced from adult stem cells. Somatic cell reprogramming to pluripotency of differentiated adult cells can be seen as more promising for therapeutic application than ESCs because of lower ethical concerns. New chemical entities for drugs may be responsible for the promotion of migration and differentiation of tissue stem cells into the affected tissues. Existing stem cell therapies are still regarded as invasive and expensive, thus cannot be widely applied. Further large-scale clinical trials are needed to apply stem cell findings into clinical applications; funding by both government and private entities should be sought. The presentation of the regulatory guidelines indicates that a critical evaluation of the guidelines for each stage of clinical trials is pertinent for evaluating success and efficiency at any given point in the planned duration. Through eradicating these problems, regenerative medicine can offer profitable solutions in medicine, limiting usage of transplants, and treatment of existing incurable diseases.

6.7 Conclusion

Regenerative medicine has emerged as a promising approach to address global healthcare challenges and promote sustainability. This rapidly evolving field combines principles from biology, medicine, and engineering to develop innovative therapies aimed at restoring, maintaining, or enhancing tissue and organ function. By harnessing the body's innate healing capabilities and leveraging advanced biotechnological interventions, regenerative medicine offers potential solutions for previously intractable conditions. While significant progress has been made in areas such as stem cell therapy, tissue engineering, and gene editing, challenges persist in translating these technologies into widespread clinical applications. As research continues to advance, regenerative medicine holds considerable potential for revolutionizing healthcare, improving patient outcomes, and contributing to more sustainable and efficacious medical practices in the future.

References

Adamič N, Vengust M (2023) Regenerative medicine in lung diseases: a systematic review. Front Vet Sci 10:1115708

Al-Ghadban S, Artiles M, Bunnell BA (2022) Adipose stem cells in regenerative medicine: looking forward. Front Bioeng Biotechnol 9:837464

Amabile G, Meissner A (2009) Induced pluripotent stem cells: current progress and potential for regenerative medicine. Trends Mol Med 15(2):59–68

Amaya-Montoya M, Pérez-Londoño A, Guatibonza-García V, Vargas-Villanueva A, Mendivil CO (2020) Cellular senescence as a therapeutic target for age-related diseases: a review. Adv Ther 37:1407–1424

Ambekar RS, Kandasubramanian B (2019) Progress in the advancement of porous biopolymer scaffold: tissue engineering application. Ind Eng Chem Res 58(16):6163–6194

Arjmand B, Abedi M, Arabi M, Alavi-Moghadam S, Rezaei-Tavirani M, Hadavandkhani M, Tayanloo-Beik A, Kordi R, Roudsari PP, Larijani B (2021) Regenerative medicine for the treatment of ischemic heart disease; status and future perspectives. Front Cell Dev Biol 9:704903

Augustine R, Gezek M, Nikolopoulos VK, Buck PL, Bostanci NS, Camci-Unal G (2024) Stem cells in bone tissue engineering: progress, promises and challenges. Stem Cell Rev Rep 20(7):1692–1731

Berniak-Woźny J, Rataj M (2023) Towards green and sustainable healthcare: a literature review and research agenda for green leadership in the healthcare sector. Int J Environ Res Public Health 20(2):908

Bizios R, Wechsler ME (2024) Concepts of tissue-biomaterial interactions: fundamentals and new directions. John Wiley & Sons

Borghesan M, Hoogaars WM, Varela-Eirin M, Talma N, Demaria M (2020) A senescence-centric view of aging: implications for longevity and disease. Trends Cell Biol 30(10):777–791

Braithwaite J, Zurynski Y, Ludlow K, Holt J, Augustsson H, Campbell M (2019) Towards sustainable healthcare system performance in the 21st century in high-income countries: a protocol for a systematic review of the grey literature. BMJ Open 9(1):e025892

Brown BN, Sicari BM, Badylak SF (2014) Rethinking regenerative medicine: a macrophage-centered approach. Front Immunol 5:510

Burdon TJ, Paul A, Noiseux N, Prakash S, Shum-Tim D (2011) Bone marrow stem cell derived paracrine factors for regenerative medicine: current perspectives and therapeutic potential. Bone Marrow Res 2011(1):207326

Chang MY, Huang TT, Chen CH, Cheng B, Hwang SM, Hsieh PC (2016) Injection of human cord blood cells with hyaluronan improves postinfarction cardiac repair in pigs. Stem Cells Transl Med 5(1):56–66

Charni M, Aloni-Grinstein R, Molchadsky A, Rotter V (2017) p53 on the crossroad between regeneration and cancer. Cell Death Differ 24(1):8–14

Childs BG, Durik M, Baker DJ, Van Deursen JM (2015) Cellular senescence in aging and age-related disease: from mechanisms to therapy. Nat Med 21(12):1424–1435

Chuangzhou G, Fang H, Yuexiu Q, Xiaoli Y (2022) Degradable regenerative medical material for promoting tissue in-situ regeneration and preparation method therefor. US20220296785

Collins MN, Ren G, Young K, Pina S, Reis RL, Oliveira JM (2021) Scaffold fabrication technologies and structure/function properties in bone tissue engineering. Adv Funct Mater 31(21):2010609

Cui H, Nowicki M, Fisher JP, Zhang LG (2017) 3D bioprinting for organ regeneration. Adv Healthc Mater 6(1):1601118

Dameshghi S, Zavaran-Hosseini A, Soudi S, Shirazi FJ, Nojehdehi S, Hashemi SM (2016) Mesenchymal stem cells alter macrophage immune responses to Leishmania major infection in both susceptible and resistance mice. Immunol Lett 170:15–26

Delorme B, Chateauvieux S, Charbord P (2006) The concept of mesenchymal stem cells. Regen Med 1(4):497–509

Deng C, Aldali F, Luo H, Chen H (2024) Regenerative rehabilitation: a novel multidisciplinary field to maximize patient outcomes. Med Rev

Dhamdar TS, Sandhya KV, Basavaraj BV (2024) Advancing innovation through biomimicry and AI: inspiration to implementation. Bionature 44(1):16–27

Di Micco R, Krizhanovsky V, Baker D, d'Adda di Fagagna F (2021) Cellular senescence in ageing: from mechanisms to therapeutic opportunities. Nat Rev Mol Cell Biol 22(2):75–95

Dominici ML, Le Blanc K, Mueller I, Slaper-Cortenbach I, Marini FC, Krause DS, Deans RJ, Keating A, Prockop DJ, Horwitz EM (2006) Minimal criteria for defining multipotent mesenchymal stromal cells. The International Society for Cellular Therapy position statement. Cytotherapy 8(4):315–317

Dong X, Gou Y, Guo M, Zhong J, Li A, Hao A, Zeng W, Haydon RC, Luu HH, Reid RR, He T (2024) A simplified noncryogenic strategy to transport mesenchymal stem cells: potential applications in cell therapy and regenerative medicine. Genes Dis 11(3):101073

Donnenberg VS, Zimmerlin L, Rubin JP, Donnenberg AD (2010) Regenerative therapy after cancer: what are the risks? Tissue Eng Part B Rev 16(6):567–575

Doudna JA (2020) The promise and challenge of therapeutic genome editing. Nature 578(7794):229–236

Durga (2021) Nanotechnology smart chip regenerative medicine. IN202141011918

Dzobo K, Thomford NE, Senthebane DA, Shipanga H, Rowe A, Dandara C, Pillay M, Motaung KS (2018) Advances in regenerative medicine and tissue engineering: innovation and transformation of medicine. Stem Cells Int 2018(1):2495848

Ehrhart J, Darlington D, Kuzmin-Nichols N, Sanberg CD, Sawmiller DR, Sanberg PR, Tan J (2016) Biodistribution of infused human umbilical cord blood cells in Alzheimer's disease-like murine model. Cell Transplant 25(1):195–199

El Ouaamari Y, Van den Bos J, Willekens B, Cools N, Wens I (2023) Neurotrophic factors as regenerative therapy for neurodegenerative diseases: current status, challenges and future perspectives. Int J Mol Sci 24(4):3866

Elkhoury K, Morsink M, Sanchez-Gonzalez L, Kahn C, Tamayol A, Arab-Tehrany E (2021) Biofabrication of natural hydrogels for cardiac, neural, and bone tissue engineering applications. Bioact Mater 6(11):3904–3923

Etulain J (2018) Platelets in wound healing and regenerative medicine. Platelets 29(6):556–568

Fatma H, Siddique HR (2024) Cancer cell plasticity, stem cell factors, and therapy resistance: how are they linked? Cancer Metastasis Rev 43(1):423–440

Fortier LA (2005) Stem cells: classifications, controversies, and clinical applications. Vet Surg 34(5):415–423

Gan QF, Lim YT, Foo CN, Yu CW, Woon CK, Cheong SK, Leong PP (2023) Incorporating insulin growth factor-1 into regenerative and personalized medicine for cardiovascular disease: a systematic review. Curr Stem Cell Res Ther 18(2):202–215

Garbayo E, Pascual-Gil S, Rodríguez-Nogales C, Saludas L, Estella-Hermoso de Mendoza A, Blanco-Prieto MJ (2020 Sep) Nanomedicine and drug delivery systems in cancer and regenerative medicine. Wiley Interdiscip Rev Nanomed Nanobiotechnol 12(5):e1637

Garg D (2022) Regenerative medicine and its potential in cardiovascular disease. J Stud Res 11(3)

Garzón I, Pérez-Köhler B, Garrido-Gómez J, Carriel V, Nieto-Aguilar R, Martín-Piedra MA, García-Honduvilla N, Buján J, Campos A, Alaminos M (2012) Evaluation of the cell viability of human Wharton's jelly stem cells for use in cell therapy. Tissue Eng Part C Methods 18(6):408–419

Giatsidis G, Dalla Venezia E, Bassetto F (2013) The role of gene therapy in regenerative surgery: updated insights. Plast Reconstr Surg 131(6):1425–1435

Gluckman E, Koegler G, Rocha V (2005) Human leukocyte antigen matching in cord blood transplantation. Semin Hematol 42(2):85–90. WB Saunders

Gubareva EA, Sjöqvist S, Gilevich IV, Sotnichenko AS, Kuevda EV, Lim ML, Feliu N, Lemon G, Danilenko KA, Nakokhov RZ, Gumenyuk IS (2016) Orthotopic transplantation of a tissue engineered diaphragm in rats. Biomaterials 77:320–335

Hirschi KK, Li S, Roy K (2014) Induced pluripotent stem cells for regenerative medicine. Annu Rev Biomed Eng 16(1):277–294

Hogan MS, Parfitt DE, Zepeda-Mendoza CJ, Shen MM, Spector DL (2015) Transient pairing of homologous Oct4 alleles accompanies the onset of embryonic stem cell differentiation. Cell Stem Cell 16(3):275–288

Hoop ED, Loeber A, Essink D (2022) Exploring, diversifying and debating sustainable health (care) approaches. Sustainability. 14(3):1698

Hosseinkhani H, Domb AJ, Sharifzadeh G, Nahum V (2023) Gene therapy for regenerative medicine. Pharmaceutics 15(3):856

Hsu MN, Chang YH, Truong VA, Lai PL, Nguyen TK, Hu YC (2019) CRISPR technologies for stem cell engineering and regenerative medicine. Biotechnol Adv 37(8):107447

Imran SAM, Hamizul MH, Khairul Bariah AA, Wan Kamarul Zaman WS, Nordin F (2022) Regenerative medicine therapy in Malaysia: an update. Front Bioeng Biotechnol 10:789644

Jafarzadeh A, PourMohammad A, Goodarzi A (2024) A systematic review of the efficacy, safety and satisfaction of regenerative medicine treatments, including platelet-rich plasma, stromal vascular fraction and stem cell-conditioned medium for hypertrophic scars and keloids. Int Wound J 21(4):e14557

Keeley D, Shetty D, Hammer JJ, Wang Z (2008) Regenerative medicine devices and foam methods of manufacture. CA2604525

Khalil R, Diab-Assaf M, Lemaitre JM (2023) Emerging therapeutic approaches to target the dark side of senescent cells: new hopes to treat aging as a disease and to delay age-related pathologies. Cells 12(6):915

Kues WA, Kumar D, Selokar NL, Talluri TR (2022) Applications of genome editing tools in stem cells towards regenerative medicine: an update. Curr Stem Cell Res Ther 17(3):267–279

Kumari DK (2024) Trajectories in life sciences. In: Today's multidisciplinary research perspectives volume-2

Kumbhar AB, Nerkar NV, Phuge AN (2024) Regenerative medicines: application to degenerative diseases and disorders. In: Biosystems, biomedical & drug delivery systems: characterization, restoration and optimization. Springer Nature Singapore, Singapore, pp 345–370

Lakey JR, Alexander M, Chan MK, Nalapko Y, Wong MB, Cox DC (2024) Advances in stem cell and regenerative therapies in autoimmune and metabolic diseases: implications for healthy longevity. Int J Transl Sci:115–126

Lapasset L, Milhavet O, Prieur A, Besnard E, Babled A, Aït-Hamou N, Leschik J, Pellestor F, Ramirez JM, De Vos J, Lehmann S (2011) Rejuvenating senescent and centenarian human cells by reprogramming through the pluripotent state. Genes Dev 25(21):2248–2253

Leferink AM, Chng YC, Van Blitterswijk CA, Moroni L (2015) Distribution and viability of fetal and adult human bone marrow stromal cells in a biaxial rotating vessel bioreactor after seeding on polymeric 3D additive manufactured scaffolds. Front Bioeng Biotechnol 3:169

Li C, Jacob V, Park MS, Dhall S, Sathyamoorthy M (2022) Primed placental tissue and uses in regenerative medicine. WO2022074523

López Cebral R, Antunes Correia De Oliveira JM, Quinteros Lopes Henriquez Da Silva TJ, Gonçalves Dos Reis RL (2020) Polysaccharide for use in regenerative medicine and/or tissue engineering. WO2020075135

Maeder ML, Gersbach CA (2016) Genome-editing technologies for gene and cell therapy. Mol Ther 24(3):430–446

Mahla RS (2016) Stem cells applications in regenerative medicine and disease therapeutics. Int J Cell Biol 2016(1):6940283

Mao AS, Mooney DJ (2015) Regenerative medicine: current therapies and future directions. Proc Natl Acad Sci 112(47):14452–14459

Martins IM, Reis RL, Azevedo HS (2016) Phage display technology in biomaterials engineering: progress and opportunities for applications in regenerative medicine. ACS Chem Biol 11(11):2962–2980

Mason C, Dunnill P (2008) A brief definition of regenerative medicine. Regen Med 3(1):1–5

Matai I, Kaur G, Seyedsalehi A, McClinton A, Laurencin CT (2020) Progress in 3D bioprinting technology for tissue/organ regenerative engineering. Biomaterials 226:119536

Memi F, Ntokou A, Papangeli I (2018) CRISPR/Cas9 gene-editing: research technologies, clinical applications and ethical considerations. Semin Perinatol 42(8):487–500. WB Saunders

Miller RR, Roubenoff R (2019) Emerging interventions for elderly patients—the promise of regenerative medicine. Clin Pharmacol Ther 105(1):53–60

Moradi M, Hood B, Moradi M, Atala A (2015) The potential role of regenerative medicine in the management of traumatic patients. J Inj Violence Res 7(1):27–35

Moradi SZ, Jalili F, Farhadian N, Joshi T, Wang M, Zou L, Cao H, Farzaei MH, Xiao J (2022) Polyphenols and neurodegenerative diseases: focus on neuronal regeneration. Crit Rev Food Sci Nutr 62(13):3421–3436

Mostepaniuk A, Akalin T, Parish MR (2023) Practices pursuing the sustainability of a healthcare organization: a systematic review. Sustainability. 15(3):2353

Murphy SV, Atala A (2014) 3D bioprinting of tissues and organs. Nat Biotechnol 32(8):773–785

Nikolova MP, Chavali MS (2019) Recent advances in biomaterials for 3D scaffolds: a review. Bioact Mater 4:271–292

O'brien FJ (2011) Biomaterials & scaffolds for tissue engineering. Mater Today 14(3):88–95

Osafune K (2021) iPSC technology-based regenerative medicine for kidney diseases. Clin Exp Nephrol 25(6):574–584

Osone S, Imamura T, Fukushima-Nakase Y, Kitamura-Masaki A, Kanai S, Imai T, Imashuku S, Kuroda H (2016) Case reports of severe congenital neutropenia treated with unrelated cord blood transplantation with reduced-intensity conditioning. J Pediatr Hematol Oncol 38(1):49–52

Oviedo NJ, Beane WS (2009) Regeneration: the origin of cancer or a possible cure? Semin Cell Dev Biol 20(5):557–564. Academic Press

Palmer AK, Kirkland JL (2016) Aging and adipose tissue: potential interventions for diabetes and regenerative medicine. Exp Gerontol 86:97–105

Park SP, Lee YJ, Lee KS, Shin HA, Cho HY, Chung KS, Kim EY, Lim JH (2004) Establishment of human embryonic stem cell lines from frozen–thawed blastocysts using STO cell feeder layers. Hum Reprod 19(3):676–684

Pedde RD, Mirani B, Navaei A, Styan T, Wong S, Mehrali M, Thakur A, Mohtaram NK, Bayati A, Dolatshahi-Pirouz A, Nikkhah M (2017) Emerging biofabrication strategies for engineering complex tissue constructs. Adv Mater 29(19):1606061

Petersdorf EW, Malkki M, Gooley TA, Martin PJ, Guo Z (2007) MHC haplotype matching for unrelated hematopoietic cell transplantation. PLoS Med 4(1):e8

Petrosyan A, Martins PN, Solez K, Uygun BE, Gorantla VS, Orlando G (2022) Regenerative medicine applications: an overview of clinical trials. Front Bioeng Biotechnol 10:942750
Phinney DG, Boregowda SV, Lee RH (2024) Method for assessing the potency of cell populations in regenerative medicine using biomarkers. WO/2024/227139
Poojari AS, Wairkar S, Kulkarni YA (2023) Stem cells as a regenerative medicine approach in treatment of microvascular diabetic complications. Tissue Cell 85:102225
Radhouani H, Da Mota Martins Gonçalves C, Gonçalves Dos Reis RL, Antunes De Oliveira JM (2019) Kefiran for use in regenerative medicine and/or tissue engineering. EP3506946
Rahat N, Sahni S, Nasim S (2024) Mapping sustainability practices in the healthcare sector: a systematic literature review and future research agenda. Int J Consum Stud 48(1):e12997
Ramadan Q, Zourob M (2021) 3D bioprinting at the frontier of regenerative medicine, pharmaceutical, and food industries. Front Med Technol 2:607648
Rattan TK, Joshi M, Vesty G, Sharma S (2022) Sustainability indicators in public healthcare: a factor analysis approach. J Clean Prod 370:133253
Reddy MS, Ponnamma D, Choudhary R, Sadasivuni KK (2021) A comparative review of natural and synthetic biopolymer composite scaffolds. Polymers 13(7):1105
Rentas S, Holzapfel NT, Belew MS, Pratt GA, Voisin V, Wilhelm BT, Bader GD, Yeo GW, Hope KJ (2016) Musashi-2 attenuates AHR signalling to expand human haematopoietic stem cells. Nature 532(7600):508–511
Rohban R, Pieber TR (2017) Mesenchymal stem and progenitor cells in regeneration: tissue specificity and regenerative potential. Stem Cells Int 2017(1):5173732
Rosemann A (2014) Why regenerative stem cell medicine progresses slower than expected. J Cell Biochem 115(12):2073–2076
Sassi S, Watanabe T, Shinoka T (2023) Scaffold and cell-based tissue engineering approaches as alternative therapy for blood vessel disease. Cardiol Res Cardiovasc Med 8:197
Sharma AK, Bury MI, Marks AJ, Fuller NJ, Meisner JW, Tapaskar N, Halliday LC, Matoka DJ, Cheng EY (2011) A nonhuman primate model for urinary bladder regeneration using autologous sources of bone marrow-derived mesenchymal stem cells. Stem Cells 29(2):241–250
Sherman JD, Thiel C, MacNeill A, Eckelman MJ, Dubrow R, Hopf H, Lagasse R, Bialowitz J, Costello A, Forbes M, Stancliffe R (2020) The green print: advancement of environmental sustainability in healthcare. Resour Conserv Recycl 161:104882
Soundarya SP, Menon AH, Chandran SV, Selvamurugan N (2018) Bone tissue engineering: scaffold preparation using chitosan and other biomaterials with different design and fabrication techniques. Int J Biol Macromol 119:1228–1239
Squillaro T, Peluso G, Galderisi U (2016) Clinical trials with mesenchymal stem cells: an update. Cell Transplant 25(5):829–848
Stone R II, Natesan S, Kowalczewski CJ, Mangum LH, Clay NE, Clohessy RM, Carlsson AH, Tassin DH, Chan RK, Rizzo JA, Christy RJ (2018) Advancements in regenerative strategies through the continuum of burn care. Front Pharmacol 9:672
Subica AM (2023) CRISPR in public health: the health equity implications and role of community in gene-editing research and applications. Am J Public Health 113(8):874–882
Suhag D (2024) Immunomodulatory biomaterials. In: Handbook of biomaterials for medical applications, volume 2: applications. Springer Nature Singapore, Singapore, pp 65–104
Takahashi K, Yamanaka S (2006) Induction of pluripotent stem cells from mouse embryonic and adult fibroblast cultures by defined factors. Cell 126(4):663–676
Takasato M, Er PX, Chiu HS, Maier B, Baillie GJ, Ferguson C, Parton RG, Wolvetang EJ, Roost MS, de Sousa C, Lopes SM, Little MH (2015) Kidney organoids from human iPS cells contain multiple lineages and model human nephrogenesis. Nature 526(7574):564–568
Travlos GS (2006) Normal structure, function, and histology of the bone marrow. Toxicol Pathol 34(5):548–565
Turinetto V, Vitale E, Giachino C (2016) Senescence in human mesenchymal stem cells: functional changes and implications in stem cell-based therapy. Int J Mol Sci 17(7):1164
Uddin F, Rudin CM, Sen T (2020) CRISPR gene therapy: applications, limitations, and implications for the future. Front Oncol 10:1387

Velikic G, Maric DM, Maric DL, Supic G, Puletic M, Dulic O, Vojvodic D (2024) Harnessing the stem cell niche in regenerative medicine: innovative avenue to combat neurodegenerative diseases. Int J Mol Sci 25(2):993

Wei L, Yan W, Shah W, Zhang Z, Wang M, Liu B, Xue Z, Cao Y, Hou X, Zhang K, Yan B (2024) Advancements and challenges in stem cell transplantation for regenerative medicine. Heliyon 10(16)

Wieringa S, McGuire H, Wang Q, Van Der Wees P, Shaw B (2024) Making sustainable healthcare decisions: three turns towards sustainable guidelines. BMJ Evid Based Med 29(4):219–222

Xu T, Rodriguez-Devora JI, Reyna-Soriano D, Bhuyan M, Zhu L, Wang K, Yuan Y (2014) Principles of bioprinting technology. In: Regenerative medicine applications in organ transplantation. Academic Press, pp 67–79

Young-Cheol KANG, Kyuboem HAN, Chun-Hyung KIM, Hong Kyu LEE, Soomin KIM, Hendra TANOTO, Yin Hua ZHANG, Zhi Jun LIN, Hui Xing CUI (2024) Composition for preservation of cells, tissues or organs, comprising isolated mitochondria, and use thereof. US20240341299

Zimmerlin L, Park TS, Zambidis ET, Donnenberg VS, Donnenberg AD (2013) Mesenchymal stem cell secretome and regenerative therapy after cancer. Biochimie 95(12):2235–2245

Zurynski Y, Herkes-Deane J, Holt J, McPherson E, Lamprell G, Dammery G, Meulenbroeks I, Halim N, Braithwaite J (2022) How can the healthcare system deliver sustainable performance? A scoping review. BMJ Open 12(5):e059207

Biomimetic Materials: Nature-Inspired Solutions for Sustainability

7

Mohammad Intakhab Alam, Md Sajid Ali, Nawazish Alam, Md. Faiyazuddin, and Sarfaraz Ahmad

Abstract

Biomimetic materials represent a transformative approach to addressing sustainability challenges by drawing inspiration from nature's time-tested solutions. These materials mimic the structures, functions, and processes of biological systems to create innovative and eco-friendly alternatives. Biomimetic structural materials, such as spider silk-inspired fibers and nacre-inspired composites, offer high strength, durability, and lightweight properties ideal for construction and engineering applications. Functional biomimetic materials, like gecko-inspired adhesives and lotus leaf-inspired self-cleaning surfaces, provide unique capabilities for industries ranging from healthcare to manufacturing. Chemical biomimetic materials, such as enzyme-inspired catalysts and photosynthesis-inspired systems, enable greener chemical processes and renewable energy solutions. Additionally, regenerative biomimetic materials, including self-healing polymers, contribute to the longevity and sustainability of products. Advances in nanotechnology have further enhanced biomimetic designs by creating nanomaterials inspired by butterfly wings and diatoms for applications in optics, filtration, and electronics. The chapter explores the principles, types, applications, and challenges of biomimetic materials, emphasizing their potential to drive sustainable development and innovation across multiple sectors. By leveraging

M. I. Alam (✉) · Md. S. Ali
Department of Pharmaceutics, College of Pharmacy, Jazan University, Jazan, Saudi Arabia
e-mail: mialam@jazanu.edu.sa

N. Alam · S. Ahmad
Department of Clinical Practice, College of Pharmacy, Jazan University, Jazan, Saudi Arabia

Md. Faiyazuddin
School of Pharmacy, Al-Karim University, Katihar, India

Md. Faiyazuddin, M. Suryawanshi (eds.), *Design and Processing of Green Materials*, Biomaterials, Bioengineering and Sustainability 4,
https://doi.org/10.1007/978-3-031-91790-5_7

nature's ingenuity, biomimetic materials offer a promising pathway to more resilient, efficient, and environmentally friendly technologies.

Keywords

Biomimetic materials · Sustainability · Bio-inspired innovation · Eco-friendly technologies · Nature-inspired design

7.1 Introduction

Biomimetic materials take inspiration from nature's evolutionary designs, replicating biological structures and processes to build new and efficient materials. These materials utilize the complex patterns created by nature over millions of years, giving sustainable and high-performance solutions (Gerola et al. 2023). Biomimetic materials span from structural and functional to chemical, each emulating distinct biological phenomena. The discipline links biological and material science, resulting to breakthroughs in numerous areas such as health, construction, and environmental engineering (Naik and Singamaneni 2017). Sustainability is a crucial consideration in material science, driven by the need to mitigate environmental impacts and conserve resources. Traditional materials often involve processes that deplete natural resources, produce significant waste, and contribute to pollution. Biomimetic materials offer a sustainable alternative by leveraging nature's efficient designs to create eco-friendly solutions. These materials are designed to minimize environmental footprint, reduce waste, and enhance the lifecycle of products (Speck and Speck 2019). By integrating sustainability into material science, biomimetic materials contribute to developing greener technologies and practices, supporting global efforts to achieve environmental sustainability and resilience.

The interdisciplinary field of biomimetics explores solutions found in nature and applies them to diverse areas such as neurophysiology, optics, electronics, and nanotechnology. Many institutions have advanced this field by fostering collaborations among biologists, engineers, and geologists to analyse natural materials' properties and responsive features, aiming to develop ecological and high-performance materials. Innovative categories like auxetic, sonomaterials, and tear-resistant materials emulate natural thermodynamic and biosensory processes to deliver sustainable applications across production, usage, and disposal phases. Biomimetic materials exemplify how integrating natural principles into material science supports global sustainability efforts and introduces groundbreaking technologies while reducing ecological impact without compromising performance (Bhushan 2009; Aizenberg and Fratzl 2009; Ciulla et al. 2023).

This chapter explores the field of biomimetic materials, detailing their principles, types, and industrial applications. The primary objective is to highlight the potential of these materials to contribute to sustainability by mimicking nature's efficient and eco-friendly designs. The review covers structural, functional, and chemical biomimetic materials, examining their unique properties and applications across various industries. Additionally, it addresses the challenges and limitations faced in the development and adoption of biomimetic materials and discusses future directions and opportunities. By providing a comprehensive overview, this review aims to foster a deeper understanding of biomimetic materials and their role in advancing sustainable material science.

7.2 Principles of Biomimicry

7.2.1 Definition and Concept of Biomimicry

Biomimicry is the practice of drawing inspiration from nature's designs and processes to solve human problems. It involves studying biological models and mimicking their forms, functions, and systems to develop innovative materials and technologies. Biomimicry emphasizes learning from nature rather than merely copying it, aiming to create sustainable and efficient solutions that align with natural principles. By understanding and emulating the strategies evolved by living organisms, biomimicry seeks to enhance human-made systems' performance, resilience, and sustainability (Aziz and El Sherif 2016).

Biomimetic materials are defined as materials inspired by natural processes, structures, and mechanisms. The term biomimetic is used interchangeably with "biomimicry" when referring to materials that blend imitation of biological and ecological systems with environmentally benign production materials and methods. For polymers, the term bio-inspired is also used when referring to synthetic polymers that mimic natural polymers or contain natural polymers. Biomimetic materials incorporate the systems and patterns of nature and are not limited to form or function. Their structures at microscopic and molecular length scales are responsible for their remarkable properties. The biological mechanisms behind these functions can be grouped into four main categories: structure-based processes, biological-based processes, structure-genetics, and structure-biochemical signaling. Focusing on the strength, self-healing, and self-cleaning properties of natural materials can lead the research into biomimetics (Lebdioui 2022; AlAli et al. 2023).

7.2.2 Historical Development and Evolution

The study of nature to derive technological principles is an ancient practice that dates back to civilizations like the Phoenicians, Etruscans, Greeks, Romans, and Chinese. These societies observed and lived in harmony with nature, drawing knowledge from plants, species, and natural processes to inform daily activities such as construction, metalwork, and ceramics. Many of these ancient applications, particularly in arts and crafts, remain relevant today. In modern times, macromolecules in polymer science have rekindled interest in ancient biomimetic practices, emphasizing their vital role in contemporary innovation. While biomimicry was indirectly present in early structures and objects, the development of new inorganic materials like concrete, steel, and glass in the twentieth century temporarily overshadowed biomimetic approaches. However, since the 1970s, biomimetic studies have been systematized, evolving into a truly interdisciplinary field (Verbrugghe et al. 2023). Historical figures like Leonardo da Vinci, who studied birds to design flying machines, exemplify early inspirations that bridged nature and technology. The twentieth century marked significant advancements in biology and material science, formalizing biomimicry as a scientific discipline. Today, biomimicry drives innovation across fields such as architecture, engineering, medicine, and environmental science, reflecting nature's ingenuity and fostering a holistic approach to addressing complex challenges (Aguilar-Planet and Peralta 2024).

7.2.3 Key Principles and Mechanisms in Nature

Nature operates on fundamental principles of efficiency, adaptability, and sustainability. Key mechanisms include hierarchical structuring, where materials are organized at multiple scales to enhance strength and functionality; self-healing, which allows organisms to repair damage and maintain functionality; and energy efficiency, where biological processes optimize resource use. Other principles involve using benign raw materials, minimizing waste through closed-loop systems, and enhancing resilience through diversity and redundancy. Understanding these principles enables the development of biomimetic materials that replicate nature's success, providing sustainable solutions for various applications (Fig. 7.1) (Sandak and Butina Ogorelec 2023; Wang et al. 2020).

7.3 Types of Biomimetic Materials

Biomimetic materials replicate natural structures, processes, and functionalities to create innovative solutions for diverse applications (Fig. 7.2). They can be broadly classified into the following categories:

7.3.1 Structural Biomimetic Materials

Structural biomimetic materials emulate the physical properties and organizational patterns found in nature to enhance strength, durability, and efficiency. Examples include materials inspired by spider silk and nacre, which offer superior mechanical properties for applications in textiles, construction, and protective coatings (Table 7.1).

7.3.1.1 Spider Silk-Inspired Fibers

Spider silk-inspired fibers are a remarkable type of biomimetic material that emulate the extraordinary mechanical properties of natural spider silk. Known for its exceptional tensile strength, elasticity, and lightweight nature, spider silk has long

Fig. 7.1 Diagrammatic presentation of different sources of biomimetic materials

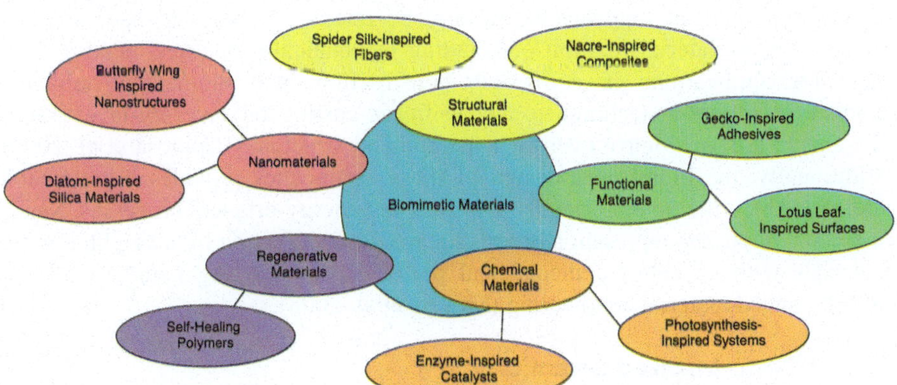

Fig. 7.2 Types of biomimetic materials

Table 7.1 Structural biomimetic materials

Material	Natural Inspiration	Properties	Applications
Spider silk-inspired fibers	Spider webs	High tensile strength, elasticity	Medical sutures, body armour
Nacre-inspired composites	Mother-of-pearl	Toughness, fracture resistance	Aerospace materials, protective coatings
Bone-inspired ceramics	Human bone	Lightweight, high strength	Implants, construction materials

fascinated scientists seeking sustainable and high-performance alternatives to synthetic fibers (Römer and Scheibel 2008; Khan et al. 2023). Spider silk is composed primarily of protein chains arranged in a hierarchical structure that combines crystalline regions for strength and amorphous regions for elasticity. Mimicking these features, synthetic spider silk is produced through techniques such as recombinant DNA technology, electrospinning, and microbial fermentation. These methods utilize engineered bacteria, yeast, or even plants to produce silk proteins, which are then spun into fibers (Rising et al. 2011). Spider silk-inspired fibers have diverse applications, ranging from biomedical sutures and tissue scaffolds to high-performance textiles and aerospace materials. The sustainable production methods and biodegradable nature of these fibers align well with the growing demand for eco-friendly materials, making them a key focus in biomimetic material science (Vollrath and Knight 2001; Hagn et al. 2011).

7.3.1.2 Nacre-Inspired Composites

Nacre-inspired composites are an exceptional class of biomimetic materials modeled after the natural structure of nacre, commonly known as mother-of-pearl. Found in mollusc shells, nacre exhibits a unique "brick-and-mortar" arrangement of aragonite platelets bonded by organic protein layers, which gives it remarkable strength, toughness, and fracture resistance (Barthelat 2010). Mimicking this structure, nacre-inspired composites are engineered using a layered assembly of ceramics and polymers to replicate the synergistic interaction between hard and soft phases. These materials exhibit outstanding mechanical properties, making them ideal for applications in aerospace, construction, and biomedical implants. Innovative fabrication techniques, such as freeze-casting and layer-by-layer assembly, enable precise control over their hierarchical structure (Espinosa et al. 2009). Additionally, the bioinspired design of these composites provides a sustainable alternative by utilizing eco-friendly materials and energy-efficient processes. Nacre-inspired composites represent a promising avenue for creating lightweight, durable materials with exceptional mechanical performance, bridging natural ingenuity with advanced material engineering (Meyers et al. 2008).

7.3.1.3 Bone-Inspired Ceramics

Bone-inspired ceramics are a remarkable type of biomimetic material designed to emulate the structure and functionality of natural bone. These materials replicate the hierarchical organization of bone, which combines a hard, mineralized outer layer with a porous, collagen-rich inner structure. Mimicking this composition, bone-inspired ceramics are often composed of hydroxyapatite, calcium phosphate, or bio-active glass, materials that closely resemble the mineral components of bone. Their porous architecture facilitates cell attachment, proliferation, and nutrient exchange, making them ideal for biomedical applications such as bone grafts, implants, and tissue engineering scaffolds. By integrating bioactive properties, these ceramics promote osteointegration, enabling the material to bond effectively with the surrounding bone tissue. Advanced fabrication techniques like 3D printing and freeze-casting are used to create these materials with controlled porosity and mechanical strength. Bone-inspired ceramics highlight the potential of biomimetic materials to merge structural functionality with biological compatibility, driving innovations in regenerative medicine and orthopaedic surgery (Hench 1991; Bose et al. 2012; Zhang and Webster 2009).

7.3.2 Functional Biomimetic Materials

Functional biomimetic materials mimic specific biological functions to achieve desired properties such as adhesion, self-cleaning, and responsiveness. These materials draw inspiration from natural phenomena like gecko feet and lotus leaves to develop innovative solutions for various applications (Table 7.2).

7.3.2.1 Gecko-Inspired Adhesives

Geckos can adhere to surfaces due to the unique microstructures on their feet. Biomimetic adhesives mimic these structures, providing strong, reversible adhesion without the need for chemical glues. These adhesives are used in medical devices, robotics, and consumer products, offering versatile and reusable bonding solutions.

Gecko-inspired adhesives are a fascinating class of biomimetic materials that replicate the remarkable adhesive capabilities of gecko feet. Geckos adhere to surfaces using millions of microscopic hair-like structures called setae, which exploit van der Waals forces to create a strong yet reversible adhesion (Autumn et al. 2002). Inspired by this mechanism, scientists have developed synthetic adhesives that mimic the hierarchical structure and functionality of gecko setae. These adhesives offer significant advantages, including exceptional grip on a variety of surfaces,

Table 7.2 Functional biomimetic materials

Material	Natural inspiration	Properties	Applications
Gecko-inspired adhesives	Gecko feet	Reusable, strong adhesion	Climbing robots, medical bandages
Lotus leaf-inspired surfaces	Lotus leaves	Superhydrophobic, self-cleaning	Self-cleaning windows, anti-fouling paint
Butterfly wing-inspired nanostructures	Butterfly wings	Iridescence, color without pigments	Anti-counterfeit technology, displays

repeatability without losing adhesion strength, and operation under dry or wet conditions (Popov et al. 2016). Gecko-inspired adhesives find applications in robotics, medical devices (such as bandages and surgical instruments), and industrial climbing or gripping tools. The development of these adhesives also contributes to sustainable practices by eliminating the need for chemical glues and adhesives, reducing environmental impact. Advances in nanotechnology and material science continue to enhance the performance and scalability of gecko-inspired adhesives, driving their potential in diverse fields (Zhou et al. 2013).

7.3.2.2 Lotus Leaf-Inspired Self-Cleaning Surfaces

Lotus leaf-inspired self-cleaning surfaces are a ground-breaking advancement in biomimetic materials, emulating the super hydrophobic properties of lotus leaves (Koch and Barthlott 2009). The surface of a lotus leaf is covered with microscopic papillae coated with a waxy substance, creating a dual-scale roughness that minimizes contact with water droplets. This unique structure enables water droplets to roll off easily, carrying dirt and contaminants with them, a phenomenon known as the "lotus effect." Synthetic self-cleaning surfaces replicate this micro- and nanoscale roughness using materials such as silicon, polymers, and coatings. These surfaces have transformative applications in fields like construction (e.g., self-cleaning windows), automotive (e.g., water-repellent coatings), and healthcare (e.g., antimicrobial surfaces). Additionally, they reduce the need for cleaning chemicals and water, making them eco-friendly solutions. Advances in fabrication techniques, such as laser etching and chemical vapor deposition, continue to improve the durability and scalability of these surfaces, enabling broader adoption across industries (Barthlott and Neinhuis 1997; Nosonovsky and Bhushan 2009).

7.3.3 Chemical Biomimetic Materials

Chemical biomimetic materials replicate natural chemical processes to create efficient and sustainable solutions. Examples include catalysts inspired by enzymes and energy systems modeled after photosynthesis, which offer eco-friendly alternatives for industrial processes and energy production (Table 7.3).

Table 7.3 Chemical biomimetic materials

Material	Natural inspiration	Mechanism	Applications
Enzyme-inspired catalysts	Enzymes	Accelerate chemical reactions	Green chemistry, pharmaceutical synthesis
Photosynthesis-inspired solar cells	Plant leaves	Convert sunlight into energy	Renewable energy, solar panels
Catalysts for CO_2 reduction	Carbon-fixing enzymes	Convert CO_2 into useful chemicals	Environmental remediation, fuel production

7.3.3.1 Catalysts Inspired by Enzymes

Catalysts inspired by enzymes, often termed biomimetic catalysts, are designed to replicate the high efficiency, specificity, and mild operating conditions of natural enzymes. Enzymes are nature's highly efficient catalysts, capable of facilitating complex biochemical reactions with remarkable precision under ambient conditions. Biomimetic catalysts emulate these properties by mimicking the active sites, structural frameworks, or catalytic mechanisms of enzymes. These synthetic catalysts often incorporate metal-organic frameworks (MOFs), biomimetic ligands, or peptide-based structures to replicate enzyme activity. For instance, heme-inspired catalysts mimic the function of cytochrome P450 enzymes in oxidation reactions, while artificial metalloenzymes emulate the catalytic roles of natural metalloproteins. These catalysts find applications in green chemistry, pharmaceuticals, and industrial processes by reducing energy consumption and waste generation. Advances in computational modeling and nanotechnology further enhance their efficiency, making them sustainable alternatives to traditional catalysts (Reetz 2013; Marchetti and Levine 2011; Turner 2009).

7.3.3.2 Photosynthesis-Inspired Energy Systems

Photosynthesis-inspired energy systems leverage the principles of natural photosynthesis to create sustainable solutions for energy conversion and storage. Mimicking the light-harvesting and energy conversion processes of chloroplasts, these systems aim to replicate the efficient capture of sunlight and its conversion into chemical energy (El-Khouly et al. 2017). Synthetic systems often incorporate artificial photoelectrochemical cells, biohybrid designs, or molecular complexes that simulate the photosynthetic reaction centers. Materials such as semiconductors paired with catalysts are employed to split water into oxygen and hydrogen, mimicking the Z-scheme electron flow in natural photosynthesis. Advances in photocatalysts and nanomaterials have significantly improved the efficiency and scalability of these systems, making them promising for renewable energy applications. Photosynthesis-inspired technologies play a pivotal role in developing artificial leaves, carbon fixation processes, and sustainable hydrogen production, paving the way for cleaner and more efficient energy solutions (Kalyanasundaram and Graetzel 2010; Barber 2009).

7.3.3.3 Catalysts for CO_2 Reduction

Catalysts for CO_2 reduction inspired by biomimetic materials mimic natural enzymatic processes to transform carbon dioxide into valuable chemicals or fuels. These catalysts emulate the efficiency and specificity of enzymes like carbonic anhydrase and RuBisCO, which play crucial roles in natural carbon fixation. Advanced biomimetic catalysts often integrate metal-organic frameworks (MOFs), nanomaterials, or synthetic molecular complexes designed to replicate the active sites of natural enzymes. For instance, catalysts based on iron, cobalt, and nickel complexes have been developed to mimic the function of metalloenzymes, enabling selective reduction of CO_2 to products like methane, methanol, or carbon monoxide. These catalysts operate under mild conditions and demonstrate enhanced reaction rates and product specificity compared to traditional approaches. By leveraging the principles of nature's catalytic mechanisms, these biomimetic catalysts hold promise for addressing climate change and promoting sustainable energy cycles through carbon capture and utilization (CCU) technologies (Xu et al. 2021; Francke et al. 2018; Qiao et al. 2014).

7.3.4 Regenerative Biomimetic Materials

Regenerative biomimetic materials are designed to emulate the natural processes of tissue repair and regeneration, offering innovative solutions for medical and biological challenges. Inspired by the self-healing properties of natural systems such as bone, skin, and plant tissues, these materials can promote cell growth, tissue integration, and repair (Liu et al. 2023). Key examples include hydrogels that mimic the extracellular matrix (ECM), enabling cell adhesion, migration, and proliferation for wound healing and cartilage repair (Geckil et al. 2010). Similarly, bioactive glasses and composites inspired by bone mineralization processes are used to stimulate bone regeneration by releasing ions that promote osteogenesis (Hench and Jones 2015). Advanced techniques like 3D bioprinting are employed to fabricate scaffolds that replicate the hierarchical structure of natural tissues, facilitating precise tissue engineering (Cui et al. 2017). Additionally, these materials often integrate bioinspired features, such as responsiveness to environmental stimuli like pH, temperature, or mechanical stress, further enhancing their functionality. The convergence of biomimicry and regenerative medicine holds immense potential for creating materials that not only repair but also restore complex biological functions, revolutionizing fields like orthopedics, dentistry, and vascular surgery (Liu et al. 2023).

7.3.4.1 Self-Healing Polymers

Self-healing polymers are innovative biomimetic materials designed to mimic the natural healing processes of living organisms, such as skin regeneration. These materials can autonomously or semi-autonomously repair damages like cracks, cuts, or tears, extending the lifecycle and reliability of the material. Inspired by biological systems that use resources like cells and proteins to repair tissue damage, self-healing polymers integrate microcapsules, vascular networks, or dynamic covalent bonds that trigger the repair process upon damage. For instance, when a crack forms, embedded microcapsules release healing agents, such as monomers or epoxies, that polymerize to seal the damage (Wang and Urban 2020; Novikov 2022).

Another approach employs reversible chemical bonds, such as hydrogen bonds, disulfide bonds, or ionic interactions, which reform when subjected to external stimuli like heat, light, or pH changes. These features are particularly advantageous in biomedical applications, such as in the design of self-healing hydrogels for wound dressings and implantable devices. Additionally, self-healing polymers have shown great potential in fields like aerospace, construction, and electronics, where durability and longevity are critical. This bioinspired innovation not only enhances material sustainability but also reduces maintenance costs and material waste (Zafeer and Bhat 2024; Sharma et al. 2019; Hia et al. 2016).

7.3.5 Bioinspired Nanomaterials

Bioinspired nanomaterials are designed with structural features ranging from 1 to 100 nanometers (nm). These features provide high surface-area-to-volume ratios, enhanced mechanical properties, and unique optical behaviours.

7.3.5.1 Butterfly Wing-Inspired Nanostructures

Butterfly wing-inspired nanostructures are remarkable biomimetic materials that replicate the intricate nanoscale patterns found on butterfly wings to achieve unique optical, structural, and functional properties. These nanostructures are designed based on the micro-ridges, multilayers, and periodic patterns that create the wings' vivid colours through structural coloration rather than pigments. This natural phenomenon, known as iridescence, results from light interference, diffraction, and scattering caused by the nanostructures (Himel et al. 2023; Potyrailo et al. 2007). In materials science, these principles have been harnessed to develop advanced nanostructured materials for diverse applications. For instance, butterfly wing-inspired photonic crystals have been used to create highly efficient sensors for detecting chemical or biological substances by analysing changes in refractive indices. Additionally, these nanostructures are being utilized in solar cells to enhance light absorption and energy efficiency by mimicking the wings' light-trapping ability. Similarly, they are applied in anti-counterfeiting measures, creating patterns and colours that are challenging to replicate (Osotsi et al. 2021). The self-cleaning and hydrophobic properties of butterfly wings, attributed to their nanoscale ridges, have also inspired innovations in water-repellent coatings. This biomimetic approach not only leads to aesthetically appealing materials but also contributes to sustainable and high-performance technologies (Bixler and Bhushan 2014; Kolle et al. 2010).

7.3.5.2 Diatom-Inspired Silica Materials

Diatom-inspired silica materials are biomimetic innovations derived from the study of diatoms (unicellular algae) that produce intricate silica-based cell walls known as frustules. These frustules exhibit unique structural properties, including high porosity, hierarchical organization, and exceptional mechanical strength, all achieved through biologically controlled silica deposition. Mimicking these natural designs has led to the creation of advanced silica materials with applications across various fields. The hierarchical porosity and large surface area of diatom-inspired silica make it ideal for catalysis, drug delivery, and biosensing. In drug delivery, these materials offer excellent biocompatibility and the ability to encapsulate and release drugs in a controlled manner. For catalysis, the porous architecture facilitates high reaction efficiency by providing access to catalytic sites. Additionally, diatom-inspired silica is utilized in water purification technologies due to its capacity for adsorbing contaminants.

The fabrication of these materials often involves bio-templating techniques or synthetic approaches inspired by diatoms' natural silica synthesis pathways (Table 7.4). Techniques like sol-gel processes have been employed to replicate the intricate designs observed in diatoms. Beyond their functional advantages,

Table 7.4 Types of biomimetic materials and their natural inspirations

Type of biomimetic material	Natural inspiration	Description	Examples
Structural biomimetic materials	Spider silk, nacre (mother-of-pearl)	Materials designed for strength, flexibility, and durability	Spider silk fibers, nacre composites
Functional biomimetic materials	Gecko feet, lotus leaves	Materials that perform specific tasks, such as adhesion or cleaning	Gecko-inspired adhesives, self-cleaning surfaces
Chemical biomimetic materials	Enzymes, photosynthesis	Materials that mimic chemical processes found in nature	Enzyme-like catalysts, solar cells
Regenerative biomimetic materials	Self-healing plants, tissues	Materials that can repair themselves when damaged	Self-healing polymers
Nanomaterials	Butterfly wings, diatom shells	Nanostructures for unique optical or filtration properties	Photonic crystals, silica nanomaterials

diatom-inspired silica materials exemplify the sustainability potential of biomimetic science by enabling eco-friendly fabrication methods and minimizing waste during synthesis (Sanchez et al. 2005; Terracciano et al. 2018).

7.4 Applications in Industry

Biomimetic materials are revolutionizing various industries by leveraging nature-inspired designs to enhance performance, efficiency, and sustainability. Below are key applications across different sectors (Tables 7.5, 7.6 and 7.7).

7.4.1 Construction and Architecture

Biomimetic materials have significant potential in construction and architecture, offering sustainable building materials and energy-efficient design solutions. These materials enhance the performance and sustainability of buildings, reducing environmental impact and promoting eco-friendly practices. The nacre-inspired composites and bone-like ceramics have improved structural durability and resilience in buildings. These materials mimic natural designs to provide lightweight yet strong alternatives for construction, reducing material use and enhancing sustainability (Tabrizian et al. 2023).

7.4.2 Sustainable Building Materials

Biomimetic materials offer sustainable alternatives to traditional construction materials. Examples include self-healing concrete, bio-inspired insulation, and

Table 7.5 Advantages of biomimetic materials for sustainability

Advantage	Description
Resource efficiency	Uses materials and designs that maximize efficiency and minimize waste
Environmental friendliness	Reduces reliance on harmful chemicals and processes
Durability and resilience	Mimics nature's ability to produce tough, long-lasting materials
Self-healing capabilities	Materials that can repair themselves, extending product lifespan
Energy efficiency	Designs inspired by nature to reduce energy consumption
Adaptability	Materials that adapt to environmental changes, reducing maintenance needs

Table 7.6 Biomimetic materials and their sustainable impact

Biomimetic material	Sustainability benefit	Example
Lotus leaf-inspired surfaces	Reduces water and chemical usage	Self-cleaning building facades
Gecko-inspired adhesives	Reusable, non-toxic	Recyclable medical tapes
Photosynthesis-inspired systems	Renewable energy generation	Solar fuel cells
Self-healing polymers	Reduces material waste and repairs damage	Long-lasting automotive parts
Nacre-inspired composites	Environmentally friendly, durable	Sustainable construction materials

Table 7.7 Applications of biomimetic materials

Industry	Biomimetic material	Application
Healthcare	Spider silk, gecko adhesives	Sutures, wound dressings
Construction	Bone-inspired ceramics, nacre composites	Strong, lightweight building materials
Energy	Photosynthesis-inspired solar cells	Renewable energy sources
Aerospace	Nacre-inspired composites	Impact-resistant materials
Textiles	Lotus leaf-inspired surfaces	Water-repellent fabrics
Environmental engineering	Diatom-inspired filters	Water purification, pollutant filtration
Electronics	Butterfly wing-inspired nanostructures	Reflective coatings, sensors

energy-efficient glazing systems, which enhance the durability, efficiency, and environmental performance of buildings.

7.4.3 Energy-Efficient Design

Biomimetic principles are applied to design energy-efficient buildings. Innovations such as passive cooling inspired by termite mounds and daylighting systems mimicking natural light diffusion reduce energy consumption and improve occupant comfort, promoting sustainability in architecture. Photosynthesis-inspired systems, such as artificial leaves, are being developed for solar energy conversion and hydrogen production. These systems mimic natural photosynthetic pathways, providing sustainable energy solutions and advancing the green energy transition (Kathpalia and Kamra Verma 2023).

7.4.4 Medical and Biomedical Applications

Biomimetic materials play a crucial role in medical and biomedical applications, providing advanced solutions for tissue engineering, regenerative medicine, and drug delivery. These materials enhance the effectiveness and safety of medical treatments, improving patient outcomes. The spider silk-inspired fibers are used for sutures and regenerative scaffolds, offering biocompatibility and high tensile strength. Additionally, self-healing polymers facilitate the repair of medical devices and implants, extending their lifespan and reducing replacement costs (Porter et al. 2013).

7.4.5 Tissue Engineering and Regenerative Medicine

Biomimetic materials are used in tissue engineering and regenerative medicine to create scaffolds that mimic the extracellular matrix, supporting cell growth and tissue regeneration. These materials are essential for developing implants, prosthetics, and treatments for damaged tissues and organs (Branković et al. 2024).

7.4.6 Drug Delivery Systems

Biomimetic materials have revolutionized drug delivery systems by mimicking natural biological mechanisms to improve therapeutic precision, safety, and efficacy. These materials facilitate targeted drug delivery by emulating cellular surfaces, such as biomimetic liposomes and exosome-like nanoparticles, enabling precise targeting of tissues or cells while evading immune detection (Peer et al. 2007; El Andaloussi et al. 2013). Inspired by controlled secretion processes in nature, biomimetic hydrogels and stimuli-responsive carriers allow for sustained and localized drug release, adapting to environmental triggers like pH, temperature, or enzymes, particularly in tumor microenvironments (Langer and Peppas 2003; Bae and Park 2011). Moreover, these materials enhance drug stability, protect sensitive molecules like proteins or nucleic acids from degradation, and enable co-delivery of multiple agents for synergistic effects in oncology and chronic disease treatments (Chen et al. 2016). Bioinspired scaffolds further support regenerative medicine by mimicking the extracellular matrix, delivering growth factors or drugs to facilitate tissue repair (Porter et al. 2013). By leveraging nature's designs, biomimetic drug delivery systems offer groundbreaking advancements, ensuring personalized, efficient, and environmentally conscious healthcare solutions.

7.4.7 Environmental Engineering

Biomimetic materials offer innovative solutions for environmental engineering, addressing challenges such as water filtration, purification, and waste management. These materials leverage nature-inspired designs to enhance the efficiency and sustainability of environmental technologies. Gecko-inspired adhesives are utilized in reusable adhesive tapes and climbing robots, while lotus leaf-inspired self-cleaning surfaces are applied in coatings for glass, textiles, and automotive exteriors to minimize maintenance and cleaning costs (Bhushan 2009).

7.4.8 Water Filtration and Purification

Nature-inspired filtration systems, such as those mimicking the structure of plant roots and animal kidneys, offer efficient and sustainable solutions for water purification and desalination. These systems improve water quality and availability, supporting environmental sustainability.

7.4.9 Waste Management Solutions

Biomimetic approaches to waste management include systems that emulate natural decomposition processes, offering sustainable methods for recycling and waste reduction. These solutions help minimize environmental impact and promote resource conservation.

7.5 Case Studies

7.5.1 Bio-Inspired Innovations in Aerospace

Biomimetic designs in aerospace engineering have led to the development of more efficient and resilient aircraft components, drawing inspiration from the flight mechanisms of birds and the structural resilience of insect wings. These innovations include advanced wing morphing technologies and lightweight composite materials inspired by avian bone structures, which enhance aerodynamic performance, reduce fuel consumption, and improve overall durability (Shyy et al. 2007). Additionally, the use of lotus-leaf-inspired self-cleaning coatings on aircraft surfaces minimizes maintenance efforts, ensuring long-term efficiency and environmental sustainability (Bhushan 2009). These bio-inspired technologies support sustainable advancements and promote eco-friendly aviation solutions.

7.5.2 Biomimetic Solutions in Automotive Industry

The automotive industry leverages biomimetic materials to develop lightweight, high-strength components and improve vehicle aerodynamics. Innovations such as shark-skin-inspired surfaces reduce drag and enhance fuel efficiency, while spider silk-inspired composites provide superior tensile strength for safer and more efficient vehicles (Bar-Cohen 2016). Furthermore, energy-efficient cooling systems modelled on termite mound ventilation and natural energy dissipation mechanisms contribute to improved sustainability in vehicle design. These biomimetic solutions pave the way for eco-friendly and high-performance transportation systems, aligning with global sustainability goals (Vincent et al. 2006).

7.5.3 Sustainable Packaging Solutions

Biomimetic materials are employed to create sustainable packaging solutions inspired by natural protective structures, such as shells, fruits, and insect cocoons. These materials mimic the strength, flexibility, and lightweight properties of biological systems, reducing material usage and environmental impact (Speck and Speck 2019). Examples include packaging films inspired by the hydrophobic properties of lotus leaves and biodegradable materials modeled after natural polymers like chitin and cellulose (Gerola et al. 2023). By improving recyclability and fostering a transition to a circular economy, biomimetic packaging supports sustainable consumption and waste reduction practices, promoting long-term environmental benefits.

7.6 Challenges and Limitations

Despite the promise of biomimetic materials in addressing contemporary challenges, their widespread adoption is hindered by several technical, economic, and ethical factors (Table 7.8). By addressing these challenges through technological innovation, cost optimization, and policy support, the field of biomimetic materials can advance toward more widespread adoption, creating a sustainable and resilient future.

7.6.1 Technical and Manufacturing Challenges

The production of biomimetic materials faces significant technical challenges, particularly in accurately replicating the intricate hierarchical structures found in nature. Advanced fabrication techniques, such as 3D printing and nanoengineering, have been developed, but these often come with high costs and technical limitations (Bhushan 2018). Scalability remains a pressing issue, as translating laboratory successes to industrial-scale production without compromising quality or functionality is complex. Innovations in materials science and process automation are essential to bridge this gap, ensuring that biomimetic materials can be produced reliably and cost-effectively at scale.

Table 7.8 Challenges and limitations of biomimetic materials

Category	Challenges and limitations	Implications
Technical challenges	Replicating complex natural structures	Requires advanced manufacturing techniques like 3D printing and nanotechnology
	Difficulty in achieving scalability in production	Limits widespread adoption and industrial applications
Material properties	Mimicking multifunctional properties of natural materials	Challenges in achieving desired strength, flexibility, and durability simultaneously
Economic feasibility	High costs associated with research and development	Impedes accessibility for small-scale industries
	Competition with traditional materials in terms of cost-efficiency	Slows market acceptance and implementation
Environmental concerns	Sustainability of sourcing bio-inspired raw materials	Potential ecological impact if sourcing disrupts natural ecosystems
	Energy-intensive manufacturing processes for biomimetic materials	Raises questions about overall environmental benefits
Regulatory and ethical issues	Lack of standardized regulations for production and application.	Slows commercialization and raises ethical concerns
	Ethical considerations in replicating natural organisms or structures	May conflict with conservation efforts and intellectual property laws.
Integration challenges	Compatibility with existing industrial systems and processes	Requires extensive retrofitting and adaptation of manufacturing setups
Performance limitations	Limited long-term performance and durability compared to synthetic alternatives	Reduces reliability for certain applications
	Lack of sufficient real-world testing for many biomimetic materials	Hinders trust and adoption in critical fields like healthcare and aerospace

7.6.2 Economic and Market Considerations

Economic feasibility and market acceptance play a pivotal role in the adoption of biomimetic materials. While their superior properties offer long-term benefits, higher initial production costs compared to conventional materials pose challenges (Speck and Speck 2019). Educating industries and consumers about the benefits of biomimetic materials, such as enhanced sustainability, durability, and performance, is critical for fostering market demand. Government subsidies and incentives for sustainable innovations could further facilitate the economic viability of biomimetic solutions in various sectors.

7.6.3 Environmental and Ethical Issues

The development and application of biomimetic materials are accompanied by environmental and ethical concerns. The sustainability of sourcing natural models must be ensured to prevent overexploitation of ecosystems. Additionally, unintended ecological impacts, such as the introduction of synthetic materials into natural habitats, must be carefully assessed (Vincent et al. 2006). Ethical considerations also arise regarding the use of biomimetic innovations inspired by biological organisms and ecosystems, particularly when intellectual property conflicts or cultural sensitivities are involved. Addressing these concerns through stringent regulations and inclusive research practices is essential for the responsible advancement of biomimetic materials.

7.6.4 Knowledge Gaps and Interdisciplinary Challenges

The limited understanding of biological systems poses a significant challenge, as fully replicating nature's complexity requires a deeper knowledge of these systems, which remains incomplete. Effective biomimetic design necessitates collaboration between biologists, engineers, material scientists, and policymakers, yet coordinating such cross-disciplinary efforts can be difficult (Gerola et al. 2023).

7.6.5 Regulatory and Standardization Issues

Regulatory frameworks for biomimetic materials are still evolving, with limited guidelines addressing safety, efficacy, and environmental impact, making approval processes more complex. Additionally, the absence of standardized testing methods for biomimetic materials hinders their commercial adoption and comparison with traditional alternatives (Chen et al. 2024).

7.6.6 Longevity and Maintenance

Durability issues also affect biomimetic materials, as they may degrade faster or require specific maintenance protocols to retain their properties, particularly in harsh conditions. Furthermore, integrating biomimetic materials seamlessly with conventional systems or technologies presents another layer of complexity (Mylo and Speck 2023).

7.7 Future Directions and Opportunities

7.7.1 Advances in Material Science and Technology

Continuous breakthroughs in material science and nanotechnology are unlocking new potentials for biomimetic materials, enabling the creation of highly functional materials that mimic nature with unprecedented precision. Innovations such as self-healing polymers, nanostructured surfaces, and responsive smart materials are paving the way for next-generation applications in healthcare, aerospace, construction, and energy storage (Bhushan 2018). The integration of artificial intelligence (AI) and machine learning into material design processes is expected to accelerate the discovery of novel biomimetic solutions tailored to specific needs, driving efficiency and sustainability.

7.7.2 Potential for Cross-Disciplinary Research

The inherently interdisciplinary nature of biomimetic materials opens pathways for collaborative research that spans diverse fields, including molecular biology, computational modeling, and advanced manufacturing (Speck and Speck 2019). For instance, insights from marine biology have inspired antifouling coatings, while robotics and bioengineering have benefited from studies on animal movement and plant mechanics. Cross-disciplinary collaborations will foster the development of multi-functional biomimetic materials that address complex challenges such as climate change, resource scarcity, and healthcare disparities.

7.7.3 Policy and Regulatory Implications

The widespread adoption of biomimetic materials necessitates robust regulatory frameworks to ensure their safety, sustainability, and ethical integrity. Policies must encourage innovation while mitigating risks associated with production, application, and end-of-life disposal. Regulatory bodies should establish clear guidelines for evaluating biomimetic materials' environmental impact and compliance with sustainability goals (Vincent et al. 2006). Incentives for industries adopting biomimetic materials, such as tax benefits or research grants, could further promote their integration into mainstream markets. Public awareness campaigns and collaborations between researchers, policymakers, and industry stakeholders will be essential in navigating the regulatory landscape and maximizing the benefits of biomimetic materials.

7.8 Conclusion

Biomimetic materials represent a groundbreaking paradigm in sustainability, bridging the wisdom of nature with modern innovation to revolutionize material science. By mimicking nature's ingenious designs, these materials not only enhance performance but also reduce environmental footprints, offering practical and eco-friendly alternatives to traditional materials. Their potential to minimize resource consumption, cut down on waste, and boost material efficiency positions them as pivotal players in addressing pressing global challenges such as climate change, resource depletion, and pollution. The integration of biomimetic materials into diverse industries is fostering the development of greener technologies and sustainable practices, supporting the global shift toward a circular economy. From healthcare and construction to energy and transportation, the applications of biomimetic materials are as diverse as they are transformative, reflecting the boundless possibilities of bio-inspired innovation. To unlock the full potential of biomimetic materials, future research must address the existing technical and economic hurdles, such as scalability, cost-efficiency, and market adoption. Additionally, a concerted effort should be made to expand their applications by exploring untapped natural models and leveraging advancements in material science, nanotechnology, and artificial intelligence. Collaborative, cross-disciplinary research will be essential in harnessing these materials for maximum societal and environmental benefit. As the field of biomimetic materials continues to evolve, it promises to redefine the intersection of technology and sustainability. By blending the principles of nature with cutting-edge science, biomimetic materials pave the way for a resilient and sustainable future, inspiring researchers and innovators worldwide to rethink and reimagine the materials of tomorrow.

References

Aguilar-Planet T, Peralta E (2024) Innovation inspired by nature: applications of biomimicry in engineering design. Biomimetics 9:523. https://doi.org/10.3390/biomimetics9090523

Aizenberg J, Fratzl P (2009) Biological and biomimetic materials. Adv Mater 21:387–388. https://doi.org/10.1002/adma.200803699

AlAli M, Mattar Y, Alzaim M, Beheiry S (2023) Applications of biomimicry in architecture, construction and civil engineering. Biomimetics 8:202. https://doi.org/10.3390/biomimetics8020202

Autumn K, Sitti M, Liang YA, Peattie AM, Hansen WR, Sponberg S, Kenny TW, Fearing R, Israelachvili JN, Full RJ (2002) Evidence for van der Waals adhesion in gecko setae. Proc Natl Acad Sci USA 99:12252–12256. https://doi.org/10.1073/pnas.192252799

Aziz MS, El Sherif AY (2016) Biomimicry as an approach for bio-inspired structure with the aid of computation. Alex Eng J 55:707–714. https://doi.org/10.1016/j.aej.2015.10.015

Bae YH, Park K (2011) Targeted drug delivery to tumors: myths, reality and possibility. J Control Release 153:198–205. https://doi.org/10.1016/j.jconrel.2011.06.001

Barber J (2009) Photosynthetic energy conversion: natural and artificial. Chem Soc Rev 38:185–196. https://doi.org/10.1039/B802262N

Bar-Cohen Y (2016) Biomimetics. CRC Press. https://doi.org/10.1201/b11230

Barthelat F (2010) Nacre from mollusk shells: a model for high-performance structural materials. Bioinspir Biomim 5:035001. https://doi.org/10.1088/1748-3182/5/3/035001

Barthlott W, Neinhuis C (1997) Purity of the sacred lotus, or escape from contamination in biological surfaces. Planta 202:1–8. https://doi.org/10.1007/s004250050096

Bhushan B (2009) Biomimetics: lessons from nature–an overview. Phil Trans R Soc A 367:1445–1486. https://doi.org/10.1098/rsta.2009.0011

Bhushan B (2018) Biomimetics: bioinspired hierarchical-structured surfaces for green science and technology, springer series in materials science. Springer International Publishing, Cham. https://doi.org/10.1007/978-3-319-71676-3

Bixler GD, Bhushan B (2014) Rice- and butterfly-wing effect inspired self-cleaning and low drag micro/nanopatterned surfaces in water, oil, and air flow. Nanoscale 6:76–96. https://doi.org/10.1039/c3nr04755e

Bose S, Roy M, Bandyopadhyay A (2012) Recent advances in bone tissue engineering scaffolds. Trends Biotechnol 30:546–554. https://doi.org/10.1016/j.tibtech.2012.07.005

Branković M, Zivic F, Grujovic N, Stojadinovic I, Milenkovic S, Kotorcevic N (2024) Review of spider silk applications in biomedical and tissue engineering. Biomimetics 9:169. https://doi.org/10.3390/biomimetics9030169

Chen G, Roy I, Yang C, Prasad PN (2016) Nanochemistry and nanomedicine for nanoparticle-based diagnostics and therapy. Chem Rev 116:2826–2885. https://doi.org/10.1021/acs.chemrev.5b00148

Chen L, Zhang Y, Chen Z, Dong Y, Jiang Y, Hua J, Liu Y, Osman AI, Farghali M, Huang L, Rooney DW, Yap P-S (2024) Biomaterials technology and policies in the building sector: a review. Environ Chem Lett 22:715–750. https://doi.org/10.1007/s10311-023-01689-w

Ciulla MG, Massironi A, Sugni M, Ensign MA, Marzorati S, Forouharshad M (2023) Recent advances in the development of biomimetic materials. Gels 9:833. https://doi.org/10.3390/gels9100833

Cui H, Nowicki M, Fisher JP, Zhang LG (2017) 3D bioprinting for organ regeneration. Adv Healthc Mater 6. https://doi.org/10.1002/adhm.201601118

El Andaloussi S, Mäger I, Breakefield XO, Wood MJA (2013) Extracellular vesicles: biology and emerging therapeutic opportunities. Nat Rev Drug Discov 12:347–357. https://doi.org/10.1038/nrd3978

El-Khouly ME, El-Mohsnawy E, Fukuzumi S (2017) Solar energy conversion: from natural to artificial photosynthesis. J Photochem Photobiol C: Photochem Rev 31:36–83. https://doi.org/10.1016/j.jphotochemrev.2017.02.001

Espinosa HD, Rim JE, Barthelat F, Buehler MJ (2009) Merger of structure and material in nacre and bone—perspectives on de novo biomimetic materials. Prog Mater Sci 54:1059–1100. https://doi.org/10.1016/j.pmatsci.2009.05.001

Francke R, Schille B, Roemelt M (2018) Homogeneously catalyzed electroreduction of carbon dioxide—methods, mechanisms, and catalysts. Chem Rev 118:4631–4701. https://doi.org/10.1021/acs.chemrev.7b00459

Geckil H, Xu F, Zhang X, Moon S, Demirci U (2010) Engineering hydrogels as extracellular matrix mimics. Nanomedicine (Lond) 5:469–484. https://doi.org/10.2217/nnm.10.12

Gerola A, Robaey Z, Blok V (2023) What does it mean to mimic nature? A typology for biomimetic design. Philos Technol 36:65. https://doi.org/10.1007/s13347-023-00665-0

Hagn F, Thamm C, Scheibel T, Kessler H (2011) pH-dependent dimerization and salt-dependent stabilization of the N-terminal domain of spider dragline silk—implications for fiber formation. Angew Chem Int Ed 50:310–313. https://doi.org/10.1002/anie.201003795

Hench LL (1991) Bioceramics: from concept to clinic. J Am Ceram Soc 74:1487–1510. https://doi.org/10.1111/j.1151-2916.1991.tb07132.x

Hench LL, Jones JR (2015) Bioactive glasses: Frontiers and challenges. Front Bioeng Biotechnol 3. https://doi.org/10.3389/fbioe.2015.00194

Hia IL, Vahedi V, Pasbakhsh P (2016) Self-healing polymer composites: prospects, challenges, and applications. Polym Rev 56:225–261. https://doi.org/10.1080/15583724.2015.1106555

Himel MH, Sikder B, Ahmed T, Choudhury SM (2023) Biomimicry in nanotechnology: a comprehensive review. Nanoscale Adv 5:596–614. https://doi.org/10.1039/D2NA00571A

Kalyanasundaram K, Graetzel M (2010) Artificial photosynthesis: biomimetic approaches to solar energy conversion and storage. Curr Opin Biotechnol 21:298–310. https://doi.org/10.1016/j.copbio.2010.03.021

Kathpalia R, Kamra Verma A (2023) Artificial photosynthesis as an alternative source of renewable energy: potential and limitations. In: Chen J-T (ed) Physiology. IntechOpen. https://doi.org/10.5772/intechopen.111501

Khan AQ, Shafiq M, Li J, Yu K, Liu Z, Zhou X, Zhu M (2023) Recent developments in artificial spider silk and functional gel fibers. SmartMat 4:e1189. https://doi.org/10.1002/smm2.1189

Koch K, Barthlott W (2009) Superhydrophobic and superhydrophilic plant surfaces: an inspiration for biomimetic materials. Phil Trans R Soc A 367:1487–1509. https://doi.org/10.1098/rsta.2009.0022

Kolle M, Salgard-Cunha PM, Scherer MRJ, Huang F, Vukusic P, Mahajan S, Baumberg JJ, Steiner U (2010) Mimicking the colourful wing scale structure of the Papilio blumei butterfly. Nature Nanotech 5:511–515. https://doi.org/10.1038/nnano.2010.101

Langer R, Peppas NA (2003) Advances in biomaterials, drug delivery, and bionanotechnology. AICHE J 49:2990–3006. https://doi.org/10.1002/aic.690491202

Lebdioui A (2022) Nature-inspired innovation policy: biomimicry as a pathway to leverage biodiversity for economic development. Ecol Econ 202:107585. https://doi.org/10.1016/j.ecolecon.2022.107585

Liu S, Yu J-M, Gan Y-C, Qiu X-Z, Gao Z-C, Wang H, Chen S-X, Xiong Y, Liu G-H, Lin S-E, McCarthy A, John JV, Wei D-X, Hou H-H (2023) Biomimetic natural biomaterials for tissue engineering and regenerative medicine: new biosynthesis methods, recent advances, and emerging applications. Military Med Res 10:16. https://doi.org/10.1186/s40779-023-00448-w

Marchetti L, Levine M (2011) Biomimetic Catalysis ACS Catal 1:1090–1118. https://doi.org/10.1021/cs200171u

Meyers MA, Chen P-Y, Lin AY-M, Seki Y (2008) Biological materials: structure and mechanical properties. Prog Mater Sci 53:1–206. https://doi.org/10.1016/j.pmatsci.2007.05.002

Mylo MD, Speck O (2023) Longevity of system functions in biology and biomimetics: a matter of robustness and resilience. Biomimetics (Basel) 8:173. https://doi.org/10.3390/biomimetics8020173

Naik RR, Singamaneni S (2017) Introduction: bioinspired and biomimetic materials. Chem Rev 117:12581–12583. https://doi.org/10.1021/acs.chemrev.7b00552

Nosonovsky M, Bhushan B (2009) Multiscale effects and capillary interactions in functional biomimetic surfaces for energy conversion and green engineering. Phil Trans R Soc A 367:1511–1539. https://doi.org/10.1098/rsta.2009.0008

Novikov AS (2022) Self-healing polymers. Polymers (Basel) 14:2261. https://doi.org/10.3390/polym14112261

Osotsi MI, Zhang W, Zada I, Gu J, Liu Q, Zhang D (2021) Butterfly wing architectures inspire sensor and energy applications. Natl Sci Rev 8, nwaa107. https://doi.org/10.1093/nsr/nwaa107

Peer D, Karp JM, Hong S, Farokhzad OC, Margalit R, Langer R (2007) Nanocarriers as an emerging platform for cancer therapy. Nature Nanotech 2:751–760. https://doi.org/10.1038/nnano.2007.387

Popov VL, Filippov AE, Gorb SN (2016) Biological microstructures with high adhesion and friction. Numerical approach. Phys-Usp 59:829–845. https://doi.org/10.3367/UFNe.2016.01.037677

Porter MM, Mckittrick J, Meyers MA (2013) Biomimetic materials by freeze casting. JOM 65:720–727. https://doi.org/10.1007/s11837-013-0606-3

Potyrailo RA, Ghiradella H, Vertiatchikh A, Dovidenko K, Cournoyer JR, Olson E (2007) Morpho butterfly wing scales demonstrate highly selective vapour response. Nature Photon 1:123–128. https://doi.org/10.1038/nphoton.2007.2

Qiao J, Liu Y, Hong F, Zhang J (2014) A review of catalysts for the electroreduction of carbon dioxide to produce low-carbon fuels. Chem Soc Rev 43:631–675. https://doi.org/10.1039/C3CS60323G

Reetz MT (2013) Biocatalysis in organic chemistry and biotechnology: past, present, and future. J Am Chem Soc 135:12480–12496. https://doi.org/10.1021/ja405051f

Rising A, Widhe M, Johansson J, Hedhammar M (2011) Spider silk proteins: recent advances in recombinant production, structure–function relationships and biomedical applications. Cell Mol Life Sci 68:169–184. https://doi.org/10.1007/s00018-010-0462-z

Römer L, Scheibel T (2008) The elaborate structure of spider silk: structure and function of a natural high performance fiber. Prion 2:154–161. https://doi.org/10.4161/pri.2.4.7490

Sanchez C, Arribart H, Giraud Guille MM (2005) Biomimetism and bioinspiration as tools for the design of innovative materials and systems. Nature Mater 4:277–288. https://doi.org/10.1038/nmat1339

Sandak A, Butina Ogorelec K (2023) Bioinspired building materials—lessons from nature. Front Mater 10:1283163. https://doi.org/10.3389/fmats.2023.1283163

Sharma S, Nandan G, Rohatgi P, Prakash R (2019) Recent advances in self-healing materials. Materials Today: Proceedings 18:4729–4737. https://doi.org/10.1016/j.matpr.2019.07.460

Shyy W, Lian Y, Tang J, Viieru D, Liu H (2007) Aerodynamics of low Reynolds number flyers, 1st edn. Cambridge University Press. https://doi.org/10.1017/CBO9780511551154

Speck O, Speck T (2019) An overview of bioinspired and biomimetic self-repairing materials. Biomimetics 4:26. https://doi.org/10.3390/biomimetics4010026

Tabrizian P, Sun H, Jargalsaikhan U, Sui T, Davis S, Su B (2023) Biomimetic nacre-like hydroxyapatite/polymer composites for bone implants. JFB 14:393. https://doi.org/10.3390/jfb14080393

Terracciano M, De Stefano L, Rea I (2018) Diatoms green nanotechnology for biosilica-based drug delivery systems. Pharmaceutics 10:242. https://doi.org/10.3390/pharmaceutics10040242

Turner NJ (2009) Directed evolution drives the next generation of biocatalysts. Nat Chem Biol 5:567–573. https://doi.org/10.1038/nchembio.203

Verbrugghe N, Rubinacci E, Khan AZ (2023) Biomimicry in architecture: a review of definitions, case studies, and design methods. Biomimetics 8:107. https://doi.org/10.3390/biomimetics8010107

Vincent JFV, Bogatyreva OA, Bogatyrev NR, Bowyer A, Pahl A-K (2006) Biomimetics: its practice and theory. J R Soc Interface 3:471–482. https://doi.org/10.1098/rsif.2006.0127

Vollrath F, Knight DP (2001) Liquid crystalline spinning of spider silk. Nature 410:541–548. https://doi.org/10.1038/35069000

Wang S, Urban MW (2020) Self-healing polymers. Nat Rev Mater 5:562–583. https://doi.org/10.1038/s41578-020-0202-4

Wang Y, Naleway SE, Wang B (2020) Biological and bioinspired materials: structure leading to functional and mechanical performance. Bioactive Materials 5:745–757. https://doi.org/10.1016/j.bioactmat.2020.06.003

Xu Y, Shrestha N, Préat V, Beloqui A (2021) An overview of in vitro, ex vivo and in vivo models for studying the transport of drugs across intestinal barriers. Adv Drug Deliv Rev 175:113795. https://doi.org/10.1016/j.addr.2021.05.005

Zafeer, Mohd K, Bhat KS (2024) Chemical approaches for fabrication of self-healing polymers. Discov Appl Sci 6:373. https://doi.org/10.1007/s42452-024-06058-y

Zhang L, Webster TJ (2009) Nanotechnology and nanomaterials: promises for improved tissue regeneration. Nano Today 4:66–80. https://doi.org/10.1016/j.nantod.2008.10.014

Zhou M, Pesika N, Zeng H, Tian Y, Israelachvili J (2013) Recent advances in gecko adhesion and friction mechanisms and development of gecko-inspired dry adhesive surfaces. Friction 1:114–129. https://doi.org/10.1007/s40544-013-0011-5

Cell-Material Interaction in Sustainable Biomedical Engineering

8

Sourav Ghosh, Amartya Sen, Abhishek Digar, Biswajit Basu, Swarnali Dutta, Swarupananda Mukherjee, and Sudarshan Singh

Abstract

This chapter provides a comprehensive analysis of the dynamic relationship between cells and their external material context, emphasizing the implications of sustainable practices in biomedical engineering. It starts with an examination of the crucial components of cellular interactions with the extracellular matrix (ECM). This includes adhesion sites, integrins, cadherins, and mechano-transduction in guiding cellular activities like adhesion, spreading, polarization, and migration. The chapter elucidates the reciprocal signalling pathways that enable communication between cells and their extracellular milieu. A notable emphasis is placed on the influence of synthetic materials on these cellular interactions, specifically how surface chemistry, including hydrophobicity, hydrophilicity, and ligand spacing, affects cellular behaviour. The discourse further investigates the significance of surface topography, particularly the implications of nanostructures on cellular guidance and the differentiation of stem cells. The mechanics of materials are analysed concerning the replication of tissue stiffness in vitro, investigating how matrix rigidity impacts cell signalling,

S. Ghosh · A. Sen · A. Digar · B. Basu (✉)
Department of Pharmaceutical Technology, School of Health and Medical Sciences, Adamas University, Kolkata, West Bengal, India

S. Dutta
Department of Pharmacology, Birla Institute of Technology, Mesra, Ranchi, Jharkhand, India

S. Mukherjee
Department of Pharmaceutical Technology, NSHM Knowledge Campus, Kolkata, West Bengal, India

S. Singh (✉)
Faculty of Pharmacy, Chiang Mai University, Chiang Mai, Thailand

Office of Research Administration, Chiang Mai University, Chiang Mai, Thailand

Md. Faiyazuddin, M. Suryawanshi (eds.), *Design and Processing of Green Materials*, Biomaterials, Bioengineering and Sustainability 4,
https://doi.org/10.1007/978-3-031-91790-5_8

receptor interactions, and stem cell differentiation. The chapter also encompasses various cellular activities reliant on interactions with the environment, such as survival, proliferation, differentiation, and protein synthesis. In conclusion, the chapter addresses the immune response elicited by implanted biomaterials, examining acute inflammation, macrophage activity, foreign body responses, and the development of a fibrotic capsule. Through a detailed analysis, the chapter emphasizes the critical importance of comprehending cell-material interactions to promote the advancement of sustainable and biocompatible materials in biomedical engineering.

Keywords

Cell-material interaction · Sustainable biomedical engineering · Inflammation · Extracellular matrix · Macrophage · Implanted biomaterial

Abbreviations

BMPR-2	Bone Morphogenetic Protein 2
BMPs	Bone Morphogenetic Proteins
BP230	Bullous Pemphigoid Antigen 1 (230 kDa)
CAMs	Cell Adhesion Molecules
ECM	Extracellular Matrix
EGFR	Epidermal Growth Factor Receptor
ES	Embryonic Stem cells
FAK	Focal Adhesion Kinase
FBGC	Foreign Body Giant Cells
FBR	Foreign Body Reaction
FGF	Fibroblast Growth Factor
GPCR	G-Protein-Coupled Receptor
IKVAV	Isoleucine-Lysine-Valine-Alanine-Valine
IL	Interleukin
MEFs	Mouse Embryonic Fibroblasts
MSCs	Mesenchymal Stem Cells
MSCs	Mesenchymal Stem Cells
PDMS	Polydimethylsiloxane
PEG	Polyethylene glycol
PNIPAAm	Poly N-isopropyl acrylamide
RGCs	Receptor Guanylate Cyclases
RGD	Arginine-Glycine-Aspartic Acid
Rho A	Ras Homolog Family Member A
ROCK	Rho-associated protein kinase
RTKs	Receptor Tyrosine Kinases
RUNX-2	Runt-Related Transcription Factor 2
SAM	Self-Assembled Monolayers
UV	Ultraviolet
VEGF	Vascular Endothelial Growth Factor

8.1 Introduction

Signals from the extracellular and intracellular environments are used by cells to regulate fundamental functions such as growth, survival, and proliferation. Cellular metabolism and the availability of nutrients have a great impact on these resolutions made by the cells. Determining the mechanism of action behind these activities may help restore cellular homeostasis in a range of diseases. A thorough grasp of how cells adhere to various textured surfaces is necessary to maximize tissue integration and reduce the production of scar tissue. Furthermore, in order to address the biophysical & biochemical needs of certain cell types as well as the physiology and anatomy of the host tissue, tissue-engineering methodologies must also construct bio-interfaces between cells and biomaterials (Feinberg et al. 2008).

Cell-ECM interaction (Fig. 8.1) is regulated by the adhesion of integrin receptors. Since these integrin-mediated adhesions serve as the anchoring to ECM that regulates cell signalling, which in turn controls activities such as cell migration or survival, they are tightly controlled in order to maintain tissue & cell homeostasis (Cox et al. 2001). Integrins mediate these junctions and some adaptor proteins like Plectin and BP230 aid in the hemidesmosomes' integrin and keratin filament fusion. Conversely, multiprotein complexes known as focal adhesion, are made up of integrins that attach either indirectly or directly to other adaptor proteins such as kindlin, talin, vinculin, and paxillin (Alberts et al. 2017). Once the cell has expanded completely to the maximum spread area, focal adhesion complexes help integrins build a stronger connection. However the different chemical or mechanical properties of ECM determine how cells behave, and elements like porosity, topography, or elasticity can influence how a cell responds to changes in these properties (Ahmad Khalili and Ahmad 2015; Van Tam et al. 2012). The crawl-like movement is called

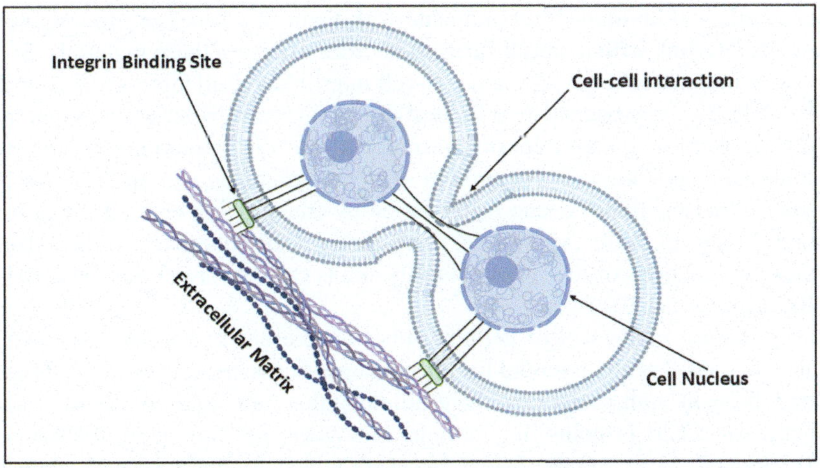

Fig. 8.1 Cell-ECM Interaction: Cell-cell interaction is done through the cytoskeleton elements and cell-ECM interact with each other through integrin binding site

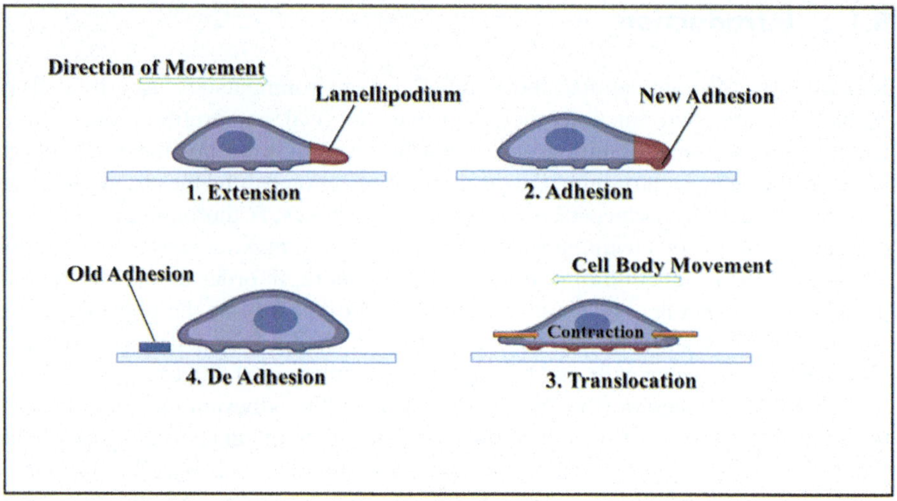

Fig. 8.2 Cell migration: cell firstly extend through lamellipodium and then new adhesion formed after that contraction happened in translocation phase and at the end de-adhesion occurs

cellular migration (Fig. 8.2), and it is mediated by the actin cytoskeleton. Together with other proteins, the actin cytoskeleton is found in the cellular cortex and is involved in both migration and cellular sculpting. Every migratory cell has to have polarity, meaning that it has a front and a rear half. If cells were not polarized, they would go in every direction (Glotzer and Hyman 1995). This polarity necessitates polarized front and rear ends of actin and tubulin filaments. Integrins play a very crucial role in the migration process because they enable lamellipodia and filopodia to adhere to the substrate in front of the cell (Lock et al. 2008).

Tissue engineering biomaterials function as synthetic extracellular matrices that sustain cells and obliquely communicate with them. It's necessary to comprehend how cells interlink with material surfaces in order to design these materials. Soluble factors, adherence to the ECM, and cell-cell interlinkages all affect cells (Bačáková et al. 2004). Two approaches exist for designing materials: [i] Substances that inhibit protein absorption or cell adhesion in order to prevent immunological reactions, thrombosis, coagulation, or the growth of microorganisms and another one is [ii] Substance that facilitates tissue regeneration by encouraging cell adhesion, migration, proliferation, differentiation, & functioning. Cell adherence to the substance mimics the functions of the ECM naturally and is crucial for cell response (Ahmad Khalili and Ahmad 2015).

Integrins are molecules of cellular adhesion present in hemidesmosomes and focal adhesions that are essential for interlinkages between cells & ECM. These are α- and β-chain transmembrane heterodimers that are able to identify certain sequences in ECM proteins like laminin, collagen, and fibronectin (Hassan et al. 2020; Barczyk et al. 2010). When integrins attach to ECM ligands, they must undergo a conformational shift from an inactive to an active state, where they split apart and reveal binding sites (Alberts et al. 2017). This process is calcium-dependent. Integrins function as mechanical sensors and receptors for bidirectional signals, connecting the cytoskeleton & the ECM. They take involvement both outside to inside signalling, which is triggered by contact with ECM components, and inside-out signalling, which is activated by talin binding to the β-chain. Numerous cellular actions such as adhesion, motility, growth, survival & proliferation, are mediated by integrins (Barczyk et al. 2010; Temenoff and Mikos 2008).

Cell adhesion molecules (CAMs), or cadherins, enable cell-to-cell attachment via desmosomes and adherens junctions. They are members of a superfamily of proteins that differ in their requirements for calcium to form stable unions according to the kind of cell. When two cadherins of the same subtype engage at their exterior domains and link to the same ligand, such as actin filaments, at their interior domains, these unions are usually homophilic and symmetric. The structure becomes stiff and able to form unions when the concentration of calcium exceeds 1 mM because it binds to the N-terminus of the cadherin's exterior domain. The structure becomes flexible and loses its ability to attach to other cadherins below 0.05 mM calcium (Alberts et al. 2017; Lodish 2008).

Integrins rapidly interact with the cytoskeleton, generating forces that facilitate their clustering and formation of focal adhesions, which are very essential for cellular motility and osteogenesis, although initial clustering does not require force, maturation does (Yu et al. 2011; Bakker et al. 2012). The maturation of focal adhesions involves a complex assembly of structural, adaptor, and signalling proteins, known as the adhesome, which consists of approximately 156 components and 690 interactions that link focal adhesions to the cytoskeleton while regulating their dynamics (Zaidel-Bar et al. 2007). As maturation progresses, these adhesions enhance their structural integrity and responsiveness to mechanical stimuli, thereby facilitating the transduction of external signals into cellular responses through integrin activation and actin regulation (Kanchanawong et al. 2010; Schoen et al. 2013).

When integrins bind to a ligand, they change shape and activate signalling proteins like FAK. This triggers mechano-transduction, leading to various cellular responses. Mechano-transduction occurs through physical and biochemical pathways, influencing cellular proliferation and differentiation (Cattaruzza et al. 2004; Miyamoto et al. 1995; Zhu and Assoian 1995). The cytoskeleton, which is anchored to adhesion sites, serves a dual function as a scaffold for solid-state biochemical reactions & a reservoir for signalling molecules. Alteration in the structure or tension of the cytoskeleton as a result of integrin signalling can modulate the release or exposure of sequestered signalling molecules, further amplifying the signal through mechanisms akin to those triggered by soluble agents such as growth factors. This process facilitates the conversion of mechanical forces into biochemical reactions within the cellular milieu (Negrutskii and Deutscher 1992).

The ECM represents a multifaceted and heterogeneous framework that envelops and provides support to cellular populations within various tissue types. It is comprised of both soluble entities, such as sequestered cytokines and growth factors, and insoluble constituents, predominantly glycoproteins and proteoglycans, which collectively establish a dense, mesh-like architecture (Rovensky 2011). The ECM plays an indispensable role in the process of tissue regeneration, as evidenced by decellularized tissues, exemplified by a trachea that was subsequently inhabited with recipient cells & successfully transplanted without necessitating administration of immunosuppressive agents (Macchiarini et al. 2008). ECM glycoproteins can be categorized into two broad classifications: structural (collagens and elastin) and adhesive. These adhesive glycoproteins, including fibronectin, tenascin, laminin, and fibulin, possess distinct binding states that engage with cell surface receptors, notably integrins. Numerous ECM components exhibit shared structural motifs, such as the arginine-glycine-aspartic acid (RGD) motif, which is critical to facilitating cell adhesion and integrin interaction. This particular motif is present in various molecules of ECM, encompassing vitronectin, fibronectin, and tenascin (Rovensky 2011).

Bidirectional communication between cells & ECM is facilitated by cell adhesion sites. This signalling is essential for many different biological processes. Because mechanical signals from the ECM are translated into intracellular signals that affect cell activity, "outside-in signalling"-the relaying of signals from the ECM to cell is crucial to maintain cell viability. "Inside-out signalling", in which cells communicate with the ECM, is critical for maintaining the ECM, as shown in the formation of fibronectin is secreted and attaches itself to activated integrins to induce strain via the actin cytoskeleton. When fibronectin is stretched, binding sites are exposed, enabling more fibronectin molecules to attach and causing fibril development, also known as fibrillogenesis. Additionally, by "pulling" on the ECM's structural elements, like bone, cells may change the ECM (Baneyx et al. 2002; Mao and Schwarzbauer 2005).

It is the interactions in cells and ECM that tissue engineer must comprehend & leverage when formulating artificial materials intended to substitute for the ECM. It is critical to acknowledge that cells typically do not engage in direct interactions with artificial materials. Whether in-vitro/in-vivo, upon encountering artificial surfaces, a branch of proteins derived from the culture media/plasma, respectively rapidly coheres with the surface in milliseconds (Atala 2012). The phenomenon of protein uptake is influenced by the characteristics of the material, including properties like surface charge, wettability (specifically hydrophobicity), & topographical features. It is this taken-up protein layer that is believed to modulate cellular interlinkages with the underlying material (Wilson et al. 2005).

8.2 Surface Chemistry

For cells to connect and spread, the chemistry at the contract between the substance and the cell is essential. Because integrins identify certain motifs in ECM proteins, such as RGD and IKVAV, the surface chemistry of a substance affects how effectively cells attach. Both the material's natural chemical composition and any further changes applied to it are included in the surface chemistry. These elements either reveal or corneal chemical or peptide signals, which influence the way sticky proteins deposit on the surface and affect cell adhesion. The role that various chemical cues play in processes like integrin clustering, stem cell differentiation, and cell adhesion are being investigated using cutting-edge methods such as chemical surface patterning.

Hydrophobicity and Hydrophilicity Surfaces can be classified as hydrophobic or hydrophilic based on their characteristics. Hydrophobic surfaces repel water, causing it to form spherical droplets with contact angles greater than 90°. On the other hand, hydrophilic surfaces facilitate water spreading, resulting in contact angles less than 90°. Super-hydrophobic surfaces exhibit extreme hydrophobicity with contact angles exceeding 150°.

For cell culture purposes, hydrophobic surfaces are generally not ideal. However, surface chemistry can be modified to increase hydrophilicity, for example, through plasma treatment to introduce polar functional groups onto the surface (Tan et al. 2010). Nonetheless, this modification may be transient, as certain materials, such as polydimethylsiloxane (PDMS), have the propensity to revert to a hydrophobic state over time. The enhancement of surface hydrophilicity fosters protein adsorption, a phenomenon that is essential for the extension & spreading of cells in tissue culture systems. Within laboratory environments, protein sourced from fetal bovine serum present in culture medium typically adheres to the surface, thereby enabling cellular attachment and proliferation through integrin-mediated signalling pathways.

Patterning Using Surface Chemistry Surface chemistry in laboratory settings, selectively regulates cell attachment on materials, forming zones for adhesion control. Poly (ethylene glycol) (PEG) is utilized for its non-fouling characteristics, inhibiting protein and cell adhesion. This technique is applied to medical device coatings to mitigate pathogen accumulation. It experiences rapid degradation. PEG derivatives, termed "PEG grafts", are under development to enhance the durability of these chemistries (Chen 1999). In mammalian cell studies, integrating PEGylated and adhesive protein-coated areas enables selective cell adhesion zones, revealing that confined cells can undergo apoptosis due to limited spreading and integrin-mediated signals. Surface chemistry also plays the main role in determining stem cells. For example, confined mesenchymal stem cells (MSCs) exhibit adipogenic tendencies in small adhesive squares, while larger squares favour osteogenesis, linked to alterations in cell shape and cytoskeletal tension mediated by Rho A and ROCK (McBeath et al. 2004). MSCs on pro-adhesive surfaces of varied shapes, flanked by PEG-treated non-adhesive zones, indicated that angular configurations heightened cytoskeletal tension, promoting osteogenesis, whereas rounded configurations diminished tension, favouring adipogenesis (Kilian et al. 2010).

Effect of Chemistry on Cell Behaviour Research indicates that surface chemistry on planar glass influences MSC lineage differentiation, with -OH promoting chondrogenesis, -NH2 aiding osteogenesis, and -CH3 maintaining stem cell markers (Curran et al. 2005). Additionally, alterations in -CH3 chain length and side chains can further influence MSC characteristics, suggesting specific -CH3 presentations may enhance MSC selection and retention from bone marrow (Curran et al. 2011).

Investigations of hydrogels with basic chemistries reveal that phosphate fosters osteogenesis, while hydrophobic t-butyl groups stimulate adipogenesis (Benoit et al. 2008). The creation of substrates in tailored surface chemistry for regulating stem cell elements could significantly impact stem cell availability for therapeutic uses.

Ligand Spacing The engagement of multiple integrins is crucial for receptor clustering and subsequent integrin-mediated signalling, influenced by the density & spacing of adhesive ligands on surfaces. Implementation of precisely controlled surface chemistries, like self-assembled monolayers, has facilitated a more thorough examination of this phenomenon. Research utilizing gold nanoparticles SAMs with RGD ligands indicated that a maximum inter-feature density of 70 nm is required for integrin clustering across various cell types (Arnold et al. 2008). Furthermore, dip-pen nanolithography has demonstrated that the arrangement and spacing of chemical modifications can significantly influence stem-cell responses (Curran et al. 2010).

Dynamic Chemistry Surface chemistries that change their states in response to light (DeForest and Anseth 2012), temperature (Da Silva et al. 2007), and electrical potential (Ng et al. 2012) have been designed to precisely regulate stem-cell fate and migration, among other aspects of cell activity. Some of these surfaces are intended to be reversible, providing an enhanced platform for study, even though many of them change states irrevocably. In one method, a "caging" group from an enzyme substrate was removed by UV radiation, allowing for targeted patterning in 3D hydrogels and regulating stem cell motility within the gel. Another technique reduced intracellular stress and promoted chondrogenesis by using light to photocleave adhesive-RGD sequences from hydrogels (Mosiewicz et al. 2013). Additionally, dynamic chemistries have been designed to change from sticky to non-adhesive states in reaction to outside stimuli. For instance, regulated cell migration was made possible by the application of 'click' chemistry to reveal or hide adhesive-RGD sites dependent on electrical potential. Cell sheets can separate thanks to thermo-responsive polymers like poly(N-isopropylacrylamide) (PNIPAAm) whose surface chemistry changes with temperature from hydrophilic to hydrophobic (Ng et al. 2012; Salinas and Anseth 2008). These dynamic surfaces can modulate cell adhesion and detachment, which is useful for harvesting cells for therapeutic applications.

8.3 Surface Topography

Topography is the physical motifs of the material surface. Topography is a helpful technique for altering cell responses because it offers a well-defined and regulated stimulus. Different pattern types with predetermined sizes may be made for the non-invasive control of cellular structure, and consequently of gene expression and cell activity. Because of these qualities, topography is a valuable tool for both translation and basic biological research.

Differential Cell Guidance by Nanostructures For more than a century, the idea of contact guidance in which cells align along topographical features has been understood (Harrison 1911). After beginning with micron-sized characteristics, the focus of recent studies has switched to comprehending how nanoscale topographies affect cellular activity. It has been discovered that nano-topographical characteristics, which are present in both the cells & ECM that surround them, affect a number of biological activities, including morphology, adhesion, gene expression, and proliferation. Research has demonstrated that distinct nanoscale characteristics, such as pillars, pits, and grooves, elicit distinct responses in cells. As an example, the growth of MSCs, osteoblasts, and fibroblasts on nanopit arrays differs from growth on flat surfaces in terms of adherence and morphological alterations. Cell behaviour is also influenced by the degree of order in nanopit patterns; more ordered patterns result in smaller adhesions and less cell spreading (Biggs et al. 2007).

Cell adhesion and proliferation are highly influenced by the height and diameter of nanopillars (Sjöström et al. 2013). While higher nanopillars tend to decrease cell adherence, shorter ones likely to increase it. Comparably, pillars with a smaller diameter promote adhesion but inhibit proliferation, while pillars with a greater diameter have the opposite effect. One of the topographies that has been investigated the most is nanogrooves. Studies have revealed that different cell types, such as neural, fibroblast, and osteoblast, prefer to align along the grooves and take on an elongated shape (Milner and Siedlecki 2007; Biela et al. 2009).

Nano Topography and Stem-Cell Differentiation Studying Embryonic stem (ES) cells on material surfaces is complex due to the need for a feeder layer, typically composed of matrigel or mouse embryonic fibroblasts (MEFs). This feeder layer may mask how nano topography affects ES cells (McFarlin et al. 2006). Nonetheless, studies have demonstrated that topography affects ES cell self-renewal and differentiation. For example, ES cells produced on feeder layer-containing grooved substrates exhibited less diffraction than those grown on flat surfaces. ES cells on nano grooves showed better spontaneous differentiation in the absence of the feeder layer (Kingham et al. 2013).

On the other hand, the response of MSCs to nano topography has been the subject of much research. On disordered nanopit topographies, MSCs have demonstrated improved osteogenic diffraction, supporting elevated intracellular tension and biochemical signalling. Furthermore, MSCs have demonstrated the ability to differentiate neurally on nano-grooved surfaces, especially when certain growth factors are present. However, additional validation is needed to ensure that this neuronal differentiation produces functional results. These results demonstrate how crucial nano topography is in controlling the activity of stem cells and how it may be used in tissue engineering to improve bone-implant integration in orthopaedic devices, among other uses (Tsimbouri et al. 2012).

8.4 Material Mechanics

Tissues and organs in mammals display considerable heterogeneity in composition, which is vital for their structural and functional efficacy. This heterogeneity, characterized by variations in cell types, extracellular matrix, and mechanical properties such as Young's modulus, significantly impacts cellular behavior and tissue functionality throughout development.

Mimicking Tissue Stiffness In Vitro It is important to replicate tissue stiffness in vitro in order to understand how the rigidity of the extracellular matrix affects tissue integrity and cellular signalling. The way cells perceive and respond to mechanical stimuli from their environment significantly influences their physiological behavior in laboratory settings (Discher et al. 2005). To achieve precision and reproducibility in creating substrates with various stiffnesses, researchers utilize synthetic polymers like polyacrylamide, polydimethylsiloxane, and polylactic acid, which can be finely tuned by adjusting cross-linking parameters during synthesis. However, these synthetic materials often lack biocompatibility, prompting the use of surface coatings of biologically relevant proteins, while naturally derived substrates like collagen, despite their biological relevance, present challenges in achieving desired precision due to production variability (Elkin et al. 2007).

Cell Behaviour and Matrix Mechanics The variations in tissue elasticity, observed within and across different kinds of tissues, show that the mechanical properties of the material used for cell cultures in a laboratory setting can result in different reactions from various types of cells. Therefore, using traditional tissue culture plastic may influence the behaviour of cells in ways that are different from what would occur in a natural biological environment. In mature cells, the stiffness of the substrate plays a critical role in influencing cell viability and proliferation rates. For example, neural cells maintained in vitro on comparatively soft substrates (within 100 to 500 Pa) exhibit greater viability than those grown on more rigid surfaces (Flanagan et al. 2002). In contrast, liver hepatocytes and stellate cells demonstrate enhanced proliferation rates on rigid substrates and enter a quiescent state or experience growth arrest when cultured on softer surfaces (Wells 2008).

How cells sense and react to the mechanical characteristics of the matrix are not completely deterministic. There is some overlap in how cells interpret these mechanical signals, suggesting that the influence of the substrate on cell behavior is not clearly defined. For instance, cell types such as neutrophils, glial cells, and myocardial cells can grow and thrive on a wide range of substrate stiffness levels. However, myocytes exhibit optimal functional performance when the substrate stiffness closely matches that of native cardiac tissue (Wells 2008). The improved function is linked to the physical connection between the cell's mechanosensory adhesions and the surrounding matrix. This is similar to how enzymes become more active at normal body temperature.

Cell mobility plays a crucial role in various body processes, including the immune response, wound healing, and the development of body structures. When cells are grown on surfaces that have the same chemical makeup but are different in stiffness, they show different levels of movement. This highlights the importance of sensing stiffness in directing cell movement, a process called mechano-axis. For instance, in fibroblast cells, the direction of movement is affected by how deformable the surface is, with cells preferring to move from softer to stiffer areas (Saha et al. 2008).

Cell Signalling Cell signalling converts extracellular signals into cellular responses via signal transduction, comprising signal reception, transduction, and response. Intermediate steps may involve intracellular signalling proteins, ligands, or signalling molecules, initiate transduction, and can function locally or distantly. Cellular signalling is categorized into four types based on the signalling molecule's reach: (1) Endocrine: Hormones enter the bloodstream to act on distant cells. (2) Paracrine: Neurotransmitters influence nearby cells. (3) Autocrine: The signalling and target cell are identical, involving cytokines. (4) Contact-dependent: This requires direct cell contact and involves membrane proteins. Signal reception occurs at the cell surface or cytoplasm, influenced by the signalling molecule's solubility. Hydrophilic molecules bind to plasma membrane receptors, whereas hydrophobic molecules, like steroids, penetrate the membrane to bind to cytosolic receptors (Cortese et al. 2009).

Receptor Types In case of cell signalling, receptors are proteins that identify and respond to ligands. There are various types of transmembrane receptors (Table 8.1).

Ion channel receptor: ionotropic receptors, also referred to as ligand-gated ion channels, consist of transmembrane proteins that form pores allowing ion passage and ligand binding to their extracellular domains, exemplified by the nicotinic acetylcholine receptor which facilitates K+ and Na+ exchange to initiate action potentials.

G-protein-coupled receptor: GPCRs are prevalent cell surface receptors crucial for many physiological functions. Their activity is initiated by ligand binding, leading to a conformational alteration that facilitates interaction with G proteins. The resultant downstream effects vary based on the specific G protein and effector protein, influencing enzyme activity and ion channel regulation. Notable examples include opioid receptors, β-adrenergic receptors, and muscarinic acetylcholine receptors.

Table 8.1 Different receptor classes, their locations, and examples

Receptor class	Location	Example of receptors	Reference
Ion channel	Cell membrane	Nicotinic acetylcholine, glutamate-gated, purine	Gibson et al. (2023)
G-protein coupled	Cell membrane	Histamine, opioid receptor, serotonin	Mangelsdorf et al. (1995)
Catalytic	Cell membrane	EGFR, FGF, VEGF, Insulin	Miller and Lappin (2020)
Nuclear	Cytoplasmic or nuclear	Thyroid hormone, androgen, estrogen retinoic acid	Mangelsdorf et al. (1995)

Catalytic receptor: catalytic or enzyme-linked receptors consist of transmembrane proteins featuring an extracellular ligand-binding domain and an intracellular enzymatic domain, initiating a cellular biochemical reaction upon ligand binding, with receptor tyrosine kinases (RTKs) and receptor guanylate cyclases (RGCs) being prominent subtypes exemplified by insulin and epidermal growth factor receptors (EGFR).

Nuclear receptor: Nuclear receptors have a conserved DNA-binding domain with zinc fingers and a ligand-binding domain that activates transcription upon hormone binding. Though their evolutionary origins are unclear, their structure enables precise physiological control. Their success led evolution to favor them as key regulators of organ function and insect development over other signaling pathways.

Stem Cell Differentiation and Substrate Stiffness Early development depends critically on the mechanical characteristics of the ECM and the interactions between pluripotent cells. During organogenesis, the differentiation processes of embryonic stem (ES) cells are influenced by the mechanical properties of the ECM. Experimental investigations have shown that aligning substrate stiffness with native tissue types effectively steers stem cell differentiation (Evans et al. 2009). When it comes to controlling stem cell differentiation, matrix mechanics is more important than chemical cues. This maintenance requires mechano-sensing and responses facilitated by focal adhesions, which generate intracellular cytoskeletal tension. This tension, influenced by substrate resistance, affects physical characteristics like adhesion strength and cell shape. In stem cells, these changes caused by cytoskeletal contractility impact signalling pathways that regulate gene expression and differentiation (Wells 2008).

For example, stem cells that round out and become tiny on softer substrates express proteins related to adipose and low-tensile neuronal cell types. Similarly, increased expression of RUNX-2 (runt-related transcription factor 2), an essential initiator of the osteogenic phenotype, and BMPR-2 (bone morphogenetic protein 2), which are critical for early-stage signalling and the development of osteoblasts, was seen in mesenchymal stem cells (MSCs) grown on substrates with a young's modulus similar to nonmineralized bone (40 kPa) (Choquet et al. 1997).

8.5 Cellular Functions Dependent on Cell-Environment Interaction

The interplay between cells and their extracellular milieu constitutes a multifaceted phenomenon that is integral to numerous cellular activities. The dynamics of cell-environment interactions are pivotal in shaping cellular processes such as viability, growth, differentiation, and protein production. Developing biomaterials that can support these tasks requires a thorough understanding of these interactions. This section elucidates how these interactions affect cell viability, growth, differentiation, and protein production.

Cell Survival The biocompatibility of materials used in biomedical applications is evaluated in large part by looking at the capacity of cells to survive in a certain environment. Biomaterials resembling the natural extracellular matrix (ECM) enhance cell survival by providing essential biochemical cues and physical support. Surface modifications, including the strategic incorporation of bioactive peptides, can significantly enhance cell adhesion and overall survival rates. Biomaterials with nano-scale features that mimic the natural ECM architecture improve cell survival by fostering favourable interactions at the cell-material interface. Understanding the interplay between structural and biochemical properties of biomaterials is vital for developing new materials for medical use. Advancements in biomaterials that closely replicate the natural ECM show promise in regenerative medicine and tissue engineering, with the potential to enhance patient outcomes (Zhu et al. 2020).

Cell Proliferation Cell proliferation, the process of cells dividing and multiplying, is vital for tissue regeneration. The characteristics of material surfaces influence this process. Micro-patterned biomaterials increase cell proliferation by guiding cell alignment and spreading. Controlled release of growth factors further enhances proliferation, aiding tissue regeneration. Optimizing surface design and incorporating bioactive molecules are crucial for biomedical applications, such as tissue engineering and regenerative medicine. Ongoing refinement of biomaterial properties is essential for improving clinical outcomes in regenerative therapies (Chen et al. 2019).

Cell Differentiation Tissue engineering is a multidisciplinary field that aims to replace or repair damaged tissues and organs through the use of engineered biomaterials. One of the key applications of biomaterials is their ability to direct and influence the process of cell differentiation. Recent studies have shown that biomaterials with mechanical properties that can be tuned, particularly those with levels of stiffness that can be adjusted, can have a substantial impact on stem cell differentiation pathways, directing them toward particular lineages like osteogenic, which is related to the formation of bones, or chondrogenic, which is related to the development of cartilage (Bose et al. 2012) (Fig. 8.3). Apart from its mechanical characteristics, the biomaterial's surface chemistry, which includes certain functional groups on its surface, can offer vital biochemical cues or signals that efficiently direct and regulate the cell's differentiation process. For instance, studies have demonstrated that biomaterials functionalized with bone morphogenetic proteins (BMPs), a class of growth factors recognized for their function in bone formation, can, in fact, effectively induce the differentiation of mesenchymal stem cells into osteoblasts—the cells responsible for bone formation—and so support bone tissue regeneration. The complex interactions among mechanical characteristics, surface chemistry, and cellular responses highlight the significance of creating biomaterials that are both biocompatible and bioactive, enabling the targeted cellular processes necessary for successful tissue engineering results. Therefore, a thorough comprehension of these elements is essential for the creation of sophisticated biomaterials that can maximize tissue regeneration and repair, which will ultimately result in enhanced therapeutic approaches in the field of regenerative medicine (Shi et al. 2022).

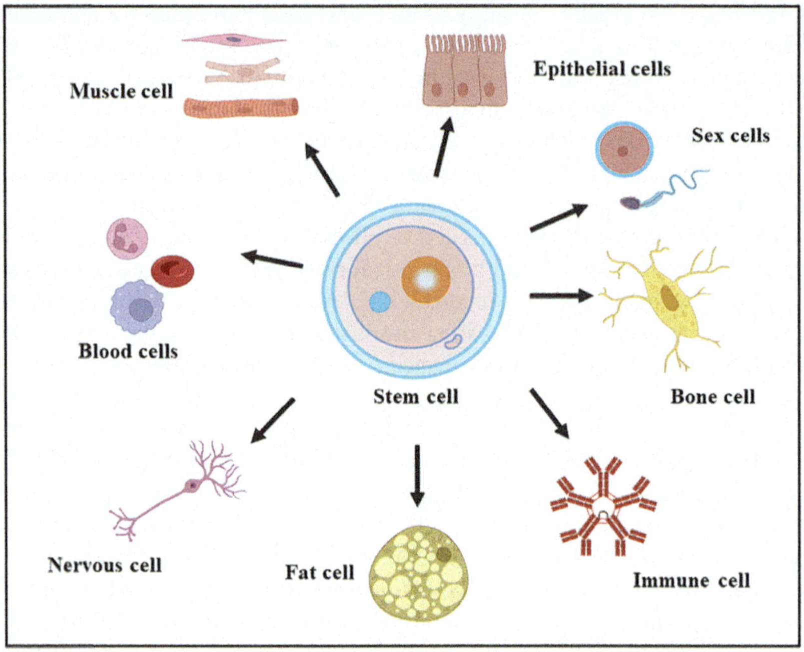

Fig. 8.3 Differentiation of stem cells into different cells

Protein Synthesis the complicated creation of different types of tissues within living organisms, as well as the preservation of cellular activities, are greatly influenced by the process of protein synthesis. The surface properties of the material, in particular, properties like electrical charge and degree of hydrophilicity, have been shown to have a significant impact on protein adsorption and the subsequent behavior that cells display in response to these materials. Recent empirical research has revealed that surfaces exhibiting hydrophilic properties combined with a moderate negative charge can significantly enhance the adsorption of fibronectin, which is recognized as a crucial ECM protein, and this enhancement subsequently facilitates improved cell adhesion and promotes the process of protein synthesis (Klimek and Ginalska 2020). Furthermore, it has been established that the controlled release of signalling molecules from the biomaterials can effectively modulate the pathways involved in protein synthesis, ultimately resulting in enhanced integration of tissue within the surrounding biological environment. The complex relationship between surface characteristics and the biological reactions they cause highlights the significance of material design in tissue engineering and regenerative medicine. Thus, developing treatment approaches targeted at enhancing tissue healing and regeneration requires a thorough understanding of these pathways (Patel et al. 2019).

8.6 Immunological Response to Implanted Biomaterial

The immune system's response to the integration of implanted biomaterials is an essential determinant that can significantly influence whether a biomedical device achieves its intended functionality or instead encounters a failure that compromises its efficacy. The characteristics and specific attributes of the biomaterial in question, alongside its dynamic interaction with the surrounding host tissue, play a pivotal role in shaping both the intensity and the nature of this immunological response. In this section, we will delve into a comprehensive exploration of the critical elements that constitute the immunological response, which encompasses a variety of phenomena such as acute inflammation, the specific responses of macrophages, the overarching foreign body reaction, the subsequent formation of a fibrotic capsule, and the associated risks pertaining to the process of cornification. Through this analysis, we aim to elucidate the complex interplay between biomaterials and the immune system, thereby enhancing our understanding of how these interactions can ultimately dictate the clinical outcomes of biomedicals.

Acute Inflammation Acute inflammation represents the body's primary and immediate reaction to the introduction of an implanted biomaterial, a physiological process that is notably marked by the activation and subsequent recruitment of various immune cells, including but not limited to neutrophils and macrophages, which play critical roles in the inflammatory response. Research conducted in this domain has revealed that the intensity and severity of acute inflammation are significantly influenced by specific characteristics of the biomaterial, particularly its surface roughness and the presence of any impurities, both of which can dramatically alter the biological response. It has been established through these studies that biomaterials exhibiting smoother surface textures and higher levels of purity are more likely to provoke a less severe inflammatory response, which is advantageous as it not only promotes the healing process but also minimizes the potential for the development of chronic inflammation, a condition that can complicate recovery. Furthermore, it is noteworthy that certain biomaterials possess the capacity to release anti-inflammatory agents into the surrounding tissues, and this property has been demonstrated to effectively mitigate the acute inflammatory response, thereby contributing to improved clinical outcomes and overall patient health. The implications of these findings underscore the importance of material selection in the design of biomaterials for medical applications, as they can significantly influence the body's response and the subsequent success of implantation. In conclusion, understanding the intricate relationship between the properties of biomaterials and the resultant inflammatory response is crucial for advancing the field of biomedical engineering and enhancing therapeutic strategies (Nakkala et al. 2021).

Macrophage Response Immune system constituents, macrophages, play a pivotal role in coordinating the immune response in response to a variety of biomaterials that are incorporated into the body. Based on the particular biochemical and physical cues that these adaptable immune cells obtain from the biomaterial in question as well as from the surrounding tissue environment, they can polarize into two distinct phenotypes: the pro-inflammatory M1 phenotype, which is linked to the initiation of inflammatory responses, and the anti-inflammatory M2 phenotype, which is associated with the resolution of inflammation and tissue repair.Recent scholarly investigations have underscored the significance of biomaterials that are meticulously engineered to promote the M2 phenotype, as such materials have the potential to significantly enhance processes related to tissue repair and integration following injury or surgical intervention. For instance, it has been shown that biomaterials with the ability to release particular cytokines, like interleukin-4 (IL-4) and interleukin-10 (IL-10), can effectively stimulate M2 polarization in macrophages. This, in turn, reduces chronic inflammation and creates an environment that is favorable for effective wound healing. Furthermore, the strategic design and fabrication of these biomaterials hold promise for improving clinical outcomes by not only mitigating adverse inflammatory responses but also by actively promoting the regeneration of damaged tissues. Thus, it is critical to continue investigating the complex relationships that exist between macrophages and biomaterials in the fields of tissue engineering and regenerative medicine, since this could lead to the development of novel treatment approaches that leverage the body's natural healing processes (Mamilos et al. 2023; Park and Kim 2024).

Foreign Body Reaction The phenomenon known as the foreign body reaction (FBR), which is defined by the intricate process of foreign body giant cell (FBGC) production, serves as a critical aspect that influences the overall long-term efficacy and success of an implant within a biological system.the phenomenon known as the foreign body reaction (FBR), which is defined by the intricate process of foreign body giant cell (FBGC) production, serves as a critical aspect that influences the overall long-term efficacy and success of an implant within a biological system. A plethora of research studies have demonstrated that various parameters, including the dimensions, geometric configuration, and surface chemical properties of the material in question, exert a substantial influence on the magnitude and severity of the foreign body reaction that ensues. It has been observed that materials possessing smaller dimensions and smoother surfaces tend to elicit a significantly reduced likelihood of triggering a pronounced foreign body reaction, whereas materials that are characterized by larger sizes and rougher textures may give rise to an increased propensity for the development of foreign body giant cells, ultimately leading to a phenomenon known as encapsulation. In light of these findings, several strategic approaches have been proposed and investigated in an effort to mitigate the adverse effects associated with the foreign body reaction, one of which includes the application of specialized coatings that serve to diminish the adsorption of proteins onto the material surface. Furthermore, the incorporation of bioactive agents that have the capacity to modulate and influence the immune response has emerged as another promising strategy aimed at alleviating the detrimental consequences of the foreign body reaction on implant success. Overall, the interplay of material characteristics and immune responses underscores the complexity of the foreign body reaction and its critical role in the field of biomaterials and implantology (Yadav and Bachhuka 2023).

Formation of Fibrotic Capsule One common and frequently anticipated result of foreign body reaction (FBR) is the formation of a fibrotic capsule around an implanted biomaterial, which can seriously impair the material's effectiveness and functionality. Research has demonstrated that the characteristics of the fibrotic capsule, including its thickness and compositional makeup, are intricately linked to the surface properties of the biomaterial itself as well as to the nature of the local immune response elicited upon implantation. Biomaterials engineered to release anti-fibrotic agents or designed with specific surface modifications that aim to attenuate immune activation have exhibited considerable potential in alleviating the degree of fibrotic encapsulation that occurs in response to their presence (Kim et al. 2023). Furthermore, the strategic employment of biodegradable materials, which are specifically formulated to degrade prior to the onset of a significant fibrotic response, emerges as an effective and innovative approach to circumventing the long-term complications that may arise from the implantation of foreign materials in biological systems. The intricate comprehension of the interplay between biomaterials and the physiological milieu highlights the need of progressing biomaterial design to minimize unfavorable reactions and augment biocompatibility. Ultimately, these insights not only contribute to the field of biomaterials science but also pave the way for more effective therapeutic interventions in regenerative medicine (Mamidi et al. 2022).

Cornification of the Inflammation Chronic inflammation surrounding an implanted biomaterial can result in a pathological condition known as cornification, a process characterized by the hardening and keratinization of the surrounding tissue, which can ultimately compromise the functionality of the implant. This occurrence presents significant challenges, particularly in applications involving soft tissues where inherent flexibility is not only desirable but often essential for maintaining normal physiological function and preventing adverse outcomes (Mehrabi et al. 2020). Research has indicated that the likelihood of cornification is markedly increased in association with biomaterials that provoke prolonged inflammatory responses, especially those that possess persistent impurities or exhibit suboptimal biocompatibility with the host tissue. Consequently, the effective prevention of cornification is heavily dependent on the strategic design and selection of biomaterials that are specifically engineered to minimize the incidence of chronic inflammation, alongside the implementation of adjunctive therapies that are both anti-inflammatory and anti-fibrotic in nature. It is feasible to improve implanted materials' performance and integration while lowering the possibility of unfavorable tissue reactions by taking care of these important parameters. Therefore, the development of safer and more efficient biomedical implants depends on a thorough understanding of the interaction between biomaterial properties and the biological response they elicit.

8.7 Conclusion

Cell-material interactions represent a highly intricate and multifarious domain within the discipline of biomedical engineering, which plays a pivotal role in determining the efficacy and overall success of various biomaterials utilized in a wide array of medical applications and interventions. Through a comprehensive understanding of the intricate mechanisms that dictate these interactions, researchers and scientists can strategically design and develop sustainable biomaterials that not only foster and encourage desirable cellular behaviours but also effectively mitigate and minimize any potential adverse immune responses that may arise. This profound and critical body of knowledge is essential for propelling the advancement of the field of sustainable biomedical engineering, thereby enhancing the overall effectiveness and outcomes of diverse medical treatments that rely on these materials. Consequently, the exploration of cell-material interactions is not merely an academic pursuit but rather a fundamental aspect of improving patient care and therapeutic efficacy in contemporary medicine.

References

Ahmad Khalili A, Ahmad MR (2015) A review of cell adhesion studies for biomedical and biological applications. Int J Mol Sci 16(8):18149–18184
Alberts B et al (2017) Biologia molecular da célula. Artmed Editora
Arnold M et al (2008) Induction of cell polarization and migration by a gradient of nanoscale variations in adhesive ligand spacing. Nano Lett 8(7):2063–2069
Atala A (2012) Regenerative medicine strategies. J Pediatr Surg 47(1):17–28
Bačáková L et al (2004) Cell adhesion on artificial materials for tissue engineering. Physiol Res 53(Suppl 1):S35–S45
Bakker GJ et al (2012) Lateral mobility of individual integrin nanoclusters orchestrates the onset for leukocyte adhesion. Proc Natl Acad Sci 109(13):4869–4874
Baneyx G, Baugh L, Vogel V (2002) Fibronectin extension and unfolding within cell matrix fibrils controlled by cytoskeletal tension. Proc Natl Acad Sci 99(8):5139–5143
Barczyk M, Carracedo S, Gullberg D (2010) Integrins. Cell Tissue Res 339:269–280
Benoit DS et al (2008) Small molecule functional groups for the controlled differentiation of human mesenchymal stem cells encapsulated in poly (ethylene glycol) hydrogels. Nat Mater 7(10):816–823
Biela SA et al (2009) Different sensitivity of human endothelial cells, smooth muscle cells and fibroblasts to topography in the nano–micro range. Acta Biomater 5(7):2460–2466
Biggs M et al (2007) The effects of nanoscale pits on primary human osteoblast adhesion formation and cellular spreading. J Mater Sci Mater Med 18:399–404
Bose S, Roy M, Bandyopadhyay A (2012) Recent advances in bone tissue engineering scaffolds. Trends Biotechnol 30(10):546–554
Cattaruzza M, Lattrich C, Hecker M (2004) Focal adhesion protein zyxin is a mechanosensitive modulator of gene expression in vascular smooth muscle cells. Hypertension 43(4):726–730
Chen C (1999) Geometric control of cell life and death. Science 30:1425
Chen Y-H et al (2019) miRNA-10a promotes cancer cell proliferation in oral squamous cell carcinoma by upregulating GLUT1 and promoting glucose metabolism. Oncol Lett 17(6):5441–5446
Choquet D, Felsenfeld DP, Sheetz MP (1997) Extracellular matrix rigidity causes strengthening of integrin–cytoskeleton linkages. Cell 88(1):39–48
Cortese B, Gigli G, Riehle M (2009) Mechanical gradient cues for guided cell motility and control of cell behavior on uniform substrates. Adv Funct Mater 19(18):2961–2968
Cox EA, Sastry SK, Huttenlocher A (2001) Integrin-mediated adhesion regulates cell polarity and membrane protrusion through the Rho family of GTPases. Mol Biol Cell 12(2):265–277
Curran J, Chen R, Hunt J (2005) Controlling the phenotype and function of mesenchymal stem cells in vitro by adhesion to silane-modified clean glass surfaces. Biomaterials 26(34):7057–7067

Curran JM et al (2010) Introducing dip pen nanolithography as a tool for controlling stem cell behaviour: unlocking the potential of the next generation of smart materials in regenerative medicine. Lab Chip 10(13):1662–1670

Curran J et al (2011) The use of dynamic surface chemistries to control msc isolation and function. Biomaterials 32(21):4753–4760

Da Silva RM, Mano JF, Reis RL (2007) Smart thermoresponsive coatings and surfaces for tissue engineering: switching cell-material boundaries. Trends Biotechnol 25(12):577–583

DeForest CA, Anseth KS (2012) Photoreversible patterning of biomolecules within click-based hydrogels. Angew Chem Int Ed Engl 51(8):1816

Discher DE, Janmey P, Wang Y-L (2005) Tissue cells feel and respond to the stiffness of their substrate. Science 310(5751):1139–1143

Elkin BS et al (2007) Mechanical heterogeneity of the rat hippocampus measured by atomic force microscope indentation. J Neurotrauma 24(5):812–822

Evans ND et al (2009) Substrate stiffness affects early differentiation events in embryonic stem cells. Eur Cell Mater 18(1):e13

Feinberg AW et al (2008) Systematic variation of microtopography, surface chemistry and elastic modulus and the state dependent effect on endothelial cell alignment. J Biomed Mater Res A 86(2):522–534

Flanagan LA et al (2002) Neurite branching on deformable substrates. Neuroreport 13(18):2411–2415

Gibson WJ et al (2023) Bifunctional small molecules that induce nuclear localization and targeted transcriptional regulation. bioRxiv

Glotzer M, Hyman AA (1995) Cell polarity: the importance of being polar. Curr Biol 5(10):1102–1105

Harrison RG (1911) On the stereotropism of embryonic cells. Science 34(870):279–281

Hassan G et al (2020) Cancer stem cell microenvironment models with biomaterial scaffolds in vitro. Processes 9(1):45

Kanchanawong P et al (2010) Nanoscale architecture of integrin-based cell adhesions. Nature 468(7323):580–584

Kilian KA et al (2010) Geometric cues for directing the differentiation of mesenchymal stem cells. Proc Natl Acad Sci 107(11):4872–4877

Kim A et al (2023) Investigating immunomodulatory biomaterials for preventing the foreign body response. Bioengineering 10(12):1411

Kingham E et al (2013) Nanotopographical cues augment mesenchymal differentiation of human embryonic stem cells. Small 9(12):2140–2151

Klimek K, Ginalska G (2020) Proteins and peptides as important modifiers of the polymer scaffolds for tissue engineering applications—a review. Polymers 12(4):844

Lock JG, Wehrle-Haller B, Strömblad S (2008) Cell–matrix adhesion complexes: master control machinery of cell migration. In: Seminars in cancer biology. Elsevier

Lodish HF (2008) Molecular cell biology. Macmillan

Macchiarini P et al (2008) RETRACTED: clinical transplantation of a tissue-engineered airway. Lancet 372(9655):2023–2030

Mamidi N et al (2022) Recent advances in designing fibrous biomaterials for the domain of biomedical, clinical, and environmental applications. ACS Biomater Sci Eng 8(9):3690–3716

Mamilos A et al (2023) Macrophages: from simple phagocyte to an integrative regulatory cell for inflammation and tissue regeneration—a review of the literature. Cells 12(2):276

Mangelsdorf DJ et al (1995) The nuclear receptor superfamily: the second decade. Cell 83(6):835–839

Mao Y, Schwarzbauer JE (2005) Fibronectin fibrillogenesis, a cell-mediated matrix assembly process. Matrix Biol 24(6):389–399

McBeath R et al (2004) Cell shape, cytoskeletal tension, and RhoA regulate stem cell lineage commitment. Dev Cell 6(4):483–495

McFarlin D et al (2006) Nanoscale topography modulates human embryonic stem cell self-renewal. In: Proceedings of the 46th Annual Meeting of The American Society for Cell Biology

Mehrabi T, Mesgar AS, Mohammadi Z (2020) Bioactive glasses: a promising therapeutic ion release strategy for enhancing wound healing. ACS Biomater Sci Eng 6(10):5399–5430
Miller EJ, Lappin SL (2020) Physiology, cellular receptor. StatPearls Publishing
Milner KR, Siedlecki CA (2007) Submicron poly (L-lactic acid) pillars affect fibroblast adhesion and proliferation. J Biomed Mater Res A 82(1):80–91
Miyamoto S et al (1995) Integrin function: molecular hierarchies of cytoskeletal and signalling molecules. J Cell Biol 131(3):791–805
Mosiewicz KA et al (2013) In situ cell manipulation through enzymatic hydrogel photopatterning. Nat Mater 12(11):1072–1078
Nakkala JR et al (2021) Immunomodulatory biomaterials and their application in therapies for chronic inflammation-related diseases. Acta Biomater 123:1–30
Negrutskii BS, Deutscher MP (1992) A sequestered pool of aminoacyl-tRNA in mammalian cells. Proc Natl Acad Sci 89(8):3601–3604
Ng CCA et al (2012) Using an electrical potential to reversibly switch surfaces between two states for dynamically controlling cell adhesion. Angew Chem Int Ed 51(31):7706–7710
Park JE, Kim DH (2024) Advanced immunomodulatory biomaterials for therapeutic applications. Adv Healthc Mater 14:e2304496
Patel S et al (2019) Messenger RNA delivery for tissue engineering and regenerative medicine applications. Tissue Eng A 25(1–2):91–112
Rovensky YA (2011) Adhesive interactions in normal and transformed cells. Springer Science & Business Media
Saha K et al (2008) Substrate modulus directs neural stem cell behavior. Biophys J 95(9):4426–4438
Salinas CN, Anseth KS (2008) The enhancement of chondrogenic differentiation of human mesenchymal stem cells by enzymatically regulated RGD functionalities. Biomaterials 29(15):2370–2377
Schoen I, Pruitt BL, Vogel V (2013) The Yin-Yang of rigidity sensing: how forces and mechanical properties regulate the cellular response to materials. Annu Rev Mater Res 43(1):589–618
Shi J et al (2022) Energy landscape decomposition for cell differentiation with proliferation effect. Natl Sci Rev 9(8):nwac116
Sjöström T et al (2013) Titanium nanofeaturing for enhanced bioactivity of implanted orthopedic and dental devices. Nanomedicine 8(1):89–104
Tan SH et al (2010) Oxygen plasma treatment for reducing hydrophobicity of a sealed polydimethylsiloxane microchannel. Biomicrofluidics 4(3):32204
Temenoff JS, Mikos AG (2008) Biomaterials: the intersection of biology and materials science, vol 1. Pearson/Prentice Hall Upper Saddle River
Tsimbouri PM et al (2012) Using nanotopography and metabolomics to identify biochemical effectors of multipotency. ACS Nano 6(11):10239–10249
Van Tam JK et al (2012) Mesenchymal stem cell adhesion but not plasticity is affected by high substrate stiffness. Sci Technol Adv Mater 13(6):064205
Wells RG (2008) The role of matrix stiffness in regulating cell behavior. Hepatology 47(4):1394–1400
Wilson CJ et al (2005) Mediation of biomaterial–cell interactions by adsorbed proteins: a review. Tissue Eng 11(1–2):1–18
Yadav TC, Bachhuka A (2023) Tuning foreign body response with tailor-engineered nanoscale surface modifications: fundamentals to clinical applications. J Mater Chem B 11(33):7834–7854
Yu C-H et al (2011) Early integrin binding to Arg-Gly-Asp peptide activates actin polymerization and contractile movement that stimulates outward translocation. Proc Natl Acad Sci 108(51):20585–20590
Zaidel-Bar R et al (2007) Functional atlas of the integrin adhesome. Nat Cell Biol 9(8):858–867
Zhu X, Assoian RK (1995) Integrin-dependent activation of MAP kinase: a link to shape-dependent cell proliferation. Mol Biol Cell 6(3):273–282
Zhu L et al (2020) ZHX2 drives cell growth and migration via activating MEK/ERK signal and induces Sunitinib resistance by regulating the autophagy in clear cell Renal Cell Carcinoma. Cell Death Dis 11(5):337

Bioinspired Scaffolds for Tissue Regeneration and Environmental Impact

9

Anas Islam, Ambareen Fatima Ahmed, Widhilika Singh, Md. Faiyazuddin, Usama Ahmad, and Faiyaz Shakeel

Abstract

Bioinspired scaffolds are revolutionizing the field of tissue regeneration by mimicking the complex structures and functionalities of natural tissues. The concepts of bioinspiration are examined in this chapter, along with how natural patterns might be adapted into artificial scaffolds to improve tissue regeneration and repair. Along with cutting-edge manufacturing techniques including electrospinning, 3D printing, and self-assembly approaches, it covers a variety of bioinspired scaffold types, such as polymer-based, ceramic-based, composite, and nanostructured scaffolds. Additionally, the chapter emphasizes how crucial scaffold functionalization is for enhancing bioactivity, cell adhesion, and proliferation. It also discusses how the manufacture of scaffolds affects the environment, highlighting the need of sustainable processes and eco-friendly materials. Examples from the fields of bone, cartilage, skin, and brain tissue engineering show how bioinspired scaffolds may be used practically and effectively. The difficulties, prospects, and possibilities of bioinspired scaffolds in promoting tissue

A. Islam
Faculty of Pharmacy, Integral University, Lucknow, India

School of Pharmaceutical Sciences, Integral University, Lucknow, India

A. F. Ahmed · W. Singh
School of Pharmaceutical Sciences, Integral University, Lucknow, India

Md. Faiyazuddin
School of Pharmacy, Al – Karim University, Katihar, Bihar, India

U. Ahmad (✉)
Faculty of Pharmacy, Integral University, Lucknow, India

F. Shakeel (✉)
Department of Pharmaceutics, College of Pharmacy, King Saud University, Riyadh, Saudi Arabia

Md. Faiyazuddin, M. Suryawanshi (eds.), *Design and Processing of Green Materials*, Biomaterials, Bioengineering and Sustainability 4,
https://doi.org/10.1007/978-3-031-91790-5_9

engineering while reducing environmental effect are covered in the chapter's conclusion.

Keywords

Bioinspired scaffolds · Tissue regeneration · Biomaterials · Environmental impact · Fabrication techniques

9.1 Introduction

Tissue regeneration represents a pivotal aspect of regenerative medicine, focusing on the restoration of damaged or diseased tissues to their original state of function (Salgado et al. 2013). Unlike conventional treatments that may rely on prosthetics or donor tissues, tissue regeneration leverages the body's inherent ability to heal, aiming to stimulate and guide this natural process (Colazo et al. 2019). Significant advancements in the principles of tissue regeneration have been achieved through the fields of stem cell biology, molecular biology, and biomaterials science. Together, these domains have made it feasible to develop complex techniques that mimic the mechanical and biological features of natural tissues, providing effective substitutes for conventional treatment strategies (Xia et al. 2018). Because chronic illnesses, traumas, and aging-related tissue deterioration are becoming more common, there is a greater need than ever for efficient tissue regeneration treatments. Irreversible tissue damage is often caused by conditions including osteoarthritis, cardiovascular disorders, and nerve injuries, which are difficult for traditional medical therapies to effectively cure (Vahidi et al. 2024; Banerjee et al. 2021). Within this particular framework, tissue regeneration serves as more than just a treatment choice, but rather a crucial need that has the capacity to reinstate functionality and enhance the overall well-being of countless individuals over the globe. Scaffolds are vital for successful tissue regeneration due to their temporary nature, which facilitates cell adhesion, proliferation, and differentiation. By providing the necessary foundation, scaffolds aid in the creation of new tissue and gradually disintegrate to make room for the regenerated structure. Scaffold design and manufacturing have advanced by using bioinspired techniques that mimic natural tissues, hence improving tissue regeneration (Yari et al. 2022; Hussain et al. 2023).

9.1.1 Importance of Bioinspired Scaffolds

Bioinspired scaffolds mark a significant progress in tissue engineering as they offer designs that closely resemble the complex architecture and functional characteristics of real tissues. These scaffolds are designed to mimic the tissues they are meant to substitute regarding their microstructure, mechanical properties, and metabolic condition. In doing so, they establish an environment that better supports cell growth and tissue development, greatly enhancing the effectiveness of regenerative treatments (Alaribe et al. 2016). The importance of bioinspired scaffolds is highlighted by their ability to address the limitations of traditional synthetic scaffolds, which often fail to adequately replicate the complexity of natural tissues (Ruffini et al. 2021). Traditional scaffolds, while useful, may lack the necessary cues for directing cell behavior, leading to suboptimal tissue regeneration. In contrast, bioinspired scaffolds are designed with a deeper understanding of the biological processes involved in tissue regeneration, enabling them to provide specific signals that promote cell adhesion, proliferation, and differentiation (Jiang et al. 2021).

Bioinspired scaffolds has the capacity to replicate the hierarchical organization seen in natural tissues, making it one of their prominent characteristics. As an example, bone tissue is made up of a complex matrix of collagen fibers, hydroxyapatite crystals, and several cellular components. In order to recreate this structure, bioinspired scaffolds for bone regeneration use composite materials that blend biopolymers and bioactive ceramics to create a scaffold that interacts with the surrounding tissue and promotes bone cell proliferation (Liu et al. 2013; Guarino et al. 2012).

In addition to mimicking the structure, bioinspired scaffolds can also be altered with bioactive substances such as growth factors, peptides, and proteins. These molecules are crucial in the signaling networks that regulate cell activity and tissue development. By incorporating such bioactive molecules, bioinspired scaffolds can actively promote tissue regeneration rather than merely providing passive support (Zhang and Wang 2019). For instance, scaffolds used in nerve regeneration may be functionalized with neurotrophic factors that enhance neuron survival and axon growth, thereby improving the chances of successful nerve repair (Qian et al. 2021). The manufacturing processes used to produce bioinspired scaffolds are equally crucial. Advanced techniques like electrospinning, 3D printing, and self-assembly enable precise manipulation of scaffold structure, producing nanoscale features critical for cell behavior and tissue integration. Moreover, the use of bioinspired design principles in these fabrication techniques ensures that the scaffolds are not only structurally similar to natural tissues but also functionally effective in guiding tissue regeneration (Chen et al. 2022a, b).

9.1.2 Environmental Impacts of Biomedical Materials

While the development of bioinspired scaffolds has brought significant advancements in tissue regeneration, it has also raised important questions regarding the environmental impact of scaffold production. The biomedical industry, like many others, faces the challenge of balancing technological innovation with environmental sustainability. The production of biomedical materials, particularly synthetic polymers and ceramics used in scaffolds, often involves processes that are energy-intensive and generate waste, contributing to environmental degradation (Riaz et al. 2024). Scaffold production raises environmental concerns due to non-renewable resource use and non-biodegradable waste generation, often utilizing synthetic polymers like PLA, PEG, and PCL. While these materials have desirable properties for tissue engineering, their production relies on petrochemical feedstocks, which are finite resources. Furthermore, the degradation of synthetic polymers can result in microplastic pollution, which poses significant risks to ecosystems and human health (Geyer et al. 2017).

Bioinspired scaffolds derived from natural polymers like collagen, chitosan, and alginate are being explored as sustainable alternatives to synthetic materials in tissue engineering due to their biodegradability and renewable nature (Arif et al. 2022; Reshmy et al. 2021). In addition to the choice of materials, the fabrication processes used to produce scaffolds also have significant environmental implications. Traditional manufacturing techniques often require the use of toxic solvents, high energy inputs, and generate substantial waste. In contrast, newer techniques such as green chemistry approaches and additive manufacturing (3D printing) offer more sustainable alternatives. These methods can reduce the use of hazardous chemicals, lower energy consumption, and minimize waste generation, thereby mitigating the environmental impact of scaffold production (Mazari et al. 2021; Krishani et al. 2023).

Life cycle assessment (LCA) evaluates the environmental impact of biomedical materials at every stage of their life cycle, from extraction to disposal. LCA helps identify key manufacturing phases with the highest environmental impact, promoting sustainable procedures and material efficiency for bioinspired scaffolds (Sukumaran and Gopi 2021; Ribeiro et al. 2013). The design of bioinspired scaffolds must consider both functional requirements for tissue regeneration and the environmental impact of the materials and their degradation products, as the emission of acidic byproducts can harm nearby ecosystems (Dey et al. 2024; Kirillova et al. 2021).

Finally, the growing emphasis on sustainability in biomedical research and development has led to the exploration of circular economy models in tissue engineering. In order to decrease waste and lower the consumption of limited resources, circular economy concepts promote the reuse, recycling, and repurposing of items. When it comes to scaffold manufacturing, this might include creating recyclable scaffolds, repurposing important scaffold components, or using leftovers from other sectors as raw materials to create scaffolds (Geissdoerfer et al. 2017).

9.2 Fundamentals of Tissue Engineering

Tissue engineering is a multidisciplinary field that uses biology, engineering, and materials science to create biological substitutes for tissue function, focusing on tissue regeneration using cells, biomaterials, and biochemical components (Chen and Liu 2016; Mathew et al. 2016). Tissue regeneration involves an intricate process of inflammation, cell proliferation, matrix deposition, and tissue remodeling. It begins with an inflammatory response, followed by the growth of stem cells, which differentiate into the required cell types, and the extracellular matrix (Tidball 2005; Xia et al. 2018; Xu et al. 2022).

A key principle of tissue regeneration is the concept of the regenerative niche, which refers to the specialized microenvironment that supports stem cell activity and tissue repair (Scadden 2006). Effective tissue engineering strategies often aim to recreate this niche, either by delivering growth factors that mimic natural signals or by designing biomaterials that replicate the physical properties of the ECM (Wade and Burdick 2012). Another crucial factor in tissue regeneration is the assimilation of the newly generated tissue with the recipient tissue. This process necessitates the development of both functioning cells and matrix, as well as the creation of vascular networks to provide nutrients and eliminate waste products (Gaharwar et al. 2020). Angiogenesis is therefore a critical component of tissue regeneration, and many tissue engineering approaches incorporate angiogenic factors or design scaffolds that promote vascularization (Lee et al. 2021; Serbo and Gerecht 2013).

Additionally, the immune system has two roles in tissue regeneration. By eliminating infections and damaged cells and generating cytokines that encourage cell division and proliferation, it may, on the one hand, aid in tissue healing. Conversely, fibrosis and scarring brought on by an overabundance or prolonged immune response might hinder the healing process. Therefore, tissue engineering strategies often include immunomodulatory approaches to balance the beneficial and detrimental effects of the immune response (Julier et al. 2017; Shanley et al. 2021).

9.2.1 Role of Scaffolds in Tissue Engineering

Scaffolds are crucial structural frameworks in tissue engineering, providing mechanical support for cells positioned on them. They mimic the physical and metabolic conditions of regenerated tissue, influencing cellular activity and matrix production. Their stiffness and flexibility can influence cell specialization, such as stem cell transformation into bone cells or adipose tissue development (Correia et al. 2012; Bružauskaitė et al. 2016; Kennedy et al. 2017; Luo 2020; Echeverria Molina et al. 2021).

Scaffolds are crucial for directing new tissue creation and spatial organization of cells. They provide mechanical support and transmit biochemical signals that regulate cellular functions. The incorporation of growth factors, cytokines, and other bioactive chemicals into the scaffold material creates a natural regenerative environment that enhances tissue formation, differentiation, and cellular proliferation. Scaffolds used in bone regeneration are often infused with bone morphogenetic proteins (BMPs), which stimulate bone growth. The scaffold's structural characteristics, including porosity, pore dimensions, and morphology, influence cellular proliferation and extracellular matrix organization. Tailored scaffold constructions may align with the anisotropic properties of certain tissues (Carotenuto et al. 2022; Chai and Leong 2007; Lee and Shin 2007; El Bialy et al. 2017; Loh and Choong 2013).

Scaffolds are essential for vascularization, which is a key process for the survival and function of tissues. They can enhance angiogenesis by integrating angiogenic factors or microchannels, and some even incorporate pre-formed vascular networks or endothelial cells to speed up the process. They are also biodegradable, slowly breaking down as new tissue develops. Nevertheless, the degradation rate needs to be meticulously controlled to guarantee sufficient support and incorporation of new tissue. Excessively fast or slow degradation can obstruct tissue integration (Zhang and King 2020; Tajvar et al. 2023; Joshi et al. 2022).

9.2.2 Types of Biomaterials Used in Scaffolds

The performance of the scaffold and the overall effectiveness of tissue regeneration are greatly influenced by the properties of the material, making the choice of biomaterials for scaffolds a crucial aspect of tissue engineering. The three main categories of scaffolds utilized in biomaterials are natural, synthetic, and composite materials. Each has its own distinct advantages and disadvantages.

9.2.2.1 Natural Biomaterials

Because of their biocompatibility, bioactivity, and capacity to encourage cell adhesion and proliferation, natural biomaterials—which are derived from biological sources—have found widespread use in the scaffolding process. Intrinsic biological signals found in these materials may promote tissue regeneration (Ullah and Chen 2020).

- **Collagen:** Collagen, the main protein in the extracellular matrix of animals, is essential for maintaining tissue structure and encouraging normal tissue function. Collagen-based scaffolds are widely used in tissue engineering because of their exceptional biocompatibility and capacity to stimulate cell adhesion and proliferation. Collagen scaffolds are helpful for several procedures, such as the regeneration of skin, bone, and cartilage. Despite its rapid degradation, collagen may be crosslinked or mixed with other substances to increase its durability and mechanical qualities (Li et al. 2021).
- **Alginate:** Because of its moderate gelation process, alginate, a polysaccharide produced from brown seaweed, is often employed to encapsulate medications and cells. Alginate-based scaffolds are used in bone and cartilage tissue engineering as well as in the administration of medications. However, since alginate lacks cell attachment sites, it must be modified or include cell-adhesive molecules in order to increase its bioactivity (Farokhi et al. 2020).
- **Chitosan:** Chitin, a material found in crab outer shells, is the source of the compound chitosan. It is highly valued because to its antibacterial, biocompatible, and biodegradable properties. Scaffolds made of chitosan have been used to bone, nerve, and skin tissue engineering. Although they may need to be chemically modified to increase mechanical strength and stability under physiological settings, their capacity to form hydrogels and films makes them adaptable for a variety of applications (Pandey et al. 2017).

9.2.2.2 Synthetic Biomaterials

The physical and chemical properties of scaffolds, including as their mechanical resilience, rate of degradation, and surface chemistry, may be more easily controlled using synthetic biomaterials. These materials may be customized to fulfill particular criteria for various tissue engineering applications. Polylactic acid (PLA), polyglycolic acid (PGA), polycaprolactone (PCL), and polyethylene glycol (PEG) are frequently used synthetic biomaterials.

- **Polylactic Acid (PLA):** PLA is a biodegradable polyester that is made from renewable materials such as maize starch or sugarcane. Its excellent mechanical qualities and biocompatibility have made it a popular choice for scaffolds used in bone and cartilage regeneration. However, PLA tends to degrade into acidic byproducts, which can lead to local inflammation. To mitigate this, PLA is often blended with other materials or used in combination with basic compounds that neutralize the acidic degradation products (Singhvi et al. 2019).
- **Polycaprolactone (PCL):** PCL is a biodegradable polyester that is characterized by its slow breakdown rate and strong mechanical properties. PCL has qualities that make it well-suited for use in applications involving bone and soft tissue engineering, particularly where extended support is required. Poly(caprolactone) (PCL) may be transformed into various scaffold structures, such as fibers, meshes, and films, by methods including electrospinning and 3D printing. Nevertheless, the water-repellent characteristics of PCL may restrict the ability of cells to adhere and grow, therefore requiring the need for altering the surface or applying a layer of biologically active substances (Malikmammadov et al. 2018).
- **Polyethylene Glycol (PEG):** PEG, a hydrophilic polymer, is often used in tissue engineering. When it comes to supplying bioactive compounds and encapsulating cells, PEG-based hydrogels are very beneficial. By adjusting the PEG's molecular weight and crosslinking density, the hydrogels' mechanical and deteriorating properties may be carefully regulated. PEG is often combined with other substances to improve bioactivity and cell interactions (Yang et al. 2022).

9.2.2.3 Composite Biomaterials

To maximize the advantages of both natural and synthetic materials, composite biomaterials are made by fusing them together. These composite materials are very versatile for a variety of tissue engineering applications because they provide improved mechanical properties, bioactivity, and controlled degradation.

- **Hydroxyapatite-Polymer Composites:** One of the main constituents of bone is calcium apatite, which occurs naturally as the mineral hydroxyapatite (HA). In order to make scaffolds for bone regeneration, HA is often mixed with polymers such as PLA, PCL, or collagen. These composites enable the scaffold to gradually integrate with the native bone tissue while providing the osteoconductivity and mechanical strength required to promote bone tissue growth (Bal et al. 2020).
- **Silk-Polymer Composites:** A protein obtained from silkworms, silk fibroin is renowned for its superior mechanical qualities and biocompatibility. Silk-based composites provide a special blend of strength, flexibility, and bioactivity when mixed with synthetic polymers like PEG or PCL, which qualifies them for use in bone, nerve, and skin tissue engineering (Sun et al. 2021).

- **Chitosan-Alginate Composites:** Alginate and chitosan are often mixed to produce hydrogels with improved bioactivity and mechanical qualities. These composites may be used to cartilage regeneration, wound healing, and medication administration. The combination of chitosan's antimicrobial properties and alginate's mild gelation process makes these composites particularly attractive for biomedical applications (Seidi et al. 2021).

9.3 Bioinspired Design Principles

Bioinspiration is a multidisciplinary approach that draws on the principles, structures, and functions observed in nature to inspire and inform the design and development of new materials, devices, and systems. The concept of bioinspiration hinges on the idea that nature, through millions of years of evolution, has optimized various biological structures and processes to achieve highly efficient, adaptable, and sustainable outcomes (Speck et al. 2017). By studying these natural systems, scientists and engineers can develop innovative solutions that mimic or are inspired by the mechanisms observed in biological organisms (Sarikaya et al. 2003).

In tissue engineering, bioinspiration imitates biological architecture to produce scaffolds that resemble genuine tissues, overcoming the difficulties in replicating their complex functional properties and hierarchical structures. Bioinspired designs can address these challenges by leveraging the inherent wisdom embedded in natural systems (Wang et al. 2020). Bioinspiration is not limited to copying nature's designs; it also involves understanding the underlying principles that govern these designs and applying this knowledge to create new solutions that are optimized for specific applications. For example, the study of spider silk, which is known for its exceptional flexibility and endurance, has led to the development of synthetic fibers with similar properties. These fibers find use in many medical contexts, including the fabrication of sutures and the construction of frameworks for tissue engineering (Omenetto and Kaplan 2010).

The concept of bioinspiration extends beyond the physical properties of materials to include the dynamic processes that occur in biological systems. For example, the creation of smart materials that can alter their characteristics in reaction to external circumstances has been influenced by the way certain plants and animals react to environmental stimuli. These substances may find utility in tissue engineering to build scaffolds that may change to meet the evolving requirements of tissues that are developing or mending (Langer and Tirrell 2004).

9.3.1 Natural Structures and Their Functionalities

Natural structures are characterized by their complexity, adaptability, and functionality, which have been refined through the process of evolution. These structures often exhibit a hierarchical organization, where different levels of structure, from the molecular to the macroscopic scale, contribute to the overall performance of the tissue or organism. Understanding the functionalities of these natural structures is crucial for designing bioinspired scaffolds that can replicate the behavior of natural tissues (Wang et al. 2020).

9.3.1.1 Hierarchical Organization

One of the key features of natural structures is their hierarchical organization, where the properties and functions of the tissue are determined by the arrangement of its components at multiple scales. For example, bone tissue exhibits a hierarchical structure that ranges from the molecular level, where collagen molecules form triple helices, to the macroscopic level, where these helices are organized into fibers, fibrils, and ultimately the entire bone matrix. This hierarchical organization gives bone its remarkable combination of strength and toughness (Reznikov et al. 2014). Skin, cartilage, and tendons are examples of natural tissues that have hierarchical structures with separate layers such as the epidermis, dermis, and hypodermis, each of which contributes to sensory and defensive roles (Gilaberte et al. 2016). Because cartilage's extracellular matrix gives it flexibility and compressive strength, it is challenging to mimic in artificial scaffolds. To produce scaffolds with exact fiber alignment and porosity, researchers may use techniques like as electrospinning and 3D printing (Sankar et al. 2018; Karamanos et al. 2021).

9.3.1.2 Adaptability and Self-Healing

Natural structures are also notable for their plasticity and capacity for self-healing. Numerous biological tissues have the ability to modify their form or function in response to environmental changes. For example, bone tissue can remodel itself in response to mechanical stress, becoming stronger in areas that experience higher loads. This adaptability is crucial for maintaining the functionality of the tissue over time (Wang et al. 2022). Adaptive and self-healing qualities for synthetic scaffolds are being researched. Materials may adapt dynamically to tissue needs by changing in reaction to outside stimuli (Beheshtizadeh et al. 2023; Mantha et al. 2019).

9.3.1.3 Biochemical Signaling

Natural structures are also characterized by their ability to engage in complex biochemical signaling, which regulates cell behavior and tissue function. The extracellular matrix (ECM), for example, is not just a structural scaffold but also a dynamic environment that interacts with cells through various signaling pathways (Gattazzo et al. 2014). In bioinspired scaffold design, it is crucial to replicate these biochemical signaling functions to create an environment that supports tissue regeneration. This can be achieved by incorporating bioactive molecules into the scaffold material or by functionalizing the scaffold surface with specific peptides or proteins that mimic the natural ECM (Carotenuto et al. 2022). For example, scaffolds that present the RGD (arginine-glycine-aspartic acid) peptide, a common motif found in ECM proteins, can enhance cell adhesion and promote tissue formation (Aye et al. 2018).

9.3.2 Translation of Natural Designs into Synthetic Scaffolds

The process of converting natural designs into synthetic scaffolds is intricate and includes both the copying of natural structures and their incorporation into useful biomaterials that promote tissue regeneration. A thorough understanding of the principles that underlie the structure and functionality of natural tissues is necessary for this translation process, as is the creation of sophisticated manufacturing methods that can faithfully reproduce these principles in artificial materials.

9.3.2.1 Mimicking Hierarchical Structures

One of the key challenges in translating natural designs into synthetic scaffolds is the replication of hierarchical structures. As mentioned earlier, natural tissues often exhibit a multi-scale organization, where the properties of the tissue are determined by the arrangement of its components at different levels. In order to recreate this structure in artificial scaffolds, scientists have devised a range of manufacturing methods that enable meticulous manipulation of scaffold design (Salvatore et al. 2021). The process of electrospinning involves drawing polymer solution into thin strands in order to produce nanoscale fibrous scaffolds (Rahmati et al. 2021).

Using computer-aided design tools, 3D printing, is a technique for producing accurate scaffolds. Microchannels and holes for cell infiltration and vascularization in bone, cartilage, and other tissues are precisely replicated, as is tissue composition and structure (Joshi et al. 2022). Self-assembly is a bioinspired process that employs materials or molecules to spontaneously organize into ordered structures such as collagen triple-helical structures and lipid bilayers are often seen in cell membranes. It is used in tissue engineering to produce nanoscale scaffolds that replicate the fibrous structure of extracellular matrix (Marin and Marchesan 2022).

9.3.2.2 Incorporating Bioactive Molecules

Bioactive substances mimic the tissue structure and biochemical signaling functions of the extracellular matrix. Attaching proteins or peptides to scaffold surfaces is one method that promotes cell migration and adherence. Similar to bone morphogenetic proteins in regenerative medicine, scaffolds that include growth factors or signaling molecules may create a microenvironment for tissue regeneration (El Bialy et al. 2017; Klimek and Ginalska 2020). A number of techniques, including encapsulating in nanoparticles or microspheres that are then implanted in the scaffold matrix, may be used to accomplish the regulated release of bioactive compounds. The availability of signaling molecules may be controlled temporally by engineering these particles to break down or release their payload in response to certain stimuli, like as variations in temperature or pH (Santo et al. 2013).

9.3.2.3 Dynamic and Responsive Scaffolds

Researchers have developed materials that can change their properties in response to external stimuli, allowing them to create scaffolds that adapt to the needs of regenerating tissue. Temperature-responsive polymers like PNIPAAm can be used for medication administration or bioactive substance discharge. Biodegradable materials, like enzymes secreted during tissue regeneration, can be used to create scaffolds that disintegrate as new tissue develops, reducing surgical extraction and inflammation (Kirillova et al. 2021; Hiruta 2022).

9.4 Types of Bioinspired Scaffolds

The variety in scaffolds depends on the type of technology applied to fabricate them (Fig. 9.1) (Adel et al. 2022).

9.4.1 Porous Scaffolds

The sponge or foam porous scaffolds possess an interconnected pore structure, which may be either directed or arranged randomly. It is very beneficial in bone regeneration process, the formation of blood vessels and deposition of the Extra Cellular Matrix (ECM). These scaffolds enhance gas and nutrient transport via the channel network by increasing their physical surface area (Nikolova and Chavali 2019).

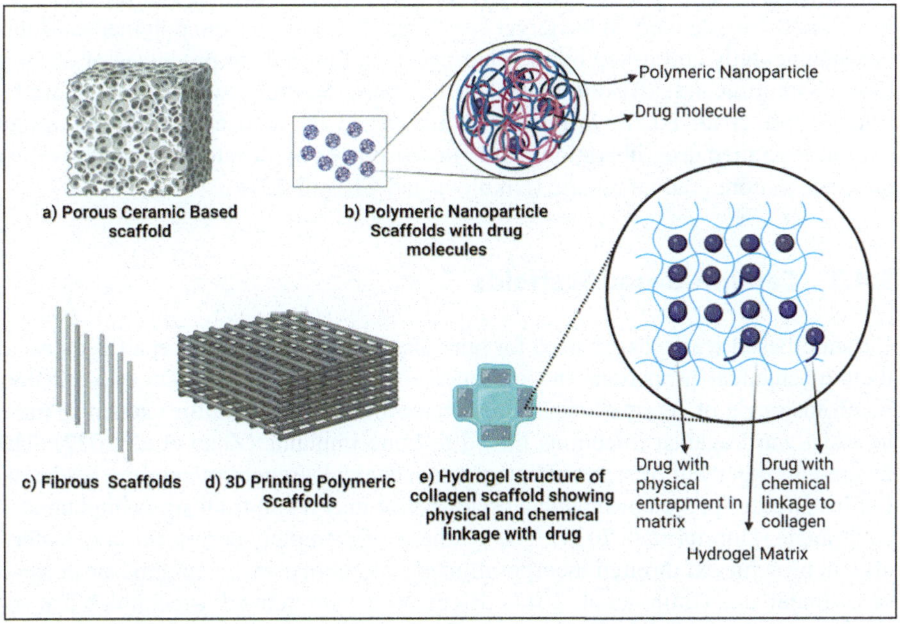

Fig. 9.1 An illustration of several kinds of scaffolds useful in tissue engineering

9.4.2 Polymer-Based Scaffolds

Polymer-based scaffolds are employed in the domains of tissue engineering and regenerative medicine to provide a microenvironment for tissue formation (Shoichet 2010). Natural polymer derived scaffolds are biocompatible with adjustable mechanical properties. They could prove efficient in treating skin disorders (Chen et al. 2022a, b). In 2021, Jadbabaei et al. introduced an innovative methodology aimed at augmenting the sodium alginate electrospinnability and subsequently fabricated alginate-PVA polymeric scaffolds intended for applications in dermal tissue engineering (Jadbabaei et al. 2021). In 2018, a group of researchers established that the scaffold made up of salmon fibrinogen and chitosan aid in wound healing with least side effects (Laidmäe et al. 2018).

'Conducting polymer' based scaffolds could restore injured or malfunctioning tissues (Hardy et al. 2013). These systems are capable of electrical responsive drug delivery. Some conducting polymers are namely polyaniline, polypyrrole, polythiophene, polyacetylene, polycarbazole, poly(p-phenylene. Nevertheless, a multitude of challenges arises when implementing conductive polymers for the purpose of electrical stimuli responsive drug delivery. These challenges encompass a restricted drug loading capacity, limitations associated with the delivery of small, charged biomolecules, degradation of electrical and mechanical properties attributed to bioactive dopants, as well as fracture and delamination from the underlying substrate material due to volumetric alterations induced by actuation upon electrical stimulation (Bansal et al. 2020; Kausar 2021). Polymer based scaffolds can provide controlled rate drug delivery at the target site. Drug distribution from polymer scaffolds is fundamentally controlled by the pharmaceutical ingredient processes of desorption, dissolution and diffusion as well as the matrix eroding, swelling, and degradation (Calori et al. 2020). Polymers characterized by reduced molecular weight exhibit expedited drug liberation, while microspheres of elevated molecular weight facilitate a more gradual release (Nikolova and Chavali 2019).

9.4.3 Ceramic-Based Scaffolds

Ceramic-based scaffolds are used for bone tissue engineering and repair because of their mechanical properties and osteogenic ability (Fig. 9.1a) (Du et al. 2018). Scaffolds made of ceramic material can have antibacterial qualities added to them to assist ward against infections linked to bone implants (Zhao et al. 2022). The layered arrangement of struts affects the mechanical strength of ceramic scaffolds. For instance, improved mechanical performance may result from struts implanted at a 45° angle (Dobriţa et al. 2023). The synthesis of extremely porous ceramic materials can be achieved through incorporation of gas bubbles into a ceramic suspension or colloidal sols (Baino et al. 2015). Bioceramics are defined as ceramics that are employed as biomaterials for the purpose of addressing defects in dental and osseous structures, securing bone grafts, and substituting compromised tissue (Yamamuro 2004).

9.4.4 Composite Scaffolds

De Groot et al. created the first tissue engineering composite in 1990 using polyurethane (PU) and poly(l-lactic acid) ((PLLA) combinations. (De Groot et al. 1990). Inorganic ceramic particles with organic polymer matrix make composite foams. Inorganic-organic composite materials represent a commendable approach to emulate bone tissue, which inherently functions as a natural composite, possessing mechanical and biological attributes that are markedly superior to those of the individual constituents (Kundu et al. 2013; Silva et al. 2010).

The major problem faced that led to development of composite scaffolds is that the bone defects are irregularly shaped, requiring versatile scaffold processing techniques and scaffolds need high strength to temporarily replace bone structure during regeneration. Poly(α-hydroxyesters) (PHE) were commonly used but lose compressive strength as porosity increased. So in order to improve mechanical strength, short hydroxyapatite fibers are added to the PHE scaffold. Even distribution of fibers is crucial for effective reinforcement, which is challenging with melt-molding. A solvent-casting technique is used to evenly distribute hydroxyapatite (HA) fibers in chloroform and a porogen in a Poly(lactic-co-glycolic acid) (PLGA) solution. And pore size is controlled by the aqueous poly(vinyl alcohol) (PVA) solution concentration and salt crystal diameter. Increasing HA content increases the scaffold's compressive strength, comparable to cancellous bone. Phase Transition technique used to make composite foams include HA powder is incorporated to a PLGA/dioxane solution and frozen to induce phase separation. The mixture is freeze-dried, resulting in composite foams with an interconnected irregular pore structure. The compressive strength of these foams is higher than pure PLGA foams. Foams having porosity about 95% and pore sizes between 30 μm to 100 μm could be produced (Thomson et al. 2000). Hence, porosity, pore size and structure can be managed by adjusting polymer concentration, HA content, solvent type, and temperature for phase-separation (Zeinali et al. 2021). The formulation of an ideal proportion between inorganic constituents and polymeric substances is crucial for promoting the development of bone tissue, while simultaneously preserving the porosity and mechanical stability of the composite materials. A varied spectrum of combinations results in composites that demonstrate outstanding efficacy within the realm of bone tissue engineering. (Kundu et al. 2013; Silva et al. 2010).

Inorganic-organic composites, such collagen with biphasic calcium phosphate (BCP) and silk fibroin with β-tricalcium phosphate (TCP) or HA, improve mechanical characteristics, cell adhesion, proliferation, and the biological response in vivo. Additionally, these composites encourage the formation of new bone, particularly when the HA to TCP ratio is raised. Polymer matrices may also include inorganic nanoparticles to enhance their mechanical and physical properties (Yan et al. 2012; Collins et al. 2021; Oliveira et al. 2006). A 2023 research devised a novel polyether-ether-ketone calcium hydroxyapatite composite with various lattice configurations to replicate natural bone structure for a quick, easy, and affordable femur bone-implant scaffold (Fig. 9.1d) (Oladapo et al. 2023).

9.4.5 Macroscopic Scaffolds

There are macroscopic pores above >100 μm in a scaffold are critical for the dispersion of cell seeding, cell motility in the three-dimensional environment, and neovascularization following scaffold implantation in vivo (Smith et al. 2009).

9.4.6 Mircoscopic Scaffolds

Encapsulated bioactive compounds can be released gradually via microspheres. Microspheres and microsphere-derived three dimensional scaffolds are produced using a variety of techniques, including solvent vapour treatment, solvent or non-solvent sintering, dispersion of oil-in-water, and selective laser sintering (SLS) (Nikolova and Chavali 2019).

9.4.7 Nanostructured Scaffolds

Nanostructured scaffolds range in size from 50–500 nm (Smith et al. 2009). To obtain nano sized 3D scaffolds uses top-down and bottom-up methods (Gritsch et al. 2017). The quick dispersion of encapsulated materials and cell penetration could be aided by the nanostructure (Wade and Burdick 2014). The integration of PLGA nanospheres, a specific category of nanoparticles, into nano-fibrous PLLA scaffolds facilitated the targeted delivery of recombinant human bone morphogenetic protein (rhBMP-7) to the cell culture milieu, thereby exerting a favorable influence on ectopic bone formation (Wei et al. 2007). The incorporation of nano-crystalline HA into natural polymer scaffolds has demonstrated an enhancement in mechanical characteristics in contrast to control scaffolds composed solely of polymers (Wei and Ma 2004).

Nanoscale drug-release scaffolds could give desired release kinetics and regulate distribution of the drug along least toxicity. So these systems could prove to be more efficient for drug's therapeutic activity. Drugs that could be added into nanofiber matrix include antibiotics like tetracycline hydrochloride, cefazolin, itraconazole, rifampin, anti inflammatory drugs like paracetamol and ibuprofen, anticoagulant heparin and proteins like bovine serum albumin (Martins et al. 2007). A study shows that carbon nanotubes and collagen scaffolds are effective substrates for stem cell proliferation and differentiation, assimilating without compromising membrane integrity. A graphene oxide-derived composite with bacterial cellulose has been developed for a resilient bone scaffold, preventing bacterial proliferation while promoting osteocyte proliferation (Bartoli et al. 2023). Nanofibrous scaffolds are lengthy, fiber-like scaffolds that have a very large length width at the nano range (Fig. 9.1c). The best technique to form nanofibrous scaffold is electrospinning (Adel et al. 2022). There exist several limitations correlated with nanofibrous scaffolds, with the lack of biological recognition on the surface of these scaffolds representing a significant drawback. In order to encourage cell-scaffold interactions, the scaffold surface must be altered to acquire more desired properties (Wei and Ma 2006).

9.5 Fabrication of Scaffolds

Scaffolds being 3D in its structure, give cells the structure and support they need to proliferate, develop and divide for the purpose of regeneration or repairing of damaged tissues or organs. While designing scaffolds following should be considered; scaffolds should be composed of biocompatible materials like polymers, ceramics or hydrogels, scaffolds should be able imitate the ECM and direct cellular activity, aid in the passage of oxygen & nutrients and eventually stimulate tissue regeneration. Physical and chemical qualities in their design influence their ability to attachment to cell, migrate, and develop tissue (Sufiyan et al. 2024). Materials and techniques utilized to fabricate scaffolds are summarized in (Fig. 9.2).

9.5.1 Materials Used in Scaffolding

The materials used in scaffolding based on their duration of use and stated as temporary or permanent. Considering the mechanical properties of the material used they become hard tissue or soft tissue. According to origin of material, they could be regarded as natural or synthetic that is further classified as polymeric, metallic, ceramic and composite materials (Perez-Puyana et al. 2020). Materials used to fabricate scaffolds are given in two categories; materials for porous solid scaffolds and materials for hydrogel scaffolds (Ma 2004).

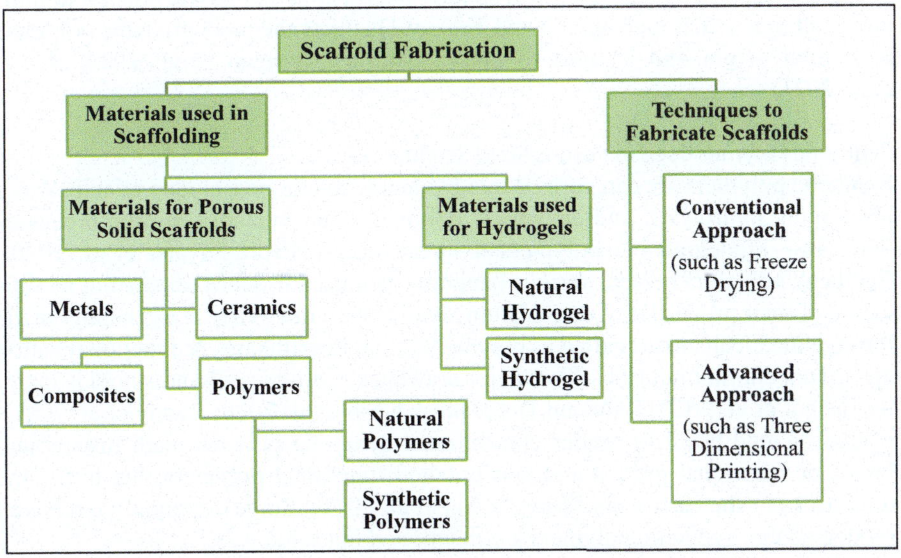

Fig. 9.2 Summary chart of materials and techniques used to fabricate scaffolds

9.5.1.1 Materials for Porous Solid Scaffolds

Materials preferred for making scaffolds in this category should be insoluble *in-vitro* aqueous condition and in *in-vivo* as well (Ma 2004).

Polymers Used in Porous Solid Scaffolding

For designing of polymer scaffolds following should be kept in consideration (Ma 2004);

- Porosity should be high
- Pore size should be optimum
- Surface area should be high
- Biodegradability
- Mechanical strength to adjust in tissue structure
- Biocompatibility
- Cellular interaction to enhance growth and migration

Two types of polymers are used to prepare porous solid scaffold, natural and synthetic polymers (Sufiyan et al. 2024).

Natural Polymer Used in Porous Solid Scaffolding

Natural polymers could be obtained from plants, animals and even from microbes. These polymers provide better biological activity and biocompatible than synthetic polymers. These are nontoxic, biodegradable in nature because upon enzymatic attack, the by-products produced by natural polymers are not toxic enough to initiate an allergic reaction. (Fig. 9.3 and Table 9.1) enlists the biodegradable polymers of natural origin and is used to make scaffolds (Sufiyan et al. 2024; Reddy et al. 2021).

Synthetic Polymer Used in Porous Solid Scaffolding

Synthetic polymers are categorized into biodegradable or non-biodegradable. Water absorption, hydrolysis, and enzymatic polymer chain breakage are the processes that cause biodegradable polymers to break down (Perez-Puyana et al. 2020). Synthetic polymers present various benefits when used in scaffolding from modification of their properties to degradation at controlled rate. They are generally made through the mechanisms called as polymerization, interlinkage, & functionalization and do not entrap impurities. The major drawback synthetic polymers hold is toxicity (Negut et al. 2020; Rahmani Del Bakhshayesh et al. 2018; Reddy et al. 2021). Chemical modifications render conventional synthetic polymer with unsaturated bonds or functional groups and hence called functional polymers that overcome drawbacks in the earlier one (Magnusson et al. 2011). Some commonly employed biodegradable synthetic polymers are mentioned in (Table 9.2).

Although the above stated synthetic polymers hold varied degrees of biodegradability, biocompatibility and mechanical properties, nevertheless, no one polymer possesses these vital characteristics at the best possible degree, altogether (Negut et al. 2020; Rahmani Del Bakhshayesh et al. 2018; Reddy et al. 2021). Nevertheless,

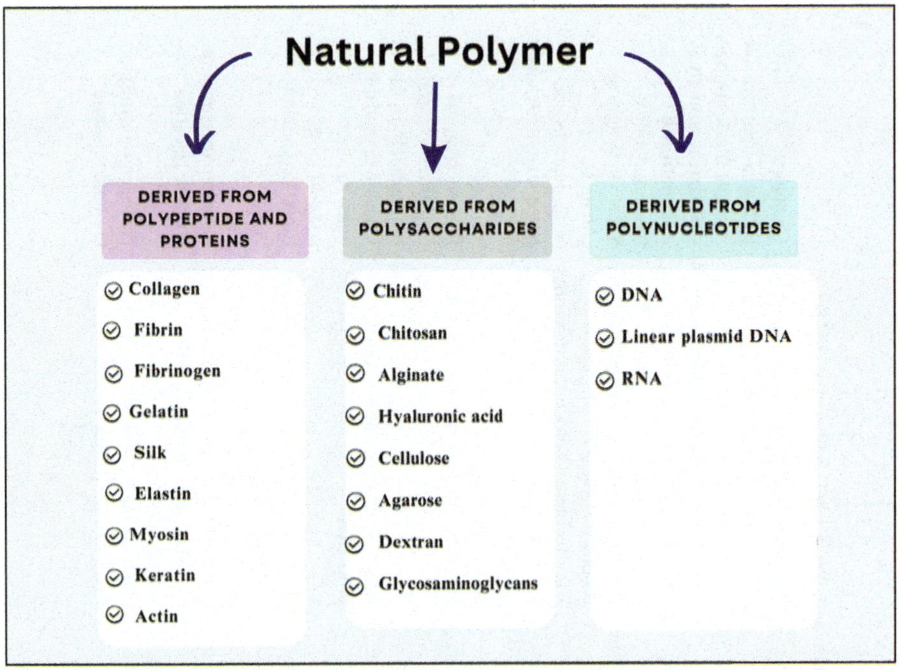

Fig. 9.3 Classification of natural polymers used in making scaffolds

when utilizing synthetic materials, it is imperative to acknowledge that the liberation of acidic byproducts will result in drop in the pH level of adjacent tissues, consequently impacting those tissues (Joshi et al. 2020).

Ceramics Used in Porous Solid Scaffolding

Certain inorganic compounds called porous bioactive glasses and calcium phosphates (CP) are also being used for bone and tissue engineering (Ma 2004). The bioactive glasses show bioactivity, elevated mechanical toughness and strength, elastic modulus, wear resistance, and rapid degradation rates can be effectively ameliorated through the integration of various ions inside the glass matrix (Kargozar et al. 2019; Du et al. 2019).

Composites Used in Porous Solid Scaffolding

Some compounds in CP group used in tissue engineering are TCP, HA and its derivatives, or their amalgamations. These are categorized as osteoconductive or osteoinductive in behavior. The drawback of these inorganic compounds is that highly porous 3D structure formation is difficult and will be mechanically brittle (Ma 2004). Result of study conducted in 2016 on controlled architectural reinforced cellulose and HA 3D construct suggest that by better mechanical properties could be imparted using natural polymers and ceramics together (Morouço et al. 2016). Another study in 2019 proves benefits of the ceramic and synthetic polymer

Table 9.1 Summary on natural polymers used in tissue engineering (Ma 2004; Shapiro and Cohen 1997; Aaliya et al. 2021)

Category	Polymer	Source	Applications	Advantages	Challenges
Natural Proteins	Collagen	Fibrous protein, major (ECM)	Soft tissue repair, tissue regeneration	Useful biological properties, component of ECM, can be fabricated into scaffolds like Collagen and glycosaminoglycan (GAG)	Pathogen transmission, immune reactions, poor mechanical properties, less controlled biodegradability
	Gelatin	Denatured collagen	Porous materials for tissue repair	Biocompatibility, processed into porous scaffolds	Same challenges as collagen
	Silk	Silkworm silk, spider silk, genetically engineered silk	Nondegradable sutures, tissue engineering	Excellent tensile properties, potential for degradation via enzymatic mechanisms, chemical modifications enhance biocompatibility	Slow degradation, potential cytotoxicity
Polysaccharides	Alginate	Seaweed	Porous and solid in state, scaffolds for tissue engineering	Biocompatibility, easily processed	Poor mechanical properties, controlled biodegradability issues
	Chitosan	Crustacean shells	Porous and solid in state, scaffolds for tissue engineering	Biocompatibility, antimicrobial properties	Variable degradation rates, requires cross-linking
	Hyaluronate	Animal tissues, bacterial fermentation	Porous and solid in state, scaffolds for tissue engineering	Hydrophilic, promotes cell migration and proliferation	Rapid degradation in vivo
Processed ECM	Small Intestinal Submucosa	Porcine small intestine submucosa	Reconstruction of various tissues	Contains type I collagen, GAGs, growth factors, promotes tissue regeneration	Pathogen transmission, immune rejection
	Decellularized Tissues	Porcine heart valves, human dermis, urinary bladder submucosa	Tissue engineering, repair applications	Contains natural macromolecules, used in heart valves, dermis, etc.	Pathogen transmission, immune rejection

Table 9.2 List of widely used synthetic polymers to make scaffolds (Reddy et al. 2021; Ma 2004; Nikolova and Chavali 2019; Xu et al. 2023)

Category	Chemical Information	Synthetic Polymer	Abbreviation	Biodegradable	Biocompatible	Properties of Polymer
Poly(α-hydroxy esters)	These polymers are derived from hydroxyl acids and are biocompatible in nature.	Polycaprolactone	PCL	Yes	Yes	Semicrystalline and aliphatic polymer, hydrophobic, good mechanical strength, degraded in body by the hydrolysis of its ester linkages, degradation rat is slow.
		Polylactide (also known as Polylactic Acid)	PLA	Yes	Yes	Semi-crystalline, hydrophobic polymer, prepared by fermentation of wheat or grain to yield lactic acid which is then polymerized. Degrade via hydrolysis of ester bonds which occurs at slow rate. Degradation products water and carbon di-oxide which is almost non toxic. Cleared through body via tricarboxylic acid (TCA) cycle.
		Polyglycolide (also known as Polyglycolic Acid)	PGA	Yes	Yes	Highly crystalline poymer,soluble in organic solvents, Degrade via hydrolysis of ester bonds. PGA degrades into glycolic acids that join with TCA giving by product water and carbon di-oxide (CO_2). PGA loses mechanical integrity within two to four weeks due to quick degradation because of hydrophilic nature.
Poly(α-hydroxy esters) copolymers	These polymers are made by combining two or more monomer of α-hydroxy esters	Poly(lactic-co-glycolic acid)	PLGA	Yes	Yes	Copolymer of PLA and PGA, high cell adhesion, controllable degradation rates, good mechanical properties, methyl side group brings hydrophobic nature similar to PLA. Degrade via hydrolysis of ester bonds

(continued)

Table 9.2 (continued)

Category	Chemical Information	Synthetic Polymer	Abbreviation	Biodegradable	Biocompatible	Properties of Polymer
Co-polyesters	A class of polymers formed from multiple types of monomers, typically containing ester functional groups	Poly(propylene fumarate)	PPF	Yes	Yes	Co-polyester derived from propylene glycol and fumaric acid. It is linear chain polyester with ester bond repeating units. Degradation is caused by hydrolysis of ester bond. It improves hydrophobicity of PLA, PGA, or PCL.
Poly(ethers)	These polymers have ether functional group in their main chain.	Polyethylene Glycol	PEG	Yes	Yes	Non-ionic, slightly immunogenic poymer. Cross-link method is used to make hydrophilic scaffolds. Cross linking affects the permeability, diffusion elasticity and degradation rate of scaffolds.
		Polyethylene Oxide	PEO	Yes	Yes	Hydrophilic polymer with minimum antigenic nature and protein binding. Photopolymerization includes adaptable mechanical property.
Poly (vinyl alcohols)	Poly vinyl derivative	Polyvinyl alcohol	PVA	Yes	Yes	PVA cause immediately solubility feature in scaffold materials.
Polyurethanes	Formed by the reaction of a polyol with a diisocyanate (DI)	Polyurethane	PU	Yes	Yes	Made by repeating units of urethane. DI used in PU formation is carcinogenic. Alternatives like lysine DI or hexamethylene DI, liberate non-toxic products upon degradation.

combination like PLLA and PCL with HA. It states that polymer is useful for drug delivery while ceramics for constructs which may cause less damage to soft tissues and have good mechanical strength. These scaffolds would reduce the chances of repeated surgeries for implants (Alizadeh-Osgouei et al. 2019).

Metalsused in Porous Solid Scaffolding

Metals may be classified as either biodegradable or non-biodegradable. The non-biodegradable metals consist of titanium, stainless steel, tantalum and alloys such as titanium-nickel and cobalt-chromium. Scaffolding incorporates biodegradable metals such as magnesium, iron, and zinc alloys (Lv et al. 2021). When manufacturing metal-based scaffolds, there are several challenges that need to be overcome. A comprehensive assessment of the kinetics of metal biodegradation is essential, particularly when considering resorbable metals. The degradation rate should be in alignment with the rate of formation of new tissue; failing to achieve this balance might lead to premature loss of mechanical integrity prior to restoration of tissue functionality (Prasad et al. 2017).

9.5.1.2 Materials for Hydrogel Scaffolds

Hydrogels (sol-gel) are hydrophilic polymers with physical (mesh structure) or chemical crosslinking (molecular level) that are soluble in water and tend to swell forming a gel-like substance when exposed to water. Because injectable hydrogels may fill irregularly shaped defects, enable minimally invasive treatments, and facilitate the integration of cells and bioactive substances, they are utilized in tissue engineering (Fig. 9.1e) (Ma 2004; Li and Mooney 2016).

Type of hydrogel scaffold is based on the type of polymer used to prepare hydrogel either natural or synthetic hydrogel (Table 9.3) (Ma 2004).

Drugs can be incorporated into scaffold system scaffold depends on the chemical interaction between polymer and drug. Drug could be trapped as a inclusion or particle in a nanofiber construct, or as a blend of drug and polymer, or could be encapsulated in tubular structures (Martins et al. 2007).

9.5.2 Techniques Used to Fabricate Scaffolds

An extensive array of novel fabrication methodologies has been thoroughly explored with particular emphasis on their applications within the realm of tissue engineering. Various prominent scaffold fabrication techniques are elucidated in this chapter. Each methodology possesses its own set of benefits; however, none can be regarded as a universally optimal scaffold fabrication method useful to each tissue types. Selection of a fabrication technique, or innovation of a new approach, is contingent upon the specific critical requirements pertinent to the tissue in question.

Conventional approaches to fabricate scaffolds: Numerous methodologies, including solvent casting accompanied by solvent casting with particulate leaching, thermally induced phase separation and freeze-drying, are employed to construct

Table 9.3 Types of hydrogels to produce respective scaffolds (Ma 2004)

S. No.	Synthetic Hydrogel	Natural Hydrogel
1.	*Polymer Used:* PEG *Advantage*: Frequently studied, biocompatible. *Synthesis*: PEG-diacrylamide and photopolymerization with cells. *Challenges*: Toxicity concerns, lack of degradability. Solutions: *Copolymers*: With PLA, PGA, and PPF. *Modifications*: Linkages in PEG backbone degradable via enzymatic action.	*Polymer used:* Alginate *Source*: Polysaccharides from seaweeds. *Cross-linking:* Using calcium sulfate (CaSO4). *Application*: Cell delivery vehicles for *in-vivo* research. *Challenges*: Uncontrolled gelation kinetics, non-uniform structure. *Solutions*: New methods for controlled, uniform cross-linking.
2.	*Polymer used:* Poly(N-isopropylacrylamide) & Poly(acrylic acid) *Type*: Network of semi-interpenetrating polymer. *Properties*: Synthetic hydrogel scaffolds.	*Polymer used*: Collagen *Forms*: Hydrogel and porous foam. *Applications*: Blood vessel, skin, cartilage, ligament, tendon repair. *Composites*: Combined with chitosan for composite hydrogels.
3.	*Polymer used:* Diblock Copolypeptide Amphiphiles *Properties*: Charged and hydrophobic segments. *Advantages*: Promising thermal stability and mechanical properties.	*Polymer Used*: Fibrin *Function*: Clotting protein, gel-forming. *Preparation*: Combining fibrinogen and thrombin. *Challenges*: Rapid degradation rate. *Applications*: Tissue engineering constructs
Artificial Proteins *Type*: Elastin-mimicking polypeptides. *Development*: Using recombinant DNA technology. *Applications*: Biomimetic hydrogels, synthetic technology for customized scaffolds.		

scaffolds in tissue engineering with regulated porosity and regarded as traditional practices (Fig. 9.4) (Dutta et al. 2017; Sachlos and Czernuszka 2003).

Advanced approaches to fabricate scaffolds: Recently, various sophisticated methodologies, including electrospinning, 3D printing, and hybrid molding techniques, innovated to create scaffolds with ability to replicate the ECM (Fig. 9.4) (Zhao et al. 2018).

9.5.2.1 Solvent Casting and Particulate Leaching

A polymer is dissolved in a volatile solvent, and salt particles or porogens are added. The mixture is cast into a mold or glass plate, evaporated, and progen leaches out, creating a porous scaffold. The scaffold is washed with water to remove salt particles, resulting in scaffolds with pore diameters ranging from 5 to 600 μm and porosity percentages between 50% and 90% (Sachlos and Czernuszka 2003; Varghese et al. 2024).

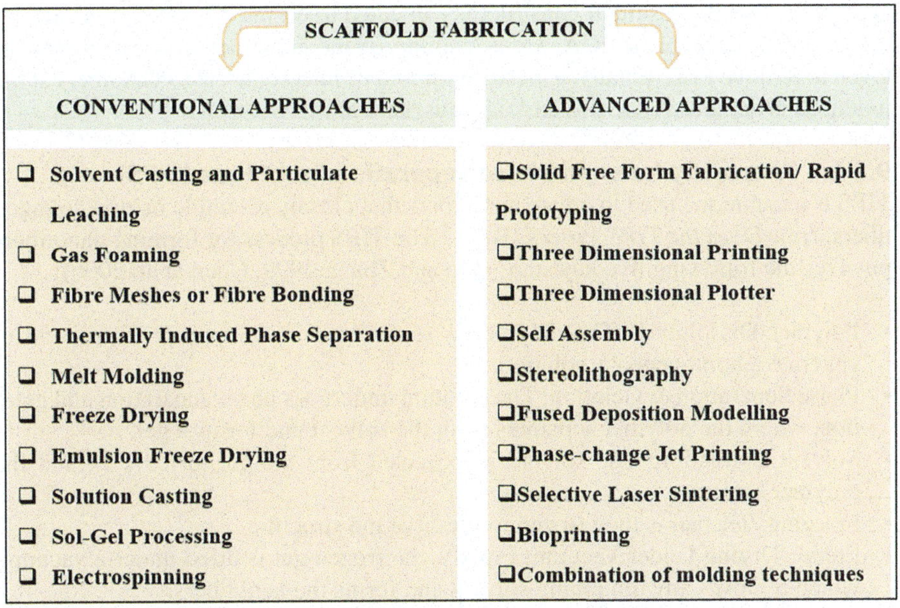

Fig. 9.4 List of conventional and advanced techniques used to fabricate scaffolds

9.5.2.2 Gas Foaming

A methodology to fabricate extremely porous scaffolds entails saturation of biodegradable polymers such as PLGA with physical blowing agents such as CO_2, nitrogen or chemical blowing agents like sodium bicarbonate ($NaHCO_3$)in lieu of organic solvents under elevated pressure conditions. Upon the rapid reduction of pressure to ambient levels, the solubility of CO_2 within the polymer diminishes, resulting in nucleation and subsequent introduction of gas bubbles or cells, typically in the size range of 100–500 μm. The evolution of gas engenders voids, thereby establishing a porous matrix within the polymer. This strategy mitigates the risks associated with organic solvents and maintains bioactive entities, yielding highly porous scaffolds with approximately 85% porosity and a pore diameter of 100 μm. Nevertheless, this approach is not without its limitations, including restricted control over pore dimensions, inadequate pore interconnectivity, and prolonged processing durations. Furthermore, it is applicable to both hydrophobic and hydrophilic polymers (Sachlos and Czernuszka 2003; Varghese et al. 2024). The anticancer gemcitabine scaffolds utilizing PLGA and CO_2 had an extremely high impregnation efficiency (>90%) (Adel et al. 2022).

9.5.2.3 Fiber Meshes or Fiber Bonding

Non-woven scaffolds made using PGA and poly(l-lactide) (PLL)were found to be insufficient in terms of mechanical strength and hence fiber bonding technique came into existence. In this methodology PLL is solubilized in highly volatile liquid like dicholoromethane (DCM) and deposited onto PGA mesh. The liquid is evaporated and the resultant construct is subjected to elevated temperature exceeding the melting point of PGA. Upon cooling of the construct, DCM treatment is given to remove PLL. Finally, a mesh composed of interconnected PGA fibers at the junctions is achieved (Sachlos and Czernuszka 2003).

9.5.2.4 Thermally Induced Phase Separation Technique (TIPS)

TIPS is a technique used to create nanofibers that closely resemble natural collagen fibers. *How Does the TIPS Process Work?* The TIPS process for forming nanofibers involves the following five key steps (Ma and Zhang 1999; Chen et al. 2006):

- Polymer Dissolution: The polymer is solubilized in an appropriate solvent to generate a homogeneous solution.
- Phase Separation & Gelation: The solution undergoes phase separation and gelation, where the polymer separates from the solvent and forms a gel.
- Solvent Extraction: The solvent is extracted from the gel to leave behind the polymer structure.
- Freezing: The gel is then frozen to preserve the structure.
- Freeze-Drying Under Vacuum: Finally, the frozen gel is dried under a vacuum, which removes any remaining solvent and forms the nanofibers.

The fibers produced using the TIPS technique has diameters ranging from 50 to 500 nanometers. These fibers exhibit a high porosity, often exceeding 98%, which gives the matrix a greater surface-to-volume ratio. The nanofibrous matrix created by TIPS has a noticeably larger surface-to-volume ratio than that of other materials, such as non-woven fabrics made with traditional textile methods or foams produced by different techniques. *What are the pros using Technique called TIPS?* One significant benefit of the TIPS technology is its adaptability, since it may be integrated with additional processing procedures, including particle leaching and 3D printing. This combination enables the creation of intricate 3D structures with well-defined pore topologies, rendering it extremely appropriate for advanced material design (Ma and Zhang 1999; Chen et al. 2006; Wei and Ma 2006; Liu and Ma 2004).

9.5.2.5 Melt Molding

In melt molding, the polymer and porogen particles are combined together. The mixture is subjected to elevated temperature exceeding the polymer's melting or glass transition temperature. Then solidification occurs where the mixture solidifies, forming the scaffold. Later extraction is performed through which the polymer porogen is extracted, leaving a network of pores in the other polymer (Sachlos and Czernuszka 2003). This technique excludes the use of organic solvents. A study was carried out using this technique where PLGA and PVA derived porous scaffolds are created using melt molding and later tested for bone growth by implanting them in rabbit skull defects. Scaffolds treated with PVA showed better bone growth than those without PVA treatment, as observed through MTT assays— (3-(4, 5-dimethylthiazolyl-2)-2, 5-diphenyltetrazolium bromide) and scanning electron microscope analysis. The enhanced bone development linked to the improved hydrophilicity of the PVA derived scaffolds, which improved their compatibility with cells (Adel et al. 2022).

9.5.2.6 Freeze Drying/Lyophilization

Freeze-drying is also called lyophilization that solidifies the solution at low-temperature and then removes the frozen solvent via sublimation, preserving the original texture. This technique allows for customization of pore structures by adjusting polymer solution parameters (concentration, viscosity) and processing conditions (lyophilization, drying parameters). Drugs can be introduced into the scaffold pores, with release rates controlled by the pore structure for continuous and steady release. In this technique the PLGA like synthetic polymers are added to organic solvents such as glacial acetic acid and benzene. The solution is allowed to freeze-dry, resulting in porous matrices. Collagen scaffolds are similarly produced by lyophilizing a collagen dispersion or solution, followed by freeze-drying. *So, how the pores are formed?* Collagen molecules are forced into the voids between the crystals when the fluid is frozen, forming ice crystals. These ice crystals are eliminated during the freeze-drying process, leaving pores behind. The freezing rate (faster freezing leads in smaller holes) and pH may be adjusted to influence the pore size. To get homogenous 3D pore configuration unidirectional solidification is required. Ethanol may be used to dehydrate collagen scaffolds, which can subsequently be dried in key locations. For lowering their solubility, antigenicity, and rate of breakdown, these scaffolds are frequently crosslinked through physical crosslinking which is acquired by dehydrothermal treatment, ultraviolet (UV) or gamma irradiation or chemical crosslinking that employs carbodiimides or bifunctional substances such as glutaraldehyde or hexamethylene DI. Chitin and alginate polymers could be lyophilized into scaffolds. This technique is time consuming and expensive (Sachlos and Czernuszka 2003; Xu et al. 2023; Varghese et al. 2024).

9.5.2.7 Emulsion Freeze Drying

According to this process, PGA is added to a DCM solution together with ultrapure water. After homogenizing the two immiscible phases that are produced, a water: oil emulsion is produced. The emulsion is quickly chilled in nitrogen and freeze-dried giving porous architecture (Sachlos and Czernuszka 2003).

9.5.2.8 Solution Casting

After dissolving PLGA in an organic solvent, such as chloroform, methanol is added to cause precipitation. To form the scaffold, the freeze-dried demineralized bone may be mixed with PLGA. The resultant composite material is then molded and heated for 24 hours at a temperature between 45 and 48 °C (Sachlos and Czernuszka 2003).

9.5.2.9 Sol-Gel Processing

The application of sol-gel techniques has been more popular in current era because of its exceptional chemical uniformity, lower processing temperatures, and capacity to regulate the shape and size of the final particles. The materials obtained from

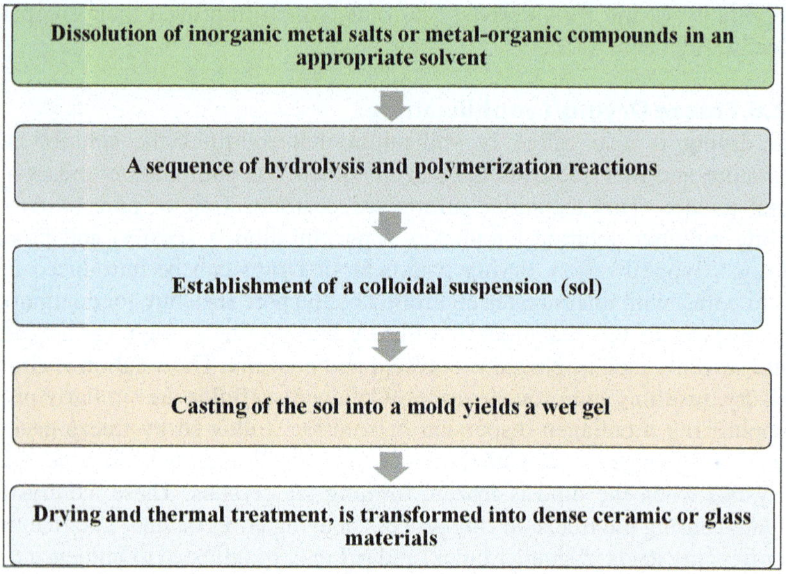

Fig. 9.5 Steps for synthesis of scaffolds through sol-gel processing

sol-gel techniques provide excellent matrices for a broad range of inorganic and organic molecules.

Hydrogels are attractive by their substantial water content and ability to interact well with living organisms, making them appropriate for a wide range of biological uses. Hydrogels could be made using naturally occurring polymers or from synthetic polymers (Fig. 9.5). Growth factors are released by diffusion, mechanical stimulation, hydrolytic disintegration, or in reaction to environmental cues. For instance, gelatin and dextran may be used to fabricate interpenetrating polymer hydrogels for drug administration, which can react to two different stimuli. By manipulating the physicochemical characteristics of the hydrogels, release process may be controlled, enabling a release duration that spans from a few days to many months. Hydrogels have a liquid form at room temperature but change into a solid-gel form at temperature of the human body. This property enables its application in a less intrusive way. Hydrogels have the ability to shield medicines, peptides, and proteins from harsh conditions, prolong the time they stay in the body, provide controlled and focused drug release, and show promise in wound healing. Sol-gel technology offers advantages for creating biological sensors, laser materials, and controlled drug delivery systems. Injectable hydrogels made of poly(N-isopropylacrylamide) that include cells, used for regenerating cartilage and nerves. Using Pluronic/heparin composite hydrogels, growth factors are delivered to promote angiogenesis, the process of creating new blood vessels. PEG-based hydrogels cross-linked with photosensitive compounds for the purpose of delivering chondrocytes and osteoblasts. Chitosan-polyol salt hydrogels that are sensitive to temperature and include bone morphogenic protein to enhance the production of bone and cartilage. PLGA modified with PEG hydrogels for long-lasting insulin release and cartilage restoration. Pluronic copolymers are used for the encapsulation of chondrocytes in the production of synthetic cartilage. Optimization of hydrogel pore size and degradation qualities is vital for ensuring hydrogels efficacy in certain applications. Additional disadvantages include the increased expenses related to the raw materials, significant reduction in size during the manufacturing process, the existence of remaining small holes, hydroxyl groups, and carbon, as well as potential health hazards caused by the prolonged duration of the organic solution's processing (Garg et al. 2012).

In scaffold-based bioprinting, characterized by a top-down methodology, an externally sourced biomaterial matrix, exemplified by a hydrogel, is utilized for printing while cellular entities are either introduced through seeding or incorporated within the bioink during the printing process (Collins et al. 2021).

9.5.2.10 Electrospinning

Electrospinning is a technique where electrical charge is used to create a mat of very fine fibers. This process was initially patented in 1902 by J.F. Cooley. A step by step process of electrospinning is described in (Fig. 9.6). A variety of materials including biodegradable polymers like PLGA and PCL, as well as PEO, PVA, collagen, silk proteins, and other peptides, may be transformed into nanofibers by the process of electrospinning. The scaffolds it produces provide an optimal structural characteristic for promoting cell proliferation and tissue development. This technology enables the production of very thin fibers with precise alignments, large surface areas, and high ratios of length to width. The pore geometry can be controlled, which enhances proliferation of cells in lab environments (in vitro) as well as in living organisms (in vivo). This method is easy to use and effective for creating both sheets and cylinder

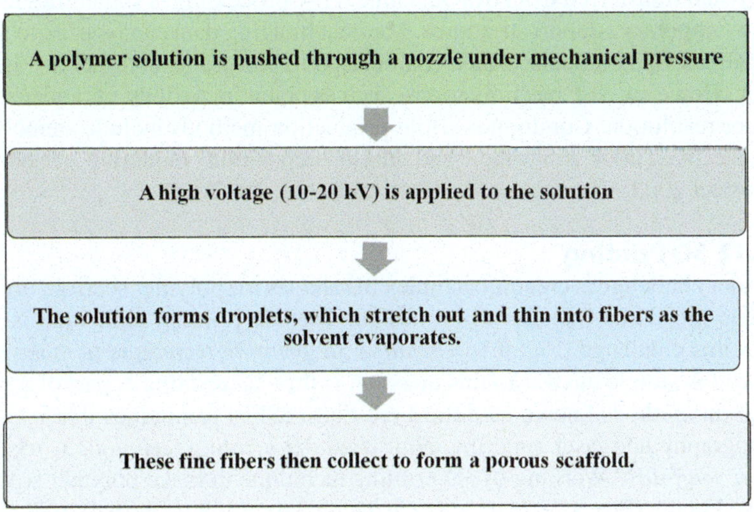

Fig. 9.6 A step by step process of electrospinning

forms. This approach is suitable for heat labile drugs since it utilizes solvents and elevated temperatures. A two-dimensional mesh with tiny holes is the main structure created by electrospinning, and it is frequently insufficient for efficient cell penetration and seeding. Seeding cells onto the electrospun matrix is the primary problem. The electrospun matrices can be improved for cell seeding by utilizing methods such as cryospinning or the use of a sacrificial biopolymer to generate holes of the appropriate size (Garg et al. 2012; Martins et al. 2007).

Electrospinning apparatus consist of syringe, needle, collector, syringe pump and high power voltage supply. Polymer solution is put in syringe. The spinneret feeds and charge the polymer solution, high voltage power source is applied on the solution of polymer ejected from the nozzle whereby an droplet (elongated in shape) is created at the needle tip called 'taylor cone' and collector made of aluminium collects the fibres that are electrospun and expelled from the feed. The mechanism working behind is that a jet emerges from the taylor cone, as the charged solution's surface tension is removed by the potential difference with the collector. The solvent is evaporated and the dried nanofibers are collected as a result of the erupting liquid jet creating fibers that proceed in the direction of the collector. So this method produces non-woven meshes, nanofibrous scaffolds (Khan and Tsukada 2014; Hutmacher 2001; Li and Tuan 2009; Lannutti et al. 2007).

9.5.2.11 Solid Free Form Fabrication (SFF)/Rapid Prototyping (RP)

SFF (Structured Fused Fabrication) is a 3D manufacturing process that creates scaffolds with customized forms and components for individual patients. This technique is crucial in tissue engineering due to its ability to regulate pore size, porosity, and distribution, enhancing the scaffold's ability to transport nutrients and oxygen. SFF involves sequential deposition of material, starting with a computer-generated model using CAD software. The model is then segmented into layers of cross-sections and transferred to an SFF machine. Post-processing is sometimes needed to remove temporary support structures. Medical imaging data can be used to generate personalized CAD models, which can be used to construct scaffolds for tissue restoration. However, SFF has limitations, such as potential toxicity of binder solutions and poor resolution. Common scaffold production methods include stereolithography, selective laser sintering, and fused deposition modeling (Sachlos and Czernuszka 2003; Varghese et al. 2024; Adel et al. 2022).

9.5.2.12 3D Printing

The major challenge is creating complex 3D shapes for polymer scaffold is essential to mimic the tissue that has to be healed. The 3D printing technique is used to address this challenge (Fig. 9.1d). *What is 3D printing technique to make polymer scaffolds?* A simple answer to this question is that 3D printing is part of a group of fabrication methods known as Solid Free-Form (SFF) techniques that include stereolithography and laser sintering. *How does 3D printing technique work to make polymer scaffolds?* Working of 3D printing technique to make polymer scaffolds is described in the (Fig. 9.7). *How to fabricate polymer scaffold with desired porosity?* Scaffolds could be formed porous by incorporating salt to the powder bed of the

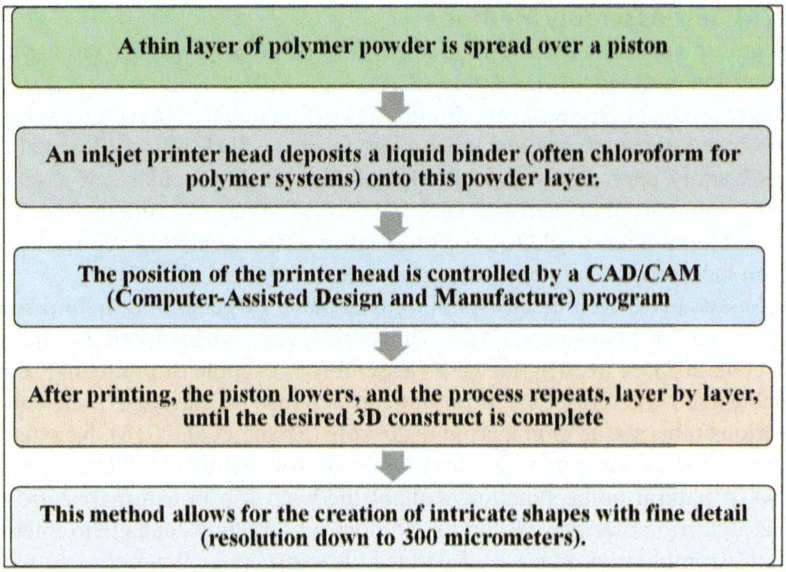

Fig. 9.7 Working of 3D printing technique to make polymer scaffolds

polymer during fabrication. PLLA and PLGA scaffolds are successfully created using 3D printing. The strength of these 3D printing-constructed scaffolds analogous to those made by conventional methods like compression molding. *What are the applications and advantages of 3D printing constructed scaffolds?* 3D printing-constructed scaffolds have been investigated for number of tissue engineering domains, including liver tissue organogenesis. Surface modifications of PLLA substrates for better cell adhesion have also been performed using 3D printing. Porous scaffolds with interconnected channels have been designed for hepatocyte (liver cell) transplantation. These scaffolds are promising for cell transplantation because their branching and porous structure may help promote blood vessel formation (angiogenesis) and tissue growth (Thomson et al. 2000).

Some 3D technologies that are frequently employed in tissue engineering include stereolithography 3D printing, FDM, SLS, Material Jetting and Bending Jetting 3D Printing (Perez-Puyana et al. 2020).

9.5.2.13 3D Plotter

This apparatus, engineered by scholars at the University of Freiburg, incorporates a dynamic extruder head (regulate along the x-, y-, and z-axis) & utilizes compressed air to expel a liquid or paste-like plotting substance. The extruder head is capable of being heated to the requisite temperature. Upon coming in contact with the substrate or an antecedent layer, the medium undergoes solidification. Figure 9.5 illustrates this apparatus (Sachlos and Czernuszka 2003).

9.5.2.14 Self-Assembly Methods

A bottom-up approach used to fabricate scaffolds out of peptides, nanostructures, and other materials is self-assembly (Zhang et al. 2018; Collins et al. 2021). This technique is capable to overcome limitations of top down approach reduced diffusion observed in large scaffolds. It is used to fabricate 3D scaffolds (Lu et al. 2013). Self assesmbly approach utilizes methodologies like molecular self-assembly of peptides employs engineered peptides that furnish functional support and facilitate bio-recognition (Nune et al. 2013). For instance, self-assembling peptide hydrogels has been incorporated into clinical applications (Gelain et al. 2021).

Self-assembly-driven electrospinning is capable of generating hybrid scaffolds characterized by honeycomb-like architectures that enhance cellular infiltration (Ding et al. 2021). Layer-by-layer self-assembly is capable of producing scaffolds, injectable nanostructures that hold significant promise for cartilage tissue constructs and various other tissue engineering endeavors (Zhang et al. 2018). New technologies have emerged known as 'scaffold-free' self assembling that allows for the replication of natural tissue function without the need for an external scaffold (Lee et al. 2017). Biomolecules, including peptides and proteins, engage in interactions and self-assemble to produce well-defined clear structures that correlate with specific functionalities. In particular environmental conditions, nano-structured fibers can be synthesized through the mechanism of molecular self-assembly (Silva et al. 2004; Zhang 2003). The diameter of fibers generated via molecular self-assembly remains considerably smaller than those fabricated through electrospinning techniques (Hartgerink et al. 2001).

9.5.2.15 Stereolithography

This method uses an UV laser beam to selectively polymerize a liquid photocurable monomer in order to produce 3D scaffolds. This process is known as stereolithography. Processing steps are as follows;

- A UV laser beam is directed over the photosensitive liquid resin surface based on data from a 3D model, typically stored in an STL file.
- The laser cures the resin layer by layer, forming the first solidified layer.
- After that the platform holding the model is brought down, and another layer of liquid resin is applied and solidified.
- This process is kept repeating till the 3D structure is complete.
- Upon the completion of the structural assembly, any residual uncured resin is meticulously removed, followed by an additional exposure to ultraviolet radiation to ensure the comprehensive curing of the product.

This method favors advantages such as it enables the formation of 3D structures with complex features and exact structural control, it is possible to attain homogeneity in the interconnectivity and pore structure and it is possible to preserve growth hormones and cell viability, which is essential for tissue engineering. Additionally, it has limits because to the time-consuming nature of the procedure, particularly as the resolution is increased. Several hardened polymers used in the procedure are non-biodegradable and might potentially provide toxicity risks. The technique is expensive due to the complexity and precision required. There is need for support structures and post-processing adds to the time and cost. A study suggests the stereolithography to fabricate calcium phosphate-based pellets and evaluate them for biocompatibility with MG-63 osteosarcoma cell lines and osseointegration in vivo. The pellets showed good biocompatibility and excellent osseointegration. Uniform pore structure and interconnectivity can be achieved by this technique (Sachlos and Czernuszka 2003; Varghese et al. 2024; Adel et al. 2022).

9.5.2.16 Fused Deposition Modelling (FDM)

FDM involves extruding a thin polymer strand in x- and y-axis directions to create a 3D physical model. The model is then lowered along the z-axis and repeated until the entire model is formed. For complex shapes, additional external structures may be created, which are usually eliminated manually upon completion, applicable on non-thermoplastic materials (Sachlos and Czernuszka 2003; Varghese et al. 2024).

9.5.2.17 Phase-Change Jet Printing

This system operates with two ink-jet print heads that deliver two different materials: one for building the model and another for supporting unconnected or overhanging features. *How does this system works?* The print heads generate and deposit molten microdroplets of material, heated above their melting point, in a drop-on-demand manner. These droplets solidify upon impact, forming beads. Overlapping beads form lines, and overlapping lines create layers. A rotary cutter arm can flatten each layer's surface to control thickness. The process repeats as the platform lowers, building the model layer by layer. After construction, the completed model remains intact after being submerged in a solution that dissolves the support material (Sachlos and Czernuszka 2003).

9.5.2.18 Selective Laser Sintering (SLS)

Using laser technology and a computationally created model as a guide, SLS is a well-known RP process that builds 3D structures by successively fusing powdered materials layer by layer. In a manner akin to stereolithography, SLS facilitates the formation of multilayered porous architectures characterized by regulated porosity and connectivity. It offers advantages such as improved pore interconnectivity and high porosity, reduced time spent processing. It also holds some drawbacks like removing uncured powder is difficult, difficulties in drawing intricacies like sharp edges. A study states that, for alveolar bone regeneration, iron oxide-doped DP-Bioglass scaffolds were produced using SLS. When the bioglass broke down, it produced ions that encouraged osteogenesis and bone mineralization while remaining non-cytotoxic (Varghese et al. 2024; Adel et al. 2022).

9.5.2.19 Bioprinting

Bioprinting is an advanced technology that constructs 3D tissues or organs using living cells or ECM components for future implantation. It utilizes a layer-by-layer sequential manufacturing approach, similar to 3D printing, to build structures from bioinks containing cells and biomaterials. Bioprinting utilizes bio inks and living cells to create 3D structures. Cells and other biomaterials are combined to create the composite bio ink used in bioprinting. Additionally, they may be manufactured from natural or synthetic polymers. Bioprinting can also be achieved through ink-jet printing, extrusion printing, and laser assisted bioprinting. Bioprinting has three process phases; pre-processing which involves obtaining bioimages via CT/MRI, creating 3D models with CAD, and selecting bioinks, processing of layer-by-layer bioprinting based on 2D stack data and post-processing where maturation in a bioreactor followed by evaluation of the construct takes place. There are two types of bioprinting approaches either scaffold based or scaffold free. In the former approach cells are deposited onto a 3D biomaterial matrix, mimicking the ECM to promote cell growth. Whereas in the later approach cells or tissues are directly patterned into structures like spheroids or honeycombs. It offers advantages such as allows the creation of tailored constructs that closely mimic the patient's tissues or organs. Layered construction leads to good pore interconnectivity, essential for tissue growth. Yet another technique imposes challenges like expensiveness and technical complexity (Varghese et al. 2024; Adel et al. 2022; Nikolova and Chavali 2019).

9.5.2.20 Combination of Molding Techniques

The process involves injecting biodegradable polymers, sometimes mixed with additives like salt particles or blowing agents, into a mold to create scaffolds with desired porosity and interconnectivity. Unlike other methods, organic solvents are not needed for injection molding, which are damaging to cell differentiation and tissue development, making it advantageous for tissue engineering. Microcellular injection involves combining particulate leaching with molding to achieve interconnected pores. By altering the kind and size of particles (example salt) or additives (example $NaHCO_3$) combined with the polymer, the size and interconnectivity of pores may be regulated. In a study PLA combined with salt particles and polyvinyl alcohol (PVA) was used to create scaffolds with controlled pore sizes. Also, $NaHCO_3$ could be used as a foaming agent to generate porous structures in microcellular injection molding. To enhance mechanical and biological qualities, scaffolds for bone tissue engineering have been developed. One example is scaffolds composed of ethylene vinyl alcohol mixed with nanoforsterite (Zhao et al. 2018).

9.6 Functionalization and Bioactivity

Scaffolds for tissue regeneration are designed considering parameters such as morphologic features, cell interactions with the matrix, and the transport of nutrients across cells. The basic structural feature of the scaffold stands on three tiers that include architecture at the macro level which constitutes anatomical features, organ, and patient specificity. Pore dimensions, porosity, interconnectivity of pores, and spatial configurations are examined at the microscopic level. The nano-scale architecture alters the surface by affixing signalling molecules that enhance cell adhesion, proliferation, and differentiation (Chia and Wu 2015). Enhancing tissue regeneration can be achieved by functionalizing bioinspired scaffolds at the nano level. This entails enhancing the biological performance of scaffold surfaces by altering them with bioactive chemicals, nanoparticles, or other stimuli. For repairing nerve, bone and osteochondral defects, such materials have significant promise for tissue regeneration applications.

9.6.1 Surface Modification Techniques

Following their introduction into the natural 3D microenvironment, artificial tissue scaffolds must not be rejected by the body. Diverse methodologies can alter the surface of the scaffold and evoke therapeutic uses. These methods replicate the microenvironment by altering the physical, chemical, and mechanical characteristics of the scaffold, therefore enhancing its biocompatibility and acceptability to the body. The objective of each of these measures was to augment the interaction amongst the scaffold and organic tissues (Chakraborty et al. 2023). To enhance the performance of scaffolds, several techniques for surface modification have been devised. These strategies may be classified into physical and chemical modifications:

Physical modifications refer to techniques that specifically target altering the topography, roughness, or wettability and porosity features of the scaffold without modifying the chemical composition of the bulk scaffolds. Several reported physical procedures to modify surface features by either sculpting at the surface level or introducing an additional active layer above the surface. Chemical changes denote procedures aimed at integrating new functional groups into the scaffold's surface (Wang et al. 2023).

Another technique is the physicochemical method for surface functionalization which includes plasma treatment.

Plasma treatment This physiochemical method alters the surface energy and chemistry of scaffolds, therefore increasing their hydrophilicity and their suitability for cell attachment. This technique entails subjecting the scaffold materials to a plasma comprising of ionized gasses, which modifies their surface energy and chemical composition, therefore improving their appropriateness for tissue engineering endeavours. The mechanism of action entails the interplay between energetic particles and the scaffold surface during plasma therapy. This specific interaction leads to polar functional groups' inclusion, enhancing the scaffolds hydrophilic characteristics. The inclusion of polar groups increases the affinity of water, so generating a surface that is more favourable and beneficial to the attachment and growth of cells. Plasma treatment alters the surface to improve compatibility with biological systems, which is crucial for tissue engineering purposes. This approach primarily modifies the surface of nanofibers. Nanofibers that have been treated with plasma are very useful in tissue engineering for the reason that they can exactly copy the extracellular matrix. This makes it easier for cells to adhere together and grow. This approach is an invaluable instrument in surmounting the constraints of hydrophobic electrospun nanofibers, hence enhancing their suitability for diverse medicinal applications (Niemczyk-Soczynska et al. 2020).

According to a study, poly(l-lactide) (PLLA) microfibrous scaffolds upon surface treatment with mild Ar or Ar-NH_3/H_2 plasmas, resulted in functionalization of surface with amine groups, while the Ar-NH_3/H_2 plasma treatment preserved the integrity of the fibrous structure and fibre roughness (Cheng et al. 2013).

Commonly used physical techniques are as following:

Physical adsorption Historically, the use of cell-adhesive proteins such as fibronectin, collagen, vitronectin, and laminin, together with extracellular matrix (ECM) analogues like chitosan and gelatin, has been a prevalent and favoured approach for improving biomaterial surfaces. Numerous biomimetic materials have been synthesised over this methodology. Biomimetic materials use biomolecular recognition to induce targeted cellular responses and guide the development of new tissues (Katti et al. 2008).

Layer-by-layer assembly (LbL) In tissue engineering, for the functionalization of scaffolds; the LbL method is an effective approach to improving their biological efficacy and therapeutic use. Through the successive layering of differently charged materials, this method enables precise management of surface characteristics, resulting in enhanced functioning. The LbL assembly technique is defined by the sequential adsorption of oppositely charged polyelectrolytes onto a substrate, facilitating the creation of uniform nanolayers. The principal driving force behind this process is electrostatic interactions, however additional factors can also add up (Zhang et al. 2018). Layer-by-layer (LbL) technology has many applications and may be used to create scaffolds with tailored properties for certain functions, including drug delivery and tissue remodelling (Borges et al. 2024). In orthopaedics, it can encapsulate pharmaceuticals or biological substances onto the surfaces of implants, therefore enabling regulated release and enhancing the process of osseointegration (Ma et al. 2023). The heparinized scaffolds demonstrated considerable promise in facilitating cell adhesion and proliferation through improved surface characteristics including increased hydrophilicity, enhanced cell adhesion, enhanced biocompatibility, regulated discharge of growth factors, and surface charge modification that impact protein adsorption and cell behaviour. This result indicates that this method may be beneficial in the advancement of cardiovascular implants (Nguyen and Tran 2022).

Langmuir-Blodgett (LB) deposition This technique stands as a widely recognized and established approach in the realm of materials fabrication on diverse substrates. This approach is especially advantageous for producing thin films with precise thickness and consistency, crucial for various applications, including scaffold functioning. The Langmuir-Blodgett method employs the transfer of a single or many layers of molecules from the interface between air and water onto a solid surface. This method is ideal for creating functional scaffolds in a variety of applications as it allows for precise control over the film's thickness and structural molecular arrangement.

Chemical modification techniques In this technique, there is a creation of chemical covalent bonds at the surface of scaffolds to alter the surface characteristics. The application of chemical functionalization allows for the formation of functional groups on the surface of scaffolds, therefore promoting the binding of diverse materials like nanoparticles, polymers, biomolecules, biologically active substances, and inorganic substances to the surface of the scaffold.

Several chemical techniques have been employed to change scaffold surfaces, including wet chemistry approaches, chemical crosslinking using crosslinking agents, and surface functionalization by chemicals with various linkers.

Wet chemistry techniques Typically, wet chemistry techniques for surface functionalization of scaffolds include the application of chemical treatments so that the functional groups such as carboxylic, amine, and phosphonate are incorporated. Although employing these techniques can greatly improve the wettability and bioactivity of scaffolds, they frequently require the use of hazardous chemicals and can compromise the structural stability of the scaffolds. Methods such as aminolysis, hydrolysis, and covalent grafting are often employed to attain the intended surface characteristics, with gelatin immobilization being especially efficient in augmenting cell adhesion and proliferation (Dulnik et al. 2022).

In the aminolysis method, ethylenediamine and 1,6 hexamethylenediamine are diamines that, when subjected to regulated conditions including concentration of diamine, temperature, duration, and pH, lead to the generation of activated amine functional groups onto the surface of scaffolds (Jeznach et al. 2019). An inherent drawback of this approach is the formation of insufficient functional groups on the surface (Bakry et al. 2018), which therefore hinders the intended cell adhesion (temporarily enhanced cell adhesion). The aminolysis process, which involves cleaving ester bonds and producing carboxylic acid and hydroxyl functional groups on scaffold surfaces, is less commonly used in tissue engineering due to its limitations. This process requires more time and demanding conditions, resulting in significant loss of essential amino groups. The hydrolysis process, on the other hand, generates hydrophilic substrates that enhance cell adhesion under acidic and alkaline conditions, requiring careful regulation of reaction parameters like temperature, time, and pH (Croll et al. 2004).

Chemical crosslinking This procedure entails the formation of chemical linkages among polymer chains, therefore enhancing the stability, mechanical robustness, and biocompatibility of the scaffolds. Gaseous crosslinking with different agents including 1-ethyl-3-(3-di-methyl-amino-propyl) carbo-di-imide hydrochloride (EDC)/N-hydroxyl succinimide (EDC/NHS), glutaraldehyde, genipin, and as well as utilizing chemical linkers for functionalization such as silane coupling agents, dopamine, and 1,6 hexamethylenediamine (Islam et al. 2021).

9.6.2 Incorporation of Bioactive Molecules

The biomaterial scaffolds may take the form of naturally occurring or artificially produced metallic, inorganic, organic, and/or hybrid materials. To enhance the accuracy in the engineered tissue process, biomaterials can undergo additional biofunctionalization using different bioactive agents. Bioactive agents encompass a wide range of substances, including naturally occurring or artificially produced cytokines, extracellular vesicles, growth factors, peptides, and small molecules which may have one or several roles. Bioactive compounds can be integrated into implants using approaches such as internal integration, electrostatic deposition, surface modification, and entrapment (Wu 2023).

The incorporation of bioactive materials on the scaffolds has implications for the treatment of bone-related disorders, inflammation, wounds, cancer, and more.

Much of the literature focuses on the application of scaffolds loaded with biologically active materials for the regeneration of bones. The ability of composite scaffolds to regenerate bone tissue may be greatly increased by including bioactive substances.

Although composite scaffolds are stronger with the ability for cell proliferation and differentiation, a biomolecule-releasing ability is much preferred. For segmental bone regeneration, gelatin-hydroxyapatite composite scaffolds filled with bone morphogenetic protein (BMP-2) were created. The findings demonstrated that scaffolds that release biomolecules can greatly help in the development of osteogens from bone marrow stem cells and fully restore the segmental bone deficiency within a 12-week timeframe. Moreover, the addition of growth factors (BMP) into composite scaffolds may result in a prolonged release, therefore expediting the progress of bone regeneration (Liu et al. 2009).

A demonstration by a recent study shows that the incorporation of bioactive glass nanoparticles to the composite scaffolds prepared with bacterial cellulose (BC) and silk fibroin (SF) an exceptional cytocompatibility and displayed the highest amount of ALP, osteocalcin, and alkaline phosphatase activity, RUNX 2 gene expression, and mineralization of bone (Niknafs et al. 2024).

Scaffolds are also loaded to release the drugs into the tissue microenvironment. To eliminate the infecting microorganisms that might have entered the body during the operative procedure for the placing of the scaffold into the body tissues, the scaffolds are laden with antibiotics. These substances are then gradually released into the surroundings to eradicate microorganisms (Kim et al. 2004).

Indeed, the scaffolds can also function as carriers for the transportation of bioactive substances such as proteins, DNA, and drugs to control biological processes within the microenvironment of the scaffold. The designing of structural scaffolds can be done to provide prolonged delivery of drugs by the incorporation of microspheres loaded with drugs into structures with regulated porosity. A recent study illustrates the fabrication of microspheres loaded with encapsulated DNA onto porous and non-porous structures and showed the release of plasmid for up to 21 days (Nof and Shea 2002).

9.6.3 Enhancement of Cell Adhesion and Proliferation

One important aspect of scaffold functionalization is the adhesion of cells to the surface of the tissues and their proliferation. This is controlled by the scaffold structure and the composition of the material which is crucial to effective tissue regeneration.

The preparation of composite scaffolds is highly beneficial in enhancing the cell adhesion properties. Alginate, a polymeric material that is biodegradable and biocompatible and broadly used in bone scaffolds has poor cell attachment due to the unavailability of cell attachment sites on its surface. When blended with gelatin, another natural polymer with peptide sequence Arg-Gly-Asp; supports adhesion of cells and their proliferation along with cell differentiation (Mostajeran et al. 2024).

Conductive nanofibrous scaffolds were fabricated by electrospinning polypyrrole /poly(ε-caprolactone)/gelatin (PPY/PCL/PG) with varying amounts of PPY added to the PCL/PG matrix. By progressively escalating the percentage of PPY (0–30%) in the blends. The composites of nanofibers with 15% PPY (PPG15) demonstrated the most stable characteristics in terms of mechanical qualities, conductivity, and biological degradation. The results of the cell proliferation assay showed that the PPG15 scaffold exhibited superior cellular attachment, proliferation, and production of cardiac-specific proteins compared to PPG30. This makes PPG15 an exceptional substrate for the management of cardiac defects and reconstruction of infarct myocardium (Kai et al. 2011).

9.7 Environmental Consideration

The environmental consequences of scaffold manufacturing are complex, including the management of waste, optimal use of resources, and consumption of energy. Cutting-edge research emphasizes novel methods for scaffold manufacturing that not only reduce environmental damage but also advance sustainability.

Optimization of Waste Utilization Experimental evidence shows that eggshell and sea-shell waste can be transformed into bioactive hydroxyapatite scaffolds, therefore mitigating the accumulation of trash in landfills and the subsequent environmental pressures (Hembrick-Holloman et al. 2020).

Vadalà et al. (2023) proposed that scaffolds specifically engineered for dark fermentation can effectively harness organic waste to generate bio-hydrogen, therefore effectively tackling waste management concerns and simultaneously producing environmentally friendly energy.

Efficiency in resource utilization Optimisation of materials: Structural manufacturing methods, particularly those incorporating calcium aluminate blends, prioritise the achievement of high porosity and mechanical strength. This emphasis on these properties can result in less material waste and enhanced resource efficiency (De Las Nieves González et al. 2019).

Energy consumption Implementing energy-efficient techniques, such as microwave-assisted synthesis, in scaffold fabrication can greatly reduce the energy consumption linked to conventional manufacturing processes (Hembrick-Holloman et al. 2020).

Although the above developments offer a promising perspective on the environmental consequences of scaffold manufacturing, there are still obstacles to expanding these environmentally friendly methods to satisfy industrial requirements without sacrificing performance or raising expenses.

9.7.1 Biodegradability and Eco-Friendly Materials

By facilitating the production of extracellular matrix (ECM) copies by cells and fibroblasts, the scaffold preserves and revitalizes the tissue's function. Biodegradability, pore interconnectivity, biocompatibility, maximal porosity, and mechanical properties are important considerations in scaffold design. In order for tissues to function properly, the extracellular matrix (ECM) gives those mechanical properties, structural support, and the ability to carry signaling chemicals. The extracellular matrix of the tissue being studied should be almost identical to the scaffold utilized to create artificial tissue (Chan and Leong 2008).

Tissue engineering depends on several essential elements, such as the careful selection of cells, their isolation and expansion, and the choice of scaffold biomaterials. Scientific research is increasingly focused on biobased macromolecules that imitate biological settings and natural matrices. Bio-based materials are well-suited for a range of tissue engineering applications by virtue of their capacity to be customized to meet the specific functional requirements of specialized tissue (Hule and Pochan 2007).

In tissue engineering applications, bio-based scaffolds establish the suitable cell-extracellular matrix environment, which influences cellular behaviour (Wheeldon et al. 2011). Composite materials provide enhanced characteristics, versatile combinations, and innovative structuring, ultimately establishing the foundation for Technological Engineering.

Biodegradable and eco-friendly biomaterials employed in scaffold construction encompass natural polymers such as collagen, chitosan, and alginate. A list of biomaterials is mentioned in the above sections.

Sustainable biomaterials are of fundamental importance in reducing the environmental impact of medical treatments. Conventional materials often depend on finite resources and entail processes that release pollutants and require substantial energy (Hutmacher 2000). Conversely, the use of natural biomaterials produced from renewable sources can greatly mitigate these effects. For example, chitosan, derived from crustacean shells, exhibits biodegradability and low toxicity, which is in line with eco-friendly efforts (Dash et al. 2011).

9.7.2 Life Cycle Assessment of Biomaterials

To assess extent of changes in the dimensions of the scaffold undergoing degradation, morphological analysis employs light microscopy and advanced methods which includes TEM and SEM (Silva et al. 2018).

Physical testing is the assessment of mechanical characteristics, such as stress, strain, and modulus, using electronic testers or rheometers. An analysis of this nature enhances understanding of the chronological development of the scaffold's function (Maisani et al. 2018). Molecular weight monitoring denotes the identification of degradation by a decrease in molecular weight. To measure this decrease, techniques such as matrix-assisted laser desorption/ionization (MALDI) or gel permeation chromatography (GPC) might be employed. (Hamid et al. 2010).

Differential scanning calorimetry (DSC) (Little et al. 2018) and X-ray diffraction (XRD) (Kamali et al. 2019) are useful techniques for studying changes in crystallinity observed during degradation processes. The identification of degradation products in biochemical analysis is achieved by the use of techniques such as NMR, FTIR, and UV-Vis spectroscopy. The mentioned analytical techniques provide valuable information about the chemical transformations occurring within the scaffold.

9.8 Challenges and Future Perspectives

Although bioinspired scaffolds have potential, there are still difficulties in achieving a desirable balance between mechanical characteristics and biological reactions impeding their clinical use. Therefore, continuous research is required to enhance scaffold designs for clinical use.

9.8.1 Technical Challenges

Material challenges The selection of materials is crucial in scaffold design since it directly affects biocompatibility, biodegradability, mechanical characteristics, and the overall effectiveness of tissue engineering applications (Hutmacher 2000). Commonly used materials include natural polymeric materials such as chitosan, collagen, and alginate, as well as synthetic polymeric substances like polylactic acid (PLA) and polycaprolactone (PCL). Composites that collectively provide the benefits of both natural as well as synthetic polymers are often used (O'Brien 2011).

The optimal scaffold material should exhibit appropriate mechanical characteristics, and degradation rates, and replicate the biological milieu of the native tissue. Achieving equilibrium among these characteristics in a single material is a formidable task. As an illustration, natural polymers demonstrate exceptional compatibility with living organisms but frequently lack the necessary mechanical strength, whereas synthetic polymers may provide adjustable mechanical characteristics but can elicit negative biological reactions (Zhang et al. 2021).

Achieving scalability and reproducibility Factors like as inconsistency in natural material origins and variations across batches can impact the performance and dependability of scaffolds. The expansion from laboratory-scale to industrial-scale manufacturing exacerbates all the previously described difficulties. Adjustments to the manufacturing line are often required during the scaling process to ensure consistency and quality, which can be both economically and technically challenging. Furthermore, industrial production on a big scale must adhere to rigorous regulatory requirements, which need thorough documentation and validation processes (Atala et al. 2012).

Precision The attainment of the intended scaffold architecture with a high level of precision is necessary to successfully direct genetic activity and organ structure. Although advanced methods like as 3D printing can provide such accuracy, they are constrained by the range of materials they can manipulate (Murphy and Atala 2014).

Pore size and interconnectivity Precise pore size and interconnectedness are essential for the efficient transport of nutrients, elimination of waste, and formation of blood vessels. Achieving the ideal porosity in scaffold design while preserving structural integrity is a complex task (Karageorgiou and Kaplan 2005).

Controlled degradation It refers to the process of scaffold deterioration aligning exactly with the rates of tissue development, therefore offering support until the new tissue reaches maximum maturity. Ensuring precise control over degradation rates can be challenging, particularly in composite materials (Tajvar et al. 2023).

9.8.2 Biological Challenges

Biocompatibility and immunogenicity Natural materials can elicit immunological responses, but synthetic materials may not facilitate ideal cell attachment and proliferation. Mitigating the risk of persistent inflammation or rejection caused by materials continues to be a major challenge (Sadowska and Ginebra 2020).

Bioactivity and Cell Signaling Scaffolds must possess both non-toxicity and bioactivity, enabling interactions between cells and the growth matrix and stimulating particular biological processes. Continued investigation is focused on improving the bioactivity of scaffolds to direct tissue regeneration (Rao et al. 2018).

Tissue-Specific Requirements Various tissues possess distinct biological, biochemical and mechanical needs. The construction of scaffolds customized for particular tissue types necessitates an insightful understanding of the unique attributes of each tissue and the capacity to accurately reproduce these complexities within the scaffold (Bakhshandeh et al. 2017).

Cell seeding and distribution The task of achieving consistent cell distribution within the scaffold and preserving optimal cell viability before and after seeding poses significant challenges. Suboptimal cell seeding might result in uneven tissue development (Bueno et al. 2007).

Vascularization Vascularization is the process of generating new microcapillaries within the scaffold to achieve successful integration of the tissue. Insufficient vascularization inhibits food delivery and causes cell death in thicker constructs, therefore restricting the size of viable engineered tissues (Grüneboom et al. 2019).

Hypersensitivity reactions Although a multitude of naturally occurring scaffold materials such as pectin, bacterial cellulose, chitosan sulfate, alginate, gellan gum, hyaluronan, β-chitin, silk fibroin, collagen, gelatin, glycosaminoglycans, and soy protein makes them cost-effective and biocompatible; they can induce minor hypersensitivity reactions in some cases (Nair and Gayathri 2023).

9.8.3 Regulatory and Ethical Considerations

The development and implementation of bioinspired scaffolds in tissue engineering are not without significant regulatory and ethical challenges. As these scaffolds transition from research laboratories to clinical applications, they must undergo a thorough assessment to ensure their safety, efficacy, and ethical acceptability. Regulatory considerations are particularly critical, given that bioinspired scaffolds often involve new materials, advanced manufacturing techniques, and complex biological interactions that do not always fit neatly into existing regulatory frameworks.

One of the primary regulatory challenges is the classification of bioinspired scaffolds. These products often blur the lines between medical devices, drugs, and biologics, each of which is subject to different regulatory pathways. For instance, scaffolds that incorporate biologically active molecules, such as stem cells or growth factors, may be classified as combination products, which require a more comprehensive evaluation involving multiple regulatory agencies (Hollister and Murphy 2011). This complexity can lead to prolonged approval processes, increasing the time and cost associated with bringing bioinspired scaffolds to market.

Another significant regulatory challenge is the lack of standardized testing protocols for bioinspired scaffolds. Unlike traditional medical devices, these scaffolds often involve dynamic interactions with living tissues, which are difficult to replicate in pre-clinical models. Therefore, there is an onus for the development of new testing methods that can accurately assess the performance of bioinspired scaffolds in a biologically relevant context. These methods should account for factors such as biodegradability, bioactivity, and the potential for immune responses, which are critical for the long-term success of these products (Shyam et al. 2023).

Ethical parameters also play a crucial role in the development and application of bioinspired scaffolds. One of the key ethical concerns is the source of biological materials used in scaffold fabrication. For example, scaffolds that incorporate animal-derived components, such as collagen or gelatin, raise questions about animal welfare and the sustainability of sourcing practices. Additionally, the use of human-derived materials, such as stem cells or ECM components, must be carefully regulated to ensure that they are obtained and used in a manner consistent with ethical guidelines (de Kanter et al. 2023).

Another ethical issue is the potential for unequal access to bioinspired scaffolds. These advanced technologies are often expensive to develop and produce, which could limit their availability to patients in low-resource settings. Ensuring equitable access to these life-saving therapies is an ethical imperative that must be addressed by policymakers, healthcare providers, and industry stakeholders.

9.8.4 Future Directions in Bioinspired Scaffold Research

The future of bioinspired scaffold research is poised to bring about significant advancements in tissue engineering, driven by ongoing innovations in materials science, fabrication techniques, and biological understanding. Several key directions are expected to shape the future landscape of this field. The creation of intelligent scaffolds that can adjust to changes in their environment is an exciting line of research. These scaffolds are designed to mimic the flexible nature of real tissues, changing their features in response to external stimuli like as mechanical strain, pH changes, or the presence of certain proteins. For example, smart scaffolds that in response to local tissue damage release growth factors; could enhance the body's natural healing processes, leading to more effective and timely regeneration of tissues (Percival et al. 2024). One potential topic for further investigation is the amalgamation of nanotechnology into bioinspired scaffolds. Nanomaterials provide distinct parameters that may be applied for the fabrication of scaffolds with improved bioactivity and mechanical robustness. For example, the addition of nanofibers or nanoparticles to scaffolds may enhance cell adhesion, proliferation, and differentiation, resulting in stronger tissue development. In addition, nanotechnology may be used to create scaffolds that possess antimicrobial characteristics, hence decreasing the likelihood of infection in implanted devices (Zhang et al. 2023).

The use of bioprinting is another promising direction for the future of bioinspired scaffolds. Bioprinting allows for the precise placement of cells, biomaterials, and bioactive molecules in a 3D structure, enabling the formation of complex, tissue-like scaffolds. This technique offers the potential to produce personalized scaffolds for patients that are tailored to the individual's anatomy and pathology, improving the likelihood of successful tissue integration and function. Moreover, combining bioprinting with imaging technologies, such as MRI or CT scans, creates exact match scaffolds as that of damaged tissue, providing a highly personalized approach to tissue engineering (Li et al. 2023). The creation of sustainable and environmentally friendly materials for scaffold construction is a crucial area for further study. Materials that are biodegradable, derived from renewable resources, and made using green chemical procedures are becoming more and more necessary as environmental effect concerns increase. Future studies will probably concentrate on creating novel biomaterials that satisfy these requirements while maintaining the biological and mechanical characteristics required for efficient tissue regeneration (Mahmud et al. 2023).

Finally, advances in computational modeling and machine learning are expected to contribute significantly in the future of bioinspired scaffold research. These tools can be used to model the behavior of scaffolds in biological environments, predict their interactions with cells and tissues, and optimize their design for specific applications. By integrating computational approaches with experimental data, researchers can accelerate the development of new scaffolds and reduce the time and cost associated with bringing these products to market (Pais et al. 2023).

References

Aaliya B, Sunooj KV, Lackner M (2021) Biopolymer composites: a review. Int J Biobased Plast 3(1):40–84

Adel IM, ElMeligy MF, Elkasabgy NA (2022) Conventional and recent trends of scaffolds fabrication: a superior mode for tissue engineering. Pharmaceutics 14(2):306

Alaribe FN, Manoto SL, Motaung SC (2016) Scaffolds from biomaterials: advantages and limitations in bone and tissue engineering. Biologia 71(4):353–366

Alizadeh-Osgouei M, Li Y, Wen C (2019) A comprehensive review of biodegradable synthetic polymer-ceramic composites and their manufacture for biomedical applications. Bioact Mater 4:22–36

Arif ZU, Khalid MY, Sheikh MF, Zolfagharian A, Bodaghi M (2022) Biopolymeric sustainable materials and their emerging applications. J Environ Chem Eng 10(4):108159

Atala A, Kasper FK, Mikos AG (2012) Engineering complex tissues. Sci Transl Med 4(160). https://doi.org/10.1126/scitranslmed.3004890

Aye SS, Li R, Boyd-Moss M, Long B, Pavuluri S, Bruggeman K, Wang Y, Barrow CR, Nisbet DR, Williams RJ (2018) Scaffolds formed via the non-equilibrium supramolecular assembly of the synergistic ECM peptides RGD and PHSRN demonstrate improved cell attachment in 3D. Polymers 10(7):690

Baino F, Novajra G, Vitale-Brovarone C (2015) Bioceramics and scaffolds: a winning combination for tissue engineering. Front Bioeng Biotechnol 3:202

Bakhshandeh B, Zarrintaj P, Oftadeh MO, Keramati F, Fouladiha H, Sohrabi-jahromi S, Ziraksaz Z (2017) Tissue engineering; strategies, tissues, and biomaterials. Biotechnol Genet Eng Rev 33(2):144–172. https://doi.org/10.1080/02648725.2018.1430464

Bakry A, Darwish MSA, El Naggar AMA (2018) Assembling of hydrophilic and cytocompatible three-dimensional scaffolds based on aminolyzed poly(l -lactide) single crystals. New J Chem 42(20):16930–16939. https://doi.org/10.1039/C8NJ03205J

Bal Z, Kaito T, Korkusuz F, Yoshikawa H (2020) Bone regeneration with hydroxyapatite-based biomaterials. Emerg Mater 3:521–544

Banerjee A, Pavane MS, Banu LH, Gopikar AS, Elizabeth KR, Pathak S (2021) Traditional medicine for aging-related disorders: implications for drug discovery. In: Stem cells and aging. Academic, pp 281–297

Bansal M, Dravid A, Aqrawe Z, Montgomery J, Wu Z, Svirskis D (2020) Conducting polymer hydrogels for electrically responsive drug delivery. J Control Release 328:192–209

Bartoli M, Piatti E, Tagliaferro A (2023) A short review on nanostructured carbon containing biopolymer derived composites for tissue engineering applications. Polymers 15(6):1567

Beheshtizadeh N, Gharibshahian M, Bayati M, Maleki R, Strachan H, Doughty S, Tayebi L (2023) Vascular endothelial growth factor (VEGF) delivery approaches in regenerative medicine. Biomed Pharmacother 166:115301

Borges J, Zeng J, Liu XQ, Chang H, Monge C, Garot C, Ren K, Machillot P, Vrana NE, Lavalle P, Akagi T, Matsusaki M, Ji J, Akashi M, Mano JF, Gribova V, Picart C (2024) Recent developments in layer-by-layer assembly for drug delivery and tissue engineering applications. Adv Healthc Mater 13(8):2302713. https://doi.org/10.1002/adhm.202302713

Bružauskaitė I, Bironaitė D, Bagdonas E, Bernotienė E (2016) Scaffolds and cells for tissue regeneration: different scaffold pore sizes—different cell effects. Cytotechnology 68(3):355–369

Bueno EM, Laevsky G, Barabino GA (2007) Enhancing cell seeding of scaffolds in tissue engineering through manipulation of hydrodynamic parameters. J Biotechnol 129(3):516–531. https://doi.org/10.1016/j.jbiotec.2007.01.005

Calori IR, Braga G, de Jesus PD, Bi H, Tedesco AC (2020) Polymer scaffolds as drug delivery systems. Eur Polym J 129:109621

Carotenuto F, Politi S, Ul Haq A, De Matteis F, Tamburri E, Terranova ML, Teodori L, Pasquo A, Di Nardo P (2022) From soft to hard biomimetic materials: tuning micro/nano-architecture of scaffolds for tissue regeneration. Micromachines 13(5):780

Chai C, Leong KW (2007) Biomaterials approach to expand and direct differentiation of stem cells. Mol Ther 15(3):467–480

Chakraborty R, Anoop AG, Thakur A, Mohanta GC, Kumar P (2023) Strategies to modify the surface and bulk properties of 3D-printed solid Scaffolds for tissue engineering applications. ACS Omega 8(6):5139–5156. https://doi.org/10.1021/acsomega.2c05984

Chan BP, Leong KW (2008) Scaffolding in tissue engineering: general approaches and tissue-specific considerations. Eur Spine J 17(S4):467–479. https://doi.org/10.1007/s00586-008-0745-3

Chen FM, Liu X (2016) Advancing biomaterials of human origin for tissue engineering. Prog Polym Sci 53:86–168

Chen VJ, Smith LA, Ma PX (2006) Bone regeneration on computer-designed nano-fibrous scaffolds. Biomaterials 27(21):3973–3979

Chen M, Jiang R, Deng N, Zhao X, Li X, Guo C (2022a) Natural polymer-based scaffolds for soft tissue repair. Front Bioeng Biotechnol 10:954699

Chen Y, Dong X, Shafiq M, Myles G, Radacsi N, Mo X (2022b) Recent advancements on three-dimensional electrospun nanofiber scaffolds for tissue engineering. Adv Fiber Mater 4(5):959–986

Cheng Q, Lee BL-P, Komvopoulos K, Yan Z, Li S (2013) Plasma surface chemical treatment of Electrospun Poly(l -Lactide) microfibrous Scaffolds for enhanced cell adhesion, growth, and infiltration. Tissue Eng A 19(9–10):1188–1198. https://doi.org/10.1089/ten.tea.2011.0725

Chia HN, Wu BM (2015) Recent advances in 3D printing of biomaterials. J Biol Eng 9:1–4

Colazo JM, Evans BC, Farinas AF, Al-Kassis S, Duvall CL, Thayer WP (2019) Applied bioengineering in tissue reconstruction, replacement, and regeneration. Tissue Eng Part B Rev 25(4):259–290

Collins MN, Ren G, Young K, Pina S, Reis RL, Oliveira JM (2021) Scaffold fabrication technologies and structure/function properties in bone tissue engineering. Adv Funct Mater 31(21):2010609

Correia C, Bhumiratana S, Yan LP, Oliveira AL, Gimble JM, Rockwood D, Kaplan DL, Sousa RA, Reis RL, Vunjak-Novakovic G (2012) Development of silk-based scaffolds for tissue engineering of bone from human adipose-derived stem cells. Acta Biomater 8(7):2483–2492

Croll TI, O'Connor AJ, Stevens GW, Cooper-White JJ (2004) Controllable surface modification of Poly(lactic- *c o* -glycolic acid) (PLGA) by hydrolysis or Aminolysis I: physical, chemical, and theoretical aspects. Biomacromolecules 5(2):463–473. https://doi.org/10.1021/bm0343040

Dash M, Chiellini F, Ottenbrite RM, Chiellini E (2011) Chitosan—A versatile semi-synthetic polymer in biomedical applications. Prog Polym Sci 36(8):981–1014. https://doi.org/10.1016/j.progpolymsci.2011.02.001

De Groot JH, Nijenhuis AJ, Bruin P, Pennings AJ, Veth RP, Klompmaker J, Jansen HW (1990) Use of porous biodegradable polymer implants in meniscus reconstruction. (1) Preparation of porous biodegradable polyurethanes for the reconstruction of meniscus lesions. Colloid Polym Sci 268:1073–1081 de Kanter AJ, Jongsma KR, Verhaar MC, Bredenoord AL (2023) The ethical implications of tissue engineering for regenerative purposes: a systematic review. Tissue Eng Part B Rev 29(2):167–187

De Las Nieves González Vicuna I, Grancianinov KJS, Dos Santos KW, Dos Santos Ortega F, De Camargo Reis Mello D, De Vasconcellos LMR, Costa MS, Pandolfelli VC, De Oliveira IR (2019) Scaffolds' production based on calcium aluminate blends and their biological properties. Res Biomed Eng 35(2):131–141. https://doi.org/10.1007/s42600-019-00015-0

Dey S, Veerendra GT, Babu PA, Manoj AP, Nagarjuna K (2024) Degradation of plastics waste and its effects on biological ecosystems: a scientific analysis and comprehensive review. Biomed Mater Devices 2(1):70–112

Ding Y, Li W, Schubert DW, Boccaccini AR, Roether JA, Santos HA (2021) An organic-inorganic hybrid scaffold with honeycomb-like structures enabled by one-step self-assembly-driven electrospinning. Mater Sci Eng C 124:112079

Dobriţa CI, Bădănoiu AI, Voicu G, Nicoară AI, Dumitru SM, Puşcaşu ME, Chiriac Ş, Ene R, Iordache F (2023) Porous bioceramic scaffolds based on akermanite obtained by 3D printing for bone tissue engineering. Ceram Int 49(22):35898–35906

Du X, Fu S, Zhu Y (2018) 3D printing of ceramic-based scaffolds for bone tissue engineering: an overview. J Mater Chem B 6(27):4397–4412

Du X, Wei D, Huang L, Zhu M, Zhang Y, Zhu Y (2019) 3D printing of mesoporous bioactive glass/silk fibroin composite scaffolds for bone tissue engineering. Mater Sci Eng C 103:109731

Dulnik J, Jeznach O, Sajkiewicz P (2022) A comparative study of three approaches to fibre's surface functionalization. J Funct Biomater 13(4):272. https://doi.org/10.3390/jfb13040272

Dutta RC, Dey M, Dutta AK, Basu B (2017) Competent processing techniques for scaffolds in tissue engineering. Biotechnol Adv 35(2):240–250

Echeverria Molina MI, Malollari KG, Komvopoulos K (2021) Design challenges in polymeric scaffolds for tissue engineering. Front Bioeng Biotechnol 9:617141

El Bialy I, Jiskoot W, Reza NM (2017) Formulation, delivery and stability of bone morphogenetic proteins for effective bone regeneration. Pharm Res 34:1152–1170

Farokhi M, Jonidi Shariatzadeh F, Solouk A, Mirzadeh H (2020) Alginate based scaffolds for cartilage tissue engineering: a review. Int J Polym Mater Polym Biomater 69(4):230–247

Gaharwar AK, Singh I, Khademhosseini A (2020) Engineered biomaterials for in situ tissue regeneration. Nat Rev Mater 5(9):686–705

Garg T, Singh O, Arora S, Murthy RS (2012) Scaffold: a novel carrier for cell and drug delivery. Crit Rev Ther Drug Carrier Syst 29(1):1–63

Gattazzo F, Urciuolo A, Bonaldo P (2014) Extracellular matrix: a dynamic microenvironment for stem cell niche. Biochim Biophys 1840(8):2506–2519

Geissdoerfer M, Savaget P, Bocken NM, Hultink EJ (2017) The circular economy–a new sustainability paradigm? J Clean Prod 143:757–768

Gelain F, Luo Z, Rioult M, Zhang S (2021) Self-assembling peptide scaffolds in the clinic. NPJ Regener Med 6(1):9

Geyer R, Jambeck JR, Law KL (2017) Production, use, and fate of all plastics ever made. Sci Adv 3(7):e1700782

Gilaberte Y, Prieto-Torres L, Pastushenko I, Juarranz Á (2016) Anatomy and function of the skin. In: Nanoscience in dermatology. Academic, pp 1–14

Gritsch L, Meng D, Boccaccini AR (2017) Nanostructured biocomposites for tissue engineering scaffolds. In: Biomedical composites. Elsevier, pp 501–542

Grüneboom A, Hawwari I, Weidner D, Culemann S, Müller S, Henneberg S, Brenzel A, Merz S, Bornemann L, Zec K, Wuelling M, Kling L, Hasenberg M, Voortmann S, Lang S, Baum W, Ohs A, Kraff O, Quick HH et al (2019) A network of trans-cortical capillaries as mainstay for blood circulation in long bones. Nat Metab 1(2):236–250. https://doi.org/10.1038/s42255-018-0016-5

Guarino V, Gloria A, Raucci MG, De Santis R, Ambrosio L (2012) Bio-inspired composite and cell instructive platforms for bone regeneration. Int Mater Rev 57(5):256–275

Hamid ZAA, Blencowe A, Ozcelik B, Palmer JA, Stevens GW, Abberton KM, Morrison WA, Penington AJ, Qiao GG (2010) Epoxy-amine synthesised hydrogel scaffolds for soft-tissue engineering. Biomaterials 31(25):6454–6467. https://doi.org/10.1016/j.biomaterials.2010.05.008

Hardy JG, Lee JY, Schmidt CE (2013) Biomimetic conducting polymer-based tissue scaffolds. Curr Opin Biotechnol 24(5):847–854

Hartgerink JD, Beniash E, Stupp SI (2001) Self-assembly and mineralization of peptide-amphiphile nanofibers. Science 294(5547):1684–1688

Hembrick-Holloman V, Samuel T, Mohammed Z, Jeelani S, Rangari VK (2020) Ecofriendly production of bioactive tissue engineering scaffolds derived from egg- and sea-shells. J Mater Res Technol 9(6):13729–13739. https://doi.org/10.1016/j.jmrt.2020.09.093

Hiruta Y (2022) Poly (N-isopropylacrylamide)-based temperature-and pH-responsive polymer materials for application in biomedical fields. Polym J 54(12):1419–1430

Hollister SJ, Murphy WL (2011) Scaffold translation: barriers between concept and clinic. Tissue Eng Part B Rev 17(6):459–474

Hule RA, Pochan DJ (2007) Polymer nanocomposites for biomedical applications. MRS Bull 32(4):354–358. https://doi.org/10.1557/mrs2007.235

Hussain Z, Mehmood S, Liu X, Liu Y, Wang G, Pei R (2023) Decoding bone-inspired and cell-instructive cues of scaffolds for bone tissue engineering. Eng Regener 5:21

Hutmacher DW (2000) Scaffolds in tissue engineering bone and cartilage. Biomaterials 21(24):2529–2543. https://doi.org/10.1016/S0142-9612(00)00121-6

Hutmacher DW (2001) Scaffold design and fabrication technologies for engineering tissues—state of the art and future perspectives. J Biomater Sci Polym Ed 12(1):107–124

Islam MM, AbuSamra DB, Chivu A, Argüeso P, Dohlman CH, Patra HK, Chodosh J, González-Andrades M (2021) Optimization of collagen chemical crosslinking to restore biocompatibility of tissue-engineered Scaffolds. Pharmaceutics 13(6):832. https://doi.org/10.3390/pharmaceutics13060832

Jadbabaei S, Kolahdoozan M, Naeimi F, Ebadi-Dehaghani H (2021) Preparation and characterization of sodium alginate–PVA polymeric scaffolds by electrospinning method for skin tissue engineering applications. RSC Adv 11(49):30674–30688

Jeznach O, Kolbuk D, Sajkiewicz P (2019) Aminolysis of various aliphatic polyesters in a form of nanofibers and films. Polymers 11(10):1669. https://doi.org/10.3390/polym11101669

Jiang S, Wang M, He J (2021) A review of biomimetic scaffolds for bone regeneration: toward a cell-free strategy. Bioeng Transl Med 6(2):e10206

Joshi SR, Pendyala GS, Shah P, Mopagar VP, Padmawar N, Padubidri M (2020) Scaffolds—the ground for regeneration: a narrative review. J Int Soc Prevent Commun Dentis 10(6):692–699

Joshi A, Choudhury S, Gugulothu SB, Visweswariah SS, Chatterjee K (2022) Strategies to promote vascularization in 3D printed tissue scaffolds: trends and challenges. Biomacromolecules 23(7):2730–2751

Julier Z, Park AJ, Briquez PS, Martino MM (2017) Promoting tissue regeneration by modulating the immune system. Acta Biomater 53:13–28

Kai D, Prabhakaran MP, Jin G, Ramakrishna S (2011) Polypyrrole-contained electrospun conductive nanofibrous membranes for cardiac tissue engineering. J Biomed Mater Res A 99A(3):376–385. https://doi.org/10.1002/jbm.a.33200

Kamali A, Oryan A, Hosseini S, Ghanian MH, Alizadeh M, Baghaban Eslaminejad M, Baharvand H (2019) Cannabidiol-loaded microspheres incorporated into osteoconductive scaffold enhance mesenchymal stem cell recruitment and regeneration of critical-sized bone defects. Mater Sci Eng C 101:64–75. https://doi.org/10.1016/j.msec.2019.03.070

Karageorgiou V, Kaplan D (2005) Porosity of 3D biomaterial scaffolds and osteogenesis. Biomaterials 26(27):5474–5491. https://doi.org/10.1016/j.biomaterials.2005.02.002

Karamanos NK, Theocharis AD, Piperigkou Z, Manou D, Passi A, Skandalis SS, Vynios DH, Orian-Rousseau V, Ricard-Blum S, Schmelzer CE, Duca L (2021) A guide to the composition and functions of the extracellular matrix. FEBS J 288(24):6850–6912

Kargozar S, Montazerian M, Fiume E, Baino F (2019) Multiple and promising applications of strontium (Sr)-containing bioactive glasses in bone tissue engineering. Front Bioeng Biotechnol 7:161

Katti D, Vasita R, Shanmugam K (2008) Improved biomaterials for tissue engineering applications: surface modification of polymers. Curr Top Med Chem 8(4):341–353. https://doi.org/10.2174/156802608783790893

Kausar A (2021) Chapter 1- Prominence of conjugated polymers. In: Conducting polymer-based nanocomposites: fundamentals and applications. Elsevier

Kennedy KM, Bhaw-Luximon A, Jhurry D (2017) Cell-matrix mechanical interaction in electrospun polymeric scaffolds for tissue engineering: implications for scaffold design and performance. Acta Biomater 50:41–55

Khan MM, Tsukada M (2014) Electrospun silk sericin nanofibers for biomedical applications. In: Silk biomaterials for tissue engineering and regenerative medicine. Woodhead Publishing, pp 125–156

Kim K, Luu YK, Chang C, Fang D, Hsiao BS, Chu B, Hadjiargyrou M (2004) Incorporation and controlled release of a hydrophilic antibiotic using poly(lactide-co-glycolide)-based electrospun nanofibrous scaffolds. J Control Release 98(1):47–56. https://doi.org/10.1016/j.jconrel.2004.04.009

Kirillova A, Yeazel TR, Asheghali D, Petersen SR, Dort S, Gall K, Becker ML (2021) Fabrication of biomedical scaffolds using biodegradable polymers. Chem Rev 121(18):11238–11304

Klimek K, Ginalska G (2020) Proteins and peptides as important modifiers of the polymer scaffolds for tissue engineering applications—a review. Polymers 12(4):844

Krishani M, Shin WY, Suhaimi H, Sambudi NS (2023) Development of scaffolds from bio-based natural materials for tissue regeneration applications: a review. Gels 9(2):100

Kundu B, Rajkhowa R, Kundu SC, Wang X (2013) Silk fibroin biomaterials for tissue regenerations. Adv Drug Deliv Rev 65(4):457–470

Laidmäe I, Ērglis K, Cēbers A, Janmey PA, Uibo R (2018) Salmon fibrinogen and chitosan scaffold for tissue engineering: in vitro and in vivo evaluation. J Mater Sci Mater Med 29:1–2

Langer R, Tirrell DA (2004) Designing materials for biology and medicine. Nature 428(6982):487–492

Lannutti J, Reneker D, Ma T, Tomasko D, Farson D (2007) Electrospinning for tissue engineering scaffolds. Mater Sci Eng C 27(3):504–509

Lee SH, Shin H (2007) Matrices and scaffolds for delivery of bioactive molecules in bone and cartilage tissue engineering. Adv Drug Deliv Rev 59(4–5):339–359

Lee JK, Link JM, Hu JC, Athanasiou KA (2017) The self-assembling process and applications in tissue engineering. Cold Spring Harb Perspect Med 7(11):a025668

Lee JH, Parthiban P, Jin GZ, Knowles JC, Kim HW (2021) Materials roles for promoting angiogenesis in tissue regeneration. Prog Mater Sci 117:100732

Li J, Mooney DJ (2016) Designing hydrogels for controlled drug delivery. Nat Rev Mater 1(12):1–7

Li WJ, Tuan RS (2009) Fabrication and application of nanofibrous scaffolds in tissue engineering. Curr Protoc Cell Biol 42(1):25

Li Y, Liu Y, Li R, Bai H, Zhu Z, Zhu L, Zhu C, Che Z, Liu H, Wang J, Huang L (2021) Collagen-based biomaterials for bone tissue engineering. Mater Des 210:110049

Li H, Zhou H, Xu C, Wei Y, Tang X (2023) 3D bioprinting and scaffold-free strategies for fabrication of multi-cellular tissues or organoids. Int J Bioprint 9(6):0135

Little H, Themistou E, Clarke SA, Cunningham E, Buchanan F (2018) Correction to: process-induced degradation of bioresorbable PDLGA in bone tissue scaffold production. J Mater Sci Mater Med 29(3):30. https://doi.org/10.1007/s10856-018-6035-7

Liu X, Ma PX (2004) Polymeric scaffolds for bone tissue engineering. Ann Biomed Eng 32:477–486

Liu Y, Lu Y, Tian X, Cui G, Zhao Y, Yang Q, Yu S, Xing G, Zhang B (2009) Segmental bone regeneration using an rhBMP-2-loaded gelatin/nanohydroxyapatite/fibrin scaffold in a rabbit model. Biomaterials 30(31):6276–6285. https://doi.org/10.1016/j.biomaterials.2009.08.003

Liu Y, Lim J, Teoh SH (2013) Development of clinically relevant scaffolds for vascularised bone tissue engineering. Biotechnol Adv 31(5):688–705

Loh QL, Choong C (2013) Three-dimensional scaffolds for tissue engineering applications: role of porosity and pore size. Tissue Eng Part B Rev 19(6):485–502

Lu T, Li Y, Chen T (2013) Techniques for fabrication and construction of three-dimensional scaffolds for tissue engineering. Int J Nanomedicine 18:337–350

Luo Y (2020) Three-dimensional scaffolds. In: Principles of tissue engineering. Academic, pp 343–360

Lv Y, Wang B, Liu G, Tang Y, Lu E, Xie K, Lan C, Liu J, Qin Z, Wang L (2021) Metal material, properties and design methods of porous biomedical scaffolds for additive manufacturing: a review. Front Bioeng Biotechnol 9:641130

Ma PX (2004) Scaffolds for tissue fabrication. Mater Today 7(5):30–40

Ma PX, Zhang R (1999) Synthetic nano-scale fibrous extracellular matrix. J Biomed Mater Res 46(1):60–72

Ma X, Zhao D, Xiang Y, Hua Y, Zhao W, Cui Y, Zhang Z (2023) Layer-by-layer self-assembly and clinical application in orthopedics. J Mater Sci Technol 147:241–268. https://doi.org/10.1016/j.jmst.2022.11.014

Magnusson JP, Saeed AO, Fernández-Trillo F, Alexander C (2011) Synthetic polymers for biopharmaceutical delivery. Polym Chem 2(1):48–59

Mahmud MZ, Islam MD, Mobarak MH (2023) The development of eco-friendly biopolymers for use in tissue engineering and drug delivery. J Nanomater 2023(1):9270064–9270015

Maisani M, Ziane S, Ehret C, Levesque L, Siadous R, Le Meins J, Chevallier P, Barthélémy P, De Oliveira H, Amédée J, Mantovani D, Chassande O (2018) A new composite hydrogel combining the biological properties of collagen with the mechanical properties of a supramolecular scaffold for bone tissue engineering. J Tissue Eng Regen Med 12(3):e1489–e1500. https://doi.org/10.1002/term.2569

Malikmammadov E, Tanir TE, Kiziltay A, Hasirci V, Hasirci N (2018) PCL and PCL-based materials in biomedical applications. J Biomater Sci Polym Ed 29(7–9):863–893

Mantha S, Pillai S, Khayambashi P, Upadhyay A, Zhang Y, Tao O, Pham HM, Tran SD (2019) Smart hydrogels in tissue engineering and regenerative medicine. Materials 12(20):3323

Marin D, Marchesan S (2022) Self-assembled peptide nanostructures for ECM biomimicry. Nano 12(13):2147

Martins A, Araújo JV, Reis RL, Neves NM (2007) Electrospun nanostructured scaffolds for tissue engineering applications. Nanomedicine 2(6):929–942

Mathew AP, Augustine R, Kalarikkal NA, Thomas S (2016) Tissue engineering: principles, recent trends and the future. In: Nanomedicine and tissue engineering: state of the art and recent trends. Apple Academic Press, pp 31–82

Mazari SA, Ali E, Abro R, Khan FS, Ahmed I, Ahmed M, Nizamuddin S, Siddiqui TH, Hossain N, Mubarak NM, Shah A (2021) Nanomaterials: applications, waste-handling, environmental toxicities, and future challenges–A review. J Environ Chem Eng 9(2):105028

Morouço P, Biscaia S, Viana T, Franco M, Malça C, Mateus A, Moura C, Ferreira FC, Mitchell G, Alves NM (2016) Fabrication of poly (ε-caprolactone) scaffolds reinforced with cellulose nanofibers, with and without the addition of hydroxyapatite nanoparticles. Biomed Res Int 2016(1):1596157–1596110

Mostajeran H, Baheiraei N, Bagheri H (2024) Effects of cerium-doped bioactive glass incorporation on an alginate/gelatin scaffold for bone tissue engineering: in vitro characterizations. Int J Biol Macromol 255:128094. https://doi.org/10.1016/j.ijbiomac.2023.128094

Murphy SV, Atala A (2014) 3D bioprinting of tissues and organs. Nat Biotechnol 32(8):773–785. https://doi.org/10.1038/nbt.2958

Nair SA, Gayathri V (2023) Scaffold materials and toxicity. In: Mohanan PV, Kappalli S (eds) Biomedical applications and toxicity of nanomaterials. Springer Nature Singapore, pp 535–558. https://doi.org/10.1007/978-981-19-7834-0_20

Negut I, Dorcioman G, Grumezescu V (2020) Scaffolds for wound healing applications. Polymers 12(9):2010

Nguyen MTN, Tran HLB (2022) Heparinization of the bovine pericardial scaffold by layer-by-layer (LBL) assembly technique. J Sci Adv Mater Dev 7(2):100405. https://doi.org/10.1016/j.jsamd.2021.11.007

Niemczyk-Soczynska B, Gradys A, Sajkiewicz P (2020) Hydrophilic surface functionalization of electrospun nanofibrous Scaffolds in tissue engineering. Polymers 12(11):2636. https://doi.org/10.3390/polym12112636

Niknafs B, Meskaraf-asadabadi M, Hamdi K, Ghanbari E (2024) Incorporating bioactive glass nanoparticles in silk fibroin/bacterial nanocellulose composite scaffolds improves their biological and osteogenic properties for bone tissue engineering applications. Int J Biol Macromol 266:131167. https://doi.org/10.1016/j.ijbiomac.2024.131167

Nikolova MP, Chavali MS (2019) Recent advances in biomaterials for 3D scaffolds: a review. Bioact Mater 4:271–292

Nof M, Shea LD (2002) Drug-releasing scaffolds fabricated from drug-loaded microspheres. J Biomed Mater Res 59(2):349–356. https://doi.org/10.1002/jbm.1251

Nune M, Kumaraswamy P, Maheswari Krishnan U, Sethuraman S (2013) Self-assembling peptide nanofibrous scaffolds for tissue engineering: novel approaches and strategies for effective functional regeneration. Curr Protein Peptide Sci 14(1):70–84

O'Brien FJ (2011) Biomaterials & scaffolds for tissue engineering. Mater Today 14(3):88–95. https://doi.org/10.1016/S1369-7021(11)70058-X

Oladapo B, Zahedi A, Ismail S, Fernando W, Ikumapayi O (2023) 3D-printed biomimetic bone implant polymeric composite scaffolds. Int J Adv Manuf Technol 126(9):4259–4267

Oliveira JM, Rodrigues MT, Silva SS, Malafaya PB, Gomes ME, Viegas CA, Dias IR, Azevedo JT, Mano JF, Reis RL (2006) Novel hydroxyapatite/chitosan bilayered scaffold for osteochondral tissue-engineering applications: Scaffold design and its performance when seeded with goat bone marrow stromal cells. Biomaterials 27(36):6123–6137

Omenetto FG, Kaplan DL (2010) New opportunities for an ancient material. Science 329(5991):528–531

Pais AI, Belinha J, Alves JL (2023) Advances in computational techniques for bio-inspired cellular materials in the field of biomechanics: current trends and prospects. Materials 16(11):3946

Pandey AR, Singh US, Momin M, Bhavsar C (2017) Chitosan: application in tissue engineering and skin grafting. J Polym Res 24:1–22

Percival KM, Paul V, Husseini GA (2024) Recent advancements in bone tissue engineering: integrating smart scaffold technologies and bio-responsive systems for enhanced regeneration. Int J Mol Sci 25(11):6012

Perez-Puyana V, Jiménez-Rosado M, Romero A, Guerrero A (2020) Polymer-based scaffolds for soft-tissue engineering. Polymers 12(7):1566

Prasad K, Bazaka O, Chua M, Rochford M, Fedrick L, Spoor J, Symes R, Tieppo M, Collins C, Cao A, Markwell D (2017) Metallic biomaterials: current challenges and opportunities. Materials 10(8):884

Qian Y, Lin H, Yan Z, Shi J, Fan C (2021) Functional nanomaterials in peripheral nerve regeneration: Scaffold design, chemical principles and microenvironmental remodeling. Mater Today 51:165–187

Rahmani Del Bakhshayesh A, Annabi N, Khalilov R, Akbarzadeh A, Samiei M, Alizadeh E, Alizadeh-Ghodsi M, Davaran S, Montaseri A (2018) Recent advances on biomedical applications of scaffolds in wound healing and dermal tissue engineering. Artif Cells Nanomed Biotechnol 46(4):691–705

Rahmati M, Mills DK, Urbanska AM, Saeb MR, Venugopal JR, Ramakrishna S, Mozafari M (2021) Electrospinning for tissue engineering applications. Prog Mater Sci 117:100721

Rao SH, Harini B, Shadamarshan RPK, Balagangadharan K, Selvamurugan N (2018) Natural and synthetic polymers/bioceramics/bioactive compounds-mediated cell signalling in bone tissue engineering. Int J Biol Macromol 110:88–96. https://doi.org/10.1016/j.ijbiomac.2017.09.029

Reddy MS, Ponnamma D, Choudhary R, Sadasivuni KK (2021) A comparative review of natural and synthetic biopolymer composite scaffolds. Polymers 13(7):1105

Reshmy R, Philip E, Madhavan A, Arun KB, Binod P, Pugazhendhi A, Awasthi MK, Gnansounou E, Pandey A, Sindhu R (2021) Promising eco-friendly biomaterials for future biomedicine: cleaner production and applications of Nanocellulose. Environ Technol Innov 24:101855

Reznikov N, Shahar R, Weiner S (2014) Bone hierarchical structure in three dimensions. Acta Biomater 10(9):3815–3826

Riaz F, Rasul I, Azeem F, Zubair M, Nadeem H, Imran M, Muzammil A, Afzal M, Siddique MH (2024) Environmental impact, health implications, and life cycle assessment of bionanocomposites. In: Advances in bionanocomposites. Elsevier, pp 387–406

Ribeiro I, Peças P, Henriques E (2013) A life cycle framework to support materials selection for ecodesign: a case study on biodegradable polymers. Mater Des 51:300–308

Ruffini A, Sandri M, Dapporto M, Campodoni E, Tampieri A, Sprio S (2021) Nature-inspired unconventional approaches to develop 3D bioceramic scaffolds with enhanced regenerative ability. Biomedicines 9(8):916

Sachlos E, Czernuszka JT (2003) Making tissue engineering scaffolds work. Review: the application of solid freeform fabrication technology to the production of tissue engineering scaffolds. Eur Cell Mater 5(29):39–40

Sadowska JM, Ginebra M-P (2020) Inflammation and biomaterials: role of the immune response in bone regeneration by inorganic scaffolds. J Mater Chem B 8(41):9404–9427. https://doi.org/10.1039/D0TB01379J

Salgado AJ, Oliveira JM, Martins A, Teixeira FG, Silva NA, Neves NM, Sousa N, Reis RL (2013) Tissue engineering and regenerative medicine: past, present, and future. Int Rev Neurobiol 108:1–33

Salvatore L, Gallo N, Natali ML, Terzi A, Sannino A, Madaghiele M (2021) Mimicking the hierarchical organization of natural collagen: toward the development of ideal scaffolding material for tissue regeneration. Front Bioeng Biotechnol 9:644595

Sankar S, Sharma CS, Rath SN, Ramakrishna S (2018) Electrospun nanofibres to mimic natural hierarchical structure of tissues: application in musculoskeletal regeneration. J Tissue Eng Regen Med 12(1):e604–e619

Santo VE, Gomes ME, Mano JF, Reis RL (2013) Controlled release strategies for bone, cartilage, and osteochondral engineering—part II: challenges on the evolution from single to multiple bioactive factor delivery. Tissue Eng Part B Rev 19(4):327–352

Sarikaya M, Tamerler C, Jen AK, Schulten K, Baneyx F (2003) Molecular biomimetics: nanotechnology through biology. Nat Mater 2(9):577–585

Scadden DT (2006) The stem-cell niche as an entity of action. Nature 441(7097):1075–1079

Seidi F, Yazdi MK, Jouyandeh M, Dominic M, Naeim H, Nezhad MN, Bagheri B, Habibzadeh S, Zarrintaj P, Saeb MR, Mozafari M (2021) Chitosan-based blends for biomedical applications. Int J Biol Macromol 183:1818–1850

Serbo JV, Gerecht S (2013) Vascular tissue engineering: biodegradable scaffold platforms to promote angiogenesis. Stem Cell Res Ther 4:1–8

Shanley LC, Mahon OR, Kelly DJ, Dunne A (2021) Harnessing the innate and adaptive immune system for tissue repair and regeneration: considering more than macrophages. Acta Biomater 133:208–221

Shapiro L, Cohen S (1997) Novel alginate sponges for cell culture and transplantation. Biomaterials 18(8):583–590

Shoichet MS (2010) Polymer scaffolds for biomaterials applications. Macromolecules 43(2):581–591

Shyam R, Reddy LVK, Palaniappan A (2023) Fabrication and characterization techniques of in vitro 3D tissue models. Int J Mol Sci 24(3):1912

Silva GA, Czeisler C, Niece KL, Beniash E, Harrington DA, Kessler JA, Stupp SI (2004) Selective differentiation of neural progenitor cells by high-epitope density nanofibers. Science 303(5662):1352–1355

Silva SS, Mano JF, Reis RL (2010) Potential applications of natural origin polymer-based systems in soft tissue regeneration. Crit Rev Biotechnol 30(3):200–221

Silva R, Singh R, Sarker B, Papageorgiou DG, Juhasz-Bortuzzo JA, Roether JA, Cicha I, Kaschta J, Schubert DW, Chrissafis K, Detsch R, Boccaccini AR (2018) Hydrogel matrices based on elastin and alginate for tissue engineering applications. Int J Biol Macromol 114:614–625. https://doi.org/10.1016/j.ijbiomac.2018.03.091

Singhvi MS, Zinjarde SS, Gokhale DV (2019) Polylactic acid: synthesis and biomedical applications. J Appl Microbiol 127(6):1612–1626

Smith IO, Liu XH, Smith LA, Ma PX (2009) Nanostructured polymer scaffolds for tissue engineering and regenerative medicine. Wiley Interdiscip Rev Nanomed Nanobiotechnol 1(2):226–236

Speck O, Speck D, Horn R, Gantner J, Sedlbauer KP (2017) Biomimetic bio-inspired biomorph sustainable? An attempt to classify and clarify biology-derived technical developments. Bioinspir Biomim 12(1):011004

Sufiyan M, Kushwaha P, Ahmad M, Mandal P, Vishwakarma KK (2024) Scaffold-mediated drug delivery for enhanced wound healing: a review. AAPS PharmSciTech 25(5):137

Sukumaran NP, Gopi S (2021) Overview of biopolymers: resources, demands, sustainability, and life cycle assessment modeling and simulation. In: Biopolymers and their industrial applications. Elsevier, pp 1–19

Sun W, Gregory DA, Tomeh MA, Zhao X (2021) Silk fibroin as a functional biomaterial for tissue engineering. Int J Mol Sci 22(3):1499

Tajvar S, Hadjizadeh A, Samandari SS (2023) Scaffold degradation in bone tissue engineering: an overview. Int Biodeterior Biodegradation 180:105599

Thomson RC, Shung AK, Yaszemski MJ, Mikos AG (2000) Polymer scaffold processing. In: Principles of tissue engineering, vol 2. Elsevier, pp 251–262

Tidball JG (2005) Inflammatory processes in muscle injury and repair. Am J Phys Regul Integr Comp Phys 288(2):R345–R353

Ullah S, Chen X (2020) Fabrication, applications and challenges of natural biomaterials in tissue engineering. Appl Mater Today 20:100656

Vadalà M, Kröll E, Küppers M, Lupascu DC, Brunstermann R (2023) Hydrogen production via dark fermentation by bacteria colonies on porous PDMS-scaffolds. Int J Hydrog Energy 48(65):25274–25284. https://doi.org/10.1016/j.ijhydene.2023.03.285

Vahidi M, Rizkalla AS, Mequanint K (2024) Extracellular matrix-surrogate advanced functional composite biomaterials for tissue repair and regeneration. Adv Healthc Mater 13:2401218

Varghese MG, Thomas A, Rupesh S, Sameer KM, Joseph D, Thomas NG (2024) Fabrication techniques for Scaffolds applied in regenerative medicine. In: Novel biomaterials for tissue engineering. IntechOpen

Wade RJ, Burdick JA (2012) Engineering ECM signals into biomaterials. Mater Today 15(10):454–459

Wade RJ, Burdick JA (2014) Advances in nanofibrous scaffolds for biomedical applications: from electrospinning to self-assembly. Nano Today 9(6):722–742

Wang Y, Naleway SE, Wang B (2020) Biological and bioinspired materials: structure leading to functional and mechanical performance. Bioact Mater 5(4):745–757

Wang L, You X, Zhang L, Zhang C, Zou W (2022) Mechanical regulation of bone remodeling. Bone Res 10(1):16

Wang Z, Wen F, Chong MSK (2023) Surface modification of tissue engineering Scaffolds. In: Gao C (ed) Polymeric biomaterials for tissue regeneration. Springer Nature Singapore, pp 227–264. https://doi.org/10.1007/978-981-99-6948-7_6

Wei G, Ma PX (2004) Structure and properties of nano-hydroxyapatite/polymer composite scaffolds for bone tissue engineering. Biomaterials 25(19):4749–4757

Wei G, Ma PX (2006) Macroporous and nanofibrous polymer scaffolds and polymer/bone-like apatite composite scaffolds generated by sugar spheres. J Biomed Mater Res A 78(2):306–315

Wei G, Jin Q, Giannobile WV, Ma PX (2007) The enhancement of osteogenesis by nano-fibrous scaffolds incorporating rhBMP-7 nanospheres. Biomaterials 28(12):2087–2096

Wheeldon I, Farhadi A, Bick AG, Jabbari E, Khademhosseini A (2011) Nanoscale tissue engineering: spatial control over cell-materials interactions. Nanotechnology 22(21):212001. https://doi.org/10.1088/0957-4484/22/21/212001

Wu G (2023) Editorial: bioactive agents for functionalization of biomaterials for precise tissue engineering. Front Bioeng Biotechnol 11:1141827. https://doi.org/10.3389/fbioe.2023.1141827
Xia H, Li X, Gao W, Fu X, Fang RH, Zhang L, Zhang K (2018) Tissue repair and regeneration with endogenous stem cells. Nat Rev Mater 3(7):174–193
Xu M, Su T, Jin X, Li Y, Yao Y, Liu K, Chen K, Lu F, He Y (2022) Inflammation-mediated matrix remodeling of extracellular matrix-mimicking biomaterials in tissue engineering and regenerative medicine. Acta Biomater 151:106–117
Xu R, Fang Y, Zhang Z, Cao Y, Yan Y, Gan L, Xu J, Zhou G (2023) Recent advances in biodegradable and biocompatible synthetic polymers used in skin wound healing. Materials 16(15):5459
Yamamuro T (2004) Bioceramics. In: Poitout DG (ed) Biomechanics and biomaterials in orthopedics. Springer, London. https://doi.org/10.1007/978-1-4471-3774-0_3
Yan LP, Oliveira JM, Oliveira AL, Caridade SG, Mano JF, Reis RL (2012) Macro/microporous silk fibroin scaffolds with potential for articular cartilage and meniscus tissue engineering applications. Acta Biomater 8(1):289–301
Yang J, Yu H, Wang L, Liu J, Liu X, Hong Y, Huang Y, Ren S (2022) Advances in adhesive hydrogels for tissue engineering. Eur Polym J 172:111241
Yari D, Ebrahimzadeh MH, Movaffagh J, Shahroodi A, Shirzad M, Qujeq D, Moradi A (2022) Biochemical aspects of scaffolds for cartilage tissue engineering; from basic science to regenerative medicine. Arch Bone Joint Surg 10(3):229–244
Zeinali R, Del Valle LJ, Torras J, Puiggali J (2021) Recent progress on biodegradable tissue engineering scaffolds prepared by thermally-induced phase separation (Tips). Int J Mol Sci 22(7):3504
Zhang S (2003) Fabrication of novel biomaterials through molecular self-assembly. Nat Biotechnol 21(10):1171–1178
Zhang F, King MW (2020) Biodegradable polymers as the pivotal player in the design of tissue engineering scaffolds. Adv Healthc Mater 9(13):1901358
Zhang D, Wang Y (2019) Functional protein-based bioinspired nanomaterials: from coupled proteins, synthetic approaches, nanostructures to applications. Int J Mol Sci 20(12):3054
Zhang S, Xing M, Li B (2018) Biomimetic layer-by-layer self-assembly of nanofilms, nanocoatings, and 3D Scaffolds for tissue engineering. Int J Mol Sci 19(6):1641. https://doi.org/10.3390/ijms19061641
Zhang X, Li L, Ouyang J, Zhang L, Xue J, Zhang H, Tao W (2021) Electroactive electrospun nanofibers for tissue engineering. Nano Today 39:101196. https://doi.org/10.1016/j.nantod.2021.101196
Zhang S, Chen X, Shan M, Hao Z, Zhang X, Meng L, Zhai Z, Zhang L, Liu X, Wang X (2023) Convergence of 3D bioprinting and nanotechnology in tissue engineering scaffolds. Biomimetics 8(1):94
Zhao P, Gu H, Mi H, Rao C, Fu J, Turng LS (2018) Fabrication of scaffolds in tissue engineering: a review. Front Mech Eng 13:107–119
Zhao C, Liu W, Zhu M, Wu C, Zhu Y (2022) Bioceramic-based scaffolds with antibacterial function for bone tissue engineering: a review. Bioact Mater 18:383–398

Bioprinting Technologies for Sustainable Tissue Engineering

10

Parth Surati, Priyanka Pawale, and Meghraj Suryawanshi

Abstract

This chapter delves into bioprinting technology (BT), a groundbreaking advancement in tissue engineering designed to create functional and durable tissue constructs for pharmaceutical and medical purposes. It offers a comprehensive overview of BT's progress, emphasizing its potential to revolutionize tissue engineering techniques while advancing the sustainable production of medical materials. The chapter explores various bioprinting approaches, such as inkjet, extrusion, and laser-based techniques, highlighting their unique features and compatible bioink options. By leveraging these innovative methods, bioprinting achieves precision and complexity in fabricating tissue structures that surpass traditional scaffold-based techniques. A critical analysis of modern bioprinting examines its current challenges, limitations, and commercial implications. Furthermore, the chapter investigates strategies and future directions to enhance bioprinting's efficiency and accessibility, ensuring its role as a transformative technology in developing eco-friendly materials for medical applications. By integrating principles of sustainability with cutting-edge bioprinting techniques, this chapter aims to provide valuable insights into the design and processing of green materials, contributing to a more sustainable future for medical and pharmaceutical sciences.

P. Surati · P. Pawale
Department of Regulatory Affairs, Northeastern University, Toronto, ON, Canada
e-mail: surati.p@northeastern.edu

M. Suryawanshi (✉)
Department of Pharmaceutics, Sandip Institute of Pharmaceutical Sciences (SIPS), Nashik, Maharashtra, India

Md. Faiyazuddin, M. Suryawanshi (eds.), *Design and Processing of Green Materials*, Biomaterials, Bioengineering and Sustainability 4,
https://doi.org/10.1007/978-3-031-91790-5_10

Keywords

Bioprinting technology · Sustainable tissue engineering · Bioinks · Pharmaceutical and medical applications · Green materials · Advanced bioprinting method

10.1 Introduction to Bioprinting Technologies for Sustainable Tissue Engineering

Bioprinting represents a cutting-edge technique in tissue engineering, enabling the creation of customized constructs with enhanced functionality compared to conventional scaffold fabrication methods. By integrating biological materials with 3D printing principles, this technology offers precise, tailored solutions for advanced tissue design (Kačarević et al. 2018; Santoni et al. 2021).

10.1.1 The Principles and Process of Bioprinting

Bioprinting is an advanced additive manufacturing technique that utilizes bioink substances composed of living cells, growth factors, and other biomolecules to construct structures layer by layer. These constructs replicate the behavior and architecture of natural tissues, making them highly effective for applications in tissue engineering, bioengineering, and materials science. The term "additive" highlights the process's precision, allowing the controlled placement of each cell type exactly where it is needed within the design (Bartolo et al. 2022).

10.1.2 Advantages and Applications of Bioprinting

Bioprinting offers numerous advantages over traditional tissue engineering methods. It enables the creation of patient-specific treatments and organ models, reduces dependence on animal testing for drug development, and ensures better compatibility between human cells and implanted biological materials. Additionally, the automation of complex processes in bioprinting minimizes human errors and ensures consistent quality across different batches of tissue constructs. In the field of regenerative medicine, bioprinting facilitates the production of personalized organs for patients, significantly accelerating drug development while reducing the need for animal trials. By streamlining the manufacturing phases of tissue constructs, this technology ensures precision and efficiency throughout the process (Xie et al. 2020).

10.1.3 Sustainability in Bioprinting

Bioprinting presents promising opportunities to advance sustainable practices in tissue engineering (TE) and regenerative medicine (RM), aligning with the growing industrial focus on sustainability. This technology supports the global shift toward greener healthcare solutions by incorporating environmentally friendly, biocompatible, and biodegradable materials in living cell-based systems. Bioprinting also fosters ethical and efficient research by reducing reliance on animal testing and expediting drug development through the creation of functional tissue models. Moreover, it addresses the critical shortage of transplantable organs by enabling the production of bio-printed tissues and organs, offering a sustainable and innovative solution to this pressing issue (Santoni et al. 2021; Dey and Ozbolat 2020).

10.1.4 Future Directions and Challenges

BT has seen rapid growth recently demonstrated by different types like inkjet printers, extrusion systems, or laser-assisted printers all having their strengths depending on various bioinks used among others. There are still some challenges; for instance, improved commercial viability while scaling up production levels may require better bioinks to enhance performance outcomes, etc... Hence researchers are continuously looking at different ways of making bioprinting work better so that it can become an enabler for environmentally friendly materials in medicine (Matai et al. 2020).

The aim is to introduce eco-friendly products (materials and practices) in pharmaceutical science too while taking advantage of the best capabilities provided by current sustainable models for advanced biomedical printing processes. This chapter provides detailed information on the recent developments made within bioprinters and their applications for engineering sustainable tissues as well as how we can design greener medical supplies for future health facilities using bioprinting technologies (Kačarević et al. 2018; Tibbetts 2021; Muskan et al. 2022).

10.2 Types of Bioprinting Technologies

Bioprinting technologies have advanced significantly in recent years, offering multiple approaches for building complex tissue structures. There are a few main categories of bioprinting that include Inkjet Bioprinting, Extrusion Bioprinting, Laser-Assisted Bioprinting, Stereolithography Bioprinting, and Emerging Hybrid Approaches; each with unique capabilities and considerations for use within the field of tissue engineering (Yilmaz et al. 2019; Highley 2019; Seol et al. 2014; Bonatti et al. 2022; Li et al. 2016).

10.2.1 Inkjet Bioprinting

The method called inkjet bioprinting is one of the earliest and most used techniques for bioprinters. In this technique, thermal or piezoelectric actuators are used to deposit droplets laden with cells on a substrate very precisely. It has high-resolution printing capabilities that allow for small structures (as small as 50 μm) to be printed.

Additionally, it also has some advantages, such as very fast speeds but low costs compared to other methods and excellent post-print cell viability; however, due to low droplet directionality, there may be problems with nozzle clogging (Kačarević et al. 2018; Yilmaz et al. 2019).

10.2.2 Extrusion Bioprinting

The extrusion-based type uses pneumatic or mechanical forces to continuously deposit material through a nozzle. It forces hard materials outwards while leaving soft ones behind, creating multi-layered products that would not have been possible otherwise. Because it works with many viscosities and cell densities for bio-inks, it is versatile enough for large-scale construction where porosity control is needed (Yilmaz et al. 2019). In addition, highly viscous materials can be printed continuously which enables scaling up to larger tissue constructs; however, the resolution is still lower than laser-based methods. This shortcoming can be addressed by combining two different techniques though (Kačarević et al. 2018; Highley 2019; Seol et al. 2014).

10.2.3 Laser-Assisted Bioprinting

Laser-assisted bioprinting employs focused laser beams to solidify small volumes of bio-inks containing living cells, precisely depositing them onto target surfaces. This method excels in accuracy, achieving micrometer-level resolution and enabling the detailed patterning of individual cells. It is particularly effective for creating intricate tissue structures while maintaining minimal impact on cell viability and accommodating high-viscosity materials. However, laser-assisted bioprinting has some drawbacks. It is more costly compared to other methods with similar throughput, and its reliance on high precision can pose challenges in certain operations, especially in unfavorable environmental conditions like low humidity. Additionally, its speed is slower than alternative techniques, making it less practical for some applications (Kačarević et al. 2018; Yilmaz et al. 2019).

10.2.4 Stereolithography Bioprinting

Stereolithography bioprinting utilizes light-sensitive materials, or photopolymers, which solidify upon exposure to specific wavelengths of light. This process enables the rapid creation of intricate 3D shapes with exceptionally high resolution. By integrating stereolithographic techniques with other methods, such as extrusion-based bioprinting, it is possible to produce hybrid constructs that incorporate both structural and cellular components. This combined approach is particularly well-suited for fabricating highly detailed and complex tissue designs (Yilmaz et al. 2019).

10.2.5 Emerging Hybrid Approaches

Recent progress in bioprinting has been fueled by efforts to integrate different bioprinting technologies, leveraging their unique strengths for enhanced outcomes. For example, researchers have successfully combined stereolithography and extrusion-based bioprinting to fabricate artificial corneal structures, achieving both high-resolution features and precise cellular placement. These innovative combinations pave the way for creating tissues with intricate architectures and customized biological properties, unlocking new possibilities in advanced tissue manufacturing (Yilmaz et al. 2019).

In the field of bioprinting, various techniques are employed, each with distinct characteristics tailored to specific applications (Fig. 10.1). **Extrusion bioprinting**, for example, offers high cell viability (85–95%) and supports high-viscosity bioinks, making it suitable for basic tissue models and drug testing, though it operates at a slower speed. Conversely, **inkjet bioprinting** is faster and more cost-effective but is limited by lower cell density and viscosity. **Laser-assisted bioprinting** enables precision cell placement with moderate resolution and medium cell density. **Stereolithography**, known for its high cell viability (>90%) and rapid processing, is ideal for creating complex scaffolds with channels. Finally, **two-photon polymerization** achieves unmatched precision (0.1–10 μm resolution) and speed, though at a significantly higher cost, making it suitable for vascularized and high-precision tissue constructs (Santoni et al. 2021).

In conclusion, each bioprinting technology has its unique advantages and limitations, making it more suitable for specific applications in tissue engineering. By understanding the capabilities and constraints of each method, scientists can select

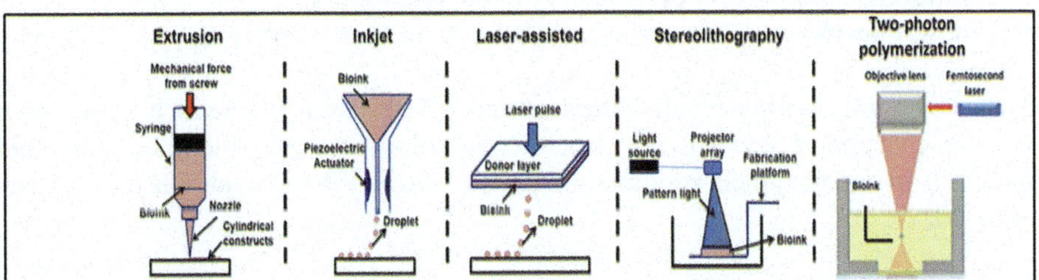

	Extrusion	Inkjet	Laser-assisted	Stereolithography	Two-photon polymerization
Speed	Slow	Fast	Medium	Fast	Very fast
Cost	Moderate	Low	High	Low	Very High
Cell viability	85-95%	80-95%	<85%	>90%	>80%
Cell density	High	Low	Medium	Medium	Medium
Resolution	100 - 500 μm	100 - 500 μm	20 -100 μm	20 -100 μm	0,1 - 10 μm
Viscosity	6-30 × 10^7 mPa·s	<10 mPa·s	1-300 mPa·s	No limitation	No limitation
Applications	Tissue models for basic cell research, drug testing and	Supplementary to other technologies	Precision cell deposition	Scaffolds and complex structures with channels	Vascularized and high precision models

Fig. 10.1 Comparative Study of Bioprinting Techniques: Inkjet, Extrusion, Laser-Assisted, Stereolithography, and Two-photon polymerization (Santoni et al. 2021)

the most appropriate technology based on their specific needs, ultimately driving the advancement of sustainable tissue engineering (Kačarević et al. 2018; Highley 2019).

10.3 Materials in Bioprinting Technologies

The success of bioprinting in creating viable tissue constructs relies heavily on the careful selection and formulation of bio-inks. These materials are essential for ensuring printability, structural integrity, and biological compatibility, all of which are critical for cell survival and the functional reproduction of native tissues. Various types of materials used in bioprinting include natural polymers, synthetic polymers, decellularized extracellular matrices (dECMs), nanocomposites, and smart stimuli-responsive materials (Bartolo et al. 2022; Yilmaz et al. 2021; Sachdev et al. 2022; Mao et al. 2020).

10.3.1 Natural Biopolymers

Natural biopolymers are often preferred in bioprinting due to their excellent biocompatibility with the human body and their ability to replicate the extracellular matrix (ECM) structure found in living tissues. Some well-known examples include:

- Animal-derived collagen: which acts as a key structural protein,
- Gelatin: A collagen derivative that easily forms gels,
- Alginate: A polysaccharide known for its ability to crosslink through gelation,
- Hyaluronic acid: A vital component of cell scaffolding,
- Fibrin: Plays a crucial role in wound healing and cell migration.

These materials promote cell adhesion and proliferation but tend to offer poor mechanical support. As such, they often require modification or blending with other materials to enhance their strength and performance (Yilmaz et al. 2021; Baranwal et al. 2022).

10.3.2 Synthetic Polymers

Synthetic polymers can be engineered to exhibit specific mechanical properties, such as rigidity or biodegradability, depending on the intended tissue engineering application, as degradation rates can vary. Common examples include:

- Polycaprolactone (PCL): Known for its slow degradation over time,
- Polylactic acid (PLA): Its stiffness makes it ideal for bone repair, especially when fast ossification is desired, while ensuring slower resorption during osteosynthesis or when using fixators like plates or screws,

- Polylactic-co-glycolic acid (PLGA): Capable of degrading at different rates based on the ratio of lactic and glycolic acids,
- Polyethylene glycol (PEG): A hydrophilic polymer that can be modified in various ways to suit different applications (Bar et al. 2014).

Although synthetic polymers lack the natural bioactive signals found in natural materials, they offer superior mechanical properties. This makes them highly adaptable for specific applications, allowing for the incorporation of tailored cues to influence cell behavior (Yilmaz et al. 2021; Sachdev et al. 2022).

10.3.3 Composite Bioinks

Composite bioinks are created by combining different materials, allowing each one to contribute its strengths while compensating for any weaknesses. Examples include:

- Blends of natural and synthetic polymers, such as PCL mixed with collagen, where the structural advantages of synthetic materials complement the biological activity of natural ones,
- Polymer-ceramic composites, like PCL combined with hydroxyapatite, which are ideal for bone tissue engineering due to the enhanced stiffness provided by ceramics compared to pure polymers,
- Nanocomposites, which incorporate nanoparticles or nanofibers to improve the ink's overall properties, both during printing and after the printed structure has been completed (Mobaraki et al. 2020; Rasouli et al. 2024). In summary, composite formulations enable the fine-tuning of mechanical, biological, and printability properties, making them suitable for regenerating different types of tissues in bioprinting (Chen et al. 2023).

10.3.4 Decellularized Extracellular Matrix (dECM)

Decellularized extracellular matrices (dECMs), derived from native tissues, are being explored as bio-inks due to their ability to preserve the complex composition and structure of whole organs. These dECMs provide an environment that greatly enhances cellular functions and supports tissue-specific differentiation (Yao et al. 2019; Cho et al. 2019).

10.3.5 Nanocomposite Materials

The incorporation of nanoparticles or nanofibers into bio-inks significantly enhances their electrical, mechanical, and other key characteristics, improving their overall performance. For example, adding carbon nanotubes or graphene oxide increases the conductivity and strength of hydrogel-based bio-inks, making them well-suited for engineering neural or cardiac tissues (Loukelis et al. 2023).

10.3.6 Smart and Stimuli-Responsive Materials

The use of smart materials in bio-inks has enabled the creation of self-transforming tissue structures that can change shape or other properties in response to external stimuli. This "4D bioprinting" technique opens new possibilities for developing tissue models that adapt to their environment and create patient-specific implants (Maan et al. 2022). In the drive for sustainability, there is a growing focus on environmentally friendly bioprinting formulations. This includes exploring plant-derived or renewable biomaterials and optimizing compositions to reduce waste and minimize environmental impact. Integrating sustainability into bioink development could lead to significant advancements in greener technologies within the medical and pharmaceutical fields (Amukarimi et al. 2022).

10.4 Application of Bioprinting Technologies for Sustainable Tissue Engineering

The application of bioprinting in tissue engineering and regenerative medicine has the potential to transform both fields. Tissue and organ transplantation have been in demand for several decades. Since demand for tissue or organs continues to grow, there is a severe shortage of tissue or organ donors (Xie et al. 2020; Bajaj et al. 2014).

10.4.1 Bone Tissue

Bone, a hard tissue, supports body tissues and organs. The body can heal minor injuries to bone tissue, but major injuries require outside stimulation to heal (Zhang et al. 2020). Obtaining bone tissue with its high mineral content and complex cellular structure remains challenging and has yet to be demonstrated. Based on the results, the cell viability decreased after printing, due to the increased viscosity caused by the higher Ma-dECM concentration. It is also possible to increase cell proliferation and osteogenic differentiation by adding an appropriate concentration of Ma-dECM. Using hybrid bio-inks (alginate and gelatin), Zhang et al. were able to print porous bone-like tissues using a simple manufacturing method. In comparison with pure PEGDMA, the printed structure showed excellent cell viability (>80%) as well as increased differentiation (Zhang et al. 2020). Xavier et al. developed bioactive hydrogels composed of nano silicates and GelMA (Xavier et al. 2015). Cell viability and mineral deposition were observed at different cell densities (Yu et al. 2020a, b).

10.4.2 Cartilage Tissue

The very limited ability of articular cartilage to self-repair makes trauma to its surface a challenging clinical problem (Huckle et al. 2003). The use of alginate-based hybrid bio-inks has been developed to engineer cartilage tissue. Bio-printed cartilage structures can be enhanced by using bio-inks based on nano-fibrillated cellulose-alginate composites. Over a seven-day culture period, this composite bio-ink demonstrated better shape fidelity and resolution compared to pure alginate (Markstedt et al. 2015). Kang and colleagues created a human-scale ear by combining bioinks such as PCL, PF127, gelatin, fibrinogen, hyaluronic acid (HA), and glycerol. During the printing process, PF-127 liquefied, and the ear construct was further cultivated to mature into tissue (Kang et al. 2016). Ni et al. developed hybrid bio-inks using silk fibrin and hydroxypropyl methylcellulose. The combination of silk fibrin with hydroxypropyl methylcellulose resulted in stronger mechanical properties due to the double network structure formed by the latter. When compared to single network bioinks, these hybrid bioinks exhibited significantly improved mechanical strength (Ni et al. 2020).

10.4.3 Osteo Cartilage

Osteochondral tissue regeneration is technically complex because it requires the simultaneous consideration of both cartilage and subchondral bone. In the past, this process was manual and lacked repeatability, making it difficult to standardize. However, 3D bioprinting has emerged as a promising solution for osteochondral tissue engineering. This technique enables the precise integration of different cells and biomaterials into a single structure (Yu et al. 2020a, b). Yu and colleagues used a multi-head 3D cell printing system to create a heterogeneous scaffold made from PCL and alginate. This scaffold allowed for the embedding of two types of progenitor cells at once, supporting the development of both cartilage and bone tissues. Additionally, the researchers developed a PDMS-based co-culture system to monitor and assess the differentiation of osteochondral tissue (Ni et al. 2020).

10.4.4 Neural Tissue

3D bioprinting techniques, including the creation of vascular networks, are also being explored for neural tissue engineering. Researchers have shown how materials like alginate, agarose, and carboxymethyl-chitosan can be used to develop a novel 3D neural mini-tissue that encapsulates neural stem cells (NSCs). These engineered constructs demonstrated high cell viability, with a uniform distribution of cells throughout the tissue. When exposed to bicuculline, the NSCs showed an enhanced calcium response, indicating their ability to differentiate into neurons and neuroglial cells (Gu et al. 2016).

For the multilayered nerve conduit, GelMA was used for the inner layer, while GelMA/PEGDA was chosen for the outer layer to provide additional mechanical strength. The inner layer was designed to foster an environment conducive to the growth of bone marrow-derived stem cells (BMSCs), ensuring the structural integrity of the conduit. This layer was seeded with PC12 cells, which showed a significantly faster attachment and proliferation rate when the conduit included BMSCs compared to when it did not. In another study, Li et al. utilized alginate and matrigel as bio-inks to encapsulate mesenchymal stem cells, encouraging neuron differentiation. These advancements in 3D bioprinting have proven to be promising for neural tissue engineering, showcasing the potential to create complex structures with viable cells that support cell growth, differentiation, and the secretion of neurotrophic factors (Li et al. 2021).

10.5 Sustainability in Bioprinting

The sustainability of bioprinting is complex to define, as it encompasses several factors, including the use of eco-friendly materials, waste reduction, ethical research practices, and the ability to address organ shortages. This section explores the role of bioprinting technologies in sustainable tissue engineering, highlighting their broad applications in medical science and the pharmaceutical industry (Whenish et al. 2022; Santoni et al. 2021; Dey and Ozbolat 2020).

10.5.1 Environmentally Friendly Materials

Bioprinting promotes sustainability by utilizing materials that are both biocompatible and biodegradable, which helps reduce the environmental impact of medical products. Many of these materials are derived from natural sources, further minimizing the ecological footprint. For instance, natural polymers such as alginate, gelatin, and hyaluronic acid are commonly used in bio-inks because they closely mimic the extracellular matrix (ECM) of native tissues, making them well-suited for use in living organisms. Additionally, synthetic polymers like polycaprolactone (PCL) and polylactic acid (PLA) are designed to degrade over time, ensuring that they do not have long-lasting environmental effects (Dey and Ozbolat 2020; Arif et al. 2023; Liang 2023).

10.5.2 Reduction of Waste

Unlike traditional manufacturing methods, which often result in significant amounts of waste, bioprinting enables the precise placement of materials, significantly reducing waste. As an additive manufacturing process, bioprinting builds objects layer by layer, using only the exact amount of material needed. This approach not only conserves resources but also lowers production costs, making it a more sustainable and efficient option for manufacturing (Santoni et al. 2021; Dey and Ozbolat 2020).

10.5.3 Ethical Research Practices

Bioprinting supports ethical research practices by reducing the need for animal testing. By enabling the creation of human-like tissues and organ models, it provides an alternative for studying biological processes and testing treatments without relying on animals. Bio-printed functional tissue models allow drug testing and disease modeling that are more effective than animal models since they are more accurate and humane at the same time. This approach not only complies with ethical principles but also accelerates drug development because human-based models provide more relevant data most often (Santoni et al. 2021; Dey and Ozbolat 2020; Arif et al. 2023).

10.5.4 Addressing Organ Shortages

Bioprinting enables the manufacturing of transplantable organs and tissues to resolve one of the healthcare's biggest challenges—donor organ shortages. Using this technology to bioprint organs or tissues specific to individual patients provides a sustainable solution towards organ shortage while minimizing reliance on donated organs along with any associated logistical or ethical problems (Santoni et al. 2021; Xie et al. 2020; Dey and Ozbolat 2020).

10.6 Challenges and Future Directions

Many challenges remain in bioprinting despite significant advancements in technology for sustainable tissue engineering.

10.6.1 Technical Challenges

- **Bioink Development**: The bioink should be printable, biocompatible, and possess certain mechanical properties which is rather challenging to achieve. Today's bio-inks often need to be optimized for those characteristics; hence, new formulations continue to be tested to make them work better (Liang 2023).
- **Scalability**: Another problem is how to scale up the bioprinting processes for commercial production while ensuring quality and consistency. This also includes making sure that the size of bio-printed tissues and organs meets clinical requirements without impairing their function or safety.
- **Resolution and Precision**: The ability to achieve high resolution and precision needed for complex tissue structures continues to pose technical difficulties.

While laser-assisted bioprinting techniques provide such accuracy, they tend to be more expensive as well as slower than other methods (Liang 2023).

10.6.2 Biological Challenges

- **Vascularization**: One of the main biological problems that scientists face in this field is creating vascular networks within big constructs since they cannot survive or work well without proper blood supply. Consequently, researchers are trying to incorporate vascular structures into 3D-printed tissues to address these concerns.
- **Cell Viability and Functionality**: Cells stay alive and perform their functions during the bioprinting process as well as after implantation. This requires fine-tuning the bio-ink formulation and printing conditions to promote cell health and activity.

10.6.3 Regulatory and Ethical Challenges

- **Regulatory Approval**: The regulations governing bio-printed tissues and organs are still being developed. Getting approval from regulatory agencies demands extensive testing to prove safety and efficacy which can take a lot of time and cost a lot of money.
- **Ethical Considerations**: Bioprinting has moral implications especially when it comes to making human beings' body parts such as skin, liver among others should be looked at critically by considering where they come from, whether people gave consent for the use or not, who will benefit from this technology among other things.

10.6.4 Future Directions

- **Advanced Bioinks**: In the future research will involve the development of advanced inks that have smart materials capable of responding to environmental changes. These could facilitate the functions and adaptation of bio-printed tissues (Tibbetts 2021).
- **Integration with Other Technologies**: Incorporating bioprinting with emerging technologies like gene editing or nanotechnology may create opportunities for elaborative constructs and functionalized biomaterials.
- **Personalized Medicine**: The integration of bioprinting with personalized medicine approaches holds great promise. By using patient-specific cells and bio-inks, bioprinting can create customized tissues and organs tailored to individual patients' needs, improving treatment outcomes and reducing the risk of rejection (Muskan et al. 2022).

10.7 Conclusion

Sustainable tissue engineering has been changed fundamentally by BT, which could result in a breakthrough for medical and pharmaceutical sciences. Bioprinting is sustainable because it uses eco-friendly materials, minimizes waste, and encourages ethical research practices. However, to unlock its full potential various technical, biological, or regulatory challenges need to be resolved. The future of bioprinting will depend on improvements in bioink development, scalability, and the creation of vascular networks which can help deal with some limitations. Moreover, the changing regulatory landscape alongside moral considerations will significantly influence what comes next in this field. Bioprinting is a new field within sustainable tissue engineering that provides innovative solutions to healthcare's biggest problems. More research should continue across different areas to use all its capabilities thus leading us toward more sustainable practices in the pharmaceutical, biotechnology, and medical industries.

References

Amukarimi S, Rezvani Z, Eghtesadi N, Mozafari M (2022) Smart biomaterials: from 3D printing to 4D bioprinting. Methods 205:191–199. https://doi.org/10.1016/j.ymeth.2022.07.006

Arif ZU, Khalid MY, Noroozi R, Hossain M, Shi HH, Tariq A, Ramakrishna S, Umer R (2023) Additive manufacturing of sustainable biomaterials for biomedical applications. Asian J Pharm Sci 18(3):100812. https://doi.org/10.1016/j.ajps.2023.100812

Bajaj P, Schweller RM, Khademhosseini A, West JL, Bashir R (2014) 3D biofabrication strategies for tissue engineering and regenerative medicine. Annu Rev Biomed Eng 16(1):247–276. https://doi.org/10.1146/annurev-bioeng-071813-105155

Bar SGK, Laurencin C, Deng M (2014) Natural and synthetic biomedical polymers. Newnes.

Baranwal J, Barse B, Fais A, Delogu GL, Kumar A (2022) Biopolymer: a sustainable material for food and medical applications. Polymers 14(5):983. https://doi.org/10.3390/polym14050983

Bartolo P, Malshe A, Ferraris E, Koc B (2022) 3D bioprinting: materials, processes, and applications. CIRP Ann 71(2):577–597. https://doi.org/10.1016/j.cirp.2022.06.001

Bonatti AF, Fortunato GM, De Maria C, Vozzi G (2022) Bioprinting technologies: an overview. In: Bioprinting. Elsevier, p 19–49. https://doi.org/10.1016/B978-0-323-85430-6.00006-6

Chen XB, Anvari-Yazdi AF, Duan X, Zimmerling A, Gharraei R, Sharma NK et al (2023) Biomaterials/bioinks and extrusion bioprinting. Bioact Mater 28:511–536. https://doi.org/10.1016/j.bioactmat.2023.06.006

Cho DW, Kim BS, Jang J, Gao G, Han W, Singh NK (2019) Decellularized extracellular matrix-based bioinks. In: 3D bioprinting. Springer, Cham. https://doi.org/10.1007/978-3-030-32222-9_6

Dey M, Ozbolat IT (2020) 3D bioprinting of cells, tissues and organs. Sci Rep 10:14023. https://doi.org/10.1038/s41598-020-70086-y

Gu Q, Tomaskovic-Crook E, Lozano R, Chen Y, Kapsa RM, Zhou Q et al (2016) Functional 3D neural mini-tissues from printed gel-based bioink and human neural stem cells. Adv Healthc Mater 5(12):1429–1438. https://doi.org/10.1002/adhm.201600095

Highley CB (2019) 3D bioprinting technologies. In: Guvendiren M (ed) 3D bioprinting in medicine. Springer, Cham. https://doi.org/10.1007/978-3-030-23906-0_1

Huckle J, Dootson G, Medcalf N, McTaggart S, Wright E, Carter A et al (2003) Differentiated chondrocytes for cartilage tissue engineering. In: Tissue engineering of cartilage and bone: Novartis Foundation symposium, vol 249. Wiley, Chichester, pp 103–117. https://doi.org/10.1002/0470867973.ch8

Kačarević ŽP, Rider PM, Alkildani S, Retnasingh S, Smeets R, Jung O, Ivanišević Z, Barbeck M (2018) An introduction to 3D bioprinting: possibilities, challenges and future aspects. Materials (Basel, Switzerland) 11(11):2199. https://doi.org/10.3390/ma11112199

Kang HW, Lee S, Ko I et al (2016) A 3D bioprinting system to produce human-scale tissue constructs with structural integrity. Nat Biotechnol 34:312–319. https://doi.org/10.1038/nbt.3413

Li J, Chen M, Fan X et al (2016) Recent advances in bioprinting techniques: approaches, applications and future prospects. J Transl Med 14:271. https://doi.org/10.1186/s12967-016-1028-0

Li Y, Cao X, Deng W, Yu Q, Sun C, Ma P et al (2021) 3D printable Sodium alginate-Matrigel (SA-MA) hydrogel facilitated ectomesenchymal stem cells (EMSCs) neuron differentiation. J Biomater Appl 35(6):709–719. https://doi.org/10.1177/0885328220961261

Liang K (2023) Tissue bioprinting: promise and challenges. Bioengineering (Basel, Switzerland) 10(12):1400. https://doi.org/10.3390/bioengineering10121400

Loukelis K, Helal ZA, Mikos AG, Chatzinikolaidou M (2023) Nanocomposite bioprinting for tissue engineering applications. Gels 9(2):103. https://doi.org/10.3390/gels9020103

Maan Z, Masri NZ, Willerth SM (2022) Smart bioinks for the printing of human tissue models. Biomol Ther 12(1):141. https://doi.org/10.3390/biom12010141

Mao H, Yang L, Zhu H, Wu L, Ji P, Yang J, Gu Z (2020) Recent advances and challenges in materials for 3D bioprinting. Prog Nat Sci Mater Int 30(5):618–634. https://doi.org/10.1016/j.pnsc.2020.09.015

Markstedt K, Mantas A, Tournier I, Martínez Ávila H, Hagg D, Gatenholm P (2015) 3D bioprinting human chondrocytes with nanocellulose–alginate bioink for cartilage tissue engineering applications. Biomacromolecules 16(5):1489–1496. https://doi.org/10.1021/acs.biomac.5b00188

Matai I, Kaur G, Seyedsalehi A, McClinton A, Laurencin CT (2020) Progress in 3D bioprinting technology for tissue/organ regenerative engineering. Biomaterials 226:119536. https://doi.org/10.1016/j.biomaterials.2019.119536

Mobaraki M, Ghaffari M, Yazdanpanah A, Luo Y, Mills DK (2020) Bioinks and bioprinting: a focused review. Bioprinting 18:e00080. https://doi.org/10.1016/j.bprint.2020.e00080

Muskan N, Gupta D, Negi NP (2022) 3D bioprinting: Printing the future and recent advances. Bioprinting 27:e00211. https://doi.org/10.1016/j.bprint.2022.e00211

Ni T, Liu M, Zhang Y, Cao Y, Pei R (2020) 3D bioprinting of bone marrow mesenchymal stem cell-laden silk fibroin double network scaffolds for cartilage tissue repair. Bioconjug Chem 31(8):1938–1947. https://doi.org/10.1021/acs.bioconjchem.0c00298

Rasouli R, Sweeney C, Frampton JP (2024) Heterogeneous and composite bioinks for 3D-bioprinting of complex tissue. Biomed Mater Devices 3:108–126. https://doi.org/10.1007/s44174-024-00171-7

Sachdev A, Acharya S, Gadodia T, Shukla SJH, Akre C, Khare M, Huse S (2022) A review on techniques and biomaterials used in 3D bioprinting. Cureus 14(8):e28463. https://doi.org/10.7759/cureus.28463

Santoni S, Gugliandolo SG, Sponchioni M, Moscatelli D, Colosimo BM (2021) 3D bioprinting: current status and trends—a guide to the literature and industrial practice. Bio-Des Manuf 5(1):14–42. https://doi.org/10.1007/s42242-021-00165-0

Seol Y-J, Kang H-W, Lee SJ, Atala A, Yoo JJ (2014) Bioprinting technology and its applications. Eur J Cardiothorac Surg 46(3):342–348. https://doi.org/10.1093/ejcts/ezu148

Tibbetts JH (2021) The future of bioprinting: multidisciplinary teams seek to create living human organs. Bioscience 71(6):564–570. https://doi.org/10.1093/biosci/biab046

Whenish R, Ramakrishna S, Jaiswal AK et al (2022) A framework for the sustainability implications of 3D bioprinting through nature-inspired materials and structures. Bio-Des Manuf 5:412–423. https://doi.org/10.1007/s42242-021-00168-x

Xavier JR, Thakur T, Desai P, Jaiswal MK, Sears N, Cosgriff-Hernandez E et al (2015) Bioactive nanoengineered hydrogels for bone tissue engineering: a growth-factor-free approach. ACS Nano 9(3):3109–3118. https://doi.org/10.1021/nn507488s

Xie Z, Gao M, Lobo AO, Webster TJ (2020) 3D bioprinting in tissue engineering for medical applications: the classic and the hybrid. Polymers 12(8):1717. https://doi.org/10.3390/polym12081717

Yao Q, Zheng YW, Lan QH, Kou L, Xu HL, Zhao YZ (2019) Recent development and biomedical applications of decellularized extracellular matrix biomaterials. Mater Sci Eng C 104:109942. https://doi.org/10.1016/j.msec.2019.109942

Yilmaz B, Tahmasebifar A, Baran ET (2019) Bioprinting Technologies in Tissue Engineering. In: Silva AC, Moreira JN, Lobo JMS, Almeida H (eds) Current applications of pharmaceutical biotechnology. Advances in biochemical engineering/biotechnology, vol 171. Springer, Cham. https://doi.org/10.1007/10_2019_108

Yilmaz B, Al Rashid A, Mou YA, Evis Z, Koç M (2021) Bioprinting: a review of processes, materials and applications. Bioprinting 23:e00148. https://doi.org/10.1016/j.bprint.2021.e00148

Yu J, Park SA, Kim WD, Ha T, Xin YZ, Lee J, Lee D (2020a) Current advances in 3D bioprinting technology and its applications for tissue engineering. Polymers 12(12):2958. https://doi.org/10.3390/polym12122958

Yu J, Lee S, Choi S, Kim KK, Ryu B, Kim C-Y, Jung C-R, Min B-H, Xin Y-Z, Park SA et al (2020b) Fabrication of a Polycaprolactone/Alginate bipartite hybrid Scaffold for Osteochondral tissue using a three-dimensional bioprinting system. Polymers 12(10):2203. https://doi.org/10.3390/polym12102203

Zhang J, Wehrle E, Adamek P, Paul GR, Qin XH, Rubert M, Müller R (2020) Optimization of mechanical stiffness and cell density of 3D bioprinted cell-laden scaffolds improves extracellular matrix mineralization and cellular organization for bone tissue engineering. Acta Biomater 114:307–322. https://doi.org/10.1016/j.actbio.2020.07.016

Recent Advancement in Biomaterial-Based Drug Delivery System

11

Manjusha Bhange, Kuldeep Vinchurkar, Ankita Bhadoriya, Sheetal Mane, and Meghraj Vivekanand Suryawanshi

Abstract

Recent advancements in biomaterial-based drug delivery systems (DDS) have significantly revolutionized the field of medicine, offering improved therapeutic efficacy, reduced side effects, and targeted delivery of drugs. This chapter highlights the latest innovations and trends in the development of biomaterials, including polymers, lipids, and inorganic materials, which are engineered to optimize drug delivery. Notably, the integration of nanotechnology has facilitated the creation of nanoparticles, liposomes, and micelles, enhancing the precision and control of drug release. Biodegradable and biocompatible materials have emerged as critical components, minimizing long-term toxicity and enhancing patient compliance. Furthermore, stimuli-responsive systems, capable of responding to environmental changes such as pH, temperature, and enzymatic

M. Bhange
Department of Pharmaceutics, Datta Meghe College of Pharmacy, DMIHER (DU), Wardha, Maharashtra, India

K. Vinchurkar (✉)
School of Pharmacy, Devi Ahilya Vishwavidyalaya, Indore, Madhya Pradesh, India

Department of Pharmaceutics, Sandip Institute of Pharmaceutical Sciences (SIPS), Affiliated To Savitribai Phule Pune University (SPPU, Pune), Nashik, India

A. Bhadoriya
Indore Institute of Pharmacy, Indore, Madhya Pradesh, India

S. Mane
School of Pharmacy, Devi Ahilya Vishwavidyalaya, Indore, Madhya Pradesh, India

M. V. Suryawanshi
Department of Pharmaceutics, Sandip Institute of Pharmaceutical Sciences (SIPS), Affiliated To Savitribai Phule Pune University (SPPU, Pune), Nashik, India

Md. Faiyazuddin, M. Suryawanshi (eds.), *Design and Processing of Green Materials*, Biomaterials, Bioengineering and Sustainability 4,
https://doi.org/10.1007/978-3-031-91790-5_11

activity, have shown promising potential in achieving controlled and sustained drug release. Advances in biofabrication techniques, such as 3D printing, have enabled the customization of drug delivery platforms tailored to individual patient needs. The incorporation of smart materials and the development of multifunctional delivery systems are also explored, offering synergistic therapeutic effects and diagnostic capabilities. Despite these advancements, challenges such as scalability, regulatory hurdles, and the need for extensive clinical validation remain. This chapter underscores the transformative impact of biomaterial-based DDS on modern therapeutics and highlights the ongoing efforts to overcome existing barriers, paving the way for next-generation drug delivery solutions.

Keywords

Biomaterials · Biodegradable materials · Drug delivery system · Innovations · Nanotechnology

11.1 Introduction

11.1.1 Overview of Drug Delivery System

Drug delivery systems (DDS) are engineered technologies for the targeted delivery and/or controlled release of therapeutic agents. These systems play a crucial role in modern medicine, enhancing the efficacy and safety of drug administration (Table 11.1). Traditional methods of drug delivery, such as oral ingestion and intravenous injection, often face challenges such as poor bioavailability, systemic side effects, and non-specific targeting. To address these issues, advanced DDS have been developed to ensure that drugs are delivered in a manner that maximizes their therapeutic potential while minimizing side effects (Zhang et al. 2013).

11.1.2 Biomaterial-Based Drug Delivery Systems

Biomaterial-based drug delivery systems represent a significant advancement over traditional drug delivery methods, offering solutions to many of the challenges faced by conventional approaches. By utilizing materials derived from natural or synthetic sources, these systems enhance the targeted delivery and controlled release

Table 11.1 Traditional vs. advanced drug delivery systems: Two different system of drug delivery are compared below

Aspect	Traditional DDS	Advanced DDS (Zhang et al. 2013)
Administration route	Oral, intravenous	Injectable, transdermal, inhalable
Bioavailability	Often low	High
Targeting	Non-specific	Specific
Side effects	High	Reduced
Release control	Immediate	Controlled/sustained

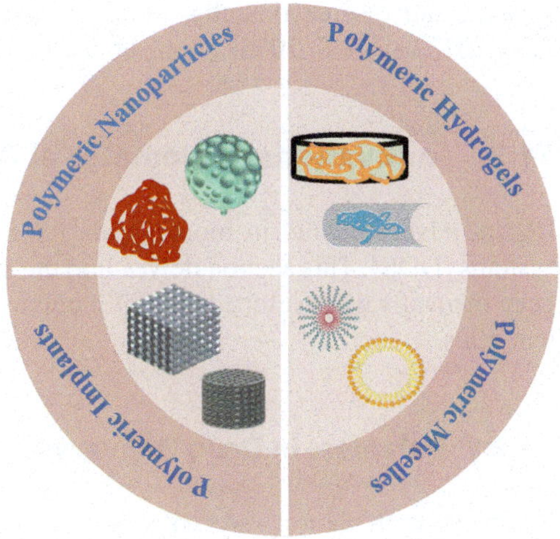

Fig. 11.1 Biomaterial-based drug delivery systems

of therapeutic agents, thereby improving the overall efficacy and safety of treatments. Biomaterials play a crucial role in various drug delivery systems, offering versatility in targeting, release kinetics, and biocompatibility. Here are several types of biomaterial-based drug delivery systems as described in Fig. 11.1.

11.1.3 Importance of Biomaterial-Based DDS

Biomaterial-based drug delivery systems incorporate materials derived from natural or synthetic sources to create carriers that can improve the delivery and efficacy of therapeutic agents. These biomaterials offer several advantages:

1. **Biocompatibility**: Biomaterials are often non-toxic and well-tolerated by the body, reducing the risk of adverse reactions.
2. **Controlled Release**: They can be engineered to release drugs at a controlled rate, maintaining therapeutic levels over extended periods.
3. **Targeted Delivery**: Biomaterials can be modified to target specific tissues or cells, enhancing the precision of treatment.
4. **Protection of Drugs**: They protect drugs from degradation before they reach the target site, improving the stability and bioavailability of the therapeutic agents.

The objective of this chapter is to provide a comprehensive overview of biomaterial-based drug delivery systems by first reviewing the different types of biomaterials, including natural and synthetic polymers, lipids, and inorganic materials, used in these systems. It will then explore various engineering strategies for designing effective drug delivery systems, such as nanoparticles, hydrogels, and microneedles. Additionally, the chapter will analyze the applications of biomaterial-based DDS in treating a range of diseases, including cancer, cardiovascular diseases, and infections. Finally, it will evaluate the current challenges and limitations of these systems and discuss potential future developments to enhance their efficacy and safety (Hickey et al. 2015; Zhang et al. 2013).

11.2 Innovations in Biomaterial Development

Recent advancements in biomaterials have significantly enhanced drug delivery systems (DDS). This section delves into the key categories of biomaterials that have been innovatively developed for DDS: polymers, lipid-based systems, and inorganic materials.

11.2.1 Polymers in Drug Delivery

Polymers, both natural and synthetic, have revolutionized the field of drug delivery by providing versatile platforms for controlled and targeted release of therapeutic agents.

11.2.2 Types of Polymers

1. **Natural Polymers**: Natural polymers are derived from biological sources and are often biocompatible and biodegradable. Their use in drug delivery systems has expanded due to their inherent biocompatibility, which minimizes the risk of adverse reactions (Table 11.2).

Table 11.2 Comparison of natural and synthetic polymers: Different types of polymers are differentiated on the basis of their features (Anselmo et al. 2019)

Feature	Natural polymers	Synthetic polymers
Source	Natural	Synthetic
Biocompatibility	High	Variable
Degradability	Biodegradable	Controllable
Mechanical properties	Less controllable	Tunable
Applications	Wound healing, tissue engineering	Nanoparticles, drug-eluting stents
	Chitosan nanoparticles: Used for the delivery of insulin orally, improving patient compliance and controlling blood glucose levels	**PLGA nanoparticles**: Used to deliver anticancer drugs like paclitaxel and doxorubicin directly to tumors, reducing systemic toxicity
	Alginate beads: Developed for encapsulating probiotics, enhancing their stability and delivery to the gut	**PEGylated proteins**: Improved pharmacokinetics and reduced immunogenicity, as seen in PEGylated interferon for hepatitis treatment
	Gelatin-based hydrogels: Employed in regenerative medicine for delivering growth factors and stem cells to damaged tissues	**PLA microparticles**: Developed for sustained release of vaccines, enhancing immune response over time

   - **Examples**: Chitosan, alginate, gelatin, hyaluronic acid.
   - **Advantages**: Biocompatibility, biodegradability, low toxicity.
   - **Applications**: Wound healing, tissue engineering, controlled drug release.
2. **Synthetic Polymers**:
   - **Examples**: Poly (lactic acid) (PLA), Poly (glycolic acid) (PGA), Poly (lactic-co-glycolic acid) (PLGA), polyethylene glycol (PEG).
   - **Advantages**: Tunable mechanical properties, controlled degradation rates, reproducibility.
   - **Applications**: Nanoparticles for cancer therapy, scaffolds for tissue regeneration, drug-eluting stents (Anselmo et al. 2019)

11.2.3 Polymer-Based Drug Delivery Systems

1. **Nanoparticles**: Enhanced drug solubility and bioavailability.
2. **Hydrogels**: Swellable matrices for controlled release.
3. **Microneedles**: Minimally invasive delivery.

Examples:

- **PLA/PLGA Nanoparticles**: Used for delivering chemotherapy drugs like paclitaxel to target cancer cells.
- **Chitosan Hydrogels**: Employed in wound dressings to provide a moist environment and controlled drug release (Wang et al. 2020; Li et al. 2020)

Advancements in both natural and synthetic polymers have propelled the field of biomaterial-based drug delivery systems forward as shown in Fig. 11.2.

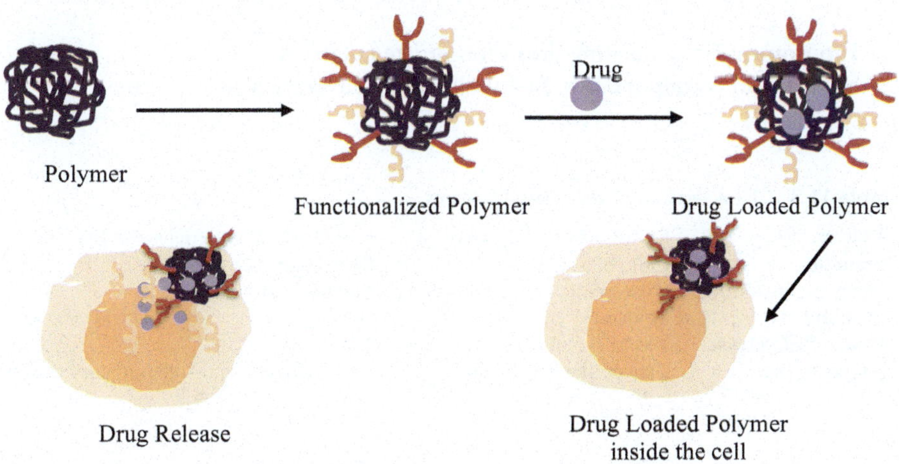

Fig. 11.2 Smart polymer based drug delivery system

Natural polymers offer biocompatibility and biodegradability, making them suitable for applications where minimal immune response is critical. Synthetic polymers provide versatility and precise control over drug release profiles, making them ideal for a wide range of therapeutic applications. Together, these innovations enhance the efficacy, safety, and patient compliance of drug therapies (Vega et al. 2021).

11.2.4 Lipid-Based Systems

Lipid-based drug delivery systems are another innovation that has improved the delivery and effectiveness of therapeutic agents, particularly for poorly water-soluble drugs (Yoo et al. 2011; Tang et al. 2021). Table 11.3 underscores various characteristics of Lipid-based drug delivery systems.

11.2.4.1 Types of Lipid-Based Systems

1. **Liposomes**:
   - **Structure**: Spherical vesicles with a lipid bilayer.
   - **Advantages**: Biocompatibility, ability to encapsulate both hydrophilic and hydrophobic drugs.
   - **Applications**: Anticancer drugs, vaccines, antifungal agents.
2. **Solid Lipid Nanoparticles (SLNs)**:
   - **Structure**: Solid lipid core stabilized by surfactants.
   - **Advantages**: Enhanced stability, controlled drug release, protection of labile drugs.
   - **Applications**: Gene delivery, neurotherapeutics, anti-inflammatory drugs.

11.2.5 Lipid-Based Drug Delivery Systems

1. **Liposomes**: Encapsulating drugs for targeted delivery.
2. **Solid Lipid Nanoparticles**: Stabilizing and delivering sensitive drugs.

Table 11.3 Characteristics of lipid-based systems

Feature	Liposomes	Solid lipid nanoparticles (SLNs)
Structure	Lipid bilayer vesicles	Solid lipid core
Drug encapsulation	Hydrophilic and hydrophobic	Mainly hydrophobic
Stability	Moderate	High
Controlled release	Yes	Yes
Applications	Cancer therapy, vaccines	Gene delivery, anti-inflammatory drugs

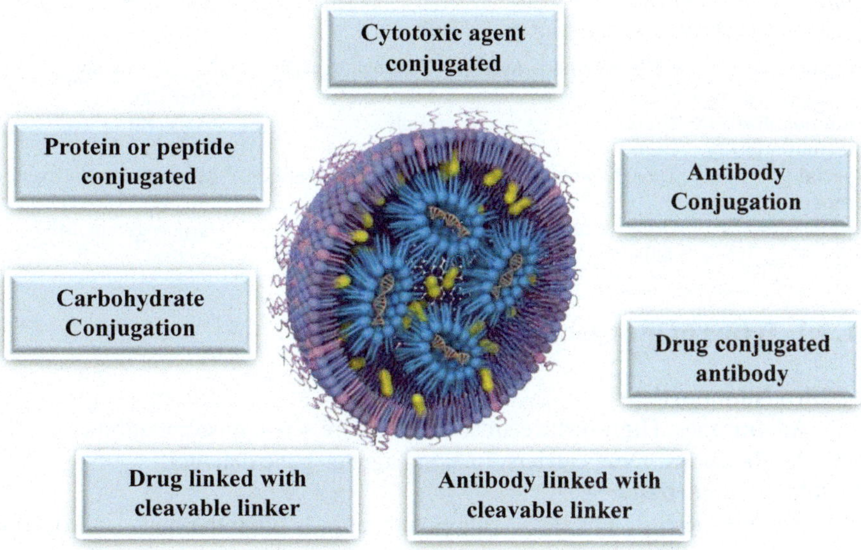

Fig. 11.3 Advancement in lipid based drug delivery system

Examples:

- **Doxil (Doxorubicin Liposomes)**: A liposomal formulation used in cancer treatment to enhance drug accumulation in tumors while reducing systemic toxicity.
- **SLNs for Insulin Delivery**: Providing a stable platform for oral delivery of insulin, improving patient compliance.

Recent advancements in lipid-based drug delivery systems have significantly improved the delivery and efficacy of therapeutic agents shown in Fig. 11.3. Liposomes, SLNs, and LDCs each offer unique advantages, from enhanced bioavailability and targeted delivery to reduced toxicity and controlled release. These innovations are transforming the landscape of drug delivery, providing more effective and safer treatments for a wide range of diseases (Torchilin et al. 2014; Li and Mooney 2016).

11.3 Inorganic Materials for Drug Delivery

Inorganic materials, including metals and metal oxides, offer unique properties that can be harnessed for innovative drug delivery systems (Table 11.4).

Table 11.4 Characteristics of inorganic materials: inorganic materials used in manufacturing of different drug delivery system (Stuart et al. 2010)

Feature	Silica nanoparticles	Gold nanoparticles	Magnetic nanoparticles
Surface area	High	Moderate	Moderate
Functionalization	High	High	Moderate
Biocompatibility	High	High	Variable
Special properties	Tunable pores	Photothermal effect	Magnetic responsiveness
Applications	Drug carriers, gene delivery	Cancer therapy, imaging	Hyperthermia, MRI contrast

11.3.1 Types of Inorganic Materials

1. **Silica Nanoparticles**:
   - **Advantages**: High surface area, tunable pore sizes, biocompatibility.
   - **Applications**: Drug carriers, imaging agents, gene delivery.
2. **Gold Nanoparticles**:
   - **Advantages**: Easily functionalized, photothermal properties, biocompatibility.
   - **Applications**: Cancer therapy, diagnostic imaging, targeted drug delivery.
3. **Magnetic Nanoparticles**:
   - **Advantages**: Magnetic responsiveness, targeted delivery through external magnetic fields.
   - **Applications**: Hyperthermia treatment, MRI contrast agents, targeted drug delivery (Makvandi et al. 2020; Stuart et al. 2010).

11.3.2 Inorganic Material-Based Drug Delivery Systems

1. **Silica Nanoparticles**: High surface area for drug loading.
2. **Gold Nanoparticles**: Photothermal properties for cancer treatment.
3. **Magnetic Nanoparticles**: External magnetic field-guided delivery.

Examples:

- **Mesoporous Silica Nanoparticles (MSNs)**: Used for targeted delivery of anti-cancer drugs, enhancing their accumulation in tumor tissues.
- **Gold Nanoparticles in Photothermal Therapy**: Converting light to heat to kill cancer cells selectively.
- **Magnetic Nanoparticles for Hyperthermia**: Heating targeted tissues to treat tumors without affecting surrounding healthy tissues.
- Some properties of Inorganic nanoparticles are shown in Fig. 11.4 (Dunncan et al. 2006; Anderson and Shive 1997)

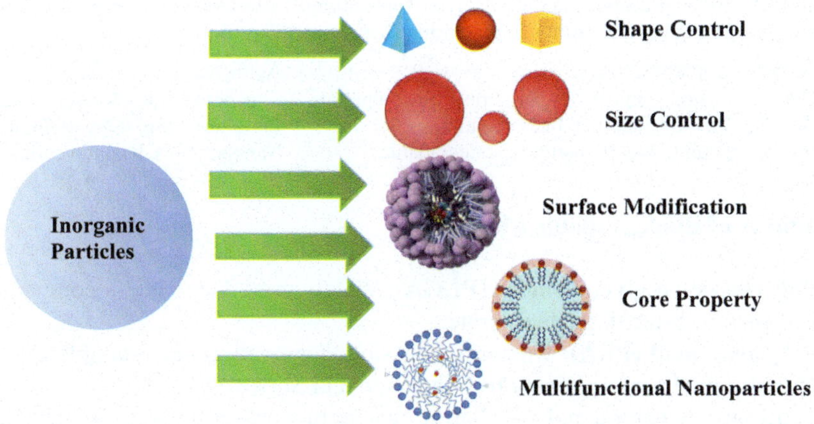

Fig. 11.4 Smart polymer based drug delivery system

11.4 Biodegradable and Biocompatible Materials

Biodegradable and biocompatible materials are pivotal in the advancement of drug delivery systems (DDS). These materials offer numerous benefits, including reduced long-term toxicity, enhanced patient compliance, and the ability to deliver drugs in a controlled and sustained manner. This section explores the importance of biodegradability and biocompatibility, along with case studies and applications that highlight their roles in modern DDS (Anderson and Shive 1997).

11.4.1 Importance of Biodegradability

Biodegradability refers to the ability of a material to be broken down by biological processes. In the context of drug delivery systems, biodegradable materials are crucial for several reasons:

1. **Reduced Long-Term Toxicity**: Biodegradable materials are broken down into non-toxic by products that can be easily eliminated from the body, minimizing the risk of long-term adverse effects (Table 11.5).
2. **Controlled Drug Release**: These materials can be engineered to degrade at a specific rate, allowing for controlled and sustained release of the therapeutic agents. This ensures a steady drug concentration within the therapeutic window over an extended period, improving treatment efficacy.
3. **Elimination of Secondary Surgeries**: Biodegradable implants and carriers eliminate the need for surgical removal after the drug has been released, enhancing patient comfort and reducing healthcare costs (Peppas and Langer 1994).

Table 11.5 Biodegradable materials and their applications: Description of biodegradable materials used for drug delivery system manufacturing (Langer and Peppas 1981)

Material	Properties	Applications
PLGA	Customizable degradation rate, biocompatible	Cancer therapy, vaccine delivery
PLA	Slow degradation, biocompatible	Long-term drug delivery implants
Chitosan	Natural, biocompatible, biodegradable	Nanoparticles, wound healing

Examples of Biodegradable Materials:

- **Poly (lactic-co-glycolic acid) (PLGA)**: Widely used due to its biocompatibility and customizable degradation rate.
- **Polylactic acid (PLA)**: Known for its slow degradation rate, suitable for long-term drug delivery applications (Anderson and Shive 1997).
- **Chitosan**: A natural polysaccharide that is biocompatible and biodegradable, often used in nanoparticle formulations for drug delivery (Bae et al. 2011; Elisseff et al. 2001).

11.4.2 Biocompatibility and Patient Compliance

Biocompatibility refers to the ability of a material to perform with an appropriate host response in a specific application. For drug delivery systems, biocompatibility is essential to ensure that the materials do not elicit adverse immune responses or toxicity.

1. **Enhanced Safety**: Biocompatible materials are less likely to cause inflammation, allergic reactions, or other immune responses, making them safer for patients.
2. **Improved Patient Compliance**: Materials that are biocompatible and biodegradable can be designed to be minimally invasive and more comfortable for patients, thereby improving adherence to treatment regimens.
3. **Versatility in Applications**: Biocompatible materials can be used in a variety of forms, including nanoparticles, hydrogels, and implants, to suit different therapeutic needs and delivery methods (Allen et al. 2013; Hua et al. 2013).

Examples of Biocompatible Materials:

- **Hyaluronic Acid**: Used in hydrogels for controlled drug release and tissue engineering (Table 11.6).
- **Silk Fibroin**: Utilized for its biocompatibility and mechanical strength in drug delivery and wound healing applications.
- **Alginate**: A natural polymer used for encapsulating cells and drugs, particularly in tissue engineering and regenerative medicine (Langer et al. 2011).

Table 11.6 Biocompatible materials and their applications: Biocompatible materials used in drug delivery systems (Hua et al. 2013)

Material	Properties	Applications
Hyaluronic acid	Hydrophilic, biocompatible	Hydrogels, tissue engineering
Silk fibroin	Strong, biocompatible	Drug delivery, wound healing
Alginate	Biocompatible, natural	Cell encapsulation, regenerative medicine

Table 11.7 Year wise advances in biofabrication techniques (Liu et al. 2016)

Year	Milestone
1980s	Emergence of tissue engineering as a field.
1990s	Development of biocompatible materials for implants.
2000	Introduction of electrospinning for scaffold creation.
2003	First demonstration of 3D printing for tissue scaffolds.
2010	Advances in microfabrication techniques for drug delivery.
2015	Development of smart hydrogels responsive to stimuli.
2020	Personalized medicine applications using 3D printing.
2023	Integration of living cells in biofabricated constructs.

11.5 Advances in Biofabrication Techniques

Biofabrication is an interdisciplinary field that combines principles from biology, materials science, and engineering to create complex biological constructs and devices. These constructs can mimic natural tissues and are used in various biomedical applications, including drug delivery systems, tissue engineering, and regenerative medicine.

The concept of biofabrication has its roots in the early twentieth century, with the development of biomaterials and early tissue engineering efforts (Table 11.7). However, significant progress began in the late 20th and early 21st centuries with advances in biotechnology and materials science (Liu et al. 2016).

Progress Timeline:

Biofabrication techniques have seen remarkable advancements that are crucial for the development of next-generation biomaterial-based drug delivery systems. The chapter on "Recent Advancement in Biomaterial-Based Drug Delivery System" discusses several cutting-edge techniques and their implications for drug delivery, providing detailed insights into how these advances are shaping the future of medicine (Liu et al. 2016).

11.5.1 Three-Dimensional (3D) Printing

3D printing, or additive manufacturing, is a breakthrough technology in biofabrication. This technique allows the precise layering of biomaterials to create complex and customized structures. In the context of drug delivery systems, 3D printing offers several significant advantages:

- **Precision and Control**: 3D printing enables the creation of drug delivery devices with intricate geometries that can control the release kinetics of therapeutic agents. This precision ensures that drugs are delivered at the desired rate and location, improving efficacy and reducing side effects.
- **Customization**: One of the major benefits of 3D printing is its ability to fabricate patient-specific drug delivery systems. Using patient imaging data, such as MRI or CT scans, devices can be tailored to fit the patient's anatomy perfectly. This customization enhances the effectiveness of treatment and minimizes discomfort.
- **Material Versatility**: 3D printing supports a wide range of materials, including biocompatible polymers, ceramics, and hydrogels, which can be used to develop drug delivery systems. This versatility allows for the incorporation of various drugs and the design of multifunctional devices that can provide structural support and therapeutic delivery (Liu et al. 2016).

11.5.2 Progressive Developments in 3D Printing

The progressive developments in 3D printing techniques and their significant implications for drug delivery systems given in detail year-by-year account.

11.5.2.1 Early Exploration and Proof of Concept

Initial explorations into using 3D printing for medical applications, focusing primarily on creating scaffolds for tissue engineering. Set the foundation for using 3D printing in drug delivery by demonstrating the ability to fabricate complex structures that could potentially house and release drugs (Sawant et al. 2012).

11.5.2.2 Advancements in Biocompatible Materials

Introduction of biocompatible polymers suitable for 3D printing. Enabled the creation of drug delivery systems that are safe for implantation in the human body (Sawant et al. 2012).

11.5.2.3 Emergence of Customization and Complexity

Advances in 3D printing technology allowed for higher resolution and more complex structures. Facilitated the development of more sophisticated drug delivery systems that could release multiple drugs at different rates (Sawant et al. 2012).

11.5.2.4 Integration with Biological Systems

Introduction of bioinks containing living cells for 3D printing. Enabled the creation of drug delivery systems that not only release drugs but also integrate with the body's tissues, improving biocompatibility and functionality (Sawant et al. 2012).

11.5.2.5 Present: Smart and Responsive Systems

Development of 3D printed hydrogels that respond to environmental stimuli (e.g., pH, temperature). These smart hydrogels offer precise control over drug release, reacting to specific physiological conditions to optimize therapeutic effects (Sawant et al. 2012).

The evolution of 3D printing techniques over the years has dramatically impacted the field of drug delivery, transforming it into a highly precise and customizable discipline. From the early days of simple scaffolds to the latest smart and responsive systems, each advancement has brought new possibilities for improving therapeutic outcomes and patient experiences. The ongoing research and development in this field continue to push the boundaries, promising even more innovative solutions in the future of medicine (Sawant et al. 2012).

11.5.3 Microfabrication

Microfabrication techniques, such as soft lithography and photolithography, have also made significant contributions to the field of drug delivery. These techniques involve the creation of micro- and nanoscale structures with high precision and repeatability:

- **Microneedles**: Microfabrication is used to produce microneedle arrays, which can deliver drugs through the skin painlessly and efficiently. These microneedles can be engineered to dissolve or remain in the skin, releasing drugs over a prolonged period.
- **Microchips and MEMS Devices**: Micro-Electro-Mechanical Systems (MEMS) devices fabricated using microfabrication techniques can be used for controlled drug release. These devices can be implanted or ingested, providing precise control over drug release profiles in response to external stimuli or pre-programmed schedules.

The progression of microfabrication techniques and their significant impact on drug delivery systems given in detail year-by-year account (Bao et al. 2020).

11.5.3.1 Foundation and Initial Developments

Introduction of photolithography for creating micro-scale structures in the semiconductor industry. Laid the groundwork for microfabrication techniques that would later be adapted for biomedical applications (Bao et al. 2020).

11.5.3.2 Early Applications in Drug Delivery

Adaptation of microfabrication techniques for creating drug delivery microchips. Enabled precise control over drug release, paving the way for programmable drug delivery systems (Bao et al. 2020).

11.5.3.3 Refinement and Commercialization

Advances in micro-electro-mechanical systems (MEMS) for drug delivery. Allowed for the development of implantable devices that could release drugs in a controlled manner over extended periods (Bao et al. 2020).

11.5.3.4 Enhanced Control and Targeting

Creation of microfabricated devices capable of multi-drug delivery. Facilitated complex treatment regimens by enabling the sequential or simultaneous release of multiple therapeutic agents (Bao et al. 2020).

11.5.3.5 Integration with Advanced Materials

Incorporation of nanomaterials into microfabricated drug delivery devices. Enhanced the targeting capabilities and efficacy of drug delivery systems by leveraging the unique properties of nanomaterials (Bao et al. 2020).

11.5.3.6 Present: Smart Systems and Real-Time Monitoring

Integration of microfabrication with smart materials for responsive drug delivery systems. These systems can adjust drug release rates in response to environmental stimuli, optimizing therapeutic outcomes (Bao et al. 2020).

Microfabrication techniques have evolved from their origins in the semiconductor industry to become pivotal in the development of advanced drug delivery systems. Each decade has seen significant innovations that have improved the precision, control, and functionality of these systems. From the initial creation of micro-scale structures to the latest smart and personalized drug delivery devices, microfabrication continues to shape the future of medicine, offering innovative solutions for more effective and patient-friendly therapies (Bao et al. 2020).

11.5.4 Electrospinning

Electrospinning is a technique that produces fine fibers from polymer solutions under the influence of an electric field. These fibers can form scaffolds with high surface area-to-volume ratios, which are beneficial for drug delivery applications:

- **Drug Encapsulation**: Drugs can be encapsulated within the electrospun fibers, providing a controlled and sustained release. This is particularly useful for wound dressings and tissue engineering scaffolds where local drug delivery is required.
- **Nanofibrous Mats**: Electrospun nanofibrous mats can be designed to deliver multiple drugs simultaneously, addressing complex conditions such as infections and chronic wounds. These mats can also incorporate growth factors and other bioactive molecules to promote tissue regeneration while providing drug therapy.

The progression of electrospinning techniques and their significant impact on drug delivery systems given in detail year-by-year account.

- Initial experiments and establishment of the electrospinning process for creating nanofibers from polymer solutions.
- Early research into the use of electrospun nanofibers for controlled drug release.

- Introduction of electrospun nanofibrous scaffolds for tissue engineering. Highlighted the dual functionality of electrospun scaffolds in providing structural support and localized drug delivery.

Development of electrospun fibers with embedded sensors for real-time monitoring of drug release and patient response (Ganesh et al. 2022).

11.5.5 Hydrogel-Based Systems

Hydrogels are hydrophilic polymer networks capable of holding large amounts of water. Advances in hydrogel technology have made them highly suitable for drug delivery:

- **Responsive Hydrogels**: Modern hydrogels can respond to various stimuli such as pH, temperature, and enzymes, releasing drugs in a controlled manner. These "smart" hydrogels enhance the precision of drug delivery, ensuring that drugs are released only under specific physiological conditions.
- **Injectable Hydrogels**: Injectable hydrogels can be used for minimally invasive delivery of drugs. Once injected, these hydrogels can solidify in situ, forming a depot that releases drugs over time. This is particularly advantageous for localized and sustained delivery of therapeutics (Gao et al. 2016).

The progression of hydrogel-based system and their significant impact on drug delivery systems given in detail year-by-year account.

- Introduction of basic hydrogel formulations for biomedical applications.
- Use of hydrogels for localized drug delivery in wound care. Showed the effectiveness of hydrogels in delivering drugs directly to wound sites, promoting healing and reducing infections.
- Introduction of biodegradable hydrogels. Enhanced the safety and effectiveness of hydrogel-based drug delivery systems by ensuring that the hydrogels degrade safely in the body.
- Advances in 3D printing of hydrogels for creating complex, patient-specific drug delivery systems. Enabled the customization of drug delivery devices to fit individual patient needs, enhancing the precision and efficacy of treatments.
- Use of hydrogels for the delivery of biologics, such as proteins and antibodies.
- Integration of biosensors within hydrogel systems for real-time monitoring of drug release and patient response.

Hydrogel-based systems have seen remarkable advancements over the past few decades, significantly impacting the field of drug delivery. From the initial formulations to the latest smart and personalized systems, each development has contributed to more effective and patient-friendly therapies (Jeong et al. 2011). The ongoing research and integration of advanced materials and technologies continue to push the boundaries, promising innovative solutions for the future of medicine (Zhao et al. 2021).

11.6 Integration of Biofabrication Techniques with Drug Delivery Systems

The integration of advanced biofabrication techniques with biomaterial-based drug delivery systems has led to several innovative solutions:

- **Multi-Functional Implants**: Combining 3D printing with drug-loaded hydrogels or electrospun fibers has resulted in implants that provide structural support while delivering drugs locally. These multi-functional implants are beneficial in orthopedic applications, where they can support bone regeneration and prevent infections.
- **Personalized Medicine**: Personalized drug delivery systems fabricated using 3D printing and microfabrication ensure that treatments are tailored to individual patients. This customization enhances therapeutic outcomes and reduces adverse effects, aligning with the goals of personalized medicine.
- **Smart Drug Delivery Devices**: The development of smart drug delivery devices using advanced biofabrication techniques allows for responsive drug release. These devices can be programmed to release drugs in response to specific physiological signals, improving the precision and efficacy of treatments (Bose et al. 2013). Table 11.8 illustrates biofabrication technique advancement in drug delivery system.

Table 11.8 Advanced biofabrication techniques with benefits and its applications: Biofabrication technique advancement in drug delivery system (Shi et al. 2010)

Solution	Biofabrication technique	Description	Benefits	Applications
Multi-functional implants	3D printing, drug-loaded hydrogels, electrospun fibers	Combines structural support with local drug delivery	Supports bone regeneration, prevents infections	Orthopedic applications[29, 30]
Personalized medicine	3D printing, microfabrication	Tailors drug delivery systems to individual patients	Enhances therapeutic outcomes, reduces adverse effects, aligns with personalized medicine goals	General medical treatments
Smart drug delivery devices	Advanced biofabrication techniques	Creates devices that respond to specific physiological signals for drug release	Provides responsive and precise drug release, improves precision and efficacy of treatments	Targeted and controlled drug delivery

The integration of advanced biofabrication techniques with biomaterial-based drug delivery systems has led to several innovative solutions, as detailed below:

11.7 Smart Materials and Multifunctional Delivery Systems

The integration of smart materials into multifunctional delivery systems has revolutionized drug delivery by enabling responsive and targeted therapeutic interventions.

Smart materials, also known as intelligent or responsive materials, have the ability to change their properties in response to external stimuli such as temperature, pH, light, magnetic fields, and more. These materials can be engineered to deliver drugs in a controlled and precise manner, enhancing the effectiveness of treatments.

Examples of Smart Materials:

1. **Thermo-responsive Polymers**:
   - **Example**: Poly(N-isopropylacrylamide) (PNIPAAm)
   - **Mechanism**: These polymers undergo a reversible phase transition at a specific temperature, making them ideal for drug delivery in response to body temperature.
   - **Application**: Used in the delivery of chemotherapeutic agents where localized heating can trigger drug release.
2. **pH-responsive Polymers**:
   - **Example**: Poly(methacrylic acid) (PMAA)
   - **Mechanism**: These polymers change their solubility in response to pH changes, making them suitable for targeted delivery in environments like the stomach or tumors.
   - **Application**: Used in oral drug delivery systems where the drug is released in the acidic environment of the stomach.
3. **Magnetic Nanoparticles**:
   - **Example**: Iron oxide nanoparticles
   - **Mechanism**: These particles can be guided to specific locations in the body using an external magnetic field and can generate heat upon exposure to an alternating magnetic field.
   - **Application**: Used in hyperthermia treatment for cancer where localized heating destroys cancer cells and releases drugs.
4. **Light-responsive Materials**:
   - **Example**: Gold nanoparticles
   - **Mechanism**: These materials change their properties upon exposure to light, enabling controlled drug release.
   - **Application**: Used in photothermal therapy where light is used to trigger drug release and heat to kill cancer cells (Ventola et al. 2014; Norman et al. 2017).

11.8 Synergistic Therapeutic Effects

Synergistic Therapeutic Effects Smart materials can be designed to deliver multiple therapeutic agents simultaneously or sequentially, providing a synergistic effect that enhances the overall therapeutic outcome (Table 11.9).

Examples:

1. **Combination Therapy for Cancer**:
   - **System**: Thermo-responsive hydrogels loaded with chemotherapy drugs and nanoparticles.
   - **Mechanism**: The hydrogel releases drugs in response to body temperature and the nanoparticles enhance the local heating effect, killing cancer cells more effectively.
   - **Result**: Increased efficacy of cancer treatment through combined chemotherapy and hyperthermia.
2. **Anti-inflammatory and Pain Relief**:
   - **System**: pH-responsive micelles loaded with anti-inflammatory drugs and analgesics.
   - **Mechanism**: The micelles release their payload in the acidic environment of inflamed tissues.
   - **Result**: Simultaneous reduction of inflammation and pain, improving patient comfort and recovery (Tenchov et al. 2021).

11.9 Diagnostic Capabilities

Smart materials can also be engineered to possess diagnostic capabilities, enabling real-time monitoring of the disease state and therapeutic response (Table 11.10).

Table 11.9 Synergistic effects of smart materials in drug delivery: Effects of smart materials alongwith their synergistic effects (Patel et al. 2023)

System	Stimulus	Drugs delivered	Mechanism	Therapeutic effect
Thermo-responsive hydrogel	Temperature	Chemotherapy drugs, nanoparticles	Release triggered by body heat	Enhanced cancer cell kill rate via chemo and hyperthermia
pH-responsive micelles	pH	Anti-inflammatory drugs, analgesics	Release in acidic inflamed tissues	Reduction of inflammation and pain
Light-responsive nanoparticles	Light	Anticancer drugs	Release upon light exposure	Targeted cancer therapy via photothermal effects

Table 11.10 Diagnostic capabilities of smart materials: Description and difference of smart materials with their diagnostic capabilities (Li and Mooney 2016)

Smart material	Diagnostic modality	Mechanism	Application
Fluorescent nanoparticles	Fluorescence imaging	Emission of light upon interaction	Cancer diagnostics, drug tracking
Gadolinium-based nanoparticles	MRI	Enhanced contrast in MRI scans	Real-time imaging of drug distribution

Examples:

1. **Fluorescent Nanoparticles**:
   - **Mechanism**: These particles can emit fluorescence in response to specific biological interactions, allowing for imaging and tracking of drug delivery.
   - **Application**: Used in cancer diagnostics to visualize tumor sites and monitor drug distribution.
2. **Magnetic Resonance Imaging (MRI) Contrast Agents**:
   - **Example**: Gadolinium-based nanoparticles
   - **Mechanism**: Enhance contrast in MRI scans, enabling precise imaging of drug delivery locations.
   - **Application**: Used to track the distribution and release of drugs in real-time (Li and Mooney 2016).

11.10 Future Outlook for Biomaterial-Based Drug Delivery Systems

Looking ahead, the future of biomaterial-based drug delivery systems is promising, with ongoing research and development focused on several key areas:

1. **Nanotechnology and Advanced Materials**:
   - Continued integration of nanotechnology will enhance the targeting capabilities and therapeutic efficacy of drug delivery systems.
   - Development of novel biomaterials with advanced properties, such as bioresorbable polymers and bioactive scaffolds, will expand the application scope.
2. **Smart and Responsive Systems**:
   - Further refinement of smart materials and stimuli-responsive platforms will enable more precise and controlled drug delivery.
   - Integration of artificial intelligence (AI) and machine learning algorithms may optimize treatment protocols based on real-time patient data.
3. **Personalized Medicine and Regenerative Therapies**:
   - Advancements in tissue engineering and regenerative medicine will lead to the development of bioactive scaffolds and organ-on-a-chip systems for personalized therapies.
   - Customization of drug delivery systems will become more commonplace, tailored to genetic, physiological, and lifestyle factors of individual patients.

4. **Translation from Bench to Bedside**:
    - Accelerated translation of innovative biomaterial-based technologies from laboratory settings to clinical practice will require robust collaboration between academia, industry, and regulatory bodies.
    - Addressing challenges such as scalability, cost-effectiveness, and regulatory approval will be crucial for widespread clinical adoption.

11.11 Conclusion

In recent years, biomaterial-based drug delivery systems have advanced significantly, revolutionizing modern therapeutics and ushering in an era of personalized, precise, and effective treatments. These systems have evolved from basic carriers to sophisticated platforms that integrate smart materials, nanotechnology, and advanced diagnostics. Smart and responsive systems, such as thermo-responsive polymers and pH-sensitive hydrogels, allow for precise drug targeting and controlled release in response to physiological cues. Furthermore, innovations in 3D printing and microfabrication enable the customization of drug delivery systems tailored to individual patient needs, optimizing treatment outcomes and minimizing side effects.

The transformative impact of these advancements on modern therapeutics is profound. Biomaterial-based systems enhance therapeutic efficacy by delivering drugs with precision and reducing systemic exposure, thereby maximizing therapeutic benefits while minimizing adverse effects. They also improve patient compliance through minimally invasive delivery methods and personalized treatment approaches, enhancing patient comfort and adherence to therapy regimens. Moreover, the integration of diagnostic capabilities into drug delivery devices enables real-time monitoring of drug distribution and patient response, facilitating timely adjustments in treatment strategies for complex diseases like cancer and diabetes.

Looking forward, the future of biomaterial-based drug delivery systems holds promise for further innovation and application in clinical settings. Ongoing research aims to refine smart materials and multi-functional delivery systems, making them more responsive, versatile, and compatible with diverse therapeutic needs. As these technologies continue to evolve and gain acceptance, they are poised to play an increasingly pivotal role in shaping the landscape of modern medicine, offering tailored solutions that improve patient outcomes and quality of life.

References

Allen TM, Cullis PR (2013) Liposomal drug delivery systems: from concept to clinical applications. Adv Drug Deliv Rev 65(1):36–48. https://doi.org/10.1016/j.addr.2012.09.037

Anderson JM, Shive MS (1997) Biodegradation and biocompatibility of PLA and PLGA microspheres. Adv Drug Deliv Rev 28(1):5–24. https://doi.org/10.1016/S0169-409X(97)00048-3

Anselmo AC, Mitragotri S (2019) Nanoparticles in the clinic: An update. Bioeng Transl Med 4(3):e10143. https://doi.org/10.1002/btm2.10143

Bae YH, Park K (2011) Targeted drug delivery to tumors: myths, reality and possibility. J Control Release 153(3):198–205. https://doi.org/10.1016/j.jconrel.2011.06.002

Bao Y, Maeki M, Ishida A, Tani H, Tokeshi M (2020) Effect of organic solvents on a production of PLGA-based drug-loaded nanoparticles using a microfluidic device. ACS Omega 7:33079–33086

Bose S, Vahabzadeh S, Bandyopadhyay A (2013) Bone tissue engineering using 3D printing. Mater Today 16(12):496–504. https://doi.org/10.1016/j.mattod.2013.11.017

Duncan R (2006) Polymer conjugates as anticancer nanomedicines. Nat Rev Cancer 6(9):688–701. https://doi.org/10.1038/nrc1958

Elisseeff J, McIntosh W, Anseth K et al (2001) Principles of tissue engineering. Tissue Eng 7(4):451–461. https://doi.org/10.1089/107632701753213046

Ganesh KN, Hillmyer M, Kaplan D, Lecommandoux S, Lutkenhaus JL, Rowan S, Schanze K, Wang S, Zhang D (2022) From biosensors to drug delivery and tissue engineering: open biomaterials research. ACS Omega 7(8):6437–6438

Gao W, Zhang Y, Zhang Q, Zhang L (2016) Nanoparticle-hydrogel: a hybrid biomaterial system for localized drug delivery. Ann Biomed Eng 44(6):2049–2061. https://doi.org/10.1007/s10439-015-1506-4

Hickey JW, Santos JL, Williford JM, Mao HQ (2015) Control of polymeric nanoparticle size to improve therapeutic delivery. J Control Release 219:536–547. https://doi.org/10.1016/j.jconrel.2015.08.033

Hua S, de Matos MBC (2013) Mets, the missing link between diabetes and cancer? Front Physiol 4:Article 302. https://doi.org/10.3389/fphys.2013.00302

Jeong B, Gutowska A, An YH (2011) Biodegradable hydrogels for drug delivery. Tech Orthop 26(3):192–202. https://doi.org/10.1097/BTO.0b013e31822953f7

Langer R (2011) Biomaterials for drug delivery and tissue engineering. MRS Bulletin 31(6):477–483

Langer R, Peppas NA (1981) Present and future applications of biomaterials in controlled drug delivery systems. Biomaterials 2(4):201–214. https://doi.org/10.1016/0142-9612(81)90044-7

Li J, Mooney DJ (2016) Designing hydrogels for controlled drug delivery. Nat Rev Mater 1:16071. https://doi.org/10.1038/natrevmats.2016.71

Li L, Nurunnabi M, Nafiujjaman M et al (2020) Recent advances of liposomes in drug delivery. J Nanobiotechnology 18(1):90. https://doi.org/10.1186/s12951-020-00643-w

Liu D, Yang F, Xiong F, Gu N (2016) The smart drug delivery system and its clinical potential. Theranostics 6(9):1306–1323

Makvandi P, Kirkby M, Zare EN, Vecchione R, Liu D, Di Natale C (2020) Engineering smart controlled release systems for precision drug delivery. Adv Sci 7(19):2001476. https://doi.org/10.1002/advs.202001476

Norman J, Madurawe RD, Moore CM, Khan MA, Khairuzzaman A (2017) A new chapter in pharmaceutical manufacturing: 3D-printed drug products. Adv Drug Deliv Rev 108:39–50. https://doi.org/10.1016/j.addr.2016.03.001

Patel N, Davis Z, Hofmann C, Vlasak J, Loughney JW, DePhillips P, Mukherjee M (2023) Development and characterization of an in vitro cell-based assay to predict potency of MRNA–LNP-based vaccines. Vaccines 11:1224

Peppas NA, Langer R (1994) New challenges in biomaterials. Science 263(5154):1715–1720. https://doi.org/10.1126/science.8134835

Sawant RR, Torchilin VP (2012) Liposomes as 'smart' pharmaceutical nanocarriers. Soft Matter 8(41):11231–11240. https://doi.org/10.1039/c2sm26077h

Shi Q, Votruba AR, Farokhzad OC, Langer R (2010) Nanotechnology in drug delivery and tissue engineering: from discovery to applications. Nano Lett 10(9):3223–3230. https://doi.org/10.1021/nl102216v

Stuart MAC, Huck WTS, Genzer J, Müller M, Ober C, Stamm M et al (2010) Emerging applications of stimuli-responsive polymer materials. Nat Mater 9(2):101–113. https://doi.org/10.1038/nmat2614

Tang Y, Wu S, Li X et al (2021) Recent progress in nanotechnology for therapy and diagnosis of atherosclerosis. Cytotechnology 73(5):493–510. https://doi.org/10.1007/s10616-021-00491-

Tenchov R, Bird R, Curtze AE, Zhou Q (2021) Lipid nanoparticles from liposomes to MRNA vaccine delivery, a landscape of research diversity and advancement. ACS Nano 15:16982–17015

Torchilin VP (2014) Multifunctional, stimuli-sensitive nanoparticulate systems for drug delivery. Nat Rev Drug Discov 13(11):813–827. https://doi.org/10.1038/nrd4333

Vega-Vásquez P, Mosier NS (2021) Advances in liposome-based drug delivery: chemistry and application. AAPS PharmSciTech 22(1):22. https://doi.org/10.1208/s12249-020-01812-7

Ventola CL (2014) Medical applications for 3D printing: current and projected uses. Pharm Ther 39(10):704–711. Retrieved from Pharmacy and Therapeutics Journal

Wang S, Li C, Qian M, Jiang H, Shi W, Chen J (2020) Recent advances in polymeric micelles for anti-cancer drug delivery. Eur J Pharm Sci 141:105122. https://doi.org/10.1016/j.ejps.2019.105122

Yoo JW, Irvine DJ, Discher DE, Mitragotri S (2011) Bio-inspired, bioengineered and biomimetic drug delivery carriers. Nat Rev Drug Discov 10(7):521–535. https://doi.org/10.1038/nrd3494

Zhang Y, Chan HF, Leong KW (2013) Advanced materials and processing for drug delivery: the past and the future. Adv Drug Deliv Rev 65(1):104–120. https://doi.org/10.1016/j.addr.2012.10.003

Zhao J, Hu G, Huang Y, Huang Y, Wei X, Shi J (2021) Polysaccharide conjugate vaccine: a kind of vaccine with great development potential. Chin Chem Lett 32:1331–1340

Polysaccharides Based Gums and Mucilage: As Potential Biomaterial

12

Hitendra S. Mahajan, Ujwaldip Vijay Deore, Meghraj Suryawanshi, and Rahul Sarode

Abstract

To interact with biological systems for a medical purpose either therapeutic or diagnostic-engineered biomaterials are used biological organisms like plants, fungi, animals, bacteria, and other life forms a material are derived or produced. Biomaterials are used for the following broad range of applications: medical implants, methods to promote the human tissues, including clips, sutures, and staples for wound recovery, and dissolvable dressings, human tissue regeneration, using a combination of biomaterial supports or scaffolds, cells, and bioactive molecules, biosensors to find out the amount and presence of specific substances and to convey that data and drug-delivery excipient that carry and/or apply drugs to a disease target. Excipients play a basic role in the formulation of medicines, which ensures the dosage, stability, and bioavailability of the active principle constituted by its excipients. The global pharmaceutical excipients market is projected to a CAGR of 5.8%; of which synthetic excipients contribute a major portion. Considering the limitations of synthetic excipients, the industry is in demand of natural excipients for pharmaceuticals, medicinal, food, and cosmeceuticals. In the present chapter a comprehensive overview of gum and mucilage as potential biomaterials for drug delivery applications is discussed.

Keywords

Gums · Mucilage · Biomaterial · Drug delivery · Biomedical applications · Tissue engineering · Gene delivery · Prebiotics

H. S. Mahajan (✉) · U. V. Deore · R. Sarode
SES'S R C Patel Institute of Pharmaceutical Education and Research, Shirpur, Maharashtra, India

M. Suryawanshi
Department of Pharmaceutics, Sandip Institute of Pharmaceutical Sciences (SIPS), Nashik, Maharashtra, India

Md. Faiyazuddin, M. Suryawanshi (eds.), *Design and Processing of Green Materials*, Biomaterials, Bioengineering and Sustainability 4, https://doi.org/10.1007/978-3-031-91790-5_12

12.1 Introduction

Plants and biological organisms like animals, bacteria, fungi, and other life forms the biomaterial are produced. To interact with biological systems for a medical purpose either medicinal or diagnostic these substances are engineered. Biomaterials are also called biologically derived materials. They are of natural or synthetic origin and are used for a medical purpose to boost, support, or restore tissue or a biological function. Biomaterials are used for the following broad range of applications; methods to promote human tissue healing, medical implants, including sutures, clips, and staples for wound recovery, and dissolvable sutures, regenerated human tissues, using a combination of biomaterial supports or scaffolds, cells, and bioactive molecules, biosensors to find out the amount and presence of specific substances and to convey that data and in drug-delivery excipient that act as carrier to deliver the drug to the targeted disease.

The major application of polysaccharides-based gums and mucilage as a potential biomaterial is as a pharmaceutical excipient as drugs cannot be administered as it is but must be formulated into a convenient dosage form by using the excipients. Excipients perform an important role in the formulation of medicines. Excipients are an extremely significant part of medicine as their weight is concerned, which ensures the stability, dosage, and bioavailability of the active principle constituted by its excipients. Excipients play various functions such as diluents & fillers, binders, suspending and viscosity agents, film-forming agents, flavouring, sweetening & coating agents, colorants, disintegrating agents, lubricants & glidants, preservatives, emulsifying agents, and other functionalities, to fulfill known functions and in certain cases, they may directly or indirectly controlling the rate and extent of drug release and absorption. Thus, excipients perform an important role not only in the formulation of medicines, but also in helpful for conserving the safety, efficacy, and stability of bioactive molecules and ensuring that proposed benefits are delivered to the patients. Pharmaceutical manufacturers can carry out cost-savings in drug development, enhance functionality, and help in drug formulation innovation by optimal use of excipients.

The global pharmaceutical excipients market reached USD 6.6 billion in 2022 and is expected to reach around USD 11.8 billion by 2032 with a CAGR of 5.8%; of which synthetic excipients contribute a major portion.

There are certain limitations while using synthetic excipients as their hostile effects will lead to toxicity and allergic reactions, their non-biodegradable nature, and their production is mostly not eco-friendly. Nowadays, the pharmaceutical industry and consumers are demanding various plant-derived materials as the choice of excipients instead of synthetic ones. Recently, many plant-derived raw materials like polysaccharides-based gums and mucilage have been exposed as excipients in pharmaceuticals, medicinal, food, and cosmeceuticals.

Excipients used in pharmaceuticals are obtained from various sources and they are classified as shown in Fig. 12.1 (Jani et al. 2005).

From this the synthetic excipients class has some disadvantages such as it has some toxicity problems; some are costly, during their synthesis some cause

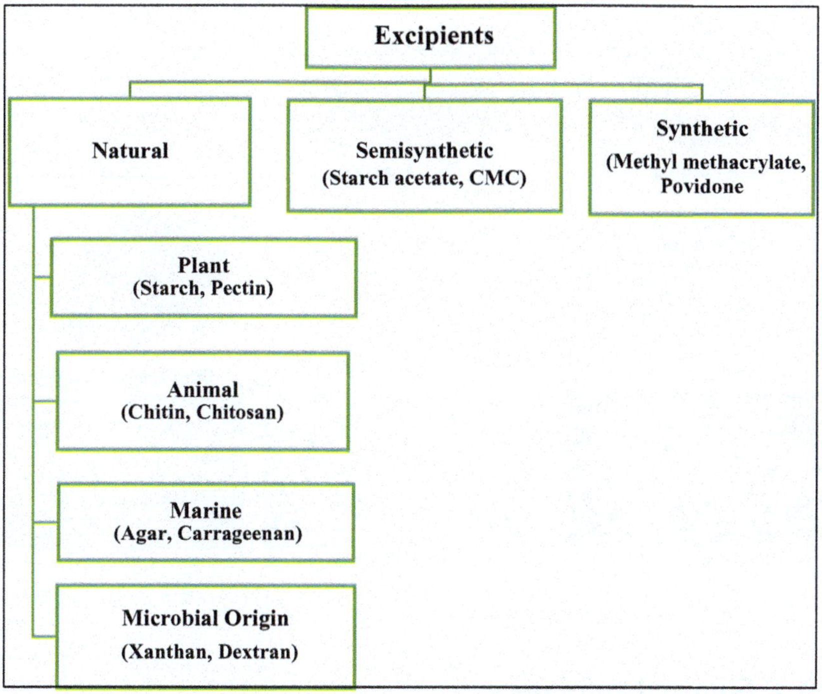

Fig. 12.1 Classification of excipients

environmental pollution, some show side effects and non-renewable sources, etc. Synthetic excipients class had reported some acute and chronic undesirable effects like irritation to the eye and skin with the workers handling substances such as methyl methacrylate and poly-methyl methacrylate. During handling of Carbomer irritation occurs to the eyes, respiratory tract, and mucous membranes. At the site of injection, the formation of subcutaneous granulomas, and adverse reactions produced by povidone. There is also evidence that after intramuscular injections povidone may accumulate in organs. Nowadays a substantial number of plant-based substances are available as pharmaceutical excipients. A wide range of gums and mucilage obtained from natural origin are used as pharmaceutical excipients such as acacia, agar, starch, alginates, pectin, gelatine, etc.

Last few decades gums and mucilage obtained from plant sources have usually been used as polymeric materials in pharmaceutical and cosmetic industries, broadly in beverages and food industries. These materials are now widely used due to their abundance in nature and have advantages such as, being bio-acceptable, economic, easily available, less toxicity, eco-friendly, as well as better public acceptance, and superior as compared to synthetic polymers in terms of biocompatibility and biodegradability.

Fig. 12.2 Plants-derived gums and mucilages

12.1.1 Gum and Mucilage

Today we have a wide range of polysaccharides as plant material obtained from natural origin. These plant materials (polysaccharides) are grouped into Gums and Mucilages (Fig. 12.2). Gums are amorphous translucent solids, partially soluble in water to yield an adhesive solution of high viscosity or colloidal solution. These may be pathological or physiological origin produced due to incision to the plant or that either exudes from the natural or artificially produced plant injuries.

So, gums are obtained by collecting the exudates from different parts of the plants mostly from the stems e.g., Acacia, tragacanth, etc.

Whereas mucilage is comparable in constitution to gums they are the normal physiological products of metabolic activity in plants and produced without incision or injury to the plant produced inside the cell.

From plants water-soluble gums, are derived which are carbohydrate polymers of high molecular weight, and easily dissolve or disperse in water to produce solutions or colloidal suspensions of varying degrees of viscosity (Jani et al. 2009).

12.1.2 Extraction and Isolation of Mucilaginous Polysaccharides

Mucilages are obtained from various parts of the plant, mostly found present in the seeds. In the general method for the extraction of mucilage the ground plant material was mixed with either enough ethanol or with benzene (ratio of 1:1) this facilitates to removal of the lipids and inactivates the enzymes if present in the material.

The decanted solution of ethanol/benzene and then the marc was soaked in a sufficient amount of distilled water and allowed to heat in a water bath at a temperature between 50 °C and 60 °C for 60 minu and kept aside at room temperature for the next 24 h, then the resultant was centrifuge at 4000–5000 rpm/min for 5–10 min. By filtration through muslin /cotton cloth, the obtained supernatant suspension was collected. The obtained supernatant was reduced to its half volume by concentrating it on a water bath. By adding ethanol, methanol, or acetone solvent in the supernatant with continuous starring this leads to precipitation of mucilage which is further isolated from concentrated supernatant. Mucilage of polysaccharide compounds is water soluble, as it forms a hydrogen bond with water. By using polar or dipolar organic solvents the polysaccharides are dehydrated, and molecular precipitation is the result of the aggregation of polysaccharide molecules by increasing intramolecular hydrogen bonding. The obtained precipitate of mucilage was then, filtered, and subjected to drying in the oven at about 40 ± 55 °C. Finally, after grinding and shifting through sieve number # 60 the free-flowing powder mucilage was obtained.

12.1.3 The Chemical Nature of Gums and Mucilage

Gums and mucilage are translucent amorphous substances and have some similarities as both are plant hydrocolloids. On hydrolysis, they produce many monosaccharide units linked by glycoside linkages. Generally, they are insoluble in oils, or organic solvents like acetone, alcohol, etc. They may or may not be soluble in water but when they are added to water, they take up water swell in water and produce a viscous colloidal solution. It gives arabinose, mannose, galactose, and glucuronic acid to hydrolysis, and many of them are associated with uronic acids. In general, gums are pathological products comprising calcium, magnesium, and potassium salts of complex substances known as 'polyuronides' whereas mucilage is are physiological product resemblance to gums, but they are generally sulphuric acid (Prajapati et al. 2013).

Mucilage and gums are hydrocolloids consisting of a few to several monosaccharide units coupled with organic acids like uranic acid (Tosif et al. 2021). Moreover, monosaccharides on hydrolysis cannot be further broken down into simpler sugar molecules as it is simple carbohydrate molecules. Hemicellulose and pectin are the main polysaccharides present in plant gums and mucilage. Hemicellulose consists of arabinoxylans whereas pectin consists of rhamnogalacturonan.

Gum and mucilage are plant hydrocolloids that contain sugars like pentose, methyl pentose, and galactose, which are linked to uranic acid residues through glycosidic linkages. The terminal carbon atom of the monosaccharide unit present at the opposite side of the carbonyl chain may occur in an oxidized form (Cai et al. 2013).

Generally, mucilage arabinoxylans are comprised of β-1,4-linked xylose backbones, they are usually replaced by 1–3 sugar residues at O-2 or/and O-3 positions.

While mucilage rhamnogalacturonan (RG-I) comprises a backbone of the repeating disaccharide of **α**-(1,2)-rhamnose and **α**-d-(1,4)-galacturonic acid (Tosif et al. 2021).

Furthermore, there is some evidence that to the hemicelluloses, rhamnogalacturonan I side chains can be covalently bound, which results in the formation of a super-macromolecular polymeric network. D-xylose, L-arabinose, and D-galactose are neutral sugars, and their proportions and types of neutral sugars vary with the source of mucilage. Due to the presence of hydroxyl and carboxyl groups as a functional group in the mucilage, they act as polyelectrolytes (Tosif et al. 2021).

12.1.4 Advantages of Gums and Mucilage

(a) **Biodegradable and biocompatible**- As gums and mucilage are easily available from nature they are biocompatible and biodegradable. They are extremely renewable sources and do not have any adverse reactions to human beings like irritation of the skin or eye or on the environment.
(b) **Non-toxic**- Chemically gums and mucilage are carbohydrates in nature and made up of repeating units of monosaccharides, and are nontoxic.
(c) **Low cost and environment friendly**- As these are plant-derived materials, so compared with the synthetic sources it is always cheaper to use and environment friendly as need not to synthesize chemically.
(d) **Local availability**- Due to the broad applications of several industries and use as an excipient in pharmaceuticals government may encourage these plants' cultivation which makes it easy and local availability, such as guar gum and tragacanth (Jani et al. 2009).

12.1.5 Disadvantages of Gums and Mucilage

(a) **Microbial contamination**- There is a chance of microbial contamination of the gums and mucilage as they are plant-derived materials and chemically all of these are polysaccharides in nature, also during their use, they get exposed to the external environment. To avoid this, it is recommended to use appropriate preservatives and suitable handling of gums and mucilage.
(b) **Variation in batch-to-batch**- Batch-to-batch variation is possible in the production of gums and mucilage as it is dependent upon various seasonal and environmental factors. This is not observed in the case of synthetic excipients.
(c) **Uncontrolled rate of hydration**- The amount of chemical constituents existing in each material may vary due to alterations in the collection methods of natural materials at different times, in regions, species, and climate conditions.
(d) **Reduced viscosity on storage**- Normally, when gums and mucilage contact with water the viscosity of the formulation will increase. Due to the complex nature of gums and mucilage, there is a reduction in viscosity after storage (Deogade et al. 2012).

12.2 Classification of Gums and Mucilage

As mentioned above natural gums and mucilage are obtained from Plants, Animals, Marine, and Microbial origin, where they perform various structural as well as metabolic functions; most of them are from plant sources among the others.

The different available gums and mucilage can be classified as follows (Jani et al. 2009).

12.2.1 According to the Sources

(a) **Marine origin/algal (seaweed) gums**: Agar, Carrageenan's, Laminarin, Alginic acid.
(b) **Plant origin**:
  (i) Tree exudates- Acacia, Tragacanth, Gum *karaya*.
  (ii) Seed gums- Starch, Amylose, Guar gum, Locust bean gum.
  (iii) Extracts- Pectin, Larch gum
  (iv) Tuber and roots- Potato starch.
(c) **Animal origin**: Chitin, Chitosan, Chondroitin sulfate, and Hyaluronic acid.
(d) **Microbial origin** (bacterial and fungal): Dextran, Xanthan, Pullulan, Curdian, Zanflo, Emulsan, Schizophyllan, Lentinan, Scleroglucan, Krestin.

12.2.2 According to the Charges

(a) **Anionic gums**: Arabic, Agar, karaya, Gellan, Tragacanth, Algin, Carrageenan.
(b) **Non-ionic seed gums**: Amylose, Tamarind, Cellulose, Guar, Locust bean, Xanthan, Arabians,, Galactomannans.

12.2.3 According to Shapes

(a) **Linear**: Algins, Amylose, Cellulose, Pectins.
(b) **Branched:**
  (i) Short branches-, Xylan, Xanthan, Galactomanan.
  (ii) Branch-on-branch- Amylopectin, Tragacanth, Gumarabic.

12.2.4 Semi-Synthetic Origin

(a) **Starch derivatives**- Starch acetate, Starch phosphates.
(b) **Cellulose derivatives**- Methylcellulose (MC), Microcrystalline Cellulose (MCC), Carboxy Methyl Cellulose (CMC), Hydroxyethyl cellulose, Hydroxypropyl- methylcellulose (HPMC).

12.2.5 According to Monomeric Units in Chemical Structure

(a) **Homoglycans**- Amylose, Cellulose, Arabinanas.
(b) **Diheteroglycans**- Carragennans, Algins, Galactomannans.
(c) **Tri-heteroglycans**- Gellan, Arabinoxylans, Xanthan.
(d) **Tetra-heteroglycans**- Psyllium seed Gum, Gum Arabic.
(e) **Penta-heteroglycans**- Tragacanth, gum ghatti (Kokate et al. 2003).

12.3 Preliminary Identification Tests for Gums and Mucilage

Extracted gums and mucilage are identified and confirmed by performing a preliminary phytochemical investigation. The tests are mentioned in Table 12.1.

The Molisch's test, Iodine test, Ruthenium test, and enzyme tests have been performed for the detection of the presence of mucilage, carbohydrate, and peroxidase enzymes (Khandelwal 2008; Lala 1981; Smith and Montgomery 1959).

12.4 Characterization of Gums and Mucilage

Characterization of extracted gums and mucilage is essential for establishing their suitability and it can be achieved by only a multiple-technique approach. It involves structural, physical, and chemical properties and toxicity. As gums and mucilage are polysaccharides in nature, the confirmation of the different sugars present in them and their structural elucidation can be carried out by various chromatographic and spectroscopic methods like TLC, HPTLC, UV, *FTIR, NMR, and* Mass spectroscopy.

Physicochemical characterization of gums and mucilage involves colour, odour, shape, taste, texture, solubility, pH, estimation of ash values, and determination of swelling index. Micrometric properties of gums and mucilage involve bulk density, tapped density, and angle of repose. For the processing and storage of particulate materials, the angle of repose of gum and mucilage is important and it can be calculated by using the fixed funnel method. On the spatial arrangement of the particles,

Table 12.1 Preliminary identification tests for gums and mucilage (Deore and Mahajan 2021)

Test	Observation	Inferences
Molisch's test: About 50 mg mucilage powder was taken in a test tube; Molisch's reagent was added over it followed by Conc. Sulphuric acid from the sides of the test tube	A Violet ring is observed at the junction of two layers	Carbohydrate present (confirms the presence of gums and mucilage)
Ruthenium test: The small amount of mucilage powder was taken on a glass slide and drops of ruthenium red solution were added then and observed under the microscope	Pink colour develops	Mucilage present

the bulk density of gum and mucilage is dependent on their density while the tapped density was calculated by subjecting the graduated measuring cylinder (containing the gums and mucilage) to mechanical tapping until constant volume or mass changes were achieved. The flow properties of gums and mucilage powder were determined by using Hausner's ratio (HR) and Carr's Index (CI). By using estimated values of bulk and tapped density the HR and CI of the gum and mucilage were calculated (Deore and Mahajan 2021; Rowe et al. 2009).

Whereas toxicity of gums and mucilage is determined by acute toxicity test (as per OECD guideline), by determination of the $LD_{50,}$ etc. The study was conducted in rats, mice, and guinea pigs of either sex (Jani et al. 2009).

12.5 Biomedical Applications of Gum and Mucilage

Drug delivery at the right time and the right place is a very prime concern to have the desired effect of the formulation whether it is taken either injectable, oral or even dermal route. While delivering the drug to the body several factors of the drug are taken into consideration pharmacokinetics, therapeutic efficacy, cellular- uptake, efficiency and toxicity, effect on metabolism, etc. Biomaterial use in drug delivery can effectively enhance its action as well as biocompatibility in any route of administration. Synthetic polymers can afford the diversity for the incorporation of desired functional groups during the chemical synthesis process with variable properties and reproducibility but lack both biodegradability and biocompatibility which are the most important properties for biomedical applications. Natural gums and mucilage have both these desired properties and this material is also manageable for required modification by chemically.

Natural polysaccharides are the ample and tenable properties of the material used for the development of several drug delivery systems, tissue engineering, and wound dressing (Fig. 12.3).

(a) **In tissue engineering as Scaffolds**

Tissue engineering is mainly intended to regenerate and restore injured and damaged tissue by different means. These natural polysaccharides have been used to form tissue engineering scaffolds that facilitate suitable 3-dimensional microenvironments for cell attachment, differentiation, and propagation. The scaffolds provide the mechanical properties to the tissues like elasticity and stiffness. Naturally, derived gums and mucilage are frequently used to prepare the scaffold as these materials are biocompatible, biodegradable, hydrophilic, and non-toxic. Additionally, these materials have biological similarities with the natural extracellular matrix. Therefore, polysaccharides-based biomaterials have developed as a suitable platform in tissue engineering (Table 12.2). Synthetically derived polymers have restricted use in tissue engineering as they lack biological similarities with the natural extracellular matrix (Sood et al. 2021).

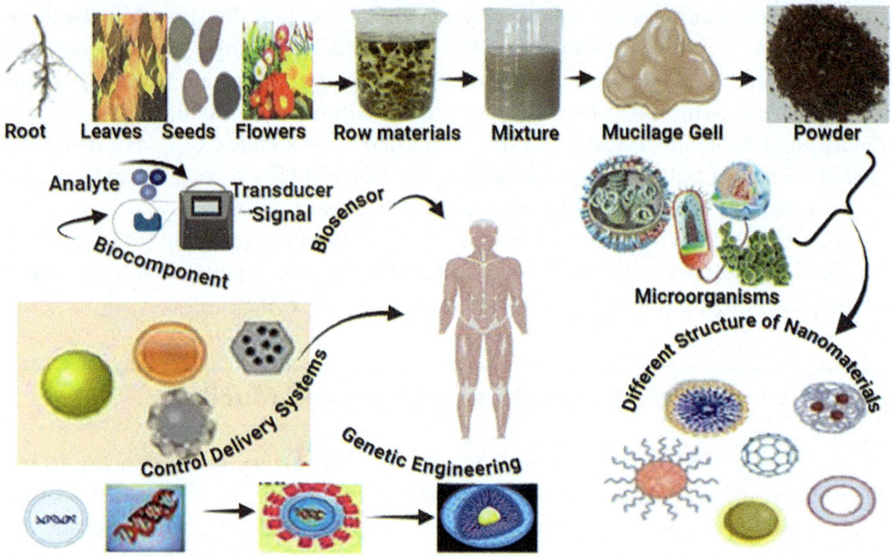

Fig. 12.3 Biomedical Applications polysaccharides

Table 12.2 Examples of different polysaccharides with their scaffold types and targeting tissues

Sr no	Components	Scaffolds	Applications	References
1.	Alginate, Silk fibroin	Hydrogels	Tissue engineering (General)	Nguyễn et al. (2019)
2.	Alginate, Collagen	3D scaffolds	Cartilage tissue engineering	Almeida et al. (2017)
3.	Chitosan	Porous scaffolds	Cartilage tissue engineering	Shamekhi et al. (2017)
4.	Chitosan, Gelatin	Hydrogels	Skin tissue engineering	Ng et al. (2016)
5.	Cellulose, Hydroxyapatite	Nanofibers and Porous scaffolds	Bone tissue engineering	Ao et al. (2017)
6.	Cellulose	Hydrogels	Tissue engineering (General)	

Natural polymers like alginates, cellulose, chitosan, gelatin, etc. are preferred due to their less or minimum toxicity, better biocompatibility, and excellent biodegradability nature. Alginates are reported to have antifungal, antibacterial, and high absorbency properties. Chitosan also is reported to have antifungal, antibacterial, mucoadhesive, and analgesic properties. These properties play an essential role and are very useful in drug delivery and tissue engineering. Alginates-based biomaterial is mainly used in wound healing, to promote the wound healing process by promoting the release of cytokines and boosting the granulation tissue formation and epithelization to promote the wound healing process. Alginates are used in different wound healing dosage forms like nano-fibre, films, hydrogels, etc. (Kumar et al. 2018).

(b) **In wound dressings**

A wound is an injury that damages or disturbs the structural and functional anatomy of the skin. In the wound healing process wound repair and tissue restoration are the key challenges. To address the current limitations in scaffold designing and biomaterial fabrication tissue engineering emphasizes the current nanotechnological-based approaches. The remedy of biocomposites and biomaterials derived from alginate, cellulose, gelatine, chitin, collagen, silk, and hyaluronate, is increasing acceptance and popularity owing to attributes like biocompatibility, bioactive nature, antimicrobial, immunomodulatory, and angiogenic effects (Table 12.3).

The significant characteristics of biocomposites and biomaterials obtained from natural sources like alginate, cellulose, chitin, gelatine, silk, etc. can be useful to produce the scaffold by tissue engineering. It can be used to produce discrete types of dressing to treat wounds and burns (Qin et al. 2022).

(c) **In mucoadhesive drug delivery systems**

Controlled release of drugs from the scaffold is possible in tissue engineering by using mucilaginous polymers. Use of mucilaginous polymers in the fabrication of scaffolds termed mucoadhesive drug delivery systems. This system facilitates local healing and eliminates the limitations of drugs like poor water solubility, low therapeutic index, etc. by adhering to the target mucosal surface. Mucilaginous polysaccharides with charge density can act as good bio-adhesive agents. According to the research report by Deore U V *et al* anionic polymers showed more bio-adhesive properties than non-ionic or cationic polymers (Deore and Mahajan 2021).

In this concern, A. Nokhodchi et al. reported that mucilage derived from the seeds of Trigonella foenum was reported to develop oral controlled-release matrix tablets of propranolol hydrochloride (Nokhodchi et al. 2008). Similarly in another report, G. K. Jani *et al* formulate up to 12 h sustained release of diclofenac tablets by using mucilage obtained from the *Hibiscus rosasinensis*. The mucilage was used as a gelling agent, disintegrating agent, and in modified-release dosage form.

Table 12.3 Examples of different polysaccharides with their scaffold in wound healing application

S. No	Components	Scaffolds	Applications	References
1.	Quince seed mucilage	Hydrogel	Skin wounds/burns	Qin et al. (2022)
2.	*Lallemantia royleana*	Silver nanoparticles	Antimicrobial, antioxidant	Tosif et al. (2021)
3.	Basil seeds mucilage	Hydrogel	Wound dressings	Tantiwatcharothai and Prachayawarakorn (2020)
4.	Balangu mucilage	Hydrogels	Wound Healing	Tosif et al. (2021)
5.	*Hibiscus* leaves mucilage	Nanofiber	Wound Healing	Tosif et al. (2021)

In another study by Deore U V *et al*, the *Cassia obtusifolia* seed mucilage was reported as a novel excipient for drug delivery. The study proved that mucilage obtained from *Cassia obtusifolia* had better bio-adhesive properties than tragacanth. The authors also extend the study by doing chemical modification of *Cassia* mucilage to improve its mucoadhesive property by thiolation (Deore et al. 2021). The authors reported that the bio-adhesive property of *Cassia* mucilage was found to be increased by thiolation (Deore and Mahajan 2018; Deore and Mahajan 2021).

(d) **In gene delivery**

In most cases, the treatment for any disorder or disease involves treating the altered signs and symptoms of the body rather than restoring or treating the defective gene which is the fundamental root cause of the disease. In the beginning, step to treat the faulty gene, there is a need to develop a vector for safely delivering the nucleic acids at the desired target site without any adverse impact. Naturally occurring sugar-containing mucopolysaccharides like Glycosaminoglycans polysaccharide compounds are mostly employed in gene delivery to reduce nonspecific interaction and improve stability and transfection efficiency.

I Fernandez Pineiro et al propose sorbitan ester-based lipid nanoparticles functionalized by glycosaminoglycans chondroitin sulphate and hyaluronic acid were used as gene delivery agents. It is observed that the addition of hyaluronic acid and chondroitin sulphate causes the stabilization of the nanosystem and protects the plasmid DNA. Such a nanosystem might transfect A549 cells without changing the cells' viability and safety profiles (Shariatinia 2019).

(e) **Application of natural polysaccharides as prebiotics**

Prebiotics are undigested food materials that escape from the digestion process (by normal human digestive enzymes) and directly reach to colon safely in an intact form where they selectively promote the growth of probiotics. Such food material stimulates the beneficial gut microflora. Polysaccharides, oligosaccharides, sugar polyols, fiber, and resistant starch are the five types of prebiotics. These undigested polysaccharides act as a carbon source for gut microflora. The human body contains different types of carbohydrate-active enzymes like carbohydrate esterases, polysaccharide lyases, glycoside hydrolases, and glycosyltransferases which utilize the undigested polysaccharides.

Plant polysaccharides are used as prebiotics to regulate the activity and composition of gut microflora. Plant polysaccharides play a crucial role in maintaining the balanced gut microflora which is vital to optimal gut health. In recent days, this material has had extensive interest, including arabinoxylan, oligofructose, inulin, and various dietary fibers. These polysaccharidal compounds boost the growth of gut probiotic microflora such as *Bifidobacterium, Lactobacillus,* etc. (Davani-Davari et al. 2019).

Zihan Qi *et al* reported that fructan polysaccharides isolated from the *P. sibiricum* did not get digested in the in-vitro saliva, gastric juice, and small intestinal digestion fluid but intestinal bacteria showed good metabolic activity to the polysaccharides isolated from the *P. sibiricum* (Qi et al. 2024).

12.6 Conclusion

Plant-derived gums and mucilage materials have unique biological potential due to their easy availability, low cost, natural abundance, ecological biodegradability, and biocompatibility. These materials are now gaining popularity as biomaterials such as additives, gelling agents, emulsifiers, and so on in the pharmaceutical and food industries.

In pharmaceuticals, it mainly acts as a versatile excipient for various formulations, not only restricted to gelling, film-forming, binding, and disintegrating agents but also showing its potential role in biomedicals, particularly in tissue engineering for drug delivery, mucoadhesive drug delivery, wound dressings, and gene delivery. Because of the polysaccharidal nature of gums and mucilage, they are used as prebiotics for gut microflora.

References

Almeida HV, Sathy BN, Dudurych I, Buckley CT, O'Brien FJ, Kelly DJ (2017) Anisotropic shape-memory alginate scaffolds functionalized with either type I or type II collagen for cartilage tissue engineering. Tissue Eng A 23(1–2):55–68. https://doi.org/10.1089/ten.tea.2016.0055

Ao C, Niu Y, Zhang X, He X, Zhang W, Lu C (2017) Fabrication and characterization of electrospun cellulose/nano-hydroxyapatite nanofibers for bone tissue engineering. Int J Biol Macromol 97:568–573. https://doi.org/10.1016/j.ijbiomac.2016.12.091

Cai M, Wang N, Xing C, Wang F, Wu K, Du X (2013) Immobilization of aluminum with mucilage secreted by root cap and root border cells is related to aluminum resistance in Glycine max L. Environ Sci Pollut Res 20:8924–8933. https://doi.org/10.1007/s11356-013-1815-6

Davani-Davari D, Negahdaripour M, Karimzadeh I, Seifan M, Mohkam M, Masoumi SJ, Berenjian A, Ghasemi Y (2019) Prebiotics: definition, types, sources, mechanisms, and clinical applications. Food Secur 8(3):92. https://doi.org/10.3390/foods8030092

Deogade UM, Deshmukh VN, Sakarkar DM (2012) Natural gums and mucilage's in NDDS: applications and recent approaches. Int J PharmTech Res 4(2):799–814. http://www.sphinxsai.com/2012/pharmAJ/PHARM/PT=38[799-814]AJ12

Deore UV, Mahajan HS (2018) Isolation and characterization of natural polysaccharide from Cassia Obtustifolia seed mucilage as film forming material for drug delivery. Int J Biol Macromol 115:1071–1078. https://doi.org/10.1016/j.ijbiomac.2018.04.174

Deore UV, Mahajan HS (2021) Isolation and structural characterization of mucilaginous polysaccharides obtained from the seeds of Cassia uniflora for industrial application. Food Chem 351:129262. https://doi.org/10.1016/j.foodchem.2021.129262

Deore UV, Mahajan HS, Surana SJ, Wagh RD (2021) Thiolated and carboxymethylated Cassia obtusifolia seed mucilage as novel excipient for drug delivery: development and characterisation. Mater Technol 36(14):857–867. https://doi.org/10.1080/10667857.2020.1800307

Jani G, Goswami J, Prajapati V, Zinzuwadia M, Joshi B, Dabhi A (2005) Studies on formulation and evaluation of new superdisintegrants for dispersible tablets. Int J Pharmaceut Excipient:37–43

Jani GK, Shah DP, Prajapati VD, Jain VC (2009) Gums and mucilages: versatile excipients for pharmaceutical formulations. Asian J Pharm Sci 4(5):309–323

Khandelwal K (2008) Preliminary phytochemical screening. Pract Pharmacogn Tech Exp 8:149–156

Kokate C, Purohit A, Gokhale S (2003) Text book of pharmacognosy. Nirali Prakashan, Pune

Kumar A, Rao KM, Han SS (2018) Application of xanthan gum as polysaccharide in tissue engineering: a review. Carbohydr Polym 180:128–144. https://doi.org/10.1016/j.carbpol.2017.10.009

Lala P (1981) Practical pharmacognosy. Lina Guha, Calcutta, p 135

Ng WL, Yeong WY, Naing MW (2016) Polyelectrolyte gelatin-chitosan hydrogel optimized for 3D bioprinting in skin tissue engineering. Int J Bioprint 2(1):53–62. https://doi.org/10.18063/IJB.2016.01.009

Nguyễn TT, Ratanavaraporn J, Yodmuang S (2019) Alginate-silk fibroin Bioink: a printable hydrogel for tissue engineering. In: 2019 12th Biomedical Engineering International Conference (BMEiCON). https://doi.org/10.1109/BMEiCON47515.2019.8990292

Nokhodchi A, Nazemiyeh H, Khodaparast A, Sorkh-Shahan T, Valizadeh H, Ford J (2008) An in vitro evaluation of fenugreek mucilage as a potential excipient for oral controlled-release matrix tablet. Drug Dev Ind Pharm 34(3):323–329. https://doi.org/10.1080/03639040701662594

Prajapati VD, Jani GK, Moradiya NG, Randeria NP, Nagar BJ, Naikwadi NN, Variya BC (2013) Galactomannan: a versatile biodegradable seed polysaccharide. Int J Biol Macromol 60:83–92. https://doi.org/10.1016/j.ijbiomac.2013.05.017

Qi Z, Gao T, Li J, Zhou S, Zhang Z, Yin M, Hu H, Liu H (2024) Structural characterization and prebiotic potential of polysaccharides from Polygonatum sibiricum. Food Sci Hum Wellness 13(4):2208–2220. https://doi.org/10.26599/FSHW.2022.9250184

Qin J, Chen F, Wu P, Sun G (2022) Recent advances in bioengineered scaffolds for cutaneous wound healing. Front Bioeng Biotechnol 10:841583. https://doi.org/10.3389/fbioe.2022.841583

Rowe RC, Sheskey P, Quinn M (2009) Handbook of pharmaceutical excipients. Libros Digitales-Pharmaceutical Press. http://repositorio.ub.edu.ar/handle/123456789/5143

Shamekhi MA, Rabiee A, Mirzadeh H, Mahdavi H, Mohebbi-Kalhori D, Eslaminejad MB (2017) Fabrication and characterization of hydrothermal cross-linked chitosan porous scaffolds for cartilage tissue engineering applications. Mater Sci Eng C 80:532–542. https://doi.org/10.1016/j.msec.2017.03.194

Shariatinia Z (2019) Pharmaceutical applications of natural polysaccharides. In: Natural polysaccharides in drug delivery and biomedical applications. Elsevier, pp 15–57. https://doi.org/10.1016/B978-0-12-817055-7.00002-9

Smith F, Montgomery R (1959) The chemistry of plant gums and mucilages. Reinhold Publishing Corporation/Chapman And Hall Ltd, New York/London

Sood A, Gupta A, Agrawal G (2021) Recent advances in polysaccharides based biomaterials for drug delivery and tissue engineering applications. Carbohydr Polym Technol Appl 2:100067. https://doi.org/10.1016/j.carpta.2021.100067

Tantiwatcharothai S, Prachayawarakorn J (2020) Property improvement of antibacterial wound dressing from basil seed (O. basilicum L.) mucilage-ZnO nanocomposite by borax crosslinking. Carbohydr Polym 227:115360. https://doi.org/10.1016/j.carbpol.2019.115360

Tosif MM, Najda A, Bains A, Kaushik R, Dhull SB, Chawla P, Walasek-Janusz M (2021) A comprehensive review on plant-derived mucilage: characterization, functional properties, applications, and its utilization for nanocarrier fabrication. Polymers 13(7):1066. https://doi.org/10.3390/polym13071066

Natural Biomaterials Used in Wound Healing and Diabetic Foot

13

Md Jasim Uddin, Amol D. Gholap, Iqra Zulfiqar, Samiha Binte Sadiq, Sharmin Lucky, and Khadiza Akter Lubna

Abstract

The potential of natural biomaterials to improve wound healing and manage diabetic foot problems is becoming more widely acknowledged. The development, as well as application of these biomaterials, are reviewed in this book chapter, emphasizing how they might be used to promote successful wound repair and address the particular difficulties associated with diabetic foot ulcers. Collagen, chitosan, alginate, and silk fibroin are important substances covered because of their inherent bioactivity, biocompatibility, and capacity to degrade naturally. These properties enhance tissue regeneration and lower infection risk. The chapter focuses on how various biomaterials encourage healing, including how they alter the wound environment, support structure, and improve cellular activity. It also looks at new developments in the creation and improvement of these biomaterials, such as innovative formulations and delivery methods. The chapter gives an overview of the benefits, drawbacks, and potential applications of natural biomaterials in wound care by combining data from current research and clinical trials. For academics, physicians, and other healthcare workers looking to enhance diabetic foot care and wound management results through creative biomaterial uses, this synthesis offers insightful information.

M. J. Uddin (✉)
Department of Pharmaceutical Technology, Faculty of Pharmacy, Universiti Malaya, Kuala Lumpur, Malaysia
e-mail: j.uddin@um.edu.my

A. D. Gholap
Department of Pharmaceutics, St. John Institute of Pharmacy and Research, Palghar, Maharashtra, India

I. Zulfiqar · S. B. Sadiq · S. Lucky · K. A. Lubna
Drug Delivery & Therapeutics Lab, Dhaka, Bangladesh

Md. Faiyazuddin, M. Suryawanshi (eds.), *Design and Processing of Green Materials*, Biomaterials, Bioengineering and Sustainability 4,
https://doi.org/10.1007/978-3-031-91790-5_13

Keywords

Natural biomaterials · Wound healing · Diabetic foot · Polymers · Drug delivery

13.1 Introduction

The human body's biggest and broadest organ, the skin, is more vulnerable to diseases, trauma, burns, cuts, fractures, or erroneous defects and may be brought on by physical trauma, burns, endogenous causes (such as diabetes, cancer, vascular disease, etc.), or pathological disorders (Pires et al. 2024). Skin damage from superficial and local wounds is a serious illness that must be treated. The biological mechanisms involved in wound recovery and repair are usually slow, intricate, multidimensional, and dynamic (Huang et al. 2024). Acute and severe wounds often take 8–12 years to heal, although simple and tiny wounds may heal by themselves, and sometimes, dust particles and bacterial infections can also interfere with the process of healing. The substance used to cover wounds and stop the infection and subsequent damage is called a wound dressing. Additionally, it facilitates quicker healing of wounds by refraining from the accumulation of germs as well as other harmful foreign elements in the barrier (Zhao et al. 2024). A substance called a wound dressing wraps wounds and guards against infection and additional harm. Along with preventing the accumulation of germs and other contagious foreign material in the barrier, it also fosters an environment that speeds up wound healing. Wounds were formerly managed and repaired over time with the use of covers made of natural materials like cotton, which absorbed fluids leaking from the wound's location and created an outer layer known as an eschar. Unfortunately, this repair technique slows down epidermal cell migration and proliferation, which in turn slows down healing. Using wet dressings to treat injuries and hasten the healing process has been recommended by researchers throughout time (Ansari and Darvishi 2024). Long-term use of moist wound dressings implies that the materials applied to expedite the process of healing have biological and physicochemical characteristics appropriate for the wound site. To treat bacterial plus microbial infections, it releases medicines and covers the surface so that it may be readily removed from the wound. To produce innovative wound dressings, a variety of polymers (synthetic as well as natural), having distinct biological, chemical, and physical properties have been studied and used. Because of their intrinsic properties, polymers derived from microorganisms, plants, and animals are a great way to heal wounds. Research has shown that because of its better compatibility, it is better than alternatives when used as wound care dressings (Pires et al. 2024). A variety of preparation techniques may be used to create these biomaterials, including film, foam, nanofiber, hydrogel, scaffold, etc. Because dressings of hydrogel for injuries made from natural biomaterials resemble extracellular matrix (ECM) in structure, they have emerged as a novel variety of wound dressing. They provide an acidic, wet environment that promotes the formation and division of the cells that are new, stops necrosis of the tissue, eases pain, as well as shields wounds. The most significant and often utilized natural polymers in wound dressings include natural biomaterials including keratin, silk, bacterial cellulose (BC), β-glucan, N-acetylglucosamine, dextran (Dex), hyaluronic acid (HA), and collagen (Col) chitin/chitosan (Ottaviano et al. 2025). One major but common consequence of persistent diabetes is the occurrence of foot ulcers, which often leads to a patient's first consultation with a vascular surgeon. The individual's absence of protective feeling and inability to automatically alter gait owing to peripheral neuropathy are the main reasons these wounds, which may stem from either acute or chronic trauma, continue to exist. This is in addition to several other common, possible factors, such as foot bony changes (amputation-related structural deformity or Charcot arthropathic), concurrent peripheral arterial disease (PAD) of the vessels in the lower extremities, and diabetic microcirculatory dysfunction that impairs healing (Zhang et al. 2024a).

As the patient's illness worsens, the chance of developing a foot ulcer expands. The incidence of Diabetic Foot Ulcers (DFUs) and their impact on the healthcare system has grown in tandem with the worldwide burden of diabetes and diabetic patients' life expectancy. In addition to having high infection rates and a higher death rate, DFUs cause substantial morbidity for the patient and are often an indication of major amputation (Yap et al. 2024). Therefore, enhancing outcomes in diabetic patients, especially those with documented peripheral neuropathy, requires careful screening for new DFUs and thorough monitoring of existing lesions. DFU is a complex issue that is best handled by a multidisciplinary team. When it comes to disorders that impact several organ systems, interdisciplinary care models are especially successful because they provide patients who may otherwise face major socioeconomic hurdles to receiving healthcare access to complete treatment. The monitoring and treatment of patients with DFU have benefited greatly from various care models in 94% of collected cases, the use of a multidisciplinary care team decreased the total number of amputations associated with DFU (Qi et al. 2024b).

This chapter's goals and purview center on comprehending the biological processes of wound healing, encompassing the phases of remodeling, cellular proliferation, hemostasis, and inflammation. Additionally, it covers the unique problems associated with DFUs, including peripheral artery disease (PAD), neuropathy, and delayed inflammation brought on by persistently high blood sugar (Qi et al. 2024a). The chapter focuses on how alginate, collagen, chitosan, and silk fibroin are examples of natural biomaterials that may promote wound healing as well as tissue regeneration. Furthermore, it describes current developments in biomaterial compositions and delivery methods, highlighting their potential to enhance chronic wound management, especially in the treatment of diabetic foot.

13.2 Biological Foundation of Healing of Wound and Diabetic Foot Problems

13.2.1 Mechanisms of Wound Healing

Healthcare providers and Clinical practices have a significant problem while treating acute and chronic wounds caused by diverse traumas. An intricate biological process of wound healing is controlled by the body's physiological response. Multiple cell types cooperate in wound healing, aided by intricate molecular signaling networks. To bring the skin back to its normal structure and function, the vascular system, cytokines, mediators, and cells all work together. The following phases are initiated by these coordinated efforts: hemostatic phase, inflammation, cellular proliferation, and remodeling phase. These activities are crucial to the body's functioning (Al Mamun et al. 2024; Peña and Martin 2024).

13.2.1.1 The Hemostasis Phase

The first phase, hemostasis, of wound healing, starts immediately upon the damage, persists for minutes to hours, and is crucial for preventing significant blood loss (Fig. 13.1). During this phase, damaged blood vessels rapidly constrict, and activated platelets aggregate with fibrin to create a clot that halts hemorrhage upon interaction with the vascular subendothelial matrix (Guo et al. 2021). The three main steps in this process are vasoconstriction, primary hemostasis, and secondary hemostasis. Vasoconstriction results in decreased blood flow from injured arteries. Platelets bind to the surface extracellular matrix (ECM) proteins during initial hemostasis, creating a platelet clog. While thrombin activation causes further aggregation and degranulation, platelet receptors help cells interact with the ECM (Yang et al. 2022b).

During secondary haemostasis, a stable thrombus ceases bleeding by turning soluble fibrinogen into insoluble fibrin. For wound healing platelets attack immune cells, and release growth factors and chemokines. The damaged vessel wall begins mending with the aid of smooth muscle and endothelial cells. To avoid thrombosis the coagulation is regulated. This congruence guarantees the clot's integrity and initiates the healing process of the wound (Wolf et al. 2021).

13.2.1.2 Inflammation

After the end of the quick phase of hemostasis (which lasts 2–5 days), the inflammation phase begins cleansing the wound and preparing the ground for tissue restoration. Platelets and injured cells begin to release cytokines as well as growth factors to recruit immune cells with neutrophils initially responding to phagocytize pathogens, dead cells, and debris. Furthermore, pro-inflammatory cytokines, proteases (elastase and cathepsin G, and VEGF) and IL-8 are released, which promote angiogenesis and amplify the inflammatory response (Wang et al. 2022). Decreasing the quantity of neutrophils, macrophages begin reducing inflammation. These cells enhance cell proliferation, facilitate tissue healing, and continue to phagocytose while secreting more growth factors and cytokines (Hong et al. 2023). Furthermore, macrophages help in wound debridement by removing apoptotic cells. This inflammatory process provides several symptoms, including pain, redness, swelling, and, heat, which is essential for both preventing infection and also preparing the wound environment for the following stage of healing (Raziyeva et al. 2021).

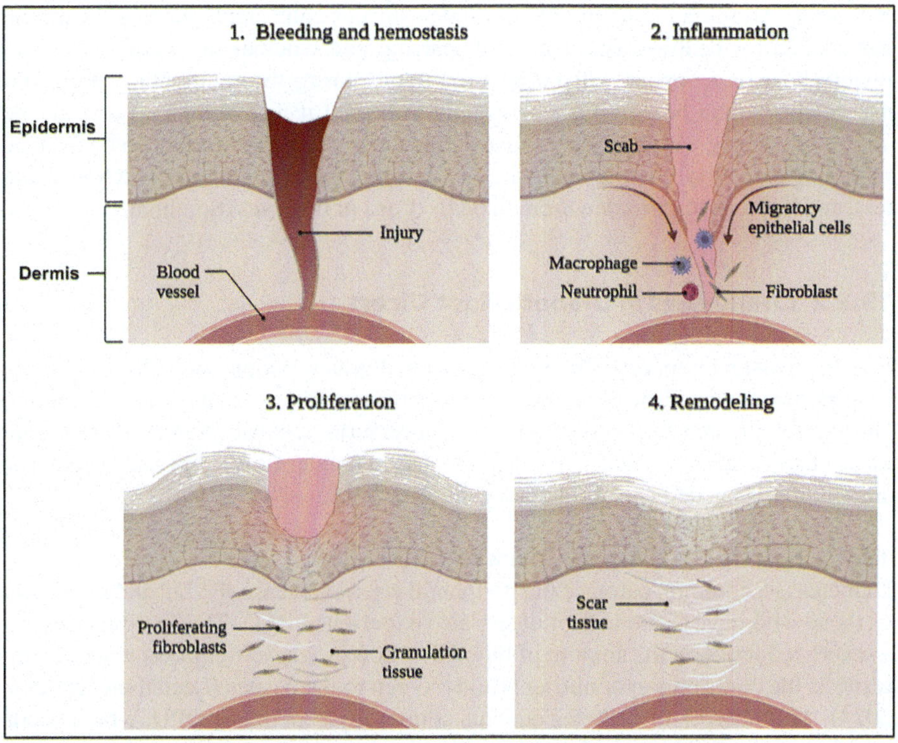

Fig. 13.1 Natural Phase of wound healing 1. Hemostasis Phase, 2. Inflammation 3. Cellular Proliferation and 4. Remodeling. (Created by using Biorender.com)

13.2.1.3 Cellular Proliferation

The cellular proliferation phase is vital for tissue restoration which lasts for 6–20 days. For tissue restoration, the 6–20-day proliferation phase of wound healing is essential. Keratinocytes, fibroblasts, and endothelial cells proliferate and migrate to produce granulation tissue, ECM, and blood vessels (Luo et al. 2021). While myofibroblasts close the wound margins, fibroblasts generate components of ECM and collagen. The new cells get nutrition and oxygen via angiogenesis and heal the wound by epithelial cells. Maintaining moistness of wounds speeds up healing; fibroblasts are mostly important as they generate collagen and rework the tissue structure (Peña and Martin 2024).

13.2.1.4 Remodeling

Depending on the degree of the damage, the final phase which is the remodeling phase, of wound healing—can span several months to up to 2 years. The main objective of the remodeling is to reinstate the natural architecture of the skin by replacing collagen III with collagen I in granulation tissue (Čoma et al. 2021). This phase is scar tissue strengthening, vascular regression, and collagen remodeling. As the tissue becomes more flexible, the skin's elasticity and tensile strength increase with time. Additionally helping to eliminate surplus cells and EMC that are no longer needed for healing. The degree of scarring and whether the wound will completely heal or reappear will be decided upon during the remodeling stage. The process of wound healing may be greatly impacted if any of these routes are not controlled. Many variables may cause the healing process of wounds to be delayed, including infection, edema, necrosis, age, trauma, chronic illnesses (such as diabetes), and immunosuppression brought on by drugs or illnesses (Spielman et al. 2023).

13.2.2 Challenges in Diabetic Foot Ulcers

Due to persistent hyperglycemia and associated issues, people with DM have a prolonged acute wound-healing process, which contributes to the pathogenesis of chronic DFUs, and fail to heal and may have serious consequences. The Wagner ulcer classification method is used to identify the extent of ulcers and whether an amputation is needed (Swaminathan et al. 2024).

13.2.2.1 Peripheral Artery Disease (PAD)

Diabetes is a common cause of the peripheral arterial disease (PAD), and about 80% of those who have diabetic foot ulcers also experience PAD. This condition causes a severe reduction in the amount of blood that flows to the extremities, which in turn restricts the contribution of nutrients and oxygen to the tissues (Leelathanalerk et al. 2024). Around 80% of diabetes patients commonly experience DFU, which occurs mainly because of the reduction of the blood flow to the extremities which restricts the supply of oxygen and nutrients to the tissue of the foot. For vasoconstriction, only high blood sugar level is not a malefactor but endothelial cell dysfunction is also a factor, which in turn reduces the synthesis of vasodilators such as nitric oxide. This condition is amenable to the development of ischemia and ulcers. As a consequence of this, ulcers could not get enough blood flow to recover properly, which would result in wounds that continue to deteriorate and even become more severe (Houghton et al. 2024).

13.2.2.2 Neuropathy

An important factor that contributes to diabetic foot ulcers is diabetic neuropathy, which damages the nerves in the periphery. About half of the people who have type 2 diabetes will eventually acquire neuropathy. A decrease in the feeling of pain is caused by sensory neuropathy, which causes small injuries such as cuts or blisters to go unrecognized, which may result in infection or ulceration. Toe deformities such as claw toes, hammer toes, and Charcot's foot may be caused by motor neuropathy, which affects the muscles of the foot. The presence of these abnormalities results in pressure points and raises the likelihood of developing ulcers (Zhu et al. 2024). The function of the sweat glands is disrupted by autonomic neuropathy, which results in excessively dry and cracked skin and also raises the risk of infection. The combination of neuropathy and poor circulation may lead to ulcers that heal slowly and are more likely to develop problems (Strand et al. 2024).

13.2.2.3 Delayed Inflammation

DFUs have a delayed or inadequate inflammatory response as a result of immune system damage caused by chronic hyperglycemia. Normally, inflammation is necessary to drive out pathogens and start the healing process of wounds (Lingyan et al. 2024). But with diabetes, this reaction is compromised, leading to persistent inflammation that obstructs the body's natural healing mechanism. DFUs often proceed to chronicity because the wound is unable to go on to the subsequent phases of healing, which include proliferation and remodeling (Mohsin et al. 2024).

13.2.2.4 Infection

Infections are a frequent and serious consequence of DFUs because of impaired immune function and decreased circulation. Patients with diabetes are more deeply infected from even very small cuts. When ischemia, neuropathy, and impaired skin barriers are present, the risk of infection is enhanced that much higher. As a result of foot infections, diabetic people experience 80% of all non-traumatic lower-limb amputations (Yang et al. 2024b). Identifying symptoms such as erythema, swelling, warmth, discomfort, and purulent discharge are necessary steps in the process of the diagnosis of an infection. The creation of biofilm, which makes infections more difficult to cure, is a consequence of bacterial colonization in DFUs. Intensive treatment, which may include hospitalization and the use of antibiotics, is often necessary for severe infections such as osteomyelitis. Infections like this may spread to tissues and bones if they are not adequately handled, which can result in amputation (Hsu et al. 2024).

13.2.3 The Role of Biomaterials in Wound Management

Controlling severe micro and macro wounds by dressing the area using biomaterials is considered is useful therapy for diabetic patients. Because of the qualities that biomaterials possess such as hydrophilicity, biocompatibility, biodegradability, and adaptability, make them well-qualified for therapeutic applications (Hosty et al. 2024). To begin, the exceptional biocompatibility and biodegradability qualities guarantee tissue repair. Additionally, it provides a warm and wet atmosphere that is advantageous for the regeneration of tissue. Finally, the dressing contains sufficient porosity, which will required for gas exchange, and it also allows cell migration and proliferation (Ansari and Darvishi 2024).

13.3 Key Natural Biomaterials

13.3.1 Collagen

Collagen is present in every human body, animal tissue, and nature and is also found in the sea.

13.3.1.1 Sources and Characteristics

Natural Sources

Collagen present in two sources natural and synthetic. Natural collagen is found in animal and herbal origin. Bovine, porcine, and marine are the forms of animal collagen. The industrial source of collagen is bovine collagen. And it has low immunogenicity and biocompatibility. The key source of bovine collagen is cattle. Type I collagen is bovine collagen and massive source of commercially available collagen. Bovine collagen cause to allergic reaction to some people (Jayaprakash et al. 2024).

Marine collagen can be found in both invertebrates and vertebrates. Marine collagen is always found in marine organisms, such as mollusks, sponges, sea urchins, octopuses, prawns, starfish, jellyfish, cuttlefish, and sea anemones. Marine collagens also have biocompatibility like bovine and porcine (Delikanlı Kıyak et al. 2024). But it has more bioavailability than bovine and porcine collagen. Porcine collagen has low antigenicity properties as like human collagen that's why it used industrial applications. The dermis and small intestinal mucosa of porcine are the sources of porcine collagen. Porcine collagen has broadly effect on wound healing, plastic and reconstructive surgery. But it has the same zoonotic risk as bovine collagen. Porcine collagen is similar to bovine collagen and also have same properties like bovine sources. Porcine is derived from pig so it has some restrictions for Middle East regions (Farooq et al. 2024).

Synthetic Sources

Natural collagen has some limitations towards immune problems so synthetic collagen is developed to improve it. Material commercially named KOD is one of the examples of synthetic collagen sources. This synthetic protein, which has 36 amino acids, self-assembles into hydrogels and triple-helix nanofibers (Sabarees et al. 2024). Another synthetic collagen source has been developed with the benefit of recombinant technology, offering premium collagens free of contaminants generated from animals. Insect and mammalian cell cultures, yeast, and mostly plant cell cultures produce it. Recombinant technologies are said to have drawbacks such as high cost, limited yield, and the absence of cofactors or enzymes in the systems (Fig. 13.2) (Josephin et al. 2024).

13.3.1.2 Applications in Wound Healing

Collagen shows a great effect on wound healing. By keeping the wound wet and shielding it from physical damage and infection, it promotes quicker and more efficient healing. Collagen reduces the scar formation faster. It can also be utilized as a delivery system for medications that promote healing and pain management, as well as in conjunction with other wound healing treatments such antibacterial agents (Khatoon et al. 2024). Additionally, by instructing the fibroblast cells to generate new collagen cells, collagen peptides can promote skin repair and rejuvenation. By accelerating the rate of cell migration and proliferation, they can also increase overall cell turnover by removing more aging, worn-out skin cells. As a crucial part of the dermal extracellular matrix (ECM), protein-based scaffolds for skin wound

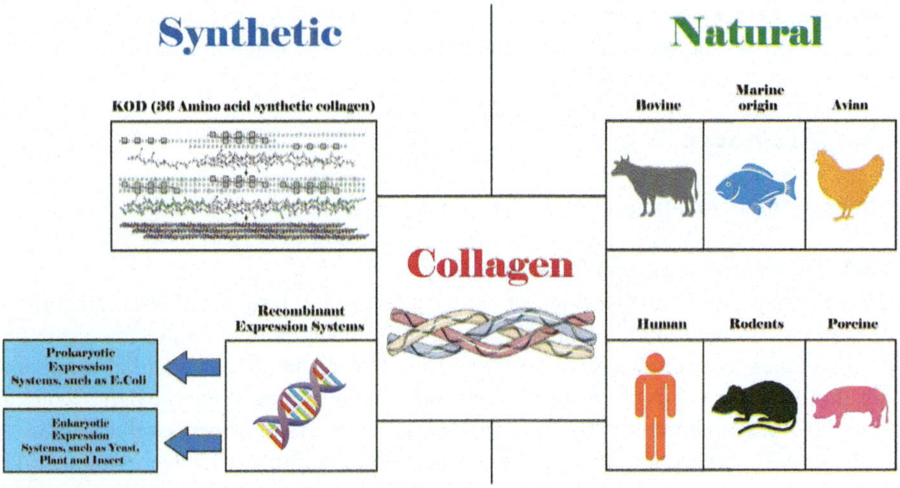

Fig. 13.2 Natural and synthetic sources of collagen (42)

healing frequently use it in a variety of forms, including powder, amorphous gels or pastes, gel-impregnated dressings, sheets, and pads (Caldera-Villalobos et al. 2025).

13.3.1.3 Clinical Efficacy and Limitations

Burns and ulcers are two examples of wounds that can significantly impair a patient's quality of life, particularly chronic wounds like pressure ulcers, diabetic foot ulcers, venous leg ulcers, and others. Diabetes mellitus patients face a common complication which is diabetic foot ulcers that are not well controlled (Liang et al. 2024a). Chronic wounds may not respond to standard care, typically due to a protracted and excessive inflammatory response, changed protease levels, and inadequate extracellular matrix. Advanced wound therapeutics are therefore being developed. Collagen is a biomaterial that is highly beneficial for the development of advance therapies. Collagen sponges have the strength to suture soft tissue and serve as a template for the creation of new tissue when they are moist. Drugs used for skin replacement have been delivered via collagen-based implants (Xiao et al. 2023).

The benefits of collagen treatment are numerous when compared to other medications. But collagen treatment also has some difficulties too. When receiving collagen treatment, a small number of patients may experience an allergic reaction. People who are allergic to shellfish or cows, for instance, may suffer from anaphylaxis (Al Hajj et al. 2024). Measles, mumps, and rubella vaccinations contain gelatine primarily made of bovine type I collagen. The majority of children who experienced anaphylaxis displayed sensitivity to these vaccines, which may have been caused by the bovine gelatine that was present in them. Collagen from animals has the potential to spread illness. There is a chance that bovine collagen will spread diseases like bovine BSE. Research has been done on the BSE risk posed by ruminant collagen and gelatine made from raw materials for human consumption. Collagens from the sea might be suitable alternatives (Jadach et al. 2024).

13.3.2 Chitosan

Chitosan another name is chitin. Chitosan is the most second abundant polymer.

13.3.2.1 Properties and Functionality

The exoskeleton of crustacea, insect cuticles, algae, and fungal cell walls all contain chitosan. Commercial chitosan, which is mostly made by chemically deacetylating chitin, comes from crustaceans. Recent opinions indicate that the market is becoming more interested in chitosan derived from fungi due to vegan desires (Rojekar et al. 2024b; Kapare et al. 2024). Chitosan has some great properties like solubility and viscosity. A number of variables, including temperature, pH, degree of acetylation, polymer molecular weight, and polymer crystallinity, affect chitosan's solubility. Homogeneous deacetylation of chitin offers the production of polymers that are soluble in acetic acid solutions, (aqueous) with DD as low as 28%, with this condition chitin is never going to be reached under heterogeneous deacetylation, such as alkali treatment at high temperatures.

Depending on the degree of deacetylation and the molecular weight of the polymer, chitosan's viscosity reduces as its molecular weight drops. Viscosity can be used to assess the durability of a polymer in a solution because it decreases with storage as a result of polymer degradation. Chitosan significantly improves wound dressing and healing (Guo et al. 2024).

13.3.2.2 Benefits in Wound Care

When paired with plasters and wound dressing materials, chitosan promotes quick wound healing by acting as an antibacterial agent against bacteria and viruses and supplying antibodies. Chitosan was shown to have the best antibacterial activity among the other nanoparticles, and the addition of graphene oxide improved the mechanical properties. Their biocompatibility is demonstrated in vitro, and they can be applied to wounds that need to be both mechanically and antibacterials patched. Chitosan is coated with copper oxide and copper nanoparticles for use as a wound dressing. Using sodium hydroxide and ascorbic acid, a straightforward chemical reduction of Cu2+ ions was used to create chitosan caps with copper oxide and copper nanoparticles. The action of the antibacterial-produced composite was evaluated using the inhibitory zone method against Gram-positive and -negative bacteria and found to be significantly greater than that of other nanomaterials, indicating that it is a superior alternative for wound dressing (Mottaghitalab et al. 2024).

The nanofiber membranes of chitosan were created and evaluated for their antibacterial activity in the treatment of chronic wounds and for their ability to minimize wound bioburden. The bactericidal activity is enhanced by the increasing concentration of polyaniline which is also efficient against both Gram-positive and -negative bacteria. This property makes the chitosan membrane unique and

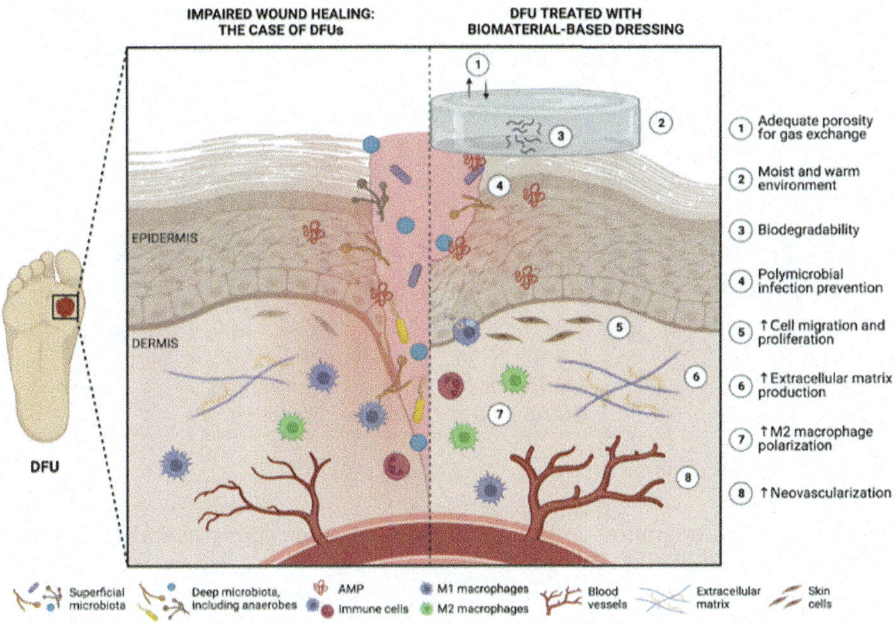

Fig. 13.3 Biomaterials in Wound Healing (Da Silva et al. 2023)

ideal for wound dressings and other healthcare applications. For antimicrobial and wound healing applications chitosan nanofibers combined with carboxymethyl chitosan nanoparticles encapsulated with an antibacterial peptide. All of the nanofibers contain concentrations of nanoparticles that were applied to a mouse's skin wound and demonstrated improved wound healing and antibacterial activity (Mawazi et al. 2024). Figure 13.3 underscores the prospective role of biomaterials in WH.

13.3.2.3 Recent Advances and Challenges

It's a serious challenge treating chronic and acute wounds. Natural polymers which are used as wound dressers are more effective than traditional ones. Sponges, foams, nanofibers, films, and hydrogels are the form of chitosan-based wound dressing. Natural polymers are generally limited by their unfavorable mechanical attributes. The combination of natural and synthetic polymers is done to enhance their physical properties, degradation patterns, and overall antimicrobial properties (Raina et al. 2022).

Natural polymers like chitosan have so many beneficial properties which allow them to be used as a therapeutic option for chronic wounds. Despite this, a natural polysaccharide may elicit the immune system to react adversely and cause irritation. Molecular weight control of the natural polysaccharides may be used to overcome such limitations. The wound-healing process in dry wounds could be affected by these properties. It may cause dehydration, thus reducing blood flow and the ability

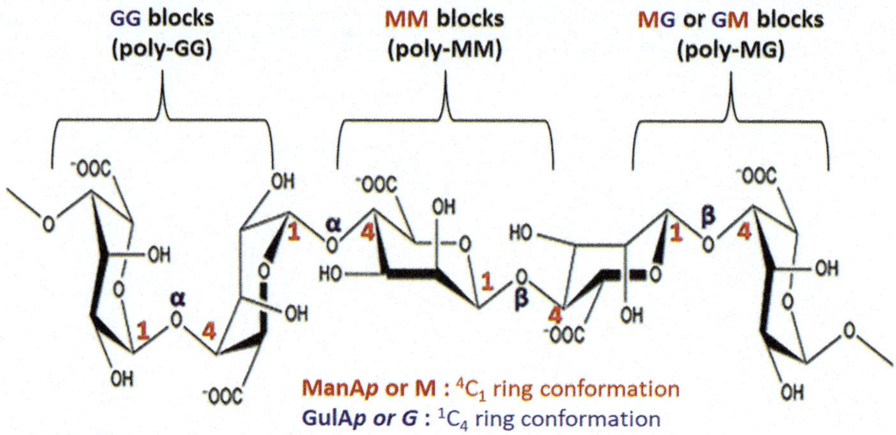

Fig. 13.4 Alginate chemical structure. (Abka-khajouei et al. 2022)

of epithelial cell migration within the wound site, interrupting new tissue creation (Sahana and Rekha 2018).

13.3.3 Alginate

In marine aquatic environments, brown algae is found and alginates are adopted from brown algae.

13.3.3.1 Structural and Biological Properties

Alginates are naturally occurring anionic polysaccharides belonging to the family of linear copolymers (without branches) found in the cell wall matrix of brown seaweed. The main structure of alginates is composed of two monomeric units: β-(1,4) linked d-mannuronic acid (ManA*p* or M) with $^4C_1$ ring conformation and α-(1,4)-linked l-guluronic acid (GulA*p* or *G*) with $^1C_4$ ring conformation (Li et al. 2024b) (Fig. 13.4).

Hydrogel-based alginates are utilized in tissue engineering, medication delivery, and wound healing.

13.3.3.2 Role in Managing Chronic Wounds

Intense burn, severe, and chronic wound infection treatment requires a longer period. In the US 6.5 million patients are severely affected by chronic wound infections along with more than 18% of diabetic patients over the age of 65 years. These diabetes patients suffer from foot ulcers, which are not easily healed or repaired. Wound healing is a complicated process that is controlled vastly by the biological system, which includes different types of cells, i.e., immune cells, keratinocytes, endothelial cells, and fibroblasts (Iqbal et al. 2024).

Alginate-based wound dressing materials have many advantages in comparison to conventional gauze dressing because the wound exudates can be easily evaporated preventing the entry of pathogenic bacteria and accelerating wound healing by providing a moist environment in the wound. ECM (extracellular matrix) is the surface of the wound areas, which is covered with endothelial cells and macrophage fibroblasts that contain new granulation tissue that occurs epithelialization by the time of propagation of the wound area (da Silva et al. 2024). ECM plays a vital role in proper and rapid healing because ECM delivers an appropriate environment for blood vessels and sustaining cells, which provides the nutrients required to restore the homeostasis and integrity of the tissue (52). ECM built functions as a suitable pliable and porous scaffold for facilitating the diffusion of nutrients, cells, and the growth factor through the wound areas. The thin films of the hydrogel are made containing sodium alginate (NaAl g) in 3 g/100 mL of water with 10% of the antiseptic povidone-iodine (PVP-I). The povidone-iodine shows a notable decrease in the wound areas to increase in re-epithelialization even after 3 days and the complete wound healing is achieved within 12 days, relative to the conventional wound healing process rapidly (Saberian et al. 2024).

13.3.3.3 Innovations and Practical Considerations

Alginate polysaccharides have valuable properties like biocompatibility, nontoxicity, biodegradability, and functional versatility with various matrices and substrates. Alginate is used as a growth stimulant plant in food, agriculture, and pharmaceutics. Alginates have so many properties that prove that they are efficient candidates to be used in various applications ranging from additives to food and beverages to scientific applications. Besides the use of alginate in the food and medicinal industries, it is explored in many other fields too (Tang et al. 2025).

A major obstacle of alginate-based hydrogels that need to be circumvented to be used as commercial products still exists even today. The major issues include introducing hazardous solvents or reagents during the synthesis process, modification, and cross-linking of the hydrogels. Therefore, there is an urgent need to explore advanced ways of chemically developing smart polysaccharide platforms with broad biomedical application properties not just for general wound healing but also for skin cancer, pressure sores, and ulcer-related wounds. Green chemistry and advanced technological devices can be used as a smart approach to achieving this (Davydova et al. 2024).

13.3.4 Silk Fibroin

A protein that extracts from the cocoon of *Bombyx Mori* silkworms is called silk fibroin.

13.3.4.1 Unique Features and Applications

Silk fibroin has so many unique features like mechanical robustness and flexibility, adjustable water solubility, and piezoelectricity. Silk fibroin also has low immunogenicity, good cell adhesion, and great permeability to oxygen. Silk fibroin has high cytocompatibility this silk fibroin is widely applied in tissue engineering. And this biomaterial used in wound healing too (De Giorgio et al. 2024).

13.3.4.2 Impact on Tissue Regeneration

Tissue engineering is the process of creating a synthetic device that mimics native tissue, hence restoring its anatomical and functional integrity while avoiding the drawbacks associated with autograft and allograft-based methods. With the advent of artificial biocompatible materials, novel scaffolds have been created that can enhance tissue regeneration when paired with cells and/or bioactive compounds (Lin et al. 2024).

Because of its improved mechanical, chemical, and bio/cytocompatibility qualities, silk fibroin has drawn interest as a viable biomaterial in regenerative medicine throughout the past 10 years. The adaptability of silk fibroin is further enhanced by the potential to create cutting-edge medical instruments from a raw aqueous solution, including films, fibers, hydrogels, 3D porous scaffolds, non-woven scaffolds, particles, or composite materials (Smita and Pramanik 2025). Because these devices can accommodate the widest range of tissue requirements, they offer a novel clinical option for the treatment of bone and cartilage, the cardiovascular system, the nervous system, skin, and pancreatic tissue regeneration, among many other biological uses. The recent review encompasses topics such as (i) the most interesting features of silk fibroin-based biomaterials, bare silk fibroin's biological nature and structural features, and comprehending the related chemo-physical properties and techniques used to produce the desired formulations of silk fibroin; (ii) the different applications of silk fibroin-based biomaterials and their related composite structures, discussing their biocompatibility and effectiveness in the medical field. For achieving bone regeneration Tissue engineering has been considered the most promising therapeutic strategy, and the earliest initial applications disclosed silk fibroin biocompatibility traits that, over time, have been demonstrated in various scientific works (Zhu et al. 2025a).

Given tissue engineering applications, it is worth mentioning that silk fibroin represents an exogenous protein for humans; hence, immunological reactions have been deeply examined. In some rare cases, delayed hypersensitivity or immunological responses were detected by the application of silk fibroin as a suture material. The most recent advancements in tissue engineering clearly indicate that the peculiar properties of silk fibroin (i.e., the as-discussed biocompatibility, environmental and mechanical stability, long degradation rate, and low immunogenicity) fulfill the major requirements for ideal scaffolds in regenerative engineering. In addition, the possibility to blend silk fibroin with other compounds can further improve material properties, specifically, those related to cell attachment, biostability, immunomodulation, and antimicrobial activity, assessing silk fibroin's higher versatility for the creation of innovative medical devices. Silk fibroin-based devices employed in recent regenerative medicine applications, schematized in (Fig. 13.5) (Wong et al. 2024).

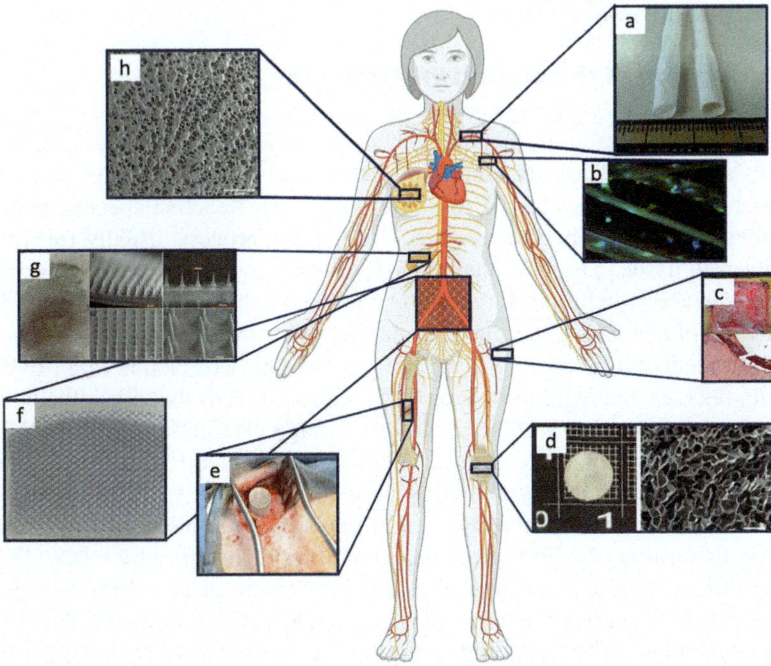

Fig. 13.5 Silk Fibroin -based materials prepared from amorphous aqueous solution and related device applications: (**a**) electro-spun silk fibroin tubes, (**b**) fibers, (**c**) films, (**d**, **e**) 3D scaffolds, (**f**, **g**) patches, and (**h**) hydrogels are some concrete applications of these devices for the regeneration of cardiovascular, neural, skin, cartilage, bone, abdominal, pancreas, and breast tissues, respectively. (De Giorgio et al. 2024)

13.3.4.3 Current Research and Developments

Silk fibroin is used in a wide range of biomedical applications these days, including disease models and tissue engineering. Because of its remarkable mechanical qualities, biocompatibility, and biodegradability, silk fibroin which has been degumming to remove sericin offers a vital range of material options for biomaterials, including scaffolds for tissue engineering, including bone, cartilage, skin, tympanic membrane, vasculature, and the oesophagus, as well as coatings for cell culture and drug delivery matrices (Jyoti et al. 2024). For a variety of biomedical uses, SF can be converted into films, hydrogels, nanoparticles, and nanofibers. Formulating a solution form of SF as a precursor is necessary to transform SF into these various forms (Zhang et al. 2024b).

13.4 Mechanisms of Action

13.4.1 Modulation of the Wound Environment

Complicated biological reactions include homeostasis (coagulation), inflammation, proliferation, and regeneration (maturation) throughout wound healing (Sawant et al. 2024a, b; Shaikh et al. 2024; Vardikar et al. 2025). Reactive species, metal ions, cytokines, and growth hormones play a role in this process, usually forming and repairing new tissue. The first phase starts within minutes post-injury, during which the body generates a blood clot to avert significant hemorrhaging. This entails the accumulation of platelets at the site of damage and the transformation of fibrinogen into fibrin, mediated by thrombin. Platelets secrete growth factors (e.g., EGF, PDGF, bFGF, β-TGF) and cytokines, such as IL-1β and α-TNF that assemble neutrophils for matrix formation within 12 h after damage. The inflammatory phase arises to counteract microbial infection (Zhu et al. 2025b). Neutrophils traverse to the site of damage, facilitated by signaling molecules such as formyl methionyl peptides and IL-1β, to commence phagocytosis. This phase is bifurcated into two segments: the first phase (24–36 h post-injury) and the secondary phase (48–72 h post-injury), during which monocytes develop into macrophages. Macrophages play a crucial role in triggering the inflammatory response, encouraging the death of neutrophils, enabling deposition, and encouraging collagen and angiogenesis through growth factor secretion. About 3 days after the injury, the proliferative phase starts and lasts for about 2 weeks. The procedure involves fibroblasts migrating to the site of the wound, where they produce essential extracellular matrix (ECM) constituents like collagen, fibronectin, along with hyaluronic acid. Fibroblasts transform into myofibroblasts, which facilitate wound contraction. Epithelial keratinocytes multiply to provide a protective barrier. The ending phase, commencing around 8 days after damage and extending up to 2 years, emphasizes the restoration and regeneration of the collagen matrix to mimic healthy tissue (Chen et al. 2024a). Nonetheless, scar tissue developed during this phase does not entirely emulate the characteristics of regular skin, attaining just around 80% of the healthy tissue's tensile strength. Smart biomaterials' moist environment permits oxygen penetration, wound exudate clearance, and growth factor or medication release to stimulate tissue proliferation and remodeling, unlike standard dressings of wounds. In wound healing and skin tissue engineering, natural biomaterials are popular due to their biodegradability, biocompatibility, and ease of manufacturing, strong tissue regeneration plus repair ability, excellent permeability as well as absorption, and good adhesion (Li et al. 2024c). Biomaterials, such as hydrogels, facilitate the maintenance of an ideal damp environment, hence facilitating autolytic debridement and augmenting cell growth. This equilibrium averts desiccation while assimilating surplus exudate, essential for cellular motility and proliferation (Li et al. 2024d).

Regulated oxygen levels are essential since transitory hypoxia promotes repair via angiogenesis, but persistent hypoxia hinders it (Nimmy et al. 2024; Patil et al. 2024; Faiyazuddin et al. 2024; Kumar et al. 2024; Mesut et al. 2024). Certain biomaterials may be engineered for oxygen transport to the wound bed, mitigating oxidative stress while guaranteeing sufficient oxygenation. Preserving a somewhat acidic environment enhances antibacterial efficacy, but a neutral pH promotes cellular proliferation and tissue regeneration. Specific biomaterials assist in regulating pH, particularly in chronic wounds, which often grow more alkaline and hinder healing (Dixit et al. 2024).

Hyperactive proteases, such as matrix metalloproteinases (MMPs), dismantle the extracellular matrix as well as obstruct the healing of wounds. Biomaterials may regulate protease levels, preserving the extracellular matrix and keeping the wound in its ideal state for healing. Certain biomaterials modulate the immune response, facilitating pro-healing (anti-inflammatory) responses and mitigating excessive inflammation that may impede recovery (Aldahish et al. 2024). Biomaterials may inhibit infection by integrating antimicrobial substances such as silver nanoparticles or peptides. This maintains a sterile wound environment, crucial for maximum healing results. Absorbent biomaterials, particularly hydrogels, are crucial in managing excessive exudate and minimizing maceration while sustaining an environment favorable to healing (Shen et al. 2025).

13.4.2 Structural Support and Integration

Considering the essential function of tissues and organs' extracellular matrix (ECM), biomaterials derived from the ECM possess significant promise in tissue repair, regeneration, and therapeutic interventions for diseases (Gholap et al. 2024c; Pardeshi et al. 2024a, b, c; Rojekar et al. 2024c). The extracellular matrix's (ECM) main function is to give tissues structural support. The extracellular matrix (ECM), which is mostly composed of collagen, elastin, and proteoglycan, is a structure with a particular mechanical strength. Cell adhesion is supported by the extracellular matrix, which makes it easier for cells to connect, disseminate, and migrate inside tissues (Chen et al. 2024b). Cell surface receptors, like integrins, interact with ECM proteins, such as fibronectin and laminin, to promote cell adherence to the extracellular matrix. Cell behavior and function are regulated by signaling molecules found in the extracellular matrix (ECM). In response to external stimuli or tissue damage, growth factors, cytokines, and other bioactive substances are released from sequester within the extracellular matrix. The initiation of intracellular signaling pathways by these substances has the potential to impact gene expression, survival, differentiation, and proliferation of cells. When developing such materials, the following essential factors must be considered (Liu et al. 2022): The materials must exhibit excellent biocompatibility, ensuring that their interaction with cells or tissues does not elicit detrimental immune responses or cytotoxic consequences (Gholap et al. 2024a, b, d; Parit et al. 2024; Hatvate et al. 2024; Rojekar et al. 2024a). Secondly, the materials must facilitate cell movement, adhesion, growth, along differentiation. Third, to meet the specific requirements of diverse tissues and applications, the materials must have modifiable properties including mechanical strength, degradation kinetics, along porosity. By changing these characteristics, the biomaterials can be made to fit different tissue kinds, clinical conditions, and therapeutic goals. Fourth, biomaterials need to control how materials or cells interact, including how well or poorly they promote (or inhibit) cell adhesion, diffusion, and migration. Hydrogels, hydrogel microspheres, electrospinning, and dECM have become more well-known as research on ECM-inspired biomaterials advances because of their similarity to the ECM and their adaptability (Golebiowska et al. 2024).

Biomimetic materials such as hydrogels and hydrogel microspheres mimic the extracellular matrix's structure and functionality (ECM) to aid in regenerative medicine as well as tissue engineering (Bhowmik et al. 2023; Gholap et al. 2023a, b, c, d, e; Pardeshi and Giram 2023). Cell adhesion, migration, and differentiation signals may be provided by these materials, which can be made utilizing collagen, gelatin, and hyaluronic acid, examples of ECM components (Ahmad et al. 2023). Hydrogels may be engineered to react to certain physiological circumstances, enabling precise modulation of drug release, whereas hydrogel microspheres provide targeted administration and regulated release of cells, growth factors, and pharmaceuticals. A method for creating nanofiber scaffolds that mimic the extracellular matrix is electrospinning, with fiber diameter and pore size influencing cellular behavior, proliferation, migration, and differentiation. By incorporating ECM components along with simulating ECM-like environments, these materials and technologies provide conducive microenvironments for cellular proliferation, tissue regeneration, and pharmaceutical delivery (Ramos-Rodriguez and Leach 2025).

13.4.3 Enhancement of Cellular Activity and Tissue Regeneration

Several million individuals throughout the globe suffer from diabetes, making it a chronic condition and a major healthcare cost (Gholap et al. 2023b; Pardeshi et al. 2023; Hatvate et al. 2023). Diabetic foot ulcers are among the most dangerous effects of diabetes (DFU). Wound healing is hindered and amputation may happen as a result of elevated blood glucose levels. It is still tough to find a comprehensive cure for DFU, despite the availability of several therapeutic alternatives (Yang et al. 2022a). One potential therapeutic option is growth factor-based therapy, such as recombinant growth factors or platelet-rich plasma (PRP), which is used to treat DFU. Improved healing results in DFU may be possible with the use of polymeric compounds to supply growth factors. Growth factors are essential in wound healing and skin tissue renewal. At present, many growth factors are clinically authorized for topical application or injection. The integration of growth factors with biomaterials preserves their characteristics and facilitates sustained release throughout time. Furthermore, this combination reduces the frequency of repeated doses of growth factors. Due to the intricate nature of the wound-healing process, multiple growth factors working together is frequently required instead of just one (Baig et al. 2022).

Managing infection and inflammation while facilitating tissue regeneration may significantly enhance the healing of wounds. The smart biomaterials provide considerable benefits in healing due to their capacity to regulate the healing process temporally and spatially (Kolimi et al. 2022). Comprehending the design of biomaterials for various phases of wound healing would enable the future customization of materials for specific wounds, enhancing their efficacy in wound care. Recent developments in nanotechnology improve drug-delivery device design and manufacturing for diabetic wound healing. To improve drug performance and maximize drug encapsulation efficiency, this research focuses on a variety of nanostructures, comprising nanoparticles, nanofibers, liposomes, and nanohydrogels (Liu et al. 2023).

Chronic wounds and diseases may have persistent inflammation due to chemokine dysregulation. Advanced therapy biomaterials affect wound healing chemokine signaling differently. Cytokine and chemokine signaling recruits different cell types at each stage of wound healing. Designing biomaterials for wound healing focuses on the dysregulation of the chemokine signaling throughout the change from inflammatory into proliferative stages. Immunomodulatory biomaterials promote tissue regeneration to boost the immune system. Biomaterials that mimic the ECM and have an immunomodulatory effect that is self-limiting could be ideal (Wang et al. 2024a). Glycosidic bonds connecting monosaccharides, or polysaccharides, are produced by animals such as chitosan and glycosaminoglycan (GAG) and plants such as alginate. These polymers replicate ECM structure and microenvironment, making them desirable tissue engineering biomaterials. GAGs, chitosan, and alginates are polysaccharides utilized in wound healing therapies. Various physiochemical features of artificial biomaterials promote wound healing, like their counterparts. PEG, PLGA, and MAA are synthetic biomaterials used in wound healing. Diabetic wound dressings employ hydrophilic ether-based polymer PEG. Both natural and manmade biomaterials may treat cutaneous wounds. These materials were initially developed using polymers that mimic the natural extracellular matrix (ECM) and replenish injured tissue. We find a more complex role for these biomaterials in wound healing when we investigate their effects on chemokine production including cellular signaling. The timing of biomaterial application during the healing of wounds also seems to affect pro-inflammatory and anti-inflammatory chemokine expression (Apte et al. 2023).

In the last 20 years, tissue engineering (in situ) has emerged as a cutting-edge technique that makes use of biomaterials to harness the body's inherent stem/progenitor cells for the regeneration of sick or damaged tissue (Safina and Embree 2022). To promote a regenerative environment, recruit resident stem cells to areas of injured or damaged tissue, and direct them towards tissue repair, immunomodulatory biomaterials are designed. Biomaterials have shown their significance in stem cell biology, differentiation, and prospective medicinal uses, particularly in wound healing. Biomaterials must be regarded as a crucial element in stem cell differentiation methodologies (Abdulghani and Mitchell 2019).

Bacterial contamination is well recognized as a substantial barrier to the healing of wounds and a primary barrier to the closure of wounds. Consequently, it is critically necessary to design as well as develop wound dressings to lower the risk of infection and enhance the quality of healing. The potential effectiveness of these functional dressings in wound closure has been demonstrated. In addition, the quick development of emerging fields like bioelectronics, 3D printing, plus smart materials will soon enable on-demand therapy and continuous monitoring of wound healing (Riha et al. 2021).

13.5 Innovations in Biomaterial Development

13.5.1 Novel Formulations and Composite Materials

13.5.1.1 Collagen

Collagen found in tissue and skin is considered a primary protein and has been used for a long time. Collagen with additional natural components such as alginate or hyaluronic acid has been used in recent advancements in collagen composite. This composite preserves the moisture of the wound works as a natural scaffold in the development of cells, and also speeds up the process of healing (Sabarees et al. 2024).

13.5.1.2 Chitosan

Chitosan, derived directly from the crustacean shells has antibiotic qualities that promote the wound healing process. Recent advancements have been made in the development of natural biomaterials based on chitosan and hydrogel. Additionally, natural materials such as honey and herbal extract have been used to improve the quality of these therapeutic materials (Saberian et al. 2024).

13.5.1.3 Silk Fibroin

Silk fibroin is renowned for its durability and also for its biodegradability. Silk fibroin is a protein derived from silkworms and is considered a protein. Wound dressings made from silk have been used to provide two essential processes such as cell attachment and angiogenesis, required in the healing of diabetic patients' wounds (De Giorgio et al. 2024).

13.5.1.4 Alginate Dressings

Alginate is a very absorbent substance that is often used for the treatment of wounds that are exudative, which is derived from seaweed. They are mainly used to facilitate the treatment of deep wounds. Additional materials such as antibacterial agents or growth hormones, are used in the development of these composite alginate dressing (Iqbal et al. 2024).

13.5.2 Advanced Delivery Systems and Techniques

The treatment of DFUs for wound healing management is a notable challenge due to the various complications that are linked with diabetes. In the last few years, the successful treatment of DFU has occurred through the development of natural biomaterials in combination with innovative delivery techniques and tactics. By reinforcing the bioavailability of medicinal compounds, these cutting-edge delivery technologies for DFU facilitate the revitalization of tissue, expedite the healing process and enable infection management. To enhance the outcomes of chronic wound healing for DFU the investigation of innovative delivery strategies mainly employed in combination with natural biomaterials is crucial (Perez-Favila et al. 2019).

In the field of wound therapy, electrospinning techniques have been used for manufacturing these nanofiber mats, and these mats are becoming are effective method of administering medicinal drugs. These techniques have a structure that is similar to that of the ECM and provide an environment that is conducive to the adherence and development of cells. For the treatment of DFU, natural biomaterials, for instance, chitosan, silk fibroin, and collagen are used because of their flexibility and biocompatibility. To transmit growth stem cells, hormones, and antimicrobial agents, in a controlled and sustained manner. Electrospinning techniques through the use of nanofiber mats can be developed to be effective (Yang et al. 2022a).

It is recommended that antimicrobial-loaded nanofibers be used to both prevent bacterial infection and encourage tissue regeneration and tissue regeneration (Borude and Mahajan 2011a, b, c; Tushar and Magar 2011a). The production of growth factors by nanofibers, such as vascular endothelial growth factor (VEGF) and epidermal growth factor (EGF), can potentially speed up angiogenesis, and stimulate cell proliferation processes. The use of these mats provides an important advantage over conventional wound dressings because these mats offer a controlled, localized mode of distribution. This is in contrast to the fact that therapeutic agents are often administered in an uncontrolled manner, leading to varying effects (Kolimi et al. 2022; Liu et al. 2023).

Hydrogels are used in the treatment of DFUs, serving as a flexible drug delivery platform for wound care. This gel is particularly made from natural materials such as chitosan and alginate, and the primary ability of hydrogels is to absorb a significant amount of water, which enables them to provide moist environments that hasten the healing process of wounds (Borude and Mahajan 2011d; Tushar and Magar 2011b; Dighe 2012a; Rajurkar 2012a, b; Tupe 2012a; Vikas and Rajurkar 2012). It is possible to inject medicinal agents (antibiotics, growth factors, or stem cells) into alginate-based hydrogels, to distribute the agents in a controlled release system. Additionally, hydrogels may be used to transport stem cells, which possess the capacity to develop into a wide variety of cell types within chronic wounds (Zheng et al. 2024). This is carried out to aid in tissue repair and regeneration. The controlled release of these therapeutic compounds from hydrogels allows for longer treatment times, which may improve patient compliance and decrease the frequency of dressing changes. Moreover, microparticle systems, produced from natural materials such as gelatin are another potential new way for delivering medicinal chemicals directly to the area where they are required (Bhagat 2012; Dattatray Tupe 2012; Dighe 2012b; Gholap 2012; Rajurkar 2012c; Tupe 2012b, c, d). These microparticles are designed to provide a healing environment that is tailored to meet the unique needs of DFU patients. Moreover, microparticle systems, produced from natural materials such as gelatin are another potential new way for delivering medicinal chemicals directly to the area where they are required. These microparticles are designed to provide a healing environment that is tailored to meet the unique needs of DFU patients. To enhance angiogenesis and tissue regeneration, microparticles that are loaded with growth factors such as oxygen transporters and VEGF or show potential performance (Sang et al. 2024). This is accomplished by the controlled release of oxygen, growth factors, or insulin into the diabetic patient. Additionally, an insulin deficit in the infected area slows down the healing process or causes ulcers, intentionally delivered insulin by using microparticle devices can improve the healing process (Rajurkar 2012d, e; Bothiraja et al. 2014). Providing a targeted and continuous release of therapeutic agents by microparticle systems provides a viable treatment for the healing process after deep fasciotomy (DFU). Another development is the use of bioactive dressings made from natural extracts. These natural extracts are derived from aloe vera, green tea, and curcumin plants, and have anti-inflammatory, antibacterial, and antioxidant properties, that help to speed up the healing process. These extracts are employed in the wound as are other biomaterials such as collagen, chitosan, or alginate; which also generate; and reduce the probability of infection (Liang et al. 2024b).

Aloe vera has anti-inflammatory properties thus it is a favorable component for wound dressings intended to reduce the chronic inflammation that is associated with DFUs. Similarly, the combination of catechins and green tea extracts has potent antioxidant and antibacterial properties that protect the wound environment against infections and also reduce oxidative stress (Yang et al. 2022a). Another combination of curcumin and turmeric has been used to enhance tissue regeneration, here, curcumin is mainly responsible for increasing the formation of collagen and reducing inflammation in the infected area. The combination of natural extracts with cutting-edge nanomaterials; creates bioactive dressings that provide dual action in wound healing. While biomaterials provide a moist and safe environment for the wound, natural substances increase the healing process and prevent infection from occurring. This technique not only enhances the therapeutic efficiency of dressings but also provides DFUs with a more thorough course of treatment (Chelu et al. 2023).

13.6 Case Studies and Recent Research Findings

Because of the serious complications in DFUs often result in persistent wounds and infections. The investigation of natural biomaterials that provide better healing conditions, encourage regeneration of tissues, and prevent infections that have been prompted by the limited efficacy of traditional therapies for DFUs in attaining full healing. According to recent studies and clinical investigations, numerous natural biomaterials have shown great efficacy in the treatment of DFUs. These biomaterials are used in drug delivery systems and have antibacterial qualities, and other therapeutic advantages, which improve the healing process. Current research results focus on the use of natural biomaterials, such as chitosan-based hydrogel, dressings made of alginate and honey composite, collagen-silver nanoparticles, honey-collagen composites, and silk fibroin scaffolds (Park et al. 2019).

Because of its inherent antibacterial characteristics, biocompatibility, and capacity to stimulate tissue regeneration chitosan has garnered noteworthy interest in the field of wound healing. The effectiveness of chitosan-based hydrogels in the treatment of DFUs was recently proven by research. The addition of antimicrobial peptides to the chitosan-based hydrogels helps mitigate the significant risk of infection often associated with DFUs (López-Cervantes et al. 2019).

Compared to patients receiving traditional therapies, those treated with chitosan-based hydrogels had quicker healing periods and a lower risk of infection. The results of this research highlight the possibility of chitosan-based hydrogels as a successful DFU therapy approach, especially in situations where infection is a complicating factor (Xiong et al. 2023).

13.7 Advantages and Limitations

13.7.1 Comparative Analysis of Biomaterials

From the previous discussion analyzing all the biomaterials can see that collagen is a natural biomaterial which is a biologically derived biomaterial. Collagen is found in every animal tissue and has a great impact on wound healing. Collagen is not only shown effective in wound healing but also has efficacy in skin mimicking, bone regeneration, biomedical implants microneedles, etc. Collagen with other wound healing agents work faster and reduce scar formation. Collagen is also used as a wound dresser for diabetic foot ulcers. However, using collagen as a wound healing agent facing some difficulties too. Chitosan is a non-toxic biopolymer and is found in aquatic environments. Chitosan is known as a wound-healing material. It has some great properties like solubility and viscosity. Chitosan extract from crustaceans. Chitosan with some wound dressing materials showed rapid wound healing. Chitosan-based wound dressing materials are also used in diabetic foot ulcers (DFU). Unlike chitosan, another naturally occurring polymer is alginate. The source of alginate is algae. Alginate contains guluronic acid in the cell wall. Alginates are one of the polymers which are applied in food, textiles pharmaceuticals, and many more. Alginates have a huge contribution to the wound healing process. Alginate-based wound dresser used in acute to chronic wound infection.

The silkworm *Bombyx Mori* is the source of silk fibroin, a naturally occurring polymer utilized as a biomaterial. The silkworm's outstanding mechanical qualities allowed it to be utilized for a long time in the textile industry. 30–20% sericin (SR) and 70–80% fibroin is present in raw silkworms. Tissue engineering makes extensive use of silk fibroin (Jyoti et al. 2024; Zhang et al. 2024b; Smita and Pramanik 2025).

13.7.2 Effectiveness in Various Clinical Settings

All the biomaterials including collagen, chitosan, alginate, and silk fibroin have excellent bioactivity, biocompatibility, and biodegradability promote tissue regeneration, and ease wound healing/infection. These biomaterials are naturally found or occur. Easy to derive and also less costly. In addition to the biomedical industry, biomaterials are utilized in many other fields. These biomaterial components are widely employed for wound dressing, medication administration, antimicrobial treatments, tissue engineering, and wound healing. Collagen has not only effects on wound healing but also has properties that are good for skin rejuvenating, people with osteoarthritis show much improved joint function and are relieved from pain after getting collagen-based treatment. Although collagen has the properties that are best for skin, hair, nails, bones, and wound healing (Sahana and Rekha 2018; da Silva et al. 2024; Saberian et al. 2024).

Chitosan-based treatment is nowadays used in cancer also. Chitosan has non-toxicity and less adverse effects than any other and chitosan derivatives so it shows a great anticancer potency. Chitosan nanoparticles are used as drug delivery vehicles for controlled-release drugs in cancer treatment. Chitosan also used in gene therapy, acts as an immunity booster agent. Chitosan can lower cholesterol and low-density lipoprotein (LDL) low-density lipoprotein, so it shows a great effect on obesity treatment. Besides chitosan-based treatment is used in nowadays tissue regenerative including skin, bone, cardiac, neural, and corneal. A biomaterial, alginate is utilized in dental tissue engineering for purposes other than wound dressing and healing. Silk fibroin hydrogels have adhesive properties which are used for dental tissue engineering, bone repair also soft tissue functions like liver & kidney as well as sustained drug release or wound healing.

13.7.3 Potential Risks and Side Effects

Biomaterial has immense effects on various biomedical sides. But still have some potential risks and side effects too. Using collagen as a wound healing agents have a high risk of anaphylaxis, as collagen sources from bovine, murine, porcine, etc. Many people are allergic to cows, shellfish, etc. so they can face anaphylaxis during collagen-based treatment. As many people are allergic to shellfish cow they can face skin rashes also. Though collagen has no major side effects. After purification chitosan is ready for use in any medical or other sectors as chitosan is found in nature and completely in raw forms. But still, chitosan has some risk for use. As a biomaterial chitosan forms in nature so it has some toxicity so needs to ensure human safety. Alginates have some major side effects including severe abdomen pain, swelling to the ankle, etc. Silk fibroin has no major side effects (Raina et al. 2022; Da Silva et al. 2023).

13.8 Future Directions and Applications

13.8.1 Emerging Trends in Biomaterial Research

Pressure ulcers, diabetic foot ulcers (DFUs), and arterial or venous ulcers are among the chronic wounds that put a heavy strain on patients, healthcare providers, and the whole healthcare system. The prevalence of chronic wounds, which pose a significant threat to public health due to their high monetary cost, lengthy treatment processes, and ongoing chronicity, has grown in recent years. To address this problem and speed up wound healing while decreasing treatment duration, the biomaterials industry has risen substantially. Polyglycolic acid, alginate, hyaluronic acid, collagen, gelatin, xanthan gum, keratin, chitosan, carboxymethylcellulose, cellulose, hemicellulose, and many more are among the multitude of other substances. Hydrogels, combinations of therapies, films, membranes, foams, gels, eco-friendly and intelligent dressings, and 3D bioprinting are all examples of biomaterials that are often employed to produce an optimal wound healing environment (Josephin et al. 2024; Yang et al. 2024b; Liang et al. 2024a).

13.8.1.1 Polymeric Biomaterials

Despite advances in wound healing, treating chronic wounds in patients with diabetic foot ulcers as well as other serious injuries remains a significant issue. An excellent wound dressing material absorbs wound exudates, aids gas exchange, and prevents microbiological infections. Additionally, they must support the synthesis of biochemical mediators necessary for wound healing, such as growth factors and cytokines. Bioactive substances from natural sources have been investigated to expedite wound healing in diabetics. Natural wound dressings have garnered attention for their ability to induce new tissue growth (Shi et al. 2023). Due to their mechanical, bioinert, and biocompatible properties, synthetic polymers are widely utilized to treat diabetic wounds. Polymeric biomaterials, both natural and synthetic, are highly effective wound dressings due to their biocompatibility, mechanical strength, and low immunogenicity. Due to their biocompatibility and healing properties, silk proteins, collagen, gelatin, and fibrin are increasingly employed in wound dressings. Gelatin is biodegradable yet weak, whereas collagen is essential for skin and wound healing. Fibrin scaffolds cell delivery and tissue regeneration, while silk proteins boost cell migration and proliferation, healing diabetic wounds. In conjunction with natural materials, synthetic polymers like polycaprolactone increase mechanical qualities (Jiang et al. 2023).

13.8.1.2 Nanofibrous Scaffolds

The unique properties of nanofibrous scaffolds have made them attractive candidates for wound care in addition to skin tissue regeneration. These scaffolds have mechanical strength, adhesion properties, biodegradability, plus compatibility with living organisms. One possible use for nanofibrous scaffolds made of polymers, both natural and artificial is in the development of drug delivery systems. These systems may enable the regulated release of bioactive substances like vitamins, growth factors, hormones, and antimicrobial agents. This can be done in two ways: either by coaxially spinning two distinct polymers containing active chemicals and pharmaceuticals or by electrospinning blended polymers. As a synthetic extracellular matrix, these scaffolds provide an ideal setting for the growth of new tissues. The increased surface-to-volume ratio of nanofibrous scaffolds makes them more effective in cell adhesion, expansion, and differentiation than other forms of scaffolds. Electrospun cellulose acetate nanofibers including nanomaterials, namely CuO/Ag nanoparticles (AgNPs), may have bacteriostatic and bactericidal properties (Zhang et al. 2022).

13.8.1.3 Smart Biomaterials with Growth Factors

"Smart" biomaterials are an emerging category of materials that react to various environmental stimuli. Biomaterials that react to pH or enzymes generated by cells might provide on-demand, localized administration of medicines to modulate cellular responses. Leveraging the physicochemical properties of smart biomaterials, efficient distribution of labile biomolecules is suggested for immunomodulation and sustained disease treatment. Intelligent and adaptive biomaterials with diverse capabilities are attractive candidates for many biological applications. To expedite the regeneration process, stimulated keratinocytes, platelets, macrophages, fibroblasts, and endothelial cells generate these throughout the reconstruction of wounds. Insulin-loaded micelles (ILM) and epidermal growth factor (EGF) were added to succinyl chitosan (OHA/SCS) hybrid hydrogels/oxidized hyaluronic acid by Zhu et al. to cure diabetic skin lesions in rats. The ILM and EGF-infused composite hydrogel demonstrated exceptional wound healing effectiveness by promoting collagen and myofibril deposition, stimulating fibroblast proliferation, and improving tissue structural integrity. Researchers Lin et al. developed a silk fiber membrane (SF) to provide insulin-like growth factor-1 to diabetic wounds. Even at low dosages (3.25 pmol), insulin-like growth factor-1 caused the development of microvessels and expanded the area of epithelial tissue in a dose-dependent manner. Consequently, intelligent biomaterials with growth factors may serve as an excellent alternative for diabetic wound dressings (Kim et al. 2021; Lin et al. 2021).

13.8.1.4 Stimuli-Responsive Materials

Smart materials, or stimuli-responsive materials, can respond to internal and external wound microenvironment changes, such as temperature, pH, metabolites, enzymes, and ultrasound fields. These materials can create responsive wound care solutions by matching material features with wound pathological circumstances. For example, liquid-state hydrogels can distribute medications but may delay delivery due to solidification transition under body temperature. Additionally, these materials produce contractile force and adapt to temperature changes, accelerating healing. DNA-crosslinked hydrogels break down in reaction to pathogen-secreted DNase and release neutrophils. Body motions and external pressure can be transformed into electrical energy by nanogenerators, revolutionizing wound care. Piezoelectric materials, sensitive to mechanical stress, activate cellular processes for tissue repair and regeneration, promoting active, non-invasive healing in dynamic wound dressings. Photodynamic treatment uses light to locally raise temperature, killing microorganisms. Effective against resistant germs, this approach has few adverse effects. Stimuli fields are often used to regulate scaffold breakdown for drug or cell treatment. Magneto-induced dynamic mechanical stimulation in wound dressing scaffolds may release regulated medicines and cutaneous cells, speeding healing (Kolipaka et al. 2024; Naghib et al. 2024).

13.8.1.5 Microparticles and Nanoparticles

Microparticles and nanoparticles have great promise because of their enormous surface area, for wound healing, adjustable features, as well as capacity to encapsulate therapeutic substances. They can be stabilized by metal shells, polymers, surfactants, and low-sensitization coatings to reduce skin irritation and multisystemic problems. Nanoparticles affect cell behavior and physiological equilibrium, such as reducing inflammation, improving epithelialization, and restoring skin hemostasis. They can modulate regeneration and immune cell activity, affecting wound healing. Metal nanoparticles, including metallic, metal oxide, and quantum dots, are effective antimicrobials and reactive oxidant species scavengers against multidrug-resistant organisms. Commercial wound dressings increasingly include silver nanoparticles, which prevent and treat wound infections. Based on the characteristics of the wound, some nanoparticles can have their drug releases externally controlled like pH or metabolites, allowing precision targeting or remote activation. Controlled medication delivery using microparticles and nanoparticles is common, containing growth hormones, antibiotics, and medications that reduce inflammation as well as genetic materials like DNA or RNA. However, wound healing and monitoring using microparticles-nanoparticles face challenges such as cytotoxicity, poor particle distribution and preservation at the location of the wound, and the complex wound microenvironment affecting nanoparticle stability and function. To properly use both nanoparticles and microparticles in the wound care, extensive preclinical research and new technical solutions are needed (Luo et al. 2024; Yang et al. 2024a).

13.8.1.6 Bioelectronic Materials

Bioelectrical signals play a crucial role in wound care, as they can be created and responded to by cutaneous cells. Techniques integrating electronics into wound care and tissue repair have been developed, including materials like carbon nanotubes, metal nanoparticles, and graphene oxides. However, uneven distribution of these materials can disrupt electrical signals, requiring long-term safety assessments. Stretchable electrodes made of conductive organic polymers can help move wounded tissue. Ion-conductive hydrogels, with their high-water content and malleability, facilitate ion mobility and charge conversion between wounds and tissues. Their solid-like appearance at the macroscopic level aids in electric field-driven healing and wound hydration. Ion-conductive hydrogels, with their lower electrical conductivity and electrochemical properties, may aid wound healing. These materials, which use electrical impulses to control cell migration, tissue regeneration, and inflammation, can monitor wound healing, carry medications, and provide electrical stimulation to expedite healing. However, addressing biocompatibility, electrical characteristics, mechanical durability, and degradation rates is crucial for successful application. A multidisciplinary approach is needed to ensure the safety, efficacy, and compatibility of these materials with physiological dynamics and medical technology (Wang et al. 2024b; Li et al. 2024a).

13.8.1.7 3D- Bioprinting Using Biomaterials

Biomaterials are defined by their inherent non-toxicity and high degree of biocompatibility. Bioprinting has made substantial use of natural ingredients for example, alginate, collagen, fibrin, and gelatin, which are similar to the extracellular matrix, as biomaterials. Natural materials have been improved by combining synthetic and natural components with specific polymers, such as poly (lactic-co-glycolic acids), polyurethane, polycaprolactone (PCL), and polyethylene glycol, which impart stable structures and adjustable mechanical properties (Guptha et al. 2024).

13.8.1.8 Hydrogel Dressings

Hydrogel dressings soothe, reduce infection risk, and increase wound moisture. These benefits come from hydrogels' unique properties. To create hydrogel dressings, the three-dimensional network scaffold made of hydrophilic polymer absorbs biological fluids and water without disintegrating. Hydrogel bandages absorb a lot of fluid from the damaged spot to comfort and prevent infection risk. They aid wound healing by keeping the environment moist. Other hydrogel properties make them therapeutic. They are appropriate drug-delivery devices due to their porosity and capacity to integrate and release drugs at the wound site. Van der Waals forces and hydrogen bonds cross-link hydrogels, allowing controlled release. Hydrogels reproduce the ECM (extracellular matrix), encouraging tissue regeneration and cell growth, and making them competitive. Hydrogel dressings are biocompatible and biodegradable because they include hyaluronic acid, dextran, chitosan, alginate, and gelatin. Manufacturing is crucial because backbone chemistry and crosslinking determine gel form, stability, mechanical behavior, and stereo complexation. The material choice may most affect hydrogel properties. Yang et al. developed hydrogen-bonding supramolecular hydrogels for postoperative anti-adhesion with improved mechanical strength and shape memory. Xin Zhao et al. created novel injectable hydrogels. Strong, moist-adhesive, anti-swelling hydrogels. They gel fast, closing infected wounds without sutures or scars. Researchers and healthcare personnel may enhance wound care and patient outcomes by understanding hydrogel dressings' therapeutic advantages and qualities (Yang et al. 2020; Jin et al. 2023).

13.8.1.9 Foam Dressings

Foam bandages are versatile hydrophilic or hydrophobic therapies for wound care. They are composed of polyurethane or silicone and may have an adhesive perimeter that insulates the wound bed while facilitating the escape of oxygen and moisture vapor. Exudate infiltrates the insulated foam across the non-adherent, semi-permeable surface. Polyurethane foams with high absorbency include hydrophilic wound contact and hydrophobic backing in two or three layers. They inhibit leakage and uniformly distribute exudate throughout the absorbent layer owing to their semipermeable backing. This design mitigates dressing contamination and maintains its integrity. Silicone foam dressings composed of rubber are pliable and conform to wounds. This function enhances wound covering and healing by maintaining proximity between the dressing and the wound bed. Foam dressings provide a more pleasant experience for nursing and dressing changes compared to gauze. The significance lies in the fact that patient comfort influences wound care and adherence to therapy. Foam dressings may be used for the management of moderate to severe exuding wounds, lower leg ulcers, and granulating wounds. They effectively treat these wounds, as per clinical practice and healthcare professionals (Hargis et al. 2024).

13.8.2 Prospective Improvements and Innovations

13.8.2.1 Enhanced Drug Delivery Systems

Currently, diabetes-related sores and ulcers have exacerbated the medical burden. A wound that is diabetic may hinder movement, result in amputations, or perhaps cause death. Recently, improved systems for delivering drugs have emerged as potential methods for improving the effectiveness of wound-healing therapies in patients with diabetes. The significance of prompt intervention and efficient wound care measures is underscored. Numerous sophisticated drug delivery methods exist, comprising hydrogels, liposomes, niosomes, transferosomes, nanoparticles, dendrimers, and nanosuspensions, which can treat chronic diabetic wounds by combining biological macromolecules with bioactive compounds. The benefits of these systems include improved wound closure, improved targeting and penetration, and continuous release of medicinal medicines. Cutting-edge methods such as stem cell therapy, gene therapy, growth factors, antibiotics, minerals, and vitamins show promise in treating diabetic wounds. The management of chronic diabetic wound healing holds great potential for the effectiveness of enhanced medication delivery systems. They offer cutting-edge techniques for improving wound healing, delivering therapeutic medications, and focusing on the unique pathophysiological characteristics of diabetic wounds (Goh et al. 2024).

13.8.2.2 Integration of Biomolecules

Integrating bioactive compounds, like peptides and proteins that stimulate cell proliferation and angiogenesis, into biomaterials, is expected to improve healing processes. This method seeks to provide a more favorable environment for tissue repair. With their superior physical structures and complex biomolecules, Decellularized biomaterials derived from the extracellular matrix of tissues and organs are naturally occurring substances rich in bioactive macromolecules and peptides, which help promote skin regeneration and wound healing. Chronic non-healing cutaneous wounds, including arterial, venous, pressure, and diabetic ulcers, provide considerable issues for patients, families, and healthcare systems (Artemi et al. 2025). Conventional therapies, such as autologous skin grafts and reconstructive surgery, are becoming ineffective owing to an aging demographic and rising morbidity rates.

This underscores the pressing need for sophisticated biomaterials that facilitate rapid and efficient skin healing, preferably resulting in scar less regeneration. Recent advancements in regenerative medicine and the engineering of tissue have shown potential in both laboratory and clinical environments for the healing of skin abnormalities. Nonetheless, first treatments mostly focused on scaffolds, cells, or growth factors, often leading to insufficient healing of wounds marked by inadequate vascularization, atypical scarring, and sensory deficits. The extracellular matrix (ECM), an essential non-cellular tissue component, is integral to development, homeostasis, inflammation, and repair. As a result, research has transitioned to developing conditions that promote endogenous tissue healing (Rathna and Kulandhaivel 2024).

13.8.2.3 Integration with Personalized Medicine

Progress in 3D bioprinting and customized biomaterials facilitates the development of wound dressings customized for each patient, personalized to the dimensions, morphology, and status of individual wounds. This tailored methodology guarantees enhanced efficacy for each distinct diabetic wound, providing precise and focused treatment. Diabetes-related foot ulcers (DFUs) are chronic lesions brought on by aberrant extracellular matrix (ECM) and malfunctioning fibroblasts. Although functional enhancements may be necessary, porous collagen-based scaffolds (CS) have therapeutic potential for the treatment of diabetic foot ulcers (DFUs). Previous research has shown that fibroblasts derived from induced pluripotent stem cells serve as an excellent supply of extracellular matrix that mimics fetal tissue, significantly facilitating scar-free repair. Similarly, the functionalization of chitosan with this revitalized extracellular matrix has promise for diabetic foot ulcer healing (Mohr et al. 2024).

13.8.3 Implications for Diabetic Foot Care and Wound Management

13.8.3.1 Innovative Wound Dressings

Biomaterial dressings are being engineered to provide a moist environment that facilitates healing plus reduces the risk of infection. These may integrate antimicrobials as well as growth factors to improve healing in chronic diabetic foot ulcers. Bioactive dressings, including hydrogels, nanofibers, or collagens, are promising materials that help expedite the process of healing. Bioactive substances like nucleic acids may also be present in the materials used to dress wounds. Small non-coding RNA molecules known as microRNAs (miRNAs) are regulators of impaired wound healing and may be targeted for diabetic foot ulcer (DFU) treatment. MiRNA treatments can be delivered to the wound site via a variety of delivery methods, such as exosomes as well as nanoparticles. By continuously applying certain miRNAs topically, these dressings hold great potential for treating diabetic wounds. Contemporary or Interactive wound dressings provide a more personalized and improved approach to wound care than passive dressings. Interactive dressings interact with the wound

(Ma et al. 2024). They usually include hydrogels, foams, and silicone to aid healing. Interactive dressings reduce wound dryness and wetness by maintaining an ideal moisture balance. Their capacity to stimulate granulation tissue production supports blood vessel and epithelial cell proliferation, closing wounds. Interactive dressings also limit exudate saturation and leakage, reduce dressing changes, and reduce infection risk. Their soft adherence lessens discomfort and tension during removal and their transparency permits wound monitoring without disturbing the wound bed. The optimum uses for interactive dressings are extremely obvious chronic wounds, difficult patients, and situations where maximizing healing is essential (Jeirani Khamene et al. 2024).

13.8.3.2 Improved Healing Outcomes

Through exudate absorption and moisture preservation, biomaterials have the potential to dramatically reduce healing times and improve outcomes for diabetic foot ulcers. Exosome-infused smart biomaterials can serve as very effective wound dressings by prolonging the duration of exosome activity and delivering active compounds to the injured surrounding tissues of diabetic wounds. This significantly increases the rate of cell migration and proliferation in the vicinity of the wound. Intelligent Biomaterials Development to Improve Diabetes Management Xiong et al. engineered hydrogels containing HA that included manganese dioxide (MnO2) nanoenzymes that could break down endogenous reactive oxygen species (H2O2) into oxygen, allowing for the controlled release of fibroblast growth factor 2 (FGF-2) and M2 macrophage-derived exosomes (M2 Exos). By eradicating bacterial infection, decreasing oxidative stress, supplying oxygen, and encouraging angiogenesis and epithelialization, this HA-based hydrogel improved wound healing in diabetic mice (Xiong et al. 2022).

13.8.3.3 Reduction of Amputations

Global statistics indicate that the average lower limb amputation for a diabetic occurs every 20 s. An infection has surfaced in a diabetic foot ulcer when there are at least two of the typical symptoms or signs of inflammation (tenderness, warmth, redness, swelling, or pain) or purulent discharge, indicating the presence of an infection in a diabetic foot ulcer (pus). Foot problems now account for more hospital stays among diabetic patients than any other condition. The primary causT (94). An excellent wound dressing material absorbs wound exudates, aids gas exchange, and prevents microbiological infections. Additionally, they must support the synthesis of biochemical mediators necessary for wound healing, such as growth factors and cytokines. Bioactive substances from natural sources have been investigated to expedite wound healing in diabetics. Natural wound dressings have garnered attention for their ability to induce new tissue growth. Due to their mechanical, bioinert, and biocompatible properties, synthetic polymers are widely utilized to treat diabetic wounds. Polymeric biomaterials, both natural and synthetic, are highly effective wound dressings due to their biocompatibility, mechanical strength, and low immunogenicity. Due to their biocompatibility and healing properties, silk proteins, collagen, gelatin, and fibrin are increasingly employed in wound dressings. Gelatin is biodegradable yet weak, whereas collagen is essential for skin and wound healing. Fibrin scaffolds cell delivery and tissue regeneration, while silk proteins boost cell migration and proliferation, healing diabetic wounds. In conjunction with natural materials, synthetic polymers like polycaprolactone increase mechanical qualities (Moschiar Almeida et al. 2025).

13.9 Conclusion

The skin can sustain wounds in a variety of ways, including burns, tears, and more. The biological, dynamic, as well as time-consuming process of natural wound healing typically takes 8–12 weeks. In the past, skin wounds were treated with sterile gauze, cloth, and tree leaves. These materials not only prolonged the wound's healing period but also infected the wound's surface with bacteria and microorganisms. Alginate, collagen, chitosan, and silk fibroin are examples of natural biomaterials that have revealed significant promise in improving the healing of wounds, particularly for complicated ailments like diabetic foot ulcers (DFUs). By providing bioactivity, biodegradability, and biocompatibility, biomaterials like these aid in tissue repair and prevention of infection. Developments in formulations of biomaterial, such as hydrogels, nanofibers, and composite materials, have made it possible to manage wounds that are chronic more successfully by fostering cellular activity, creating a moist wound environment, and carefully administering therapeutic agents. Significant benefits, including increased durability, hydrophilicity, antibacterial characteristics, and reduced toxicity, are provided by biomaterial-based wound dressings in comparison to conventional ones. But for wider clinical applications, problems like scalability, raw material scarcity, and high synthesis costs need to be resolved. To make these treatments more widely available, future studies should concentrate on streamlining delivery methods, investigating cutting-edge technologies like 3D bioprinting and microfluidics, and cutting expenses. These biomaterials can be used by medical practitioners to enhance healing rates and lower the risk of complications such as infections that can result in amputations in cases of diabetic foot. Continuous innovation in formulations of biomaterial as well as delivery systems will be essential to transforming wound care management as medicine and technology advance. To provide customized solutions for various wound types, interdisciplinary research in the future should concentrate on tackling present issues and incorporating new technologies.

References

Abdulghani S, Mitchell G (2019) Biomaterials for in situ tissue regeneration: a review. Biomol Ther 9:750. https://doi.org/10.3390/biom9110750

Abka-khajouei R, Tounsi L, Shahabi N et al (2022) Structures, properties and applications of alginates. Mar Drugs 20:364. https://doi.org/10.3390/md20060364

Ahmad K, Shaikh S, Chun HJ et al (2023) Extracellular matrix: the critical contributor to skeletal muscle regeneration—a comprehensive review. Inflamm Regen 43:58. https://doi.org/10.1186/s41232-023-00308-z

Al Hajj W, Salla M, Krayem M et al (2024) Hydrolyzed collagen: exploring its applications in the food and beverage industries and assessing its impact on human health—a comprehensive review. Heliyon 10:e36433. https://doi.org/10.1016/j.heliyon.2024.e36433

Al Mamun A, Shao C, Geng P et al (2024) Recent advances in molecular mechanisms of skin wound healing and its treatments. Front Immunol 15. https://doi.org/10.3389/fimmu.2024.1395479

Aldahish A, Shanmugasundaram N, Vasudevan R et al (2024) Silk fibroin nanofibers: advancements in bioactive dressings through electrospinning Technology for Diabetic Wound Healing. Pharmaceuticals 17:1305. https://doi.org/10.3390/ph17101305

Ansari M, Darvishi A (2024) A review of the current state of natural biomaterials in wound healing applications. Front Bioeng Biotechnol 12. https://doi.org/10.3389/fbioe.2024.1309541

Apte A, Liechty KW, Zgheib C (2023) Immunomodulatory biomaterials on chemokine signaling in wound healing. Front Pharmacol 14. https://doi.org/10.3389/fphar.2023.1084948

Artemi G, Augello A, Sciandra F et al (2025) BPS2025 – novel 3D-printed wound healing transwell system for reproducible analysis of skin dressing. Biophys J 124:166a–167a. https://doi.org/10.1016/j.bpj.2024.11.907

Baig MS, Banu A, Zehravi M et al (2022) An overview of diabetic foot ulcers and associated problems with special emphasis on treatments with antimicrobials. Life 12:1054. https://doi.org/10.3390/life12071054

Bhagat V (2012) Gold nanorods: a promising drug delivery vehicle. Inven Rapid NDDS [ISSN 0976-3864] 1:1–4

Bhowmik DD, Deshmukh AY, Hatvate NT (2023) Quintessential impact of dendrimer bioconjugates in targeted drug delivery. In: Verma SK, Das AK (eds) Bioconjugated materials part 2 applications in drug delivery, vaccine formulations and important conjugates for cancer therapy. Elsevier, Amsterdam, pp 257–302

Borude SS, Mahajan AM (2011a) POEMS syndrome: a review. Pharmacologyonline 1:265–274

Borude SS, Mahajan AM (2011b) Recent updates on Q fever: diagnosis and treatment. Pharmacologyonline 1:254–264

Borude SS, Mahajan AM (2011c) Smart crystals technology: a review. Pharmacologyonline 1:238–243

Borude SS, Mahajan AM (2011d) Aquasomes: a potential drug delivery carrier. Pharmacologyonline 1:230–237

Bothiraja C, Gholap KS, Pawar AP (2014) Investigation of ethyl cellulose microsponge gel for topical delivery of eberconazole nitrate for fungal therapy. Ther Deliv 5:781–794. https://doi.org/10.4155/tde.14.43

Caldera-Villalobos M, Ramos-Montañez DG, Cabrera-Munguía DA et al (2025) Biocompatible hydrogels comprised of collagen, chitosan and polyurethane with potential applications for wound healing and controlled drug release. Polym Int 74:9–19. https://doi.org/10.1002/pi.6590

Chelu M, Musuc AM, Popa M, Calderon Moreno J (2023) Aloe vera-based hydrogels for wound healing: properties and therapeutic effects. Gels 9:539. https://doi.org/10.3390/gels9070539

Chen W, Liu K, Liao X et al (2024a) Harmonizing thickness and permeability in bone tissue engineering: a novel silk fibroin membrane inspired by spider silk dynamics. Adv Mater 36:e2310697. https://doi.org/10.1002/adma.202310697

Chen Z, Du C, Liu S et al (2024b) Progress in biomaterials inspired by the extracellular matrix. Giant 19:100323. https://doi.org/10.1016/j.giant.2024.100323

Čoma M, Fröhlichová L, Urban L et al (2021) Molecular changes underlying hypertrophic scarring following burns involve specific deregulations at all wound healing stages (inflammation, proliferation and maturation). Int J Mol Sci 22:897. https://doi.org/10.3390/ijms22020897

Da Silva J, Leal EC, Carvalho E, Silva EA (2023) Innovative functional biomaterials as therapeutic wound dressings for chronic diabetic foot ulcers. Int J Mol Sci 24:9900. https://doi.org/10.3390/ijms24129900

da Silva CM, Reis RL, Correlo VM, Jahno VD (2024) The efficient role of sodium alginate-based biodegradable dressings for skin wound healing application: a systematic review. J Biomater Sci Polym Ed 35:397–414. https://doi.org/10.1080/09205063.2023.2289247

Dattatray Tupe (2012) The threat of photoaging: clinical conditions, evaluation and treatment. Inven Rapid Cosmeceuticals [ISSN 0976-3864] 1:1–3

Davydova GA, Chaikov LL, Melnik NN et al (2024) Polysaccharide composite alginate–pectin hydrogels as a basis for developing wound healing materials. Polymers (Basel) 16:287. https://doi.org/10.3390/polym16020287

De Giorgio G, Matera B, Vurro D et al (2024) Silk fibroin materials: biomedical applications and perspectives. Bioengineering 11:167. https://doi.org/10.3390/bioengineering11020167

Delikanlı Kıyak B, İnan Çınkır N, Çelebi Y et al (2024) Advanced technologies for the collagen extraction from food waste—a review on recent progress. Microchem J 201:110404. https://doi.org/10.1016/j.microc.2024.110404

Dighe SB (2012a) Role of sebum in human health & acne. Inven Rapid Cosmeceuticals [ISSN 0976-3864] 1:1–3

Dighe SB (2012b) Dandruff: pathogenesis and treatment. Inven Rapid Cosmeceuticals [ISSN 0976-3864] 1:1–4

Dixit A, Mahajan A, Saxena R et al (2024) Engineering sulfated polysaccharides and silk fibroin based injectable IPN hydrogels with stiffening and growth factor presentation abilities for cartilage tissue engineering. Biomater Sci 12:2067–2085. https://doi.org/10.1039/D3BM01466E

Faiyazuddin M, Gholap AD, Maqsood S et al (2024) Prospects and ethics of ChatGPT in biomedical research and medical practice: an emphasis on nanotechnology. J Biomed Nanotechnol 20:1661–1678. https://doi.org/10.1166/jbn.2024.3925

Farooq S, Ahmad MI, Zheng S et al (2024) A review on marine collagen: sources, extraction methods, colloids properties, and food applications. Collagen Leather 6:11. https://doi.org/10.1186/s42825-024-00152-y

Gholap AD (2012) Translational nanomedicine: status assessment and opportunities. Int J Res Pharm Biomed Sci 4:1454–1463

Gholap AD, Said PR, Pawar RD, Ambore GS, Hatvate NT (2023a) Importance of carbohydrate-drug conjugates in vaccine development: a detailed review. In: Verma SK, Das AK (eds) Bioconjugated materials part 2 applications in drug delivery, vaccine formulations and important conjugates for cancer therapy. Elsevier, Amsterdam, pp 191–256

Gholap AD, Gupta SJ, Kamandar PA, Banchhod GV, Hatvate NT (2023b) Antibody-drug conjugates for cancer therapy: an up-to-date review on the chemistry and pharmacology. In: Verma SK, Das AK (eds) Bioconjugated materials part 2 applications in drug delivery, vaccine formulations and important conjugates for cancer therapy. Elsevier, Amsterdam, pp 105–190

Gholap AD, Hatvate NT, Khuspe PR et al (2023c) Proteomics in oncology: retrospect and prospects. In: Kulkarni S (ed) Novel technologies in biosystems, biomedical & drug delivery. Springer Nature Singapore, Singapore, pp 243–269

Gholap AD, Sawant SD, Sayyad SF et al (2023d) Food biopackaging for human benefits: status and perspectives. In: Ahmed S (ed) Biobased packaging materials. Springer Nature Singapore, Singapore, pp 1–35

Gholap AD, Sayyad SF, Hatvate NT et al (2023e) Drug delivery strategies for avobenzone: a case study of photostabilization. Pharmaceutics 15:1008. https://doi.org/10.3390/pharmaceutics15031008

Gholap AD, Gupta J, Kamandar P et al (2024a) Harnessing nanovaccines for effective immunization—a special concern on COVID-19: facts, Fidelity, and future prospective. ACS Biomater Sci Eng 10:271–297. https://doi.org/10.1021/acsbiomaterials.3c01247

Gholap AD, Khuspe PR, Pardeshi SR et al (2024b) Achieving optimal health with host-directed therapies (HDTs) in infectious diseases—a new horizon. Adv Ther 2400169. https://doi.org/10.1002/adtp.202400169

Gholap AD, Pardeshi SR, Hatvate NT et al (2024c) Environmental implications and nanotechnological advances in octocrylene-enriched sunscreen formulations: a comprehensive review. Chemosphere 358:142235. https://doi.org/10.1016/j.chemosphere.2024.142235

Gholap AD, Uddin MJ, Faiyazuddin M et al (2024d) Advances in artificial intelligence for drug delivery and development: a comprehensive review. Comput Biol Med 178:108702. https://doi.org/10.1016/j.compbiomed.2024.108702

Goh M, Du M, Peng WR et al (2024) Advancing burn wound treatment: exploring hydrogel as a transdermal drug delivery system. Drug Deliv 31. https://doi.org/10.1080/10717544.2023.2300945

Golebiowska AA, Intravaia JT, Sathe VM et al (2024) Decellularized extracellular matrix biomaterials for regenerative therapies: advances, challenges and clinical prospects. Bioact Mater 32:98–123. https://doi.org/10.1016/j.bioactmat.2023.09.017

Guo B, Dong R, Liang Y, Li M (2021) Haemostatic materials for wound healing applications. Nat Rev Chem 5:773–791. https://doi.org/10.1038/s41570-021-00323-z

Guo Y, Qiao D, Zhao S et al (2024) Biofunctional chitosan–biopolymer composites for biomedical applications. Mater Sci Eng R Rep 159:100775. https://doi.org/10.1016/j.mser.2024.100775

Guptha PM, Kanoujia J, Kishore A et al (2024) A comprehensive review of the application of 3D-bioprinting in chronic wound management. Expert Opin Drug Deliv 21:1573–1594. https://doi.org/10.1080/17425247.2024.2355184

Hargis A, Yaghi M, Bermudez NM, Gefen A (2024) Foam dressings for wound healing. Curr Dermatol Rep 13:28–35. https://doi.org/10.1007/s13671-024-00422-2

Hatvate NT, Sawant SD, Khemka PN, Vora LK (2023) Unlocking the potential of nanobiohybrids to combat environmental pollution. In: Lens P (ed) Nanobiohybrids for advanced wastewater treatment and energy recovery. IWA Publishing, pp 169–190

Hatvate NT, Khuspe PR, Mandhare TA et al (2024) Decoding of Nanoneutraceuticals and their delivery. In: Mohanta YK (ed) Sustainable green nanotechnology. CRC Press, New York, pp 284–309

Hong Y-K, Chang Y-H, Lin Y-C et al (2023) Inflammation in wound healing and pathological scarring. Adv Wound Care 12:288–300. https://doi.org/10.1089/wound.2021.0161

Hosty L, Heatherington T, Quondamatteo F, Browne S (2024) Extracellular matrix-inspired biomaterials for wound healing. Mol Biol Rep 51:830. https://doi.org/10.1007/s11033-024-09750-9

Houghton JSM, Saratzis AN, Sayers RD, Haunton VJ (2024) New horizons in peripheral artery disease. Age Ageing 53. https://doi.org/10.1093/ageing/afae114

Hsu L, Li L, Poon LY (2024) Analysis of risk factors of infection in diabetic foot patients. Int Wound J 21:e14411. https://doi.org/10.1111/iwj.14411

Huang Q, Zeng Y, Qiu Y et al (2024) Applications and prospects of new transdermal drug delivery system based on metal-organic frameworks for diabetic wound healing. Dyes Pigments 222:111865. https://doi.org/10.1016/j.dyepig.2023.111865

Iqbal Y, Amin F, Usman Y, Farrukh Sarfraz M (2024) Alginate-based hydrogels with inorganic nanomaterials: a promising approach for wound healing and bone tissue regeneration. Eur Polym J 212:113057. https://doi.org/10.1016/j.eurpolymj.2024.113057

Jadach B, Mielcarek Z, Osmałek T (2024) Use of collagen in cosmetic products. Curr Issues Mol Biol 46:2043–2070. https://doi.org/10.3390/cimb46030132

Jayaprakash S, Mohamad Abdul Razeen Z, Naveen Kumar R et al (2024) Enriched characteristics of poultry collagen over other sources of collagen and its extraction methods: a review. Int J Biol Macromol 273:133004. https://doi.org/10.1016/j.ijbiomac.2024.133004

Jeirani Khamene A, Yazdian F, Pourmadadi M et al (2024) Fabrication and characterization of an innovative g-C3N4/calcium/aloe vera enriched PVA-bacterial cellulose wound dressing: a novel approach to diabetic wound management. Inorg Chem Commun 168:112879. https://doi.org/10.1016/j.inoche.2024.112879

Jiang P, Li Q, Luo Y et al (2023) Current status and progress in research on dressing management for diabetic foot ulcer. Front Endocrinol (Lausanne) 14. https://doi.org/10.3389/fendo.2023.1221705

Jin S, Newton MAA, Cheng H et al (2023) Progress of hydrogel dressings with wound monitoring and treatment functions. Gels 9:694. https://doi.org/10.3390/gels9090694

Josephin A, Intan N, Basari B, Rahman SF (2024) Extraction of collagen and hydroxyapatite from fish for bone scaffold: a review. AIP Conf Proc 3080:090006

Jyoti TN, Rudrakshi C, Prabhuji MLV (2024) A new horizon in biocompatible membrane: the role of collagen and silk fibroin in tissue regeneration-case series. J Biol Med 8:018–024. https://doi.org/10.17352/jbm.000043

Kapare HS, Pagar S, Kamandar P et al (2024) Exploring modified chitosan-based gene delivery technologies for therapeutic advancements. Int J Biol Macromol 260:129581. https://doi.org/10.1016/j.ijbiomac.2024.129581

Khatoon F, Narula AK, Sehgal P (2024) Efficacy of collagen based biomaterials in diabetic foot ulcer wound healing. Eur Polym J 217:113345. https://doi.org/10.1016/j.eurpolymj.2024.113345

Kim J, Lee K-M, Han SH et al (2021) Development of stabilized dual growth factor-loaded hyaluronate collagen dressing matrix. J Tissue Eng 12. https://doi.org/10.1177/2041731421999750

Kolimi P, Narala S, Nyavanandi D et al (2022) Innovative treatment strategies to accelerate wound healing: trajectory and recent advancements. Cells 11:2439. https://doi.org/10.3390/cells11152439

Kolipaka T, Pandey G, Abraham N et al (2024) Stimuli-responsive polysaccharide-based smart hydrogels for diabetic wound healing: design aspects, preparation methods and regulatory perspectives. Carbohydr Polym 324:121537. https://doi.org/10.1016/j.carbpol.2023.121537

Kumar D, Kumar S, Gholap AD et al (2024) A comprehensive review on the biomass-mediated synthesis of silver nanoparticles: opportunities and challenges. J Environ Chem Eng 13:115133. https://doi.org/10.1016/j.jece.2024.115133

Leelathanalerk A, Piankarnka N, Srimongkon P et al (2024) Quality of life and utility of patients with peripheral artery disease (PAD): a systematic review and meta-analysis. Sci Rep 14:28823. https://doi.org/10.1038/s41598-024-80387-1

Li J, Xie Y, Liu G et al (2024a) Bioelectret materials and their bioelectric effects for tissue repair: a review. ACS Appl Mater Interfaces 16:38852–38879. https://doi.org/10.1021/acsami.4c07808

Li Q-Q, Xu D, Dong Q-W et al (2024b) Biomedical potentials of alginate via physical, chemical, and biological modifications. Int J Biol Macromol 277:134409. https://doi.org/10.1016/j.ijbiomac.2024.134409

Li Z, Tan G, Xie H, Lu S (2024c) The application of regenerated silk fibroin in tissue repair. Materials (Basel) 17:3924. https://doi.org/10.3390/ma17163924

Li Z, Xu S, Xu Z et al (2024d) Enhancing cellular behavior in repaired tissue via silk fibroin-integrated triboelectric nanogenerators. Microsyst Nanoeng 10:68. https://doi.org/10.1038/s41378-024-00694-5

Liang C-W, Cheng H-Y, Lee Y-H et al (2024a) Efficacy and safety of collagen derivatives for osteoarthritis: a trial sequential meta-analysis. Osteoarthr Cartil 32:574–584. https://doi.org/10.1016/j.joca.2023.12.010

Liang X, Huang C, Liu H et al (2024b) Natural hydrogel dressings in wound care: design, advances, and perspectives. Chin Chem Lett 35:109442. https://doi.org/10.1016/j.cclet.2023.109442

Lin M-J, Lu M-C, Chan Y-C et al (2021) An insulin-like growth Factor-1 conjugated Bombyx mori silk fibroin film for diabetic wound healing: fabrication, physicochemical property characterization, and dosage optimization in vitro and in vivo. Pharmaceutics 13:1459. https://doi.org/10.3390/pharmaceutics13091459

Lin D, Li M, Wang L et al (2024) Multifunctional hydrogel based on silk fibroin promotes tissue repair and regeneration. Adv Funct Mater 34. https://doi.org/10.1002/adfm.202405255

Lingyan L, Liwei X, Han Z et al (2024) Identification, influencing factors and outcomes of time delays in the management pathway of diabetic foot: a systematic review. J Tissue Viability 33:345–354. https://doi.org/10.1016/j.jtv.2024.04.007

Liu C, Pei M, Li Q, Zhang Y (2022) Decellularized extracellular matrix mediates tissue construction and regeneration. Front Med 16:56–82. https://doi.org/10.1007/s11684-021-0900-3

Liu W, Zu L, Wang S et al (2023) Tailored biomedical materials for wound healing. Burn Trauma 11. https://doi.org/10.1093/burnst/tkad040

López-Cervantes J, Escárcega-Galaz AA, Sánchez-Machado DI et al (2019) Characterization and efficacy of chitosan membranes in the treatment of skin ulcers. Egypt J Basic Appl Sci 6:195–205. https://doi.org/10.1080/2314808X.2019.1694734

Luo R, Dai J, Zhang J, Li Z (2021) Accelerated skin wound healing by electrical stimulation. Adv Healthc Mater 10. https://doi.org/10.1002/adhm.202100557

Luo Z, Bian F, Cao X et al (2024) Bio-inspired self-adhesive microparticles with melanin nanoparticles integration for wound closure and healing. Nano Today 54:102138. https://doi.org/10.1016/j.nantod.2023.102138

Ma J, Fang Y, Yu H et al (2024) Recent advances in living algae seeding wound dressing: focusing on diabetic chronic wound healing. Adv Funct Mater 34. https://doi.org/10.1002/adfm.202308387

Mawazi SM, Kumar M, Ahmad N et al (2024) Recent applications of chitosan and its derivatives in antibacterial, anticancer, wound healing, and tissue engineering fields. Polymers (Basel) 16:1351. https://doi.org/10.3390/polym16101351

Mesut B, Al-Mohaya M, Yeşilkaya E et al (2024) Demystifying the potential of lipid-based nanocarriers in targeting brain malignancies. Naunyn Schmiedeberg's Arch Pharmacol 397:9243–9279. https://doi.org/10.1007/s00210-024-03212-6

Mohr AE, Ortega-Santos CP, Whisner CM et al (2024) Navigating challenges and opportunities in multi-omics integration for personalized healthcare. Biomedicines 12:1496. https://doi.org/10.3390/biomedicines12071496

Mohsin F, Javaid S, Tariq M, Mustafa M (2024) Molecular immunological mechanisms of impaired wound healing in diabetic foot ulcers (DFU), current therapeutic strategies and future directions. Int Immunopharmacol 139:112713. https://doi.org/10.1016/j.intimp.2024.112713

Moschiar Almeida B, Evans R, Kayssi A (2025) Fundamentals of wound care for amputation prevention. Semin Vasc Surg 38:54–63. https://doi.org/10.1053/j.semvascsurg.2025.01.001

Mottaghitalab F, Khodadadi Yazdi M, Reza Saeb M et al (2024) Green and sustainable hydrogels based on quaternized chitosan to enhance wound healing. Chem Eng J 492:152288. https://doi.org/10.1016/j.cej.2024.152288

Naghib SM, Amiri S, Mozafari MR (2024) Stimuli-responsive chitosan-based nanocarriers for drug delivery in wound dressing applications: a review. Carbohydr Polym Technol Appl 7:100497. https://doi.org/10.1016/j.carpta.2024.100497

Nimmy S, Ibrahim SA, Nasr MHA (2024) Manipulating the rhizosphere microbiome for plant health. In: Microbiome engineering, 1st edn. CRC Press, Florida, p 19

Ottaviano L, Buoso S, Zamboni R et al (2025) Natural protein films from textile waste for wound healing and wound dressing applications. J Funct Biomater 16:20. https://doi.org/10.3390/jfb16010020

Pardeshi S, Giram P (2023) Preparation, antibacterial and antiviral activity measurements and detection methods. In: Deshmukh K (ed) Antibacterial and antiviral functional materials. ACS Publications, pp 33–64

Pardeshi S, More M, Togre N, Rebello N, Giram P (2023) Dendrimers based antibacterial and antiviral materials. In: Antibacterial and antiviral functional materials. ACS Publications, pp 139–169

Pardeshi S, Gholap A, Kapare H et al (2024a) Coating approaches of functionalized magnetic nanoparticles for Theranostic applications. In: Functionalized magnetic nanoparticles for theranostic applications. Wiley, pp 131–171

Pardeshi S, Pillai D, Giram P et al (2024b) Nanocarriers for ocular drug delivery. In: Nanocarrier drug delivery systems. De Gruyter, Germany, pp 469–500

Pardeshi SR, Gholap AD, Hatvate NT et al (2024c) Advances in dorzolamide hydrochloride delivery: harnessing nanotechnology for enhanced ocular drug delivery in glaucoma management. Discov Nano 19:199. https://doi.org/10.1186/s11671-024-04154-x

Parit S, Manchare A, Mundhe P et al (2024) Antibody-drug conjugates: a promising breakthrough in cancer therapy. Int J Pharm 659:124211. https://doi.org/10.1016/j.ijpharm.2024.124211

Park KH, Kwon JB, Park JH et al (2019) Collagen dressing in the treatment of diabetic foot ulcer: a prospective, randomized, placebo-controlled, single-center study. Diabetes Res Clin Pract 156:107861. https://doi.org/10.1016/j.diabres.2019.107861

Patil M, Dhumal D, Rojekar S et al (2024) Nanocarriers for antimicrobial therapy. In: Nanocarrier drug delivery systems. De Gruyter, Germany, pp 561–602

Peña OA, Martin P (2024) Cellular and molecular mechanisms of skin wound healing. Nat Rev Mol Cell Biol 25:599–616. https://doi.org/10.1038/s41580-024-00715-1

Perez-Favila A, Martinez-Fierro ML, Rodriguez-Lazalde JG et al (2019) Current therapeutic strategies in diabetic foot ulcers. Medicina (B Aires) 55:714. https://doi.org/10.3390/medicina55110714

Pires PC, Damiri F, Zare EN et al (2024) A review on natural biopolymers in external drug delivery systems for wound healing and atopic dermatitis. Int J Biol Macromol 263:130296. https://doi.org/10.1016/j.ijbiomac.2024.130296

Qi X, Ge X, Chen X et al (2024a) An Immunoregulation hydrogel with controlled hyperthermia-augmented oxygenation and ROS scavenging for treating diabetic foot ulcers. Adv Funct Mater 34. https://doi.org/10.1002/adfm.202400489

Qi X, Shi Y, Zhang C et al (2024b) A hybrid hydrogel with intrinsic immunomodulatory functionality for treating multidrug-resistant Pseudomonas aeruginosa infected diabetic foot ulcers. ACS Mater Lett 6:2533–2547. https://doi.org/10.1021/acsmaterialslett.4c00392

Raina N, Rani R, Pahwa R, Gupta M (2022) Biopolymers and treatment strategies for wound healing: an insight view. Int J Polym Mater Polym Biomater 71:359–375. https://doi.org/10.1080/00914037.2020.1838518

Rajurkar VG (2012a) Potential applications of Superparmagnetic nanoparticles technology. Inven Rapid NDDS [ISSN 0976-3791] 1:1–3

Rajurkar VG (2012b) Recent technologies in Nanomedicine for treatment of cancer. Inven Rapid Pharm Tech 4:1–3

Rajurkar VG (2012c) Potential applications of capillary Electromigration methods for the improvement of separation efficiency. Inven Rapid Pharm Anal Qual Assur [ISSN 0976-3813] 1:1–3

Rajurkar VG (2012d) Lipid Nanocapsules: a novel promising tool as drug delivery system. Inven Rapid NDDS [ISSN 0976-3791] 1:1–4

Rajurkar VG (2012e) New threat of clinical toxicities evoked from Nanocarrier systems. Inven Rapid NDDS [ISSN 0976-3791] 1:1–4

Ramos-Rodriguez DH, Leach JK (2025) Decellularized cell-secreted extracellular matrices as biomaterials for tissue engineering. Small Sci 5. https://doi.org/10.1002/smsc.202400335

Rathna RP, Kulandhaivel M (2024) Advancements in wound healing: integrating biomolecules, drug delivery carriers, and targeted therapeutics for enhanced tissue repair. Arch Microbiol 206:199. https://doi.org/10.1007/s00203-024-03910-y

Raziyeva K, Kim Y, Zharkinbekov Z et al (2021) Immunology of acute and chronic wound healing. Biomol Ther 11:700. https://doi.org/10.3390/biom11050700

Riha SM, Maarof M, Fauzi MB (2021) Synergistic effect of biomaterial and stem cell for skin tissue engineering in cutaneous wound healing: a concise review. Polymers (Basel) 13:1546. https://doi.org/10.3390/polym13101546

Rojekar S, Gholap AD, Togre N et al (2024a) Current status of mannose receptor-targeted drug delivery for improved anti-HIV therapy. J Control Release 372:494–521. https://doi.org/10.1016/j.jconrel.2024.06.002

Rojekar S, Kapare HS, Vishwakarma N et al (2024b) Chitosan scaffolds: expanding horizons in biomedical applications. Carbohydr Polym 323:121394. https://doi.org/10.1016/j.carbpol.2023.121394

Rojekar S, Parit S, Gholap AD et al (2024c) Revolutionizing eye care: exploring the potential of microneedle drug delivery. Pharmaceutics 16:1398. https://doi.org/10.3390/pharmaceutics16111398

Sabarees G, Vishvaja S, Raghuraman S et al (2024) Collagen-based nanofibers: revolutionizing therapeutics for impaired wound healing. Int J Polym Mater Polym Biomater:1–29. https://doi.org/10.1080/00914037.2024.2402364

Saberian M, Safari Roudsari R, Haghshenas N et al (2024) How the combination of alginate and chitosan can fabricate a hydrogel with favorable properties for wound healing. Heliyon 10:e32040. https://doi.org/10.1016/j.heliyon.2024.e32040

Safina I, Embree MC (2022) Biomaterials for recruiting and activating endogenous stem cells in situ tissue regeneration. Acta Biomater 143:26–38. https://doi.org/10.1016/j.actbio.2022.03.014

Sahana TG, Rekha PD (2018) Biopolymers: applications in wound healing and skin tissue engineering. Mol Biol Rep 45:2857–2867. https://doi.org/10.1007/s11033-018-4296-3

Sang F, Liu C, Yan J et al (2024) Polysaccharide- and protein-based hydrogel dressings that enhance wound healing: a review. Int J Biol Macromol 280:135482. https://doi.org/10.1016/j.ijbiomac.2024.135482

Sawant SD, Pardeshi SR, Khuspe PR et al (2024a) The role of inulin in altering the animal microbiome: a comprehensive review. Microbiome Eng 1:79–102. https://doi.org/10.1201/9781003394662

Sawant SD, Pardeshi SR, Khuspe PR et al (2024b) Omics tools for determining microbial composition: a systematic review. Microbiome Eng 1:35–50. https://doi.org/10.1201/9781003394662

Shaikh MS, Faiyazuddin M, Khan MS et al (2024) Chikungunya virus vaccine: a decade of progress solving epidemiological dilemma, emerging concepts, and immunological interventions. Front Microbiol 15. https://doi.org/10.3389/fmicb.2024.1413250

Shen C, Zhou Z, Li R et al (2025) Silk fibroin-based hydrogels for cartilage organoids in osteoarthritis treatment. Theranostics 15:560–584. https://doi.org/10.7150/thno.103491

Shi S, Wang L, Song C et al (2023) Recent progresses of collagen dressings for chronic skin wound healing. Collagen Leather 5:31. https://doi.org/10.1186/s42825-023-00136-4

Smita SS, Pramanik K (2025) Development of Silk Fibroin/Gelatin/PCL tri-polymeric complex nanofibrous two-dimensional mat for epithelial tissue regeneration. Chem Eng J 503:158251. https://doi.org/10.1016/j.cej.2024.158251

Spielman AF, Griffin MF, Parker J et al (2023) Beyond the scar: a basic science review of wound remodeling. Adv Wound Care 12:57–67. https://doi.org/10.1089/wound.2022.0049

Strand N, Anderson MA, Attanti S et al (2024) Diabetic neuropathy: pathophysiology review. Curr Pain Headache Rep 28:481–487. https://doi.org/10.1007/s11916-024-01243-5

Swaminathan N, Awuah WA, Bharadwaj HR et al (2024) Early intervention and care for diabetic foot ulcers in low and middle income countries: addressing challenges and exploring future strategies: a narrative review. Health Sci Rep 7. https://doi.org/10.1002/hsr2.2075

Tang L, Xie S, Wang D et al (2025) Astragalus polysaccharide/carboxymethyl chitosan/sodium alginate based electroconductive hydrogels for diabetic wound healing and muscle function assessment. Carbohydr Polym 350:123058. https://doi.org/10.1016/j.carbpol.2024.123058

Tupe AKD (2012a) Cubosomes: emerging drug delivery system. Inven Rapid NDDS [ISSN 0976-3791] 1:1–3

Tupe AKD (2012b) Pharmacosomes: an emerging vesicular system. Inven Rapid NDDS [ISSN 0976-3791] 1:1–3

Tupe D (2012c) Potential applications of botanical extract. Inven Rapid Ethnopharmacol [ISSN 0976-3805] 1:1–4

Tupe D (2012d) Artificial viral envelopes: emerging vehicle for gene delivery. Pharmatutor-Art-1439 1:1–3

Tushar V Magar DSM (2011a) Advances in the field of nanooncology: an overview. Inven Impact NDDS [ISSN 0976-3791] 4:1–3

Tushar V Magar DSM (2011b) Elastic vesicles as drugs carriers through the skin. Inven Rapid NDDS [ISSN 0976-3791] 1:1–2

Vardikar A, Das U, Mandal S et al (2025) Surface engineered multimodal magnetic nanoparticles for neurodegenerative diseases. In: Targeted therapy for the central nervous system. Elsevier, USA, pp 121–153

Vikas G Rajurkar SBD (2012) Phonophoresis: ultrasound enhanced transdermal drug delivery. Inven Rapid NDDS [ISSN 0976-3791] 1:1–3

Wang Z, Qi F, Luo H et al (2022) Inflammatory microenvironment of skin wounds. Front Immunol 13. https://doi.org/10.3389/fimmu.2022.789274

Wang Y, Vizely K, Li CY et al (2024a) Biomaterials for immunomodulation in wound healing. Regen Biomater 11. https://doi.org/10.1093/rb/rbae032

Wang Y, Wu H, Xiao A et al (2024b) Combined amniotic membrane and self-powered electrical stimulator bioelectronic dress promotes wound healing. ACS Appl Mater Interfaces 16:15809–15818. https://doi.org/10.1021/acsami.3c18547

Wolf SJ, Melvin WJ, Gallagher K (2021) Macrophage-mediated inflammation in diabetic wound repair. Semin Cell Dev Biol 119:111–118. https://doi.org/10.1016/j.semcdb.2021.06.013

Wong K, Tan XH, Li J et al (2024) An in vitro macrophage response study of silk fibroin and silk fibroin/Nano-hydroxyapatite scaffolds for tissue regeneration application. ACS Biomater Sci Eng 10:7073–7085. https://doi.org/10.1021/acsbiomaterials.4c00976

Xiao H, Chen X, Liu X et al (2023) Recent advances in decellularized biomaterials for wound healing. Mater Today Bio 19:100589. https://doi.org/10.1016/j.mtbio.2023.100589

Xiong Y, Chen L, Liu P et al (2022) All-in-one: multifunctional hydrogel accelerates oxidative diabetic wound healing through timed-release of exosome and fibroblast growth factor. Small 18:e2104229. https://doi.org/10.1002/smll.202104229

Xiong Y, Feng Q, Lu L et al (2023) Immunomodulatory hydrogels: advanced regenerative tools for diabetic foot ulcer. Adv Funct Mater 33. https://doi.org/10.1002/adfm.202213066

Yang Y, Zhao X, Yu J et al (2020) H-bonding supramolecular hydrogels with promising mechanical strength and shape memory properties for postoperative antiadhesion application. ACS Appl Mater Interfaces 12:34161–34169. https://doi.org/10.1021/acsami.0c07753

Yang L, Rong G-C, Wu Q-N (2022a) Diabetic foot ulcer: challenges and future. World J Diabetes 13:1014–1034. https://doi.org/10.4239/wjd.v13.i12.1014

Yang Y, Du Y, Zhang J et al (2022b) Structural and functional design of electrospun nanofibers for hemostasis and wound healing. Adv Fiber Mater 4:1027–1057. https://doi.org/10.1007/s42765-022-00178-z

Yang A-L, Sun S-B, Qu L-Y et al (2024a) Polysaccharide hydrogel containing silver nanoparticle@catechol microspheres with photothermal, antibacterial and anti-inflammatory activities for infected-wounds repair. Int J Biol Macromol 265:130898. https://doi.org/10.1016/j.ijbiomac.2024.130898

Yang S, Hu L, Zhao Y et al (2024b) Prevalence of multidrug-resistant bacterial infections in diabetic foot ulcers: a meta-analysis. Int Wound J 21:e14864. https://doi.org/10.1111/iwj.14864

Yap MH, Cassidy B, Byra M et al (2024) Diabetic foot ulcers segmentation challenge report: benchmark and analysis. Med Image Anal 94:103153. https://doi.org/10.1016/j.media.2024.103153

Zhang Y, Li J, Habibovic P (2022) Magnetically responsive nanofibrous ceramic scaffolds for on-demand motion and drug delivery. Bioact Mater 15:372–381. https://doi.org/10.1016/j.bioactmat.2022.02.028

Zhang W, Liu L, Cheng H et al (2024a) Hydrogel-based dressings designed to facilitate wound healing. Mater Adv 5:1364–1394. https://doi.org/10.1039/D3MA00682D

Zhang Y, Li M, Chang J et al (2024b) Silk fibroin–gelatine haemostatic sponge loaded with thrombin for wound haemostasis and tissue regeneration. Burn Trauma 12. https://doi.org/10.1093/burnst/tkae026

Zhao J, Liu J, Hu Y et al (2024) Research advances in hydrogel-based wound dressings for diabetic foot ulcer treatment: a review. J Mater Sci 59:8059–8084. https://doi.org/10.1007/s10853-024-09493-9

Zheng Y, Pan C, Xu P, Liu K (2024) Hydrogel-mediated extracellular vesicles for enhanced wound healing: the latest progress, and their prospects for 3D bioprinting. J Nanobiotechnol 22:57. https://doi.org/10.1186/s12951-024-02315-9

Zhu J, Hu Z, Luo Y et al (2024) Diabetic peripheral neuropathy: pathogenetic mechanisms and treatment. Front Endocrinol (Lausanne) 14. https://doi.org/10.3389/fendo.2023.1265372

Zhu J, Du Y, Backman LJ et al (2025a) Cellular interactions and biological effects of silk fibroin: implications for tissue engineering and regenerative medicine. Small 21:e2409739. https://doi.org/10.1002/smll.202409739

Zhu S, Zhang Q, Xu X et al (2025b) Recent advances in silk fibroin-based composites for bone repair applications: a review. Polymers (Basel) 17:772. https://doi.org/10.3390/polym17060772

Life Cycle Assessment of Biomaterials: Towards Sustainable Production and Consumption

14

Sonia Gupta, Dinesh Kumar, Monika Sharma,
and Rajni Tanwar

Abstract

Life Cycle Assessment (LCA) is a critical analytical methodology for assessing the environmental performance of biomaterials across their entire life cycle, encompassing production, usage, and end-of-life stages. This chapter investigates LCA as a holistic tool for pinpointing environmental hotspots and opportunities to optimize energy efficiency, resource utilization, and waste management within biomaterial systems. Although LCA faces challenges such as data constraints, regional variability, and methodological intricacies, it remains instrumental in advancing sustainable biomaterial production, guiding policy development, and facilitating market integration. The integration of LCA into biomaterial innovation promotes a transition toward circular economy principles by emphasizing material reuse, recycling, and valorization. By ensuring transparency in communicating LCA outcomes, stakeholder trust is strengthened, encouraging widespread adoption of biomaterials across industries. This chapter highlights the surging demand for sustainable alternatives and underscores the necessity of refining LCA methodologies, enriching data repositories, and addressing associated trade-offs to unlock the full potential of biomaterials in contributing to global sustainability objectives.

Keywords

Life cycle assessment · Biomaterials · Sustainable production · Circular economy · Environmental impacts · Resource efficiency · Waste management · Methodological complexities · Policy integration · Market adoption

S. Gupta (✉) · M. Sharma
Swami Devi Dyal Institute of Pharmacy, Golpura Barwala, Panchkula, Haryana, India

D. Kumar · R. Tanwar
School of Pharmacy, Desh Bhagat University, Mandi Gobindgarh, Punjab, India

Md. Faiyazuddin, M. Suryawanshi (eds.), *Design and Processing of Green Materials*, Biomaterials, Bioengineering and Sustainability 4,
https://doi.org/10.1007/978-3-031-91790-5_14

14.1 Introduction

The escalating challenges of environmental sustainability, climate change, and resource depletion have necessitated the exploration of sustainable alternatives to conventional fossil-derived materials. Biomaterials, derived from renewable biological resources such as plants, algae, and agricultural waste, have emerged as promising substitutes for petrochemical-based materials. These materials, including bioplastics, biopolymers, and bio-composites, have demonstrated utility across diverse industries such as packaging, construction, healthcare, and automotive, reflecting their adaptability and potential to drive a transition toward a sustainable future. A key tool for evaluating the environmental impact of biomaterials is LCA, a systematic framework that assesses the environmental implications of a product or process across its entire life cycle, from raw material extraction to end-of-life disposal or recycling. LCA provides essential insights into the advantages and limitations of biomaterials, offering a comprehensive perspective on their role in sustainable production and consumption. The pressing environmental issues, alongside the adverse ecological impacts of synthetic materials—manifested as pollution of air, water, and land—highlight the urgency of adopting new strategies for environmental conservation. Addressing these challenges necessitates the development and implementation of innovative, sustainable materials to mitigate life-threatening environmental risks (Geyer et al. 2017).

In recent decades, heightened environmental awareness has spurred significant interest among researchers in bio-based materials, including biofibers, biopolymers, and bio-composites, which are increasingly recognized for their potential to replace synthetic materials. Bio-based materials are derived from renewable biological resources, with their building blocks sourced from living organisms (Wu et al. 2020). Among these, natural fiber-reinforced biopolymers, often termed "green composites," have gained attention as environmentally degradable materials capable of breaking down under natural conditions such as exposure to air, light, heat, or microorganisms. Although natural fibers generally exhibit lower stiffness and strength compared to synthetic counterparts, they offer numerous advantages, including biodegradability, accessibility, non-toxicity, combustibility, and cost-effectiveness (Mazerolles et al. 2019). Classified based on their origin as plant, animal, or mineral fibers, these materials are widely employed as reinforcements in polymer matrices to create bio-based and polymer composites tailored for specific applications. Researchers have extensively documented the favorable properties of natural fibers, such as low density, ease of processing, and sustainability, prompting their use as substitutes for synthetic materials. Industrial adoption of bio-based materials has accelerated, with applications ranging from aerospace components—such as wing boxes, pilot control panels, and thermal insulators—to consumer goods, including furniture, railway interiors, gardening products, packaging, construction materials, and sports equipment. Notably, companies like Boeing have pioneered efforts to recycle biomaterials from decommissioned aircraft, exemplifying the innovative use of these materials. This growing integration of bio-based materials across industries underscores their transformative potential in advancing environmental sustainability (Dunlap et al. 2024).

Objectives and Scope of LCA Are

(i) **Identify Environmental Impacts**: Evaluate the potential environmental impacts associated with all stages of a product's life cycle, from raw material extraction to disposal.
(ii) **Support Decision-Making**: Provide insights to help organizations and policymakers make informed decisions to minimize environmental impacts.
(iii) **Improve Product Sustainability**: Identify opportunities to improve the environmental performance of products or processes.
(iv) **Compare Alternatives**: Assess and compare the environmental footprints of different products, systems, or services.
(v) **Enhance Transparency:** Foster accountability by documenting and communicating the environmental impacts throughout the product life cycle.

A well-structured LCA framework ensures a consistent study and the attainment of the desired objectives. It provides the context for subsequent assessment steps, enabling a clear and comprehensive outcome aligned with the study's purpose. LCA is founded on systems thinking, which examines all interconnected processes as a whole rather than in isolation. This holistic approach helps prevent unintended consequences, such as improving one stage of a product's life cycle at the expense of another. Over the years, LCA methodologies have advanced, guided by internationally recognized standards like ISO 14040 and ISO 14044, solidifying its role as a globally accepted tool for environmental assessment (Snowdon et al. 2017a; Taherymoosavi et al. 2016).

14.1.1 Key Phases of LCA

14.1.1.1 Life Cycle Inventory (LCI)

The LCI phase of a LCA involves compiling a detailed inventory of various flows within the defined system boundaries that based on the functional unit. The LCI phase is often the most time-consuming and labor-intensive part of an LCA, as the accuracy and reliability of data are essential for producing representative and meaningful results. The LCI flows include emissions to air, energy consumption, raw material, water, and land use. The final quantification of inputs and outputs within these boundaries, known as the life cycle inventory. Developing a robust LCI requires extensive data collection. To facilitate this process, literature values and established LCI databases are frequently used, including resources such as the U.S. National Renewable Energy Laboratory (NREL) database, the Ecoinvent database, and the European Reference Life Cycle Database (ELCD) (Deng 2014).

14.1.1.2 The Life Cycle Impact Assessment (LCIA)

The LCIA phase is a crucial part of LCA that assesses the potential environmental impacts of the inputs and outputs identified during the LCI phase. This phase involves classifying and characterizing data to associate inventory flows with specific environmental impact categories, effectively translating raw data into meaningful environmental indicators. LCIA typically includes the following steps:

(i) **Classification**: Assigning inventory data to relevant environmental impact categories, such as climate change, ozone depletion, acidification, eutrophication, and human toxicity.
(ii) **Characterization**: Quantifying the contributions of inventory flows to each impact category using scientifically validated models and characterization factors. For example, greenhouse gas emissions are converted into carbon dioxide equivalents for the climate change category.
(iii) **Normalization and Weighting**: Normalizing results to enable comparison across categories and applying weighting factors to prioritize impacts based on their relative importance.

This phase aids to identify significant environmental hotspots within a product's life cycle moreover, offering a scientific foundation for enhancing environmental performance. However, LCIA results have limitations, such as uncertainties in characterization models and the exclusion of certain environmental or social impacts beyond the selected categories. Consequently, the findings should be interpreted in the context of the study's overall scope and objectives.

Various LCIA methodologies, including CML 2001, TRACI, and ReCiPe, translate inventory data into standardized equivalence units, which are then aggregated to determine overall environmental impact. Most LCA studies adopt midpoint modeling due to its strong scientific foundation. This study applies the TRACI methodology, which covers a wide range of impact categories addressing critical environmental concerns (Falano et al. 2014; Faruk et al. 2014).

LCIA methodologies are generally classified into two approaches: midpoint and endpoint.

(i) **Midpoint methods**: This method categorize environmental flows into specific impact categories, such as climate change, acidification, and eutrophication that represent distinct environmental processes. These categories are independent and are not meant to be compared or weighted against each other.
(ii) **Endpoint methods**: This method consolidate midpoint indicators into broader categories of environmental damage, typically focusing on three main areas: ecosystem health, human health, and resource depletion. This study employs the midpoint approach using the TRACI methodology, providing a detailed and scientifically robust evaluation of individual environmental impacts (Falano et al. 2014).

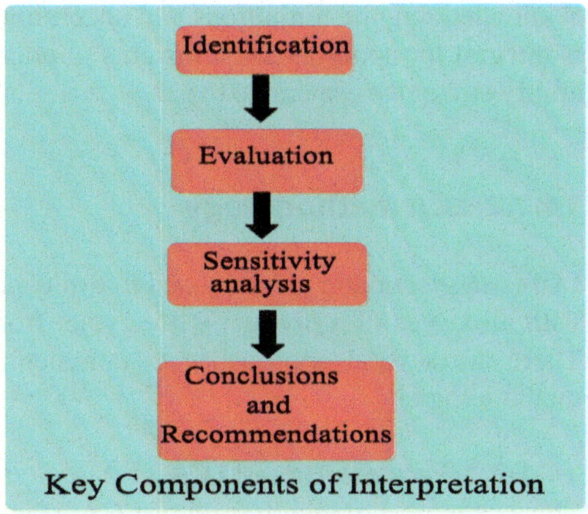

Fig. 14.1 Key contents of interpretation

14.1.1.3 Interpretation

The interpretation phase is the concluding stage of LCA, where the results from the LCI and LCIA are analyzed, synthesized, and evaluated to derive meaningful conclusions and actionable recommendations. This phase ensures that the outcomes align with the study's defined goal and scope and are relevant for the intended audience or decision-making process (Fig. 14.1).

Key Components of Interpretation

(i) **Identification of Significant Issues**: In this component the key contributors to environmental impacts determined and hotspots across the life cycle stages, processes, or impact categories are analyze. Moreover, critical flows or activities that drive the most significant impacts also identify.
(ii) **Evaluation of Study Consistency and Completeness:** During this the verification of data, assumptions, and methodologies align with the defined goal and scope performed. It also make check for gaps, inconsistencies, or limitations that could affect the reliability of the results.
(iii) **Sensitivity and Uncertainty Analysis**: Variations in input data, assumptions, or modeling choices that influence the results have been subjected for assesement. It also Quantify uncertainties to enhance the robustness and transparency of conclusions.
(iv) **Conclusions and Recommendations**: Further in interpretation conclusions based on the identified significant issues and their implications for decision-making have been formulated. This also provide practical recommendations for improving the environmental performance of the product, process, or system.

Role of Interpretation

Interpretation serves to validate the relevance and reliability of LCA findings, ensuring they are meaningful for stakeholders and aligned with the study's objectives. It also facilitates the communication of results in a clear and actionable manner while acknowledging the limitations and uncertainties inherent in the analysis. This phase is integral for applying LCA insights to policy-making, product design, or sustainability strategies (Peters 2015).

14.1.2 LCA Methodology

LCA is a systematic method used to evaluate the environmental impacts associated with all stages of a product's life cycle, from raw material extraction to disposal. There are several approaches to conducting an LCA, each with its focus and methodology:

14.1.2.1 Process-Based LCA

This is the most common form of LCA. It involves a detailed accounting of all processes in the product life cycle, such as raw material extraction, manufacturing, transportation, use, and disposal. It also provides high level of detail after specific product comparisons. Due to more time-consuming and have system boundary issues some processes might be omitted.

14.1.2.2 Hybrid LCA

This kind of methodology combines process-based LCA and I-O LCA to leverage the strengths of both approaches. By using both processes it provides more comprehensive and detailed outcomes additionally, addresses boundary issues.

14.1.2.3 Attributional LCA (ALCA)

Focuses on describing the environmental impacts directly attributable to a product system over its life cycle. It provides a snapshot of the current system. This method provides straightforward results for comparative studies but potential changes in the system has not accounted e.g., market dynamics.

14.1.2.4 Input-Output (I/O) LCA

It is based on economic input-output tables, it assesses environmental impacts by analyzing the economic transactions between sectors. It is associated with merits such as Comprehensive, avoids boundary issues by including all economic activities but it is less specific to individual products that relies on aggregated data.

14.1.2.5 Streamlined LCA

A simplified version of LCA focusing on key processes and impacts to reduce complexity and time. With strengths i.e. quicker and less resource-intensive. But, this method is less comprehensive.

14.1.2.6 Social and Economic LCA

Social and economic impacts have been explores while this methodology. It gives a broader perspective on sustainability along with multiple methodological challenges.

14.1.2.7 Consequential LCA (CLCA)

The environmental consequences of changes in a system are under Examination, such as introducing a new product or technology. However, It is a quite useful for policy analysis and strategic decision-making but requires assumptions about future market conditions and results are scenario-dependent (Ramesh 2016; Teuber et al. 2016).

14.1.3 LCA Models

LCA models are computational tools and frameworks designed to facilitate the analysis of environmental impacts across the life cycle of products or systems. These models implement different methodologies and databases to provide comprehensive results. Below are some commonly used (Table 14.1).

14.1.4 Biomaterial Life Cycle Stage

14.1.4.1 Raw Material Extraction for Biomaterials

The raw material extraction phase is critical in the life cycle of biomaterials, as it involves sourcing biological feedstocks that serve as the foundation for subsequent processing and manufacturing stages. Various types of feedstocks, their environmental impacts, and ethical considerations also have effect on it. Biomaterials can be derived from diverse biological sources, each with unique characteristics and applications (Table 14.2).

14.1.4.2 Environmental Impacts of Raw Material Extraction

The cultivation or collection of feedstocks for biomaterials can have significant environmental consequences, as illustrated in Table 14.3 on the environmental impacts of raw material extraction.

14.1.4.3 Ethical Considerations

The extraction of raw materials for biomaterials presents both opportunities and challenges in terms of environmental and ethical sustainability. A balanced approach, incorporating sustainable agricultural practices, resource-efficient technologies, and ethical considerations, is essential to maximize the benefits of biomaterials while minimizing their negative impacts (Table 14.4).

14.1.4.4 Production and Processing

In LCA, production and processing focus on transforming raw materials into biomaterials through key processes such as fermentation and polymerization. Fermentation, often used in producing bioethanol or bioplastics, requires careful management of microbial cultures, substrates, and energy inputs. Similarly, polymerization—especially in creating biodegradable polymers like polylactic acid (PLA)—can be energy-intensive, involving steps such as heating and catalysis

Table 14.1 Different types of LCA models

LCA models	Included Features
SimaPro	Includes a variety of impact assessment methods (e.g., ReCiPe, Eco-indicator 99) Supports process-based and hybrid LCA Extensive databases like Ecoinvent and Agri-footprint Suitable for both academic research and industrial applications
GaBi (Ganzheitliche Bilanzierung)	Large, robust database covering various industries Advanced modeling options for product systems and supply chains Customizable reports for decision-making Often used in industrial sectors such as automotive, electronics, and energy
Open LCA	Free and supports various LCA databases (e.g., Ecoinvent, ELCD, USLCI) User-friendly interface with strong visualization tools Compatible with a wide range of impact assessment methods Ideal for small and medium-sized businesses, as well as academic users
Umberto	Graphical modeling of product systems Integrates with databases like Ecoinvent Supports process optimization and scenario analysis Used in process industries and for carbon footprint analysis
CMLCA	Based on the CML baseline methodology Provides robust data for environmental impact assessments Strong in methodological transparency Primarily used for academic research
Brightway2	Open-source and customizable Handles large datasets and complex scenarios Supports Monte Carlo uncertainty analysis Preferred for research requiring custom modeling and scripting capabilities
GREET (Greenhouse Gases, Regulated Emissions, and Energy Use in Transportation)	Specialized for transportation sector analysis Focus on energy and emissions Includes well-to-wheels and cradle-to-grave analysis Policy analysis and technology assessment in transportation

(continued)

Table 14.1 (continued)

LCA models	Included Features
EASETECH (Environmental Assessment System for Environmental Technologies)	Strong focus on waste and resource management Includes life cycle inventory (LCI) and impact assessment options Customizable system boundaries and scenarios Used for policy-making and technology optimization in waste management
Ecochain	User-friendly interface Real-time collaboration and reporting Integration with supply chain data Suitable for businesses looking to improve sustainability performance

Table 14.2 Raw material extraction for biomaterials

Biomaterials	Examples	Applications
Agricultural Crops	Corn, sugarcane, soybeans, wheat, and other starch-rich or oilseed crops	Bioplastics (e.g., polylactic acid, bio-polyethylene), biofuels, and biodegradable packaging
Waste Products	Agricultural residues (e.g., corn stover, rice husks), food waste, and industrial by-products (e.g., glycerol, lignin)	Bioenergy, bio-composites, and bio-based chemicals
Algae	Microalgae and macroalgae (seaweed)	Bioplastics, biofuels (e.g., biodiesel, bioethanol), and nutraceuticals
Forestry Products	Wood, bamboo, and cellulose fibers	Biodegradable polymers, paper-based materials, and bio-based construction materials
Animal-Derived Feedstocks	Chitin from shellfish, collagen from animal hides, and keratin from wool	Medical biomaterials (e.g., sutures, implants), and cosmetics

Witik et al. (2011), Pelletier et al. (2021), Smith et al. (2023), Jones et al. (2022)

(Wang and Zhao 2023; Rodrigue et al. 2020; Zhang et al. 2021). These processes typically lead to significant energy consumption and waste generation, contributing to carbon emissions, water pollution, and resource depletion. However, innovations in green production are mitigating these impacts. For example, microbial metabolic engineering improves fermentation efficiency, reducing substrate usage and emissions Moreover, enzymatic polymerization has emerged as a sustainable alternative, operating under mild conditions with lower energy demands and fewer hazardous by-products Waste valorization techniques, like utilizing fermentation by-products as bioenergy sources, further enhance resource efficiency (Rodrigue et al. 2020; Zhang et al. 2021). These advancements are essential for achieving environmentally sustainable biomaterial production while maintaining industrial scalability.

Table 14.3 Environmental impacts of raw material extraction

Feed stock	Impacts	Mitigation
Land use	Large-scale cultivation of crops for biomaterials can lead to deforestation, habitat destruction, and competition for arable land	Utilizing marginal lands or integrating agroforestry practices
water use	Biomaterial feedstocks like sugarcane and corn are water-intensive, potentially exacerbating water scarcity in regions with limited freshwater resources	Adopting drip irrigation and rainwater harvesting techniques
Pesticide and fertilizer inputs	Excessive use of agrochemicals contributes to soil degradation, water pollution (eutrophication), and greenhouse gas emissions (e.g., nitrous oxide)	Transitioning to organic farming or precision agriculture
Biodiversity	Monoculture farming reduces biodiversity, disrupting ecosystems and increasing vulnerability to pests and diseases	Promoting crop diversification and habitat conservation

Table 14.4 Ethical considerations

Biomaterial	Action	Issues included
Land-use change	The conversion of forests, grasslands, or wetlands to agricultural land for biomaterial production can lead to significant carbon emissions, loss of biodiversity, and disruption of local ecosystems	Deforestation Displacement of Indigenous Communities
Food vs. non-food crops	The allocation of arable land for biomaterial feedstocks raises concerns about food security, particularly in regions where agricultural land is limited	Competition with Food Production Equity Concerns

14.1.4.5 Transportation and Distribution in LCA

This stage contribute significantly to environmental impacts, particularly through carbon emissions from fuel combustion in various modes of transport, including trucks, ships, and airplanes. This critical stages encompassing the logistics of moving raw materials to production facilities and delivering final products to consumers. For instance, heavy reliance on road and air freight generates substantial CO_2 emissions, as these modes have higher carbon footprints compared to rail or maritime shipping Additionally, packaging materials used during transportation, such as single-use plastics, cardboard, and pallets, contribute to solid waste and environmental degradation. In LCA, these impacts are quantified to assess the sustainability of the entire supply chain. Innovations in sustainable logistics, such as optimizing route planning, adopting low-carbon fuels (e.g., biodiesel, electric vehicles), and implementing reusable or biodegradable packaging, are helping reduce the carbon footprint and waste associated with transportation and distribution. For example, transitioning to electric. These strategies are vital for enhancing the environmental performance of biomaterial supply chains.

14.1.4.6 Use Phase in LCA

The use phase of biomaterials evaluates their performance, durability, and environmental impacts during application. For instance, bioplastics in packaging provide functional benefits such as light weighting and biodegradability, while biomaterials in healthcare, like bioresorbable implants, offer compatibility and reduced waste. Product longevity plays a critical role in environmental performance; durable biomaterials can extend product life and reduce resource demand. However, this phase may generate microplastics or require specific conditions, impacting their overall sustainability Specific innovations for instances, enhanced polymer stabilization, biodegradable coatings enhance product lifespan and reduces bad environmental effects during use (Rodrigue et al. 2020).

14.1.4.7 End-of-Life (EOL) Management in LCA

This Stage of LCA makes management for biomaterials. It Includes disposal methods i.e. biodegradation, recycling, and composting. Biodegradation and composting offer low-impact solutions, returning materials to the environment as organic matter, but they require specific conditions to be effective. Recycling, while reducing resource extraction, faces challenges due to contamination and limited infrastructure for biomaterials (Wang et al. 2022; Zhao et al. 2022). Comparatively, landfill disposal results in methane emissions and long degradation times, while incineration generates energy but also releases CO_2 Advancing biomaterial recycling and circularity presents opportunities to close the loop, but it requires innovations in material design, improved sorting technologies, and policy support to overcome logistical and economic barriers (Zhao et al. 2022).

14.1.5 Application of LCA

LCA serves as a pivotal analytical framework for evaluating environmental impacts across diverse sectors, thereby facilitating informed decision-making to support sustainable development (Fig. 14.2).

Within the energy sector, LCA provides a comparative analysis of renewable energy sources, such as solar and wind, relative to conventional fossil fuels. This analysis underscores the environmental advantages of renewables, particularly their role in mitigating greenhouse gas emissions and reducing resource depletion, while also identifying pathways to enhance efficiency in conventional energy systems through cleaner technologies (Chen et al. 2023; Ramírez et al. 2018). In the domain of waste management, LCA plays a crucial role in advancing recycling, reuse, and resource recovery initiatives, aligning with the principles of a circular economy. It further informs policy development by offering a scientific basis for comparing the environmental performance of various waste management technologies. In manufacturing and product design, LCA facilitates the evaluation of production processes, enabling the identification of inefficiencies and guiding the adoption of sustainable practices to minimize emissions and optimize resource use. This integration not only supports compliance with environmental regulations but also meets the growing consumer demand for environmentally friendly products, thereby enhancing market competitiveness (Ramírez et al. 2018). In agriculture and food systems, LCA provides valuable insights into the environmental impacts of farming practices, including the use of fertilizers and pesticides, as well as livestock

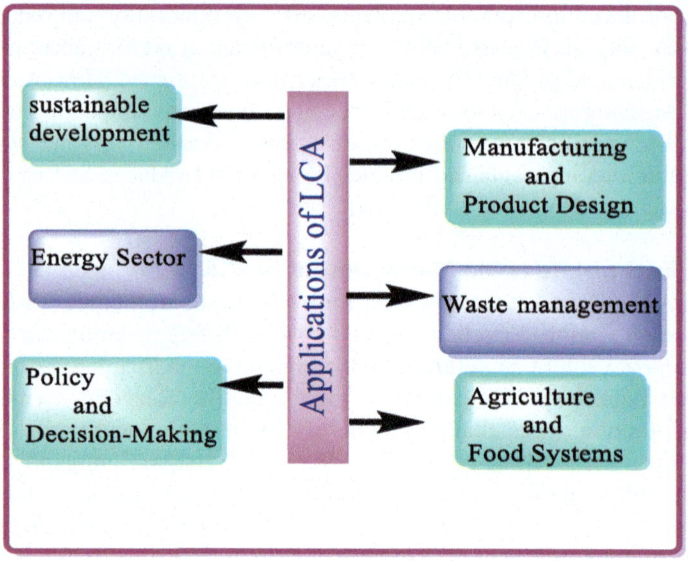

Fig. 14.2 Characteristics of effective indicators and metrics

production. It promotes the adoption of sustainable supply chain practices, such as organic farming and plant-based diets, while informing policy interventions aimed at reducing food waste and improving resource efficiency (Chen et al. 2023).

14.2 Case Studies of Biomaterials

LCA is an analytical methodology designed to comprehensively evaluate the environmental impacts associated with materials and processes across their entire life cycle, encompassing stages from raw material extraction to end-of-life disposal (Zhang et al. 2021). This systematic approach facilitates the assessment of energy consumption, emissions, and waste generation, offering insights into the sustainability of various materials. Biomaterials, including bioplastics, bio-based composites, and biofuels, have emerged as promising alternatives to conventional petrochemical-based materials due to their renewable origins and potential environmental benefits (Chen et al. 2023). Through LCA, researchers can quantify the ecological advantages and identify areas for improvement in the production and utilization of these biomaterials. The following sections present case studies analyzing the life cycle performance of different biomaterials, supported by evidence from recent literature. These studies underscore the role of LCA in advancing sustainable material innovations and guiding their integration into various industrial applications.

1. Ramírez et al. (2018) Demonstrated the potential of polyurethane-coated hydrangea stems as an effective bio-based insulation material for enhancing energy

efficiency in buildings. Their study revealed that while conventional insulation materials such as polyurethane and polystyrene displayed low compressive stresses of 0.3 MPa and 0.1 MPa, respectively, the biomaterial exhibited a significantly higher compressive stress of approximately 0.5 MPa. This improvement corresponds to a 1.7-fold increase over polyurethane and a remarkable fivefold increase over polystyrene, highlighting the superior mechanical properties of the bio-based material. Additionally, the thermal conductivity of polyurethane-coated hydrangea stems exceeded that of polystyrene and uncoated polyurethane by 18.5% and 16.3%, respectively, underscoring their enhanced thermal insulation capability. These findings emphasize the potential of bio-based materials to outperform conventional options, offering a sustainable and efficient alternative for building insulation (Choi et al. 2023).

2. Choi et al. (2023) Conducted a restoration project on a century-old edifice in South Korea, utilizing bio-based materials to enhance its energy efficiency and acoustic performance. The study incorporated repurposed coffee waste, sourced from local cafes, as a sustainable resource for the building renovation. The results indicated a significant 9% reduction in total heating energy consumption, decreasing from 18,221 kWh to 16,634 kWh. This reduction not only highlights the environmental benefits of employing bio-based materials but also underscores the potential for economic savings in operational energy costs. Computational simulations further analyzed the building's acoustic properties, revealing a notable 7.39% reduction in mid-frequency sound pressure levels, from 75.15 dB to 70.1 dB. Moreover, coffee waste panels exhibited a sound absorption coefficient approximately five to nine times greater than the conventional ceiling materials. These findings demonstrate the efficacy of bio-based materials in achieving substantial improvements in the energy efficiency and acoustic quality of heritage architecture (Niyasom and Tangboriboon 2021).
3. Niyasom and Tangboriboon (2021) Investigated the incorporation of eggshell powder, sourced from Thailand, as an additive to conventional concrete. The study revealed that concrete enriched with eggshell powder demonstrated a water absorption rate of 1.62 ± 0.16%, signifying a 40% reduction compared to standard concrete without eggshell supplementation. This reduction highlights the enhanced durability, water resistance, and chemical resilience of eggshell-infused concrete. Furthermore, mechanical testing showed that the maximum load and compressive strength of the eggshell-enriched concrete were 55.68 ± 1.64 kN and 22.08 ± 0.66 MPa, respectively, compared to 49.35 ± 5.81 kN and 19.42 ± 2.25 MPa for the control specimens. These results underscore the superior mechanical properties of eggshell-derived bio-based materials, which contribute to improved durability and sustainability in concrete applications. The enhanced physical and mechanical characteristics of eggshell concrete offer the potential to reduce maintenance costs for concrete structures, promoting environmental stewardship and advancing sustainable construction practices (Brundtland 2024).

14.3 Sustainability Indicators and Metrics

Sustainability indicators and metrics serve as essential tools for assessing, monitoring, and communicating the effectiveness of initiatives targeted at fostering sustainable development. These tools provide both quantitative and qualitative insights into the performance of systems across the interconnected dimensions of environmental, social, and economic sustainability (Brundtland 2024). Indicators represent specific variables or measures designed to capture changes and trends in sustainability, while metrics standardize the quantification of these indicators, ensuring consistent and comparable evaluations (Fig. 14.3). Together, they facilitate evidence-based decision-making, enable tracking of progress over time, and support the integration of sustainability principles into policy and practice (Smith and Zhang 2023).

14.3.1 Sustainability Indicators

The following table categorizes sustainability indicators into environmental, social, and economic dimensions, highlighting key metrics to assess sustainability at various levels such as corporate, community, or national scales (Table 14.5) (Smith and Zhang 2023).

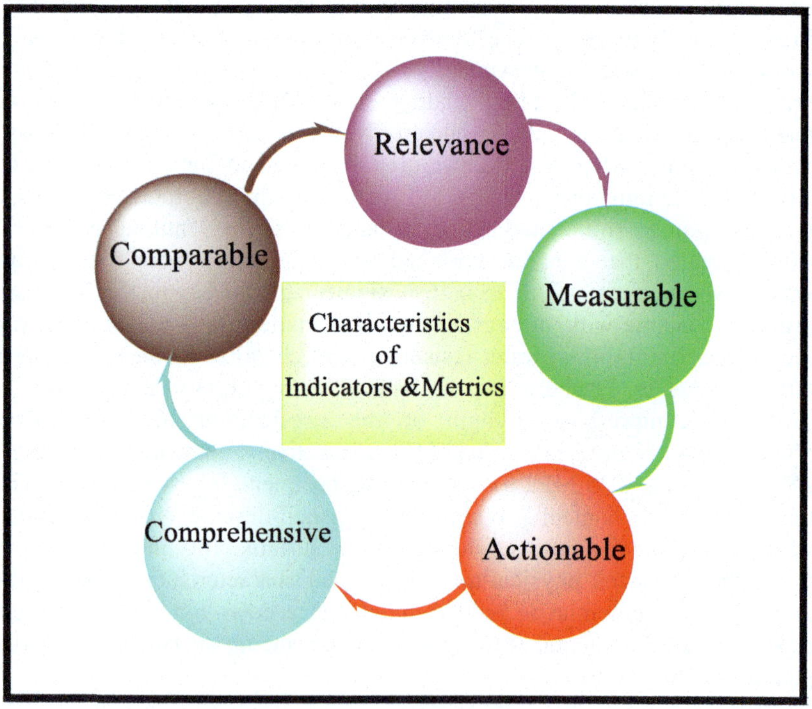

Fig. 14.3 Characteristics of effective indicators and metrics

Table 14.5 Sustainability indicators

Indicator category	Specific indicators	Measurement units/ methods	Metric
Environmental			
Resource Efficiency	Energy consumption per unit output	kWh/unit, MJ/unit	CO_2 equivalent (tons/ year)
	Water usage	Liters, m^3	% of total energy consumed
	Material efficiency (e.g., waste to product ratio)	%	Number of protected species or land area conserved (ha)
Climate Impact	Greenhouse gas (GHG) emissions	CO_2e (tons/year)	
Biodiversity	Land-use change	Hectares affected	
Pollution	Air and water quality (e.g., particulate matter, BOD levels)	μg/m^3, mg/L	
Social			
Employment and Labor	Job creation, workforce diversity	jobs created, % women/ underrepresented groups	% of workforce from underrepresented groups
Health and Safety	Workplace injuries, occupational illnesses	Incident rate per 1000 workers	Workplace incidents per 1000 workers
Education and Skills	Access to education and training programs	% employees trained, training hours	% employees receiving skill-based training
Community Engagement	Stakeholder involvement in decision-making	community consultations	
Equity	Income inequality (e.g., Gini coefficient)	Index	
Economic			
Financial Performance	Revenue growth, cost savings from sustainable practices	% change, monetary values	% of total revenue from sustainable products/ services
Innovation	R&D expenditure in sustainable technologies	% of total budget, $ amount	Material input per unit output (e.g., kg/unit)
Market Access	Share of sustainable products in the market	% of sales	
Infrastructure	Investments in renewable energy and green infrastructure	invested, % share	

Brundtland (2024), Smith and Zhang (2023)

14.3.2 Key Considerations for Effective Use

(i) **Contextualization:** Sustainability indicators should be tailored to align with the specific priorities and challenges of the organization or region in which they are applied.

(ii) **Data Availability:** Accurate and reliable data sources are essential for ensuring the precision and consistency of sustainability measurements.

(iii) **Integration:** Indicators must not only provide valuable insights for decision-making but also be integrated within broader sustainability frameworks, such as the Sustainable Development Goals (SDGs), to ensure holistic progress.

14.3.3 Foremost Challenges

(i) **Data Availability:** The lack of consistent or complete datasets often impedes the accurate measurement of sustainability indicators.
(ii) **Integration:** A key challenge lies in balancing the environmental, social, and economic dimensions of sustainability, which are often interdependent and complex.
(iii) **Standardization:** The absence of universally accepted metrics across industries or regions poses a challenge to ensuring comparability and consistency in sustainability assessments.

14.3.4 Future Prospects

Advancements in technologies such as the Internet of Things (IoT) and Artificial Intelligence (AI) are facilitating more precise and real-time monitoring of sustainability metrics, enabling organizations to track and respond to changes dynamically. Furthermore, increasing regulatory pressures and growing consumer demand for sustainable practices are encouraging the widespread adoption of standardized frameworks globally. By effectively implementing sustainability indicators and metrics, organizations can not only meet their sustainability objectives but also enhance their competitive edge through improved transparency and resource management (Lee and Kim 2024; Global Reporting Initiative 2024; International Organization for Standardization (ISO) 2023; United Nations Global Compact 2024).

14.4 Challenges in LCA of Biomaterials

Biomaterials often have diverse and geographically distributed supply chains, making it difficult to track inputs and outputs accurately (Fig. 14.4).

14.4.1 Lack of Reliable and Standardized Data

Accurate environmental assessments of biomaterials are hindered by the absence of consistent and standardized data, especially for stages like cultivation, transportation, and energy usage. This lack of data consistency complicates the reliability of LCA findings.

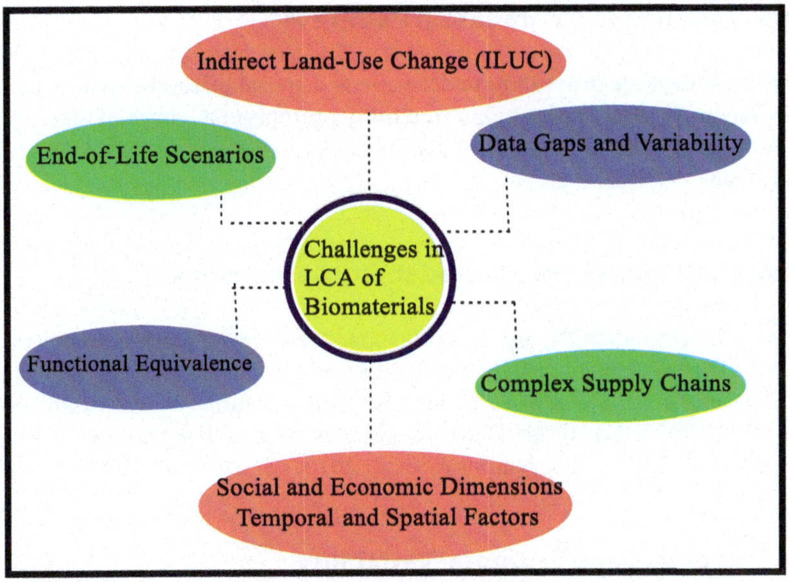

Fig. 14.4 Characteristics of effective indicators and metrics

14.4.2 High Variability in Agricultural Practices

The variability in agricultural practices, such as fertilizer use and irrigation methods, significantly influences LCA results. This makes it challenging to establish universal environmental impact data for biomaterials sourced from different regions or under different conditions.

14.4.3 Environmental Impact of Cultivation

Cultivating biomaterials can contribute to deforestation or the conversion of natural habitats, resulting in carbon emissions and biodiversity loss. These impacts are difficult to quantify accurately and add complexity to sustainability assessments (Fig. 14.4).

14.4.4 Ensuring Functional Equivalence

Comparing biomaterials with conventional materials in LCA is complicated by the need to ensure that the biomaterials are functionally equivalent. Differences in material performance and properties complicate the direct comparison of their environmental impacts.

14.4.5 Variation in Degradation Pathways

Biomaterials degrade differently under various disposal methods, such as composting, incineration, or recycling. This variability presents challenges in understanding their full environmental impact and complicates LCA assessments focused on end-of-life stages.

14.4.6 Challenges in Biodegradability Assessment

Assessing biodegradability and its corresponding environmental impact is a complex task. The breakdown of biomaterials depends on factors such as environmental conditions and time, which can vary greatly (United Nations Global Compact 2024; Geyer et al. 2017; Wu et al. 2020; Mazerolles et al. 2019; Dunlap et al. 2024; Snowdon et al. 2017b; Taherymoosavi et al. 2016; Gholap et al. 2024).

14.4.7 Regional and Temporal Variability

Environmental impacts, such as greenhouse gas emissions and water use, can vary significantly across regions and time periods. Site-specific data is essential for accurate LCA but is often difficult to obtain or inconsistent.

14.4.8 Incorporating Socio-economic Factors

Including socio-economic factors such as job creation, fair trade, and community impacts into LCA adds complexity. These factors are essential for a comprehensive evaluation of sustainability but require additional data and considerations (Geyer et al. 2017; Wu et al. 2020; Mazerolles et al. 2019; Dunlap et al. 2024).

14.5 Opportunities for Sustainable Production and Consumption

LCA plays a crucial role in identifying environmental hotspots within production processes, guiding efforts to enhance energy efficiency, water conservation, and waste reduction (Snowdon et al. 2017b). LCA aids in advancing sustainable practices by supporting the design of biomaterials that are reusable, recyclable, or biodegradable, reducing the reliance on fossil-based resources (Snowdon et al. 2017b; Taherymoosavi et al. 2016; Gholap et al. 2024; Gholap et al. 2025).

14.5.1 Advancing Sustainable Practices

LCA helps design biomaterials that can be reused or biodegraded, thus reducing dependence on fossil-based materials and fostering more sustainable product life-cycles (Gholap et al. 2024).

14.5.2 Promoting Circular Economy

Insights from LCA guide the development of environmental regulations and industry standards that integrate biomaterials into the circular economy, focusing on waste reduction and material life extension (Gholap et al. 2025).

14.5.3 Policy and Standards Development

LCA reveals environmental trade-offs in production, driving technological innovations, alternative feedstocks, and the creation of sustainable biomaterials, which can inform the development of new policies and standards.

14.5.4 Driving Innovation

LCA facilitates the communication of environmental benefits to consumers, businesses, and policymakers, thereby driving the wider adoption of sustainable biomaterials (Gholap et al. 2024; Gholap et al. 2023).

14.5.5 Stakeholder Awareness and Engagement

The use of advanced digital tools and databases enhances the accuracy and accessibility of LCA, empowering stakeholders to make informed decisions regarding sustainable production and resource management (Sharmila et al. 2024).

14.5.6 Integration with Digital Tools

Digital technologies, including AI and IoT, improve the precision and real-time monitoring of sustainability metrics, further enhancing decision-making in biomaterial production (Gholap et al. 2025).

14.5.7 Carbon Sequestration Potential

LCA can quantify the carbon sequestration benefits of plant-based biomaterials, providing a comprehensive view of their environmental impact across their entire lifecycle (Gholap et al. 2023).

14.5.8 Market Opportunities

As demand for sustainable materials grows across industries such as packaging and construction, LCA-validated biomaterials present significant market opportunities, supporting their widespread adoption.

By leveraging LCA, organizations and industries can improve their sustainability efforts, foster innovation, and promote the use of biomaterials that contribute to a more sustainable future (Hassan et al. 2025).

14.6 Conclusion

LCA is an indispensable methodology for advancing the sustainable production and consumption of biomaterials. Through a comprehensive evaluation of environmental impacts across all stages of a material's life cycle, LCA identifies critical areas for enhancing energy efficiency, optimizing resource use, and minimizing waste generation. This systemic approach provides a foundation for actionable strategies, enabling advancements in production technologies and resource management practices. Despite challenges such as data inconsistency, regional variability, and the inherent complexity of assessing multifunctional systems, LCA delivers essential insights that inform innovation, regulatory frameworks, and market transformations. Its integration into biomaterial development not only supports the adoption of circular economy principles—focusing on material reuse, recycling, and valorization—but also facilitates transparency in communicating environmental trade-offs. This transparency plays a pivotal role in fostering stakeholder trust and encouraging the broader acceptance of biomaterials across industries and consumers. As global demand for sustainable alternatives continues to rise, the refinement of LCA methodologies, alongside the establishment of robust and regionally nuanced datasets, will be imperative. These advancements will ensure that biomaterials achieve their potential in contributing to sustainability objectives, driving progress toward environmentally responsible consumption and production patterns.

References

Brundtland GH (2024) The role of sustainability indicators in global policy development. Sustain Rep 35(1):22–35

Chen M, Wang Y, Liu Z (2023) Circular economy in biomaterials: challenges and innovations in recycling processes. J Environ Manag 332:117386. https://doi.org/10.1016/j.jenvman.2023.117386

Choi JY, Nam J, Yuk H, Yun BY, Lee S, Lee JK, Kim S (2023) Proposal of retrofit of historic buildings as cafes in Korea: recycling biomaterials to improve building energy and acoustic performance. Energ Buildings 287:112988. https://doi.org/10.1016/j.enbuild.2023.112988

Deng Y (2014) Life cycle assessment of biobased fibre-reinforced polymer composites. Levenscyclusanalyse van biogebaseerde, vezelversterkte polymeercomposieten. Science, Engineering & Technology, Leuven

Dunlap J, Schramski JR, Li G et al (2024) Identifying uncertainty in the global warming impacts of biomaterials: an analysis of biosuccinic acid. Int J Life Cycle Assess 29:1137–1149. https://doi.org/10.1007/s11367-024-02290-1

Falano T, Jeswani HK, Azapagic A (2014) Assessing the environmental sustainability of ethanol from integrated biorefineries. Biotechnol J 9(7):753–765. https://doi.org/10.1002/biot.201300246

Faruk O, Bledzki AK, Fink H-P, Sain M (2014) Progress report on natural fiber reinforced composites. Macromol Mater Eng 299(1):9–26. https://doi.org/10.1002/mame.201300008

Geyer R, Jambeck JR, Law KL (2017) Production, use, and fate of all plastics ever made. Sci Adv 3(7):e1700782. https://doi.org/10.1126/sciadv.1700782

Gholap AD, Gupta J, Kamandar P, Bhowmik DD, Rojekar S, Faiyazuddin M, Hatvate NT, Mohanto S, Ahmed MG, Subramaniyan V, Kumarasamy V (2023) Harnessing nanovaccines for effective immunization – a special concern on COVID-19: facts, fidelity, and future prospective. ACS Biomater Sci Eng 10(1):271–297. https://doi.org/10.1021/acsbiomaterials.3c01247

Gholap AD, Pardeshi SR, Hatvate NT, Dhorkule N, Sayyad SF, Faiyazuddin M, Khalid M (2024) Environmental implications and nanotechnological advances in octocrylene-enriched sunscreen formulations: a comprehensive review. Chemosphere 358:142235. https://doi.org/10.1016/j.chemosphere.2024.142235

Gholap AD, Sawant SD, Pardeshi SR, Khuspe PR, Sayyad SF, Chavan M, Hatvate NT, Faiyazuddin M, Uddin MJ (2025) Omics tools for determining microbial composition: a systematic review. Microbiome Eng:35–50. https://doi.org/10.1201/9781003394662

Global Reporting Initiative (2024) GRI standards for sustainability reporting: an overview. GRI

Hassan J, Faiyazuddin M, Uddin MJ (2025) Marine drug delivery systems. In: Handbook of research in marine pharmaceutics. Apple Academic Press, pp 49–135. https://doi.org/10.1201/9781003537403

International Organization for Standardization (ISO) (2023) ISO 14040: sustainability and environmental performance metrics. ISO Rep Sustain Stand 2023(9):13–17

Jones R, Patel M, Singh R (2022) Valorization of industrial fermentation by-products for circular bioeconomy. Bioresour Technol 346:126545. https://doi.org/10.1016/j.biortech.2022.126545

Lee C, Kim H (2024) Technological innovations enhancing sustainability measurement in corporate settings. Sustain Dev Technol 40(2):99–112

Mazerolles T, Heuzey MC, Soliman M, Martens H, Kleppinger R, Huneault MA (2019) Development of co-continuous morphology in blends of thermoplastic starch and low-density polyethylene. Carbohydr Polym 206:757–766. https://doi.org/10.1016/j.carbpol.2018.11.038

Niyasom S, Tangboriboon N (2021) Development of biomaterial fillers using eggshells, water hyacinth fibers, and banana fibers for green concrete construction. Constr Build Mater 283:122627. https://doi.org/10.1016/j.conbuildmat.2021.122627

Pelletier N, Audsley E, Brodt S (2021) Sustainability in the production of biobased materials: a life cycle perspective. J Clean Prod 280:124837. https://doi.org/10.1016/j.jclepro.2020.124837

Peters JF (2015) Pyrolysis for biofuels or biochar? A thermodynamic, environmental and economic assessment. University Rey Juan Carlos

Ramesh M (2016) Kenaf (Hibiscus cannabinus L.) fibre-based bio-materials: a review on processing and properties. Prog Mater Sci 78–79:1–92. https://doi.org/10.1016/j.pmatsci.2015.11.001

Ramírez JPC, Cea M, Santín K, Valdés G, Hunter R, Navia R (2018) Characterization and application of a natural polymer obtained from Hydrangea macrophylla as a thermal insulation biomaterial. Compos Part B 132:10–16. https://doi.org/10.1016/j.compositesb.2017.07.086

Rodrigue J-P, Slack B, Notteboom T (2020) The geography of transport systems, 5th edn. Routledge

Sharmila S, Saranya A, Arulprakasajothi M, Saranya R, Srimanickam B, Abel SK, Shakeel F, Faiyazuddin M (2024) Expediting the bioactivity of zinc sulfide nanoparticles with copper oxide as a nanocomposite. BMC Chem 18(1):204. https://doi.org/10.1186/s13065-024-01320-1

Smith J, Zhang R (2023) Advancements in the application of sustainability metrics in business practices. J Environ Manag 182:45–62

Smith J, Taylor A, Lee K (2023) Advances in sustainable polymerization techniques: reducing energy and waste in biomaterial manufacturing. Green Chem 25(8):4567–4578. https://doi.org/10.1039/D3GC01256D

Snowdon MR, Mohanty AK, Misra M (2017a) Miscibility and performance evaluation of biocomposites made from polypropylene/poly(lactic acid)/poly(hydroxybutyrate). Biomacromolecules 18(7):2259–2270. https://doi.org/10.1021/acs.biomac.7b00433

Snowdon MR, Mohanty AK, Misra M (2017b) Miscibility and performance evaluation of biocomposites made from polypropylene/poly(lactic acid)/poly(hydroxybutyrate). Compos A: Appl Sci Manuf 61:9–26

Taherymoosavi S, Joseph S, Munroe P (2016) Characterization of organic compounds in a mixed feedstock biochar generated from Australian agricultural residues. J Anal Appl Pyrolysis 120:441–449. https://doi.org/10.1016/j.jaap.2016.06.017

Teuber L, Osburg VS, Toporowski W et al (2016) Wood polymer composites and their contribution to cascading utilisation. J Clean Prod 110:9–15. https://doi.org/10.1016/j.jclepro.2015.04.009

United Nations Global Compact (2024) Sustainable here is the list of references in APA style

Wang Y, Zhao H (2023) Energy-efficient strategies in enzymatic polymerization for biobased materials. Renew Sust Energ Rev 168:113345. https://doi.org/10.1016/j.rser.2022.113345

Wang F, Liu J, Zhou X (2022) Reducing carbon emissions in last-mile logistics: the role of electric vehicles and route optimization. Sustain Cities Soc 84:104000. https://doi.org/10.1016/j.scs.2022.104000

Witik RA, Payet J, Michaud V et al (2011) Assessing the life cycle costs and environmental performance of lightweight materials in automobile applications. Compos A: Appl Sci Manuf 42(10):1694–1709. https://doi.org/10.1016/j.compositesa.2011.07.024

Wu C, Li Y, Sun J, Lu Y, Tong C, Wang L, Yan Z, Pang J (2020) Novel konjac glucomannan films with oxidized chitin nanocrystals immobilized red cabbage anthocyanins for intelligent food packaging. Food Hydrocoll 98:105245. https://doi.org/10.1016/j.foodhyd.2019.105245

Zhang Y, Li X, Chen M (2021) Sustainable packaging solutions in global logistics: A life cycle assessment perspective. J Environ Manag 288:112427. https://doi.org/10.1016/j.jenvman.2021.112427

Zhao X, Li J, Zhang Y (2022) Environmental impacts of end-of-life management for bioplastics: a comparative assessment. Resour Conserv Recycl 187:106629. https://doi.org/10.1016/j.resconrec.2022.106629

Patents, Clinical Trials and Dosage Form Based on Biomaterials

15

Mohd Sayeed Shaikh, Rupesh R. Kurhade, Shaikh Shahbaz A. Majeed, Mujtaba Md Shaikh, Paresh R. Mahaparale, Mayura A. Kale, Mohd Usman Mohd Siddique, Md. Faiyazuddin, and Yahya E. Choonara

Abstract

Biomaterials such as hydrogels, nanoparticles, and biodegradable polymers are being increasingly utilized to develop advanced dosage forms that aim to improve the efficiency and safety of drug delivery. Our analysis highlights a significant rise in patent filings for novel biomaterial-based technologies, reflecting innovation in smart drug delivery systems that respond to physiological stimuli and enhance targeted therapeutic effects. Concurrently, clinical trials have

M. S. Shaikh (✉) · S. S. A. Majeed
Y. B. Chavan College of Pharmacy, Dr. Rafiq Zakaria Campus, Aurangabad, Maharashtra, India

R. R. Kurhade · P. R. Mahaparale
Government College of Pharmacy Aurangabad, Aurangabad, Maharashtra, India

M. Md. Shaikh
Government College of Pharmacy Aurangabad, Aurangabad, Maharashtra, India

Aurangabad College of Pharmacy Osmanpura, Aurangabad, Maharashtra, India

M. A. Kale
Government College of Pharmacy Karad, Karad, Maharashtra, India

M. U. M. Siddique
Department of Pharmaceutical Chemistry, Shri Vile Parle Kelavani Mandal's Institute of Pharmacy, Dhule, Maharashtra, India

Md. Faiyazuddin
School of Pharmacy, Al-Karim University, Katihar, Bihar, India

Y. E. Choonara (✉)
Wits Advanced Drug Delivery Platform, Faculty of Health Sciences, School of Therapeutic Sciences, Department of Pharmacy and Pharmacology, University of the Witwatersrand, Parktown, Johannesburg, South Africa
e-mail: yahya.choonara@wits.ac.za

Md. Faiyazuddin, M. Suryawanshi (eds.), *Design and Processing of Green Materials*, Biomaterials, Bioengineering and Sustainability 4,
https://doi.org/10.1007/978-3-031-91790-5_15

progressively incorporated these advanced biomaterials, demonstrating their potential in diverse therapeutic areas, including cancer treatment, diabetes management, and regenerative medicine. These trials reveal promising outcomes related to controlled release, targeted delivery, and reduced side effects. Despite these advancements, several challenges remain, including complex regulatory pathways, high production costs, and scalability issues. The review emphasizes that while the integration of biomaterials into dosage forms offers substantial improvements in therapeutic interventions, the path to widespread clinical adoption requires overcoming these barriers. Future research should focus on addressing these challenges, optimizing biomaterial properties for specific applications, and ensuring cost-effective production methods. By synthesizing data from patents and clinical trials, this review provides a comprehensive overview of the current state of biomaterial-based dosage forms and identifies key areas for future exploration and development. The chapter critically examines the evolving role of biomaterials in drug delivery systems by analyzing recent trends in patents, clinical trials, and dosage forms.

Keywords

Biomaterial · Patents · Clinical trials · Dosage forms · Regulatory pathways

15.1 Introduction

Biomaterials are specially engineered to interact with biological systems, performing therapeutic and diagnostic functions in medicine. They have transformed healthcare by enabling the development of novel drug delivery systems, tissue engineering frameworks, and medical devices aimed at restoring, supporting, or improving biological functions (Chong et al. 2023; Ajmal et al. 2022). In the last 5 years, biomaterial science has advanced dramatically, expanding its applications beyond traditional implants to include innovative, smart systems. These advanced biomaterials now allow for controlled drug release and tissue regeneration, ushering in a new era of biomedical solutions that are responsive and multifunctional (Ajmal et al. 2022; Deng et al. 2022). Recent advances in biomaterials have resulted in innovative solutions such as hyaluronic acid-based injectable hydrogels for sustained drug delivery, which effectively treat osteoarthritis (Jin et al. 2020; Hu et al. 2023). Similarly, biodegradable polycaprolactone nanofibers to promote tissue regeneration and reduce scarring in wound healing applications (Khunová et al. 2022). The global biomaterials market is expected to expand significantly, reaching $297.5 billion by 2028, owing to an aging population and a growing demand for biocompatible medical devices (Patshina et al. 2021). New developments in biomaterials include 3D-printed patient-specific implants made of bioactive glasses and self-healing polymers designed for drug delivery purposes. This field's research and patents are increasingly focused on smart delivery systems that respond to biological triggers (Elumalai et al. 2023; Kumar et al. 2022). Notably, Suarato et al. reported a pH-responsive chitosan-based delivery system for cancer treatment, while Panova and colleagues designed FDA-approved temperature-sensitive liposomes for controlled antibiotic release (Suarato et al. 2016; Panova et al. 2021).

Biomaterials have recently transformed medical treatments, making significant advances in tissue engineering, drug delivery, and regenerative medicine. Innovations such as stimuli-responsive hydrogels for controlled drug release, bioactive scaffolds for tissue repair, and nanostructured materials for targeted therapies have demonstrated significant promise in improving patient outcomes (Palmese et al. 2020; Zhong et al. 2023; Sun and Tan 2013). Injectable hydrogels derived from modified hyaluronic acid have demonstrated potential in cartilage regeneration, with clinical trials indicating long-term therapeutic benefits. Similarly, bioactive glass scaffolds containing strontium and zinc have shown increased bone formation in dental and orthopedic applications (Chen et al. 2021; Zhu et al. 2023). Furthermore, the development of self-healing biomaterials such as citrate-based elastomers has resulted in longer-lasting medical implants with improved mechanical properties (Zhu et al. 2017; Turner et al. 2023; Du et al. 2015). Biodegradable polymeric nanoparticles, particularly those based on PLGA, are demonstrating efficacy for targeted cancer therapy by increasing drug bioavailability while minimizing toxicity (Malathi et al. 2023; Xu et al. 2016; Valcourt et al. 2019). Furthermore, 3D-printed biomaterial scaffolds containing growth factors show great promise for wound healing (Metwally et al. 2024). The combination of biomaterials with technologies such as artificial intelligence and 3D bioprinting is creating new opportunities in personalized medicine and regenerative therapies (Chen et al. 2022).

The incorporation of biomaterials into pharmaceutical development has transformed drug delivery systems, thereby improving therapeutic outcomes and patient compliance. This chapter investigates the role of patents, clinical trials, and biomaterial-based dosage innovations, such as chitosan nanoparticles for targeted cancer therapy, hyaluronic acid hydrogels for extended drug release, and PLGA microspheres for protein delivery. These innovations have prompted major pharmaceutical companies to file numerous patent applications to secure biomaterial-based drug delivery technologies (Puluhulawa et al. 2022; Sheikh et al. 2022; Virmani et al. 2023). Clinical trials with biomaterial-based pharmaceuticals have yielded promising results, particularly in tissue engineering and regenerative medicine. For example, a cellulose-based wound dressing infused with growth factors has shown promise in Phase III trials, and silk fibroin-based oral insulin delivery systems are being studied (Tabatabaei Hosseini et al. 2024; Park et al. 2002; Jung et al. 2017). The combination of biomaterials science and pharmaceutical innovation has revolutionized drug delivery and therapy approaches. Patent growth has been particularly notable in biodegradable polymers and advanced delivery systems, as evidenced by Novartis' 2023 patent for peptide hydrogels designed for sustained protein release and Moderna's 2022 lipid nanoparticle patents for mRNA treatments (Ahn et al. 2021; Wang et al. 2023a). These developments demonstrate the industry's commitment to precision medicine and targeted delivery methods. Clinical trials with biomaterial-based formulations have yielded encouraging results in a variety of therapeutic areas. Notably, the novel hydrogel-based neural tissue matrix for treating Parkinson's disease, whereas the investigated bioactive glass scaffolds for bone regeneration. These studies highlight the successful implementation of biomaterial innovations in clinical settings, particularly in regenerative medicine (Mukherjee et al. 2020; Aldhaher et al. 2023).

The integration of biomaterials has significantly accelerated recent advances in dosage form innovation. 3D-printed tablets with biocompatible polymers for personalized medicine, mucoadhesive buccal films with natural polysaccharides, and injectable in-situ forming implants are all examples of innovations (Elisetti et al. 2023). The FDA's approval of Zynrelef® in 2021, an extended-release formulation based on a bioabsorbable polymer matrix, demonstrates successful commercialization in this area (NCT05863221 2023). The synergy of these advancements has resulted in next-generation therapeutic solutions, such as the trial of a chitosan-based nasal delivery platform for peptide therapeutics, which demonstrates the effective clinical translation of complex biologics. This chapter explores critical aspects of biomaterial-based pharmaceutical development in light of recent scientific and regulatory advances, focusing on both challenges and opportunities. It focuses on emerging trends such as stimuli-responsive materials, bioprinting technologies, and artificial intelligence-assisted biomaterial design, all of which are transforming pharmaceutical innovation (Georgieva et al. 2023; Desai et al. 2023).

15.2 Biomaterials Overview

Biomaterials are substances specifically engineered to interact with biological systems for medical purposes, including both therapeutic and diagnostic applications (Davidson Hernandez and Reyes-Romero 2019). They play a crucial role in the restoration of bodily functions, tissue repair, and the development of innovative medical devices that significantly improve patient outcomes. As a multidisciplinary field, biomaterials encompass various disciplines such as medicine, biology, chemistry, and materials science, contributing to advancements in healthcare and regenerative medicine (Chen and Liu 2016). Various types of biomaterials used in medicine and allied sciences are well illustrated by Fig. 15.1.

15.2.1 Types of Biomaterials

Biomaterials can be generally classified into several categories based on their origin and properties, including metals, ceramics, polymers, composites, and hydrogels. Each category exhibits distinct characteristics, making them suitable for various medical applications, such as implants, drug delivery systems, and tissue engineering materials (Spoială et al. 2023).

Metallic biomaterials are made up of metals and their alloys, such as stainless steel, titanium, and cobalt-based alloys, are widely used in medical applications due to their excellent mechanical properties, including high strength and fatigue resistance. These metals are used in orthopedic implants, dental implants (Tapscott and Wottowa 2023), and cardiovascular devices like stents and heart valves (Davis et al.

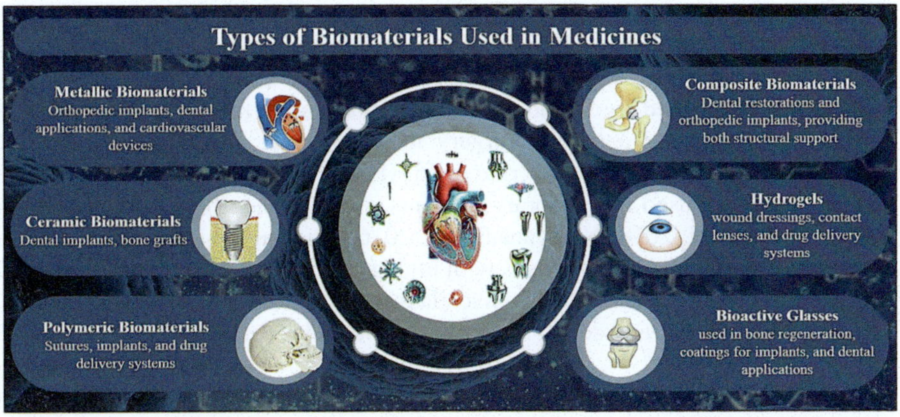

Fig. 15.1 Types of biomaterials

2022). They are often utilized in orthopedic implants, dental applications, and cardiovascular devices. However, they can suffer from corrosion and biocompatibility issues, which limit their applications in some contexts (Adhikari et al. 2018). Ceramic biomaterials, including calcium phosphates and bioactive glasses, are favored for their high compressive strength and biocompatibility, making them ideal for bone repair and dental applications (Hou et al. 2022). These materials are used in dental implants, bone grafts, and coatings for metallic implants to enhance biocompatibility. These materials can promote bone growth and integration with surrounding tissues. Despite these advantages, ceramics can be brittle, which may lead to mechanical failures in load-bearing applications (Tavoni et al. 2021). Polymeric biomaterials are versatile biomaterials that can be either synthetic or natural and are increasingly used in medical devices, drug delivery systems, and scaffolds for tissue engineering (Guo and Ma 2014). They offer advantages such as ease of fabrication, tunable mechanical properties, and biodegradability (Satchanska et al. 2024). Examples include polylactic acid (PLA) and polyurethane, which are commonly utilized in sutures, implants, and drug delivery systems (Boni et al. 2018).

Composite biomaterials combine two or more different biomaterials to enhance specific properties and functionalities. For example, reinforced polymers may incorporate ceramic fillers to improve strength and wear resistance (Voskerician 2017). These advanced materials are often used in dental restorations and orthopedic implants, providing both structural support and biological compatibility (Zhang et al. 2020). Another examples are bone cement (a mixture of polymer and ceramic), and fiber-reinforced composites (used in orthopedic and dental applications). These materials can provide tailored mechanical properties and improved biocompatibility (Vallittu 2018). Hydrogels, which are three-dimensional networks of hydrophilic polymers, have gained attention for their biocompatibility and ability to retain large amounts of water. They are particularly effective in drug delivery and tissue engineering because their properties can be tailored to control the release of therapeutic agents. However, their mechanical strength can be a limiting factor in certain applications (Thang et al. 2023). Polyethylene glycol (PEG) hydrogels, hyaluronic acid, and polyvinyl alcohol (PVA) hydrogels, these are used in wound dressings, contact lenses, and drug delivery systems (Liu and Chen 2024). Biodegradable biomaterials are designed to degrade within the body after fulfilling their purpose, reducing the need for a second surgery to remove the implant. Examples are Polylactic acid (PLA), polyglycolic acid (PGA), and their copolymers like PLGA. These are used in temporary implants, sutures, and drug delivery systems (Wawrzyńska et al. 2018; Narayanan et al. 2016). Bioactive glasses are a type of glass-ceramic material that interacts with biological tissues to promote healing and regeneration. Example is 45S5 bioactive glass. These are used in bone regeneration, coatings for implants, and dental applications (Nogueira et al. 2024).

15.2.1.1 Natural vs. Synthetic Biomaterials

Natural biomaterials possess several advantages, notably their biocompatibility, which allows them to integrate well with biological tissues without provoking significant immune responses (Tripathi et al. 2023). E.g. Collagen, silk fibroin, and decellularized tissue matrices. These materials support cell growth and tissue regeneration. They are also inherently biodegradable, which means they can be absorbed by the body without the need for subsequent removal surgeries, potentially leading to cost savings in healthcare (Schneider et al. 2023). Moreover, because they are derived from renewable sources, natural biomaterials contribute to sustainability in medical practices. Despite their benefits, natural biomaterials have certain drawbacks. One primary concern is the potential for immunogenic responses, which can vary among patients, leading to complications in their medical application (Kelly et al. 2017). Additionally, natural biomaterials may exhibit inconsistent quality and durability compared to their synthetic counterparts, which can limit their performance in some applications. Furthermore, there is a risk of leaching unwanted compounds into bodily fluids, which may pose safety concerns (Reddy et al. 2021).

Synthetic biomaterials are often favored for their reproducibility and tailored properties, allowing precise control over their mechanical and chemical characteristics. This customization enables them to meet specific requirements for various applications, catering to a broad range of medical needs (Prete et al. 2023). Furthermore, synthetic biomaterials usually demonstrate enhanced mechanical strength and durability, making them well-suited for load-bearing applications in orthopedics and dentistry (Al-Shalawi et al. 2023). Nonetheless, synthetic biomaterials also have disadvantages. They may lack adequate biocompatibility due to the absence of natural cell adhesion sites, which often necessitates chemical modifications to promote cellular interactions. Additionally, some synthetic materials can leach unwanted substances into the surrounding tissues, raising concerns regarding patient safety. Lastly, synthetic biomaterials carry certain risks associated with disease transmission, especially when sourced from biological materials (Reddy et al. 2021). Generally used biomaterials in medicine are presented in Table 15.1, with their potential application area.

Table 15.1 Commonly used biomaterials in medicine

S. No.	Type of biomaterial	Examples	Common applications	References
1.	Natural biomaterials	Collagen	Wound healing and tissue engineering	Chattopadhyay and Raines (2014)
		Chitosan	Drug delivery systems and wound dressings	Che et al. (2024)
		Alginate	Scaffolds for tissue engineering and drug delivery	Farshidfar et al. (2023)
2.	Synthetic biomaterials	Polylactic Acid (PLA)	Surgical sutures and drug delivery devices	Alhulaybi (2023)
		Polyethylene (PE)	Hip socket liners and surgical implants	García-Rey and García-Cimbrelo (2010)
		Polymethylmethacrylate (PMMA)	Dental implants and bone cements	(Frazer et al. 2005)
3.	Metallic biomaterials	Stainless steel	Bone plates and vascular stents	Fu et al. (2020)
		Titanium and its alloys	Hip and knee implants, orthopedic applications	Tapscott and Wottowa (2023)
		Cobalt-chromium alloys	Artificial heart valves and joint prostheses	Meek et al. (2008)
4.	Ceramic biomaterials	Hydroxyapatite	Bone grafts and coatings for metal implants	Arcos and Vallet-Regí (2020)
		Calcium phosphate	Artificial bones, teeth, and ligaments	Mishchenko et al. (2023)
		Zirconia	Dental implants and hip joint replacements	Gul et al. (2024)
5.	Hybrid biomaterials	Composite materials (e.g., plastic-ceramic combinations)	Used in various implants to enhance mechanical properties	Victor and Muthu (2014)
		Bioceramics	Drug delivery and bone repair applications	Arcos and Vallet-Regí (2013)
6.	Hydrogels	Polyvinyl Alcohol (Polyethylene Glycol (PEG) Hydrogels, Hyaluronic Acid, Polyvinyl Alcohol (PVA) Hydrogels)	Drug delivery systems and tissue engineering	Wang et al. (2021)
		Fibrin	Artificial blood clots and wound healing	Bayer (2022)

(continued)

Table 15.1 (continued)

S. No.	Type of biomaterial	Examples	Common applications	References
7.	Bioactive glasses	45S5 bioactive glass	Bone regeneration, coatings for implants, dental applications	Skallevold et al. (2019)

15.2.2 Properties and Functions

Successful integration with biological tissues depends on the ability of a biomaterial to perform its intended function without triggering adverse reactions (Ramakrishna and Huang 2003). Since materials that are suitable for one application may not be appropriate for another, requiring specific evaluations, biocompatible materials support tissue integration, foster cell adhesion and growth, and minimize immune or inflammatory responses to reduce post-implantation complications. The mechanical properties of biomaterials, such as tensile strength, elasticity, and ductility, are critical for their effective use in medical devices and implants, as they must match the mechanical properties of surrounding tissues to avoid stress concentration and failure (Wu et al. 2023; Olamide Omiyale et al. 2023). Key functions include providing structural support to damaged tissues or organs, evenly distributing mechanical loads to avoid material failure, and ensuring implant durability to reduce the need for replacement surgeries, with a particular emphasis on matching the Young's modulus to that of biological tissues for optimal performance (Olamide Omiyale et al. 2023). Degradation and bioactivity are critical in the design of resorbable biomaterials, with degradation referring to the breakdown process influenced by the material's composition and environment, and bioactivity reflecting the material's ability to support cell adhesion and tissue regeneration (Wei et al. 2020a; Vach Agocsova et al. 2023; Feng et al. 2006). These materials should degrade in harmony with tissue healing, providing temporary support, enhancing healing, and reducing the need for implant removal surgery (Davison et al. 2014).

15.3 Patents Related to Biomaterials

Patents are critical in biomaterials because they protect intellectual property, encourage innovation, and promote investments in the development of new technologies, all while ensuring inventors receive recognition, rewards, and business opportunities (EBIN.PUB n.d.; Haynes 2005). Protecting intellectual property is critical in the biomaterials sector because it prevents competitors from copying innovations, with patents providing legal protection for a set period of time, usually 20 years (Saha and Bhattacharya 2011). This exclusivity encourages investment by ensuring financial returns on novel biomaterials, thereby supporting future medical advancements (EBIN.PUB n.d.). Patents motivate investment, provide market exclusivity, generate revenue through licensing or sales, and provide legal recourse in the event of infringement, ensuring that inventors are compensated for their innovations (Williams 2017). Patents promote innovation in the biomaterials field by protecting inventions, encouraging collaboration and knowledge sharing among researchers, and attracting investment, all of which lead to the development of groundbreaking technologies (PDF n.d.). This facilitates research and development, public disclosure, the establishment of new markets and industries, and economic growth (Wang et al. 2014; Huebsch and Mooney 2009). Biomaterial patenting show an increasing emphasis on engineered materials for specific biomedical applications, with a particular emphasis on biodegradable, bioactive, self-healing, and programmable materials to meet the growing demand for functional and environmentally friendly solutions in drug delivery and tissue engineering (Bhat and Kumar 2013; Tu et al. 2019). Emerging biomaterials patenting technologies, such as additive manufacturing for customized structures, bioactive materials for improved implant integration, and nanotechnology for enhanced drug delivery systems, are accelerating biomaterial evolution and opening up new possibilities in healthcare, including regenerative medicine and innovative drug formulations (Kravanja and Finšgar 2022; Hakim et al. 2024). Patent databases reveal that biomaterials innovation is primarily concentrated in China, the United States, Europe, and Japan, with a notable increase in patents for biodegradable and bioactive materials, indicating a collaborative effort to improve medical devices and treatments (Wu and Lee 2007). Academic institutions, biotech companies, and research organizations have all contributed to a steady increase in patent filings over the last decade (Hernández-Melchor et al. 2023). Multinational corporations, academic institutions, and specialized biotech firms are significant stakeholders, with a focus on drug delivery systems, tissue engineering, regenerative medicine, and implantable devices, particularly those involving biodegradable polymers, nanotechnology, and 3D printing (Guan et al. 2017; Serrano et al. 2023). Significant biomaterials patents have had a significant impact on biotechnology and medical fields, with the Harvard Oncomouse case being particularly important in addressing the patentability of transgenic animals, sparking ethical and legal debates, and the Tomatoes and Broccoli cases shaping the patentability of plants produced through biological processes, influencing the protection of biotechnological innovations in Europe (New Developments in Biotechnology 1989; ABG IP n.d.).

15.3.1 Case Studies of Significant Patents

The listed patents represent significant advancements in biomaterials and biotechnology, each shaping its respective field. Table 15.2, compiles case studies examples of significant patents in biomaterials. The Cohen-Boyer patents (1976) laid the groundwork for genetic engineering, enabling the production of therapeutic proteins like insulin (Cook-Deegan and Heaney 2010). The Harvard Oncomouse patent (1988) established the precedent for patenting transgenic animals (Gilbert and Walter 2001), revolutionizing cancer research. BRCA gene patents (1994–1995) led

Table 15.2 Case studies of significant patents in biomaterials

Sr. No.	Patent name	Description	Year	Impact on field	References
1.	**Cohen-boyer Patents**	Related to recombinant DNA technology, covering methods for splicing genes to produce proteins	1976	Established foundational techniques for genetic engineering, leading to commercial applications like insulin and other therapeutic proteins	Cook-Deegan and Heaney (2010)
2.	**Harvard oncomouse Patent**	Patent for the genetically modified mouse designed for cancer research, enhancing cancer susceptibility studies	1988	Set precedent for the patentability of transgenic animals, profound implications for genetic research models	Gilbert and Walter (2001)
3.	**BRCA Gene Patents (Myriad)**	Patents covering isolated BRCA1 and BRCA2 genes associated with breast cancer susceptibility	1994-1995	Led to a Supreme Court ruling that naturally occurring DNA cannot be patented, reshaping the genetic patent landscape	Cartwright-Smith (2014)
4.	**Tomatoes and Broccoli Cases**	Addressing the patentability of biotechnologically enhanced breeding methods for fruits and vegetables	2010, 2013	Clarified regulations surrounding the patenting of products obtained through biological processes, influencing biotechnological innovation	Minssen and Nordberg (2015)
5.	**Brüstle Case**	Clarified the patentability of human embryonic stem cells and the conditions required for their patentability	2011	Influenced the understanding of what constitutes a human "embryo" and how stem cell technologies can be patented in Europe	Nielen et al. (2013)
6.	**Chitosan as Biomaterial Patent**	Patent covering the use of chitosan for biomedical applications, particularly in drug delivery systems and wound dressings	2000	Enhanced the use of biopolymer-based materials, driving innovation in drug delivery and tissue engineering applications	Fatimi (2022)

(continued)

Table 15.2 (continued)

Sr. No.	Patent name	Description	Year	Impact on field	References
7.	**Fungal Mycelium Biomaterial Patents**	Patents on utilizing fungal mycelium as a sustainable alternative for packaging and construction materials	2009-2018	Promoted the exploration of eco-friendly materials, highlighting the potential of fungal mycelium in replacing petroleum-based products	(Alemu et al. 2022)
8.	**Biodegradable Polymer Patents**	Patents focused on the development of biodegradable polymers for various biomedical applications	2005-2015	Contributed to advancements in environmentally friendly medical devices, addressing sustainability concerns in healthcare materials	Hosseini et al. (2021)
9.	**Polymer-Extracellular Matrix (ECM) Patents**	Patents that focus on the composite technology utilizing ECMs for enhancing biocompatibility in implants	2015	Encouraged innovation in regenerative medicine by improving integration of synthetic materials with biological tissues	Xing et al. (2020)

to a landmark Supreme Court decision, affirming that naturally occurring DNA cannot be patented (Cartwright-Smith 2014). The Tomatoes and Broccoli cases (2010, 2013) clarified the patentability of biotechnological processes, influencing agricultural innovation (Minssen and Nordberg 2015). The Brüstle case (2011) set standards for patenting human embryonic stem cells in Europe (Nielen et al. 2013). In biomaterials, the chitosan patent (2000) advanced drug delivery and wound care, while the hydrogel patents (2010–2013) pioneered smart drug delivery systems (Fatimi 2022). Fungal mycelium patents (2009–2018) highlighted sustainable alternatives for packaging and construction (Alemu et al. 2022), and biodegradable polymer patents (2005–2015) addressed environmental concerns in medical devices (Hosseini et al. 2021). Lastly, the Polymer-ECM patents (2015) improved biocompatibility in implants, enhancing regenerative medicine (Xing et al. 2020). These patents collectively demonstrate the evolution and impact of innovative technologies in both biotechnology and biomaterials.

15.4 Clinical Trials Involving Biomaterials

15.4.1 Overview of Clinical Trials

A cornerstone of evidence-based medicine is that clinical trials give critical data as to the safety and efficacy of new medical interventions, which also involve the use of biomaterials. In biomaterials study, clinical trials link together the developments of laboratories and practice in hospitals so that new materials or devices have not just to escape factory for wide affordable production without meeting stringent safety and performance standards (Holzapfel et al. 2013). The unique properties of biomaterials, such as their interaction with living tissues, degradation profiles, and potential long-term effects, necessitate a specialized approach to clinical trial design and execution (Williams 2008).

15.4.1.1 Preclinical Studies

Preclinical studies are required prior to human trials, and include in vitro tests and animal experiments to assess the biomaterial's safety, biocompatibility, and effects. In vitro testing evaluates cytotoxicity, cell adhesion, and proliferation, as demonstrated by osteoblast culture studies for bone scaffolds (Bose et al. 2012). Animal models provide insights into tissue integration, long-term behavior, and systemic effects, emphasizing the need for models that mimic human physiology (Navarro et al. 2008). Biomechanical testing ensures that materials used in load-bearing applications match native tissue properties (Hutmacher 2000), whereas degradation studies investigate material breakdown and the impact of byproducts, such as PLGA scaffolds (Li and Vert 2002).

15.4.1.2 Phases of Clinical Trials

The clinical trial process in biomaterials typically follows a structured pathway, progressing through several phases, each with specific purposes and greater complexity (Jaalouk et al. 2024). This process systematically evaluates the safety, efficacy, and optimal use of new biomaterials while minimizing participant risks.

(a) **Phase I Trials**

Phase I trials are the first step in human testing, with 20–80 participants evaluating biomaterials' safety, tolerability, and preliminary efficacy. To ensure safety, key aspects include monitoring for adverse reactions and tissue responses, determining optimal dosing for drug-eluting biomaterials that deliver agents such as growth factors or antibiotics, and assessing biocompatibility through tissue integration and inflammatory response analysis. Furthermore, these trials aim to improve surgical techniques for implantable biomaterials in order to standardise procedures and reduce variability. While safety is the primary objective, early efficacy indicators, such as initial bone formation in novel scaffolds, have been observed (Kearney et al. 2021; Lüscher 2020).

(b) **Phase II Trials**

Phase II trials, which typically involve 100–300 participants, Aim to further evaluate the efficacy of biomaterials while prioritizing safety (Kearney et al. 2021; Lüscher 2020). These trials provide strong evidence on the material's efficacy by comparing it to existing treatments or placebos, as demonstrated by a study of bioresorbable vascular scaffolds versus metallic stents in patients with coronary artery disease (Kearney et al. 2021; Lüscher 2020). During this phase, biomaterial application protocols are optimized, including surgical techniques and methods for non-implantable materials. Subgroup analyses also identify specific patient populations who are most likely to benefit, taking into account factors such as age and comorbidities (Collins et al. 2014). Dose-response relationships are evaluated to achieve the best balance of efficacy and safety, particularly in drug-eluting biomaterials (Lee et al. 2011). Furthermore, Phase II trials reveal the biomaterial's impact on patient function and quality of life using validated patient-reported outcome measures, providing comprehensive insights into therapeutic benefits (Smith et al. 2019).

(c) **Phase III Trials**

Phase III trials are large-scale, involving hundreds or thousands of subjects, and are typically designed as randomized controlled, superiority, or equivalence trials to provide conclusive evidence on the safety and efficacy of biomaterials when compared to standard treatments (Kearney et al. 2021; Lüscher 2020). These trials emphasize randomization, in which participants are assigned to treatment groups at random in order to reduce bias and accurately determine the effectiveness of the biomaterial (Schulz and Grimes 2002). The International Consortium of Orthopaedic Registries (ICOR) has demonstrated that Phase III trials, which are conducted across multiple research centers, aim for generalizability by addressing geographical and population-based differences (Sedrakyan et al. 2011). Holzapfel and colleagues emphasize the importance of long-term follow-up, particularly for implantable biomaterials, in assessing durability and extended safety (Holzapfel et al. 2013). Comprehensive safety monitoring is carried out by independent data safety monitoring boards to ensure participant safety, while economic evaluations, including cost-effectiveness analysis, guide healthcare decision-making and reimbursement (Demets et al. 2006). Furthermore, these trials assess quality of life impacts using patient-reported outcomes, providing a broader view of the intervention's benefits (Patrick et al. 2007). The ABSORB III trial, which included 2008 patients from 193 sites and compared a bioresorbable vascular scaffold to a metallic stent for coronary artery disease, exemplifies the rigorous and multifaceted nature of Phase III biomaterials research (Ellis et al. 2015).

(d) **Phase IV Trials**

Phase IV trials, also known as post-marketing surveillance studies, are carried out following regulatory approval to assess long-term performance and detect rare adverse events in real-world settings (Kearney et al. 2021; Lüscher 2020). These trials involve large patient populations, often numbering in the thousands, in order to uncover infrequent complications that were not identified in earlier phases (Sedrakyan et al. 2011). They typically last several years, especially for biomaterials intended for long-term implantation, to ensure a thorough assessment of safety and efficacy (Holzapfel et al. 2013). Phase IV trials collect real-world data to assess effectiveness across diverse populations and clinical settings, providing important insights into routine clinical use (Sherman et al. 2016). The large scale of these studies allows for the detection of rare adverse events that may only occur with widespread application, ensuring biomaterials' long-term safety (Resnic and Normand 2012). They also compare biomaterial performance to alternative treatments in clinical practice, developing their position in the treatment landscape (Sox and Greenfield 2009). These trials also help tailor patient selection criteria, identifying those who are most likely to benefit or face increased risks, allowing for more personalized treatment plans (Collins et al. 2014). The International Consortium of Orthopaedic Registries (ICOR) is a notable example of Phase IV biomaterial research, as it aggregates global data to track long-term trends and identify potential issues with joint implants (Sedrakyan et al. 2011).

15.4.1.3 Regulatory Considerations

Regulatory considerations for biomaterials and medical devices are complex and ever-changing, with global agencies tailoring their frameworks to ensure safety and efficacy while allowing for faster clinical integration of innovative products. This section provides an overview of key regulatory aspects of clinical trials involving biomaterials, highlighting the roles and requirements of various agencies, including the United States Food and Drug Administration (FDA), whose Center for Devices and Radiological Health (CDRH) regulates the majority of biomaterials (White 2018; Food and Drug Administration 2018).

15.4.1.4 Classification of Medical Devices

The FDA classifies medical devices into three categories based on risk level and necessary regulatory controls. Most novel biomaterials fall into Class II or III categories due to their potential risks and the level of clinical evidence required to demonstrate their safety and efficacy (Jarow and Baxley 2015).

Class I: Low-risk devices
Class II: Moderate-risk devices
Class III: High-risk devices

15.4.1.5 Premarket Approval (PMA)

Most innovative Class III biomaterials require premarket approval (PMA), which necessitates extensive preclinical and clinical trial data to ensure their safety and efficacy for intended use (Food and Drug Administration 2019). Key components include nonclinical studies, clinical investigations, manufacturing details, labeling, and post-approval requirements, which frequently involve randomized controlled trials with long-term follow-up (Maisel 2004). Alternatively, the 510(k) clearance process, which is less stringent than PMA, applies if the device is substantially equivalent to an existing one and requires evidence of comparable safety and efficacy (Food and Drug Administration 2020). Furthermore, before conducting significant risk clinical trials, researchers must obtain an Investigational Device Exemption (IDE) to ensure the safety of human subjects. Many modern biomaterials are combination products, which combine drugs or biologics with base materials (Food and Drug Administration 2021). The FDA's Office of Combination Products assesses their primary mode of action. In the European Union, the Medical Device Regulation (EU MDR), which goes into effect in May 2021, replaces the Medical Devices Directive (MDD) and includes stricter pre-market controls, mandatory Unique Device Identification (UDI) for traceability, enhanced post-market surveillance, a new review process for high-risk devices, and increased transparency through the European Database on Medical Devices (European Parlament, Council of the European Union 2017; Pane et al. 2017). International standards are critical in biomaterials research and development, including ISO 10993, which addresses biological evaluation of medical devices, ISO 13485, which outlines quality management system requirements for medical devices, and ASTM F2150, which provides guidelines for testing biomaterial scaffolds in tissue-engineered products, with adherence to these standards frequently mandated or strongly recommended by global regulatory agencies (ANSI/AAMI/ISO 10993-1:2018 2020; International Organization for Standardization 2016; Schuh and Funk 2019).

Several challenges are shaping the regulatory landscape for biomaterials, including the rise of personalized medicine, which is forcing traditional frameworks to be reevaluated in order to accommodate tailored implants and 3D-printed products (Di Prima et al. 2016). As nanotechnology becomes more integrated into biomaterials, regulators are developing new guidelines to ensure the safety and efficacy of the materials (Lombardo et al. 2020). Furthermore, adaptive clinical trial designs are gaining traction, with the potential to accelerate biomaterial development. There is also an increasing emphasis on incorporating real-world evidence into regulatory decision-making to supplement traditional clinical trial data (Pallmann et al. 2018). Finally, efforts to harmonize regulatory requirements around the world are underway, which could speed up the approval process for biomaterials in various regions (International Regulatory Co-operation 2016).

15.5 Biomaterials in Clinical Trials

Biomaterials play an important role in modern medicine, providing innovative solutions for tissue repair, drug delivery, and medical device development, with ongoing research aimed at determining their safety and efficacy through clinical trials. These materials are being studied in various categories based on their composition and intended application, including synthetic polymers such as poly(lactic-co-glycolic acid) (PLGA), polycaprolactone (PCL), and polyethylene glycol (PEG), which are valued for their tunable properties in tissue engineering and drug delivery (Gentile et al. 2014). Natural polymers like collagen, chitosan, and alginate are known for their biocompatibility and ability to mimic the extracellular matrix (Nezhad-Mokhtari et al. 2019). Ceramics, particularly bioactive glasses and calcium phosphate-based materials, for bone regeneration and dental use (Baino et al. 2018). Biodegradable metals, such as magnesium alloys, are being investigated for orthopedic and cardiovascular applications; composite materials combining polymers and ceramic items to capitalize on the strengths of both; hydrogels, which are being investigated for drug delivery, tissue engineering, and wound healing; and nanoparticles, including liposomes, polymeric, and inorganic varieties, for targeted drug delivery and imaging applications (Zhao et al. 2017; Ramakrishna et al. 2001; Caló and Khutoryanskiy 2015).

15.5.1 Successful and Failed Trials

The journey from biomaterial development to clinical application is often long and complex, with both successes and failures along the way. Here, we examine some notable examples of successful and failed clinical trials involving biomaterials.

15.5.2 Successful Trials

(a) **Drug-Eluting Stents**

Significant reductions in restenosis rates were demonstrated by the SIRIUS trial that assessed a sirolimus-eluting Cypher stent compared to bare-metal stents (Moses et al. 2003). This was a great success and lead to the wide spread usage of drug-eluting stents in percutaneous coronary interventions (Moses et al. 2003).

(b) **Artificial Corneas**

The AlphaCor artificial cornea, which has a poly(2-hydroxyethyl methacrylate) hydrogel composition was tested in clinical trials for patients with corneal blindness (Hicks et al. 2003). The FDA approved the device in 2002 and has used it to repair vision for patients with corneal damage that are too severe (Hicks et al. 2003).

(c) **Bone Graft Substitutes**

The clinical trials for Recombinant Human Bone Morphogenetic Protein-2 (rhBMP-2) in an absorbable collagen sponge resulted in the successful union of autogenous bone graft and rhBMP-2-treated groups that were approved by FDA, specifically aimed at spinal fusion surgeries (Burkus et al. 2002).

(d) **Hydrogel-Based Wound Dressings**

Compared to standard care, a novel hydrogel-based wound dressing consisting of an antimicrobial silver nanoparticle provided faster healing rates and decreased infection in chronic wounds in a clinical trial (Moura et al. 2013).

15.5.3 Failed Trials

(a) **Metal-on-Metal Hip Implants**

However, once implanted in actual subjects, the performance of large-diameter metal-on-metal hip implants failed to live up to initial promises with substantially higher-than-expected failure rates and adverse tissue reactions. In 2010, the ASR XL Acetabular System manufactured by DePuy Orthopaedics was recalled after clinical data revealed high rates of revision (Rahman et al. 2016).

(b) **Silicone Breast Implants**

Silicone breast implants from earlier generations were highly controversial over rupture and accompanied by legal challenges on claims of risks. While newer versions have been approved, the history of silicone implants should give us all pause about only approving them based upon short term safety data (Institute of Medicine 1999).

(c) **Bioabsorbable Stents**

Due to increased rates of target lesion failure when compared with metallic drug-eluting stents as seen in the ABSORB III trial evaluating a completely bioresorbable vascular scaffold, this product was withdrawn from sale (Kereiakes et al. 2017).

(d) **Synthetic Trachea**

Clinical trials of tissue-engineered tracheal replacements using synthetic scaffolds seeded with autologous stem cells faced significant challenges and ethical concerns due to poor outcomes and complications (Elliott et al. 2012).

15.5.4 Case Studies

15.5.4.1 Examples of Recent Clinical Trials

(a) **Bioresorbable Vascular Scaffolds (BVS)**

The ABSORB III trial, a large-scale randomized controlled trial, compared the safety and efficacy of the Absorb bioresorbable vascular scaffold (BVS) to the everolimus-eluting XIENCE metallic stent in patients with coronary artery disease, enrolling 2008 patients from 193 sites. The trial aimed to evaluate the potential benefits of a dissolvable scaffold as an alternative to permanent metallic stents, with the goal of reducing long-term complications. After 3 years, the BVS group had higher rates of target lesion failure (13.4% vs. 10.4%, $p = 0.06$) and device thrombosis (2.3% vs. 0.7%, $p = 0.01$), than the metallic stent group. However, subgroup analysis revealed that optimal implantation techniques and vessel sizes may improve BVS outcomes (Caixeta et al. 2019).

(b) **3D-Printed Titanium Spinal Implants**

Mobbs et al. (2017) conducted a clinical trial to investigate the use of 3D-printed titanium implants for cervical and lumbar spinal fusion in 20 patients undergoing anterior cervical discectomy and fusion (ACDF) or transforaminal lumbar interbody fusion (TLIF). The study discovered that all patients had successful fusion 12 months after surgery, with custom implants providing a better anatomical fit and reducing operative time, and that no implant-related complications occurred during the follow-up period (Mobbs et al. 2017).

(c) **Hydrogel-Based Drug Delivery for Age-Related Macular Degeneration**

The LADDER trial (Campochiaro et al. 2019) investigated into a novel port delivery system (PDS) that used a refillable hydrogel implant to provide sustained ranibizumab to patients with neovascular age-related macular degeneration (nAMD). The PDS demonstrated similar effectiveness to monthly intravitreal injections, with the highest dose formulation having a median time to first refill of 15.8 months. Furthermore, the implant significantly reduced the treatment burden, with 80% of patients not requiring refills for at least 6 months (Campochiaro et al. 2019).

(d) **Silk Fibroin Scaffolds for Tympanic Membrane Repair**

Levin et al. (2010) conducted a significant phase 1 clinical trial using silk fibroin scaffolds to repair tympanic membrane perforations, involving 30 patients with chronic perforations, and discovered that 92% of patients experienced complete closure of the perforation within 6 months, with no adverse immune reactions observed, indicating excellent biocompatibility, and 87% of patients improved their hearing (Levin et al. 2010).

15.5.4.2 Lessons Learned and Outcomes

(a) **Importance of Long-Term Follow-Up**

The ABSORB III trial highlighted the critical importance of long-term follow-up in biomaterials research. While initial results at 1 year were promising, the 3-year data revealed unexpected complications.

Lesson learned: Clinical trials for biomaterials, especially those designed for long-term implantation, should include extended follow-up periods to fully assess safety and efficacy (Kereiakes et al. 2017).

(b) **Optimization of Implantation Techniques**

Subgroup analyses from the ABSORB III trial indicated that outcomes with BVS could be improved with optimal implantation techniques and appropriate patient selection.

Lesson learned: The success of a biomaterial can be highly dependent on the technique used for implantation. Clinical trials should include standardized protocols and consider the learning curve associated with new technologies (Kereiakes et al. 2017).

(c) **Potential of Personalized Medicine in Biomaterials**

The success of 3D-printed titanium implants in spinal fusion (Mobbs et al. 2017) demonstrated the potential of personalized medicine in biomaterials research.

Lesson learned: Custom-designed implants can offer improved anatomical fit and potentially better clinical outcomes. Future trials should explore the balance between standardization and personalization in biomaterial design (Mobbs et al. 2017).

(d) **Addressing Unmet Clinical Needs**

The LADDER trial (Campochiaro et al. 2019) showed how innovative biomaterial-based solutions can address significant unmet needs in clinical practice, such as reducing treatment burden in chronic conditions.

Lesson learned: Biomaterials research should focus on addressing key clinical challenges, with patient quality of life and healthcare system efficiency as important considerations alongside traditional efficacy and safety endpoints (Campochiaro et al. 2019).

(e) **Translational Challenges in Biomaterials**

The silk fibroin scaffold trial (Levin et al. 2010) demonstrated successful translation of a novel biomaterial from bench to bedside. However, the journey from preclinical promise to clinical success remains challenging for many biomaterials.

Lesson learned: Successful translation requires close collaboration between material scientists, clinicians, and regulatory bodies. Early consideration of manufacturing scalability, sterilization methods, and regulatory requirements is crucial (Levin et al. 2010).

(f) **Importance of Post-Market Surveillance**

The experience with metal-on-metal hip implants highlighted the importance of robust post-market surveillance for implantable biomaterials (Drummond et al. 2015).

Lesson learned: Clinical trials should be complemented by comprehensive post-market surveillance programs to detect rare complications and long-term issues that may not be apparent in initial studies (Drummond et al. 2015).

15.6 Dosage Forms Based on Biomaterials

Biomaterial-based dosage forms play an important role in improving drug efficacy and patient compliance by allowing for tailored drug release and absorption mechanisms, including solid, liquid, and semi-solid forms, each with unique advantages in drug delivery (Trucillo 2024).

15.6.1 Types of Dosage Forms

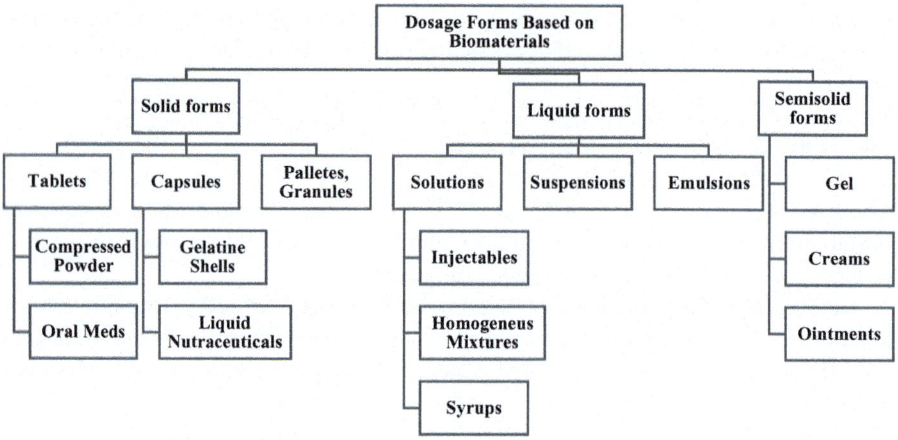

15.6.1.1 Solid Dosage Forms

Solid dosage forms such as tablets, capsules, and nanoparticles are popular due to their convenience and stability, with nanoparticles offering innovative encapsulation techniques to improve drug bioavailability and enable targeted therapies, while tablets remain a popular compressed form containing APIs and excipients. Biomaterials can be used as binders (e.g., cellulose derivatives), disintegrants (e.g., starch), or coating agents (e.g., alginate) (Adepu and Ramakrishna 2021). Capsules are solid dosage forms in which APIs are enclosed within a shell. They can be either hard gelatin capsules or soft gelatin capsules. Biomaterials such as hydroxypropyl methylcellulose (HPMC) or alginate are often used for capsule shells, especially for creating vegetarian or vegan capsules (Majee et al. 2017). Pellets are small, spherical particles, while granules are irregularly shaped particles. Both can be used for oral administration. Biomaterials can be used to form matrix systems or as coating materials to control the release of the drug (Al-Hashimi et al. 2018).

15.6.1.2 Liquid Dosage Forms

Solutions are homogeneous mixtures in which APIs are dissolved in solvents. Biomaterials such as polysaccharides (e.g., xanthan gum) or proteins (e.g., gelatin) can be used to enhance the stability or solubility of the solution (Wang et al. 2016). Suspensions are heterogeneous mixtures in which solid particles are dispersed in a liquid medium. Biomaterials can act as suspending agents (e.g., cellulose derivatives) or stabilizers (e.g., chitosan) (Kostag and El Seoud 2021). Emulsions are mixtures of two immiscible liquids, typically oil and water, stabilized by emulsifying agents. Biomaterials like lecithin or surfactants derived from natural sources are used to stabilize emulsions (Kupikowska-Stobba et al. 2024).

15.6.1.3 Semi-Solid Dosage Forms

Gels are semi-solid systems where the API is dispersed in a three-dimensional network of a polymer. Biomaterials such as carbopol, alginate, or chitosan are used to form the gel network. Creams are emulsions of oil and water, while ointments are greasy, semi-solid preparations used for topical application. Natural oils, waxes, and cellulose derivatives can be used as the base or emulsifying agents in creams and ointments (Slavkova et al. 2023).

15.7 Delivery Systems

Biomaterial-based delivery systems are designed to improve the efficacy, safety, and patient compliance of drug therapies. These systems offer controlled, sustained, or targeted release of active pharmaceutical ingredients (APIs). Some of delivery systems are.

15.7.1 Drug-Eluting Implants

Biodegradable implants are devices implanted in the body that gradually degrade and release the drug over time. Common biomaterials include polylactic acid (PLA), polyglycolic acid (PGA), and their copolymers like PLGA (poly(lactic-co-glycolic acid)) are used. Biodegradable Implants are used for localized, long-term treatment, such as in cancer therapy, pain management, and cardiovascular diseases (Makadia and Siegel 2011; Liang et al. 2023). Non-Biodegradable implants do not degrade in the body but are designed to release the drug over an extended period. Common materials include silicone, polyethylene, and titanium. Often used for chronic conditions where long-term drug release is required, such as in contraceptive devices and certain ophthalmic implants (Magill et al. 2023).

15.7.2 Injectable Hydrogels

In Situ Forming Hydrogels are liquid formulations that gel upon injection into the body, providing a depot for sustained drug release. Materials like natural polymers like gelatin, hyaluronic acid, chitosan, and synthetic polymers like polyethylene glycol (PEG) derivatives. Used for localized drug delivery in wound healing, tissue engineering, and cancer therapy (Madan et al. 2009). Pre-Formed Hydrogels that are pre-formed outside the body and then injected. They can encapsulate drugs and release them in a controlled manner. Materials Used are Alginate, polyvinyl alcohol (PVA), and other biocompatible polymers. Applied in soft tissue fillers, drug delivery for localized infections, and injectable scaffolds for regenerative medicine (Stapleton et al. 2020; Rizwan et al. 2017).

15.7.3 Microparticles and Nanoparticles

Microparticles are small particles, typically ranging from 1 to 1000 micrometers in size, used to encapsulate drugs for controlled release. Materials Used are polymers like PLA, PLGA, and chitosan. Used for vaccine delivery, sustained-release formulations, and targeted drug delivery (da Silva et al. 2023). Nanoparticles are ultra-small particles, typically less than 100 nanometers in size, designed for targeted and controlled drug delivery. Materials used are lipids, polymers like PLGA, and inorganic materials like gold and silica. Used for targeted cancer therapy, gene delivery, and delivery of poorly soluble drugs (Mitchell et al. 2020).

15.7.4 Liposomes and Niosomes

Liposomes are spherical vesicles composed of one or more phospholipid bilayers, used to encapsulate drugs. Materials Used are Natural phospholipids, cholesterol. Used for targeted drug delivery, reducing drug toxicity, and improving the bioavailability of drugs, such as in chemotherapy (Nsairat et al. 2022; Akbarzadeh et al. 2013). Niosomes are similar to liposomes but made from non-ionic surfactants, offering a more stable and cost-effective alternative. Materials used are non-ionic surfactants like Span and Tween. Used for topical drug delivery, targeted drug delivery, and vaccines (Karim et al. 2010).

15.7.5 Polymeric Micelles

Polymeric micelles are nanoscale carriers formed by the self-assembly of amphiphilic block copolymers in aqueous solutions. Materials used are block copolymers like PEG-PLA, PEG-PCL (polycaprolactone). Used for the delivery of hydrophobic drugs, enhancing solubility and bioavailability, especially in cancer therapy (Assiri et al. 2024; Patel et al. 2023).

15.8 Design Considerations

To ensure effectiveness, safety, and ease of use, biomaterial-based delivery systems must be carefully designed, taking into account factors such as stability, release profiles, and patient compliance. Biomaterial stability throughout the product's shelf life is critical, necessitating the use of appropriate container materials and an understanding of drug-container interactions to prevent degradation (Zothanpuii et al. 2020). To achieve therapeutic goals, release profiles must be optimized, with controlled-release systems that use advanced polymers for precise pharmacokinetics. Stability considerations include physical stability, which ensures the structural integrity of nanoparticles, microparticles, or implants over time; chemical stability, which protects the drug from degradation caused by oxidation, hydrolysis, or photolysis; and biological stability, which preserves the drug's bioactivity and prevents enzymatic breakdown within the delivery system (Yusuf et al. 2023; Ioele et al. 2022). Drug release profiles can be tailored for controlled, sustained, or targeted delivery. Controlled release ensures a consistent therapeutic effect by releasing the drug at a predetermined rate using mechanisms such as diffusion, degradation, or swelling in biomaterials (Geraili et al. 2021). Sustained release extends drug release over time, reducing dosing frequency and improving patient adherence; it is frequently used with biodegradable polymers that degrade gradually (Baryakova et al. 2023). Targeted release directs the drug to the site of action, reducing systemic side effects, by using targeting ligands or stimuli-responsive materials that activate in specific environments, such as pH or temperature (Wang et al. 2023b). For example PLA/PLGA Microspheres polymers provide a biodegradable matrix for the sustained release of drugs. The degradation rate can be tailored by adjusting the ratio of lactic acid to glycolic acid. Hydrogels can be designed to swell in response to specific physiological conditions, providing controlled release of encapsulated drugs (Swider et al. 2018). Improving patient compliance can be accomplished by reducing dosing frequency through controlled release systems, minimizing side effects by targeting specific drug delivery sites, and alleviating pain with user-friendly devices such as fine-needle injections or painless transdermal patches, while also offering non-invasive routes such as oral or nasal administration; additionally, designing portable, easy-to-store systems with clear instructions enhances convenience and adhere to treatment regimens (Baryakova et al. 2023). For example, Transdermal Patches provide a non-invasive means of delivering drugs through the skin, offering sustained release and ease of use. Pre-Filled Syringes and Auto-Injectors devices simplify the injection process, making it easier for patients to self-administer medications with minimal training (Basu et al. 2011). Orally Disintegrating Tablets (ODTs) dissolve quickly in the mouth without the need for water, enhancing ease of administration, especially for patients with difficulty swallowing (Ghourichay et al. 2021).

15.9 Innovations and Developments

With the use of biodegradable and bio-absorbable polymers like poly(lactic-co-glycolic acid) (PLGA), which offer biocompatibility and customizable properties for controlled-release formulations that improve therapeutic effects (Makadia and Siegel 2011). Recent developments in biomaterial-based dosage forms have greatly improved drug delivery systems. Smart polymers that respond to environmental stimuli, such as pH and temperature, are enabling targeted drug delivery for optimal treatment outcomes and advancements in polymeric nanoparticles, micelles, and hydrogels have addressed issues with drug stability and bioavailability, especially for poorly soluble drugs (Hwang et al. 2020; Xu et al. 2013). With the development of personalized drug delivery systems to improve efficacy and compliance, the integration of digital health technologies to monitor medication intake and improve adherence, and bioprinting technology, which holds great promise for creating patient-specific tissues and implants, as well as potential solutions to organ shortages and personalized therapeutic interventions, the future of these dosage forms is focused on personalized medicine (Su et al. 2024; Yeung et al. 2023; Janarthanan et al. 2020). All of these developments collectively mark a transformative period in biomaterials and drug delivery aimed at improving clinical outcomes and patient care.

15.10 Integration of Patents, Clinical Trials, and Dosage Forms

15.10.1 Interplay Between Patents and Clinical Trials

The pharmaceutical industry's development paradigm fundamentally rests on the complex relationship between patent protection and clinical research progression (Gurgula 2020). Strategic patent filing decisions directly influence clinical trial timelines, with companies typically securing intellectual property protection before initiating significant clinical investments (Li and Bergan 2020). Recent industry analyses reveal that pharmaceutical companies spend approximately $2.6 billion on research and development per approved drug, making patent protection crucial for investment recovery (DiMasi et al. 2016). Patent protection strategies increasingly incorporate anticipated clinical development pathways. Statistical analyses demonstrate that companies file an average of 3.2 patent applications per compound before phase I trials begin (Seng 2021). This front-loaded approach provides multiple layers of protection while clinical evidence accumulates (Gadiya et al. 2023). Pharmaceutical companies commonly employ continuation applications to maintain patent prosecution flexibility as clinical data emerges (Harrington et al. 2017). Modern patent portfolios frequently include multiple patent families covering various aspects of drug development. Companies file an average of 8.7 patents per approved drug, spanning compound structure, synthetic routes, polymorphs, and formulation technologies (Ni et al. 2015). This comprehensive approach creates robust barriers to competition while clinical trials progress (Markey et al. 2024).

15.10.1.1 How Patents Influence Clinical Research and Vice Versa

Patent specifications significantly shape clinical trial designs through detailed disclosure requirements and claim scope considerations. Researchers must carefully balance protecting intellectual property while generating sufficient clinical data to support patent claims (Timpe et al. 2020). Recent studies indicate that 65% of phase II trial protocols undergo modifications based on patent prosecution feedback

(Trenfield and Basit 2020). Clinical trial results frequently necessitate patent strategy adjustments. Approximately 40% of pharmaceutical patents face amendments during prosecution based on emerging clinical data (Xu and Zhu 2021). These modifications particularly affect dosing regimens and treatment indication claims (Yang 2023). The USPTO's increasing emphasis on clinical evidence supporting therapeutic claims has led to more comprehensive early-phase trials (Thakur et al. 2020). Second medical use patents face particularly stringent requirements for clinical evidence. Recent USPTO guidelines require substantial clinical data supporting new therapeutic applications (Brougher and Fazzolare 2014). This requirement has driven a 35% increase in phase II trials specifically designed to support new use patents (Gadiya et al. 2023). Clinical trial design evolution reflects patent strategy considerations. Companies increasingly incorporate pharmacogenomic endpoints in clinical trials to support personalized medicine patents (Oliver et al. 2013). Recent analysis shows a 50% increase in biomarker-focused patent applications supported by clinical trial data since 2020 (Oliver et al. 2013).

15.11 Impact on Dosage Form Development

Dosage form innovation represents a crucial intersection between patent strategy and clinical development. Novel drug delivery systems often receive independent patent protection, extending product lifecycles beyond compound patents (Gubby 2020). Market analysis indicates that innovative formulation patents extend market exclusivity by an average of 4.8 years (Moorkens et al. 2020). Companies increasingly focus on developing patent-protected delivery technologies. Recent studies show that 45% of pharmaceutical patents filed in 2023 included claims covering novel formulation approaches (Gadiya et al. 2023). This trend reflects growing recognition of delivery technology's commercial importance.

15.11.1 Role of Clinical Trial Outcomes in Shaping Dosage Forms

Clinical trial results directly inform dosage form optimization through iterative development processes. Patient compliance data from phase III trials frequently drives formulation refinements (Allen 2010). Recent studies indicate that 70% of controlled-release formulations undergo modifications based on clinical pharmacokinetic data (Allen 2010). Bioavailability studies particularly influence modified release technology development. Clinical trials comparing different formulation approaches have increased by 55% since 2020 (Olorunsola et al. 2023). This trend reflects growing emphasis on optimizing drug delivery for patient benefit. Population pharmacokinetic studies increasingly guide formulation development. Recent analyses show that 80% of modified release formulations undergo design optimization based on population PK modelling (Allen 2010). This data-driven approach enables more precise targeting of release profiles.

15.11.2 Patent Protection and Commercialization

Successful commercialization requires strategic alignment between patent protection and clinical validation of dosage forms. Companies increasingly file formulation patents before phase II trials, securing broad protection for potential delivery innovations (Ranganathan et al. 2018; Gupta et al. 2020). Market analysis indicates that patents covering novel delivery systems generate 40% higher licensing revenues compared to compound patents alone (Olsen et al. 2023). Regulatory authorities increasingly demand comparative bioavailability studies between formulations. This requirement has led to more comprehensive patent applications incorporating multiple delivery technologies. Recent studies show that 65% of formulation patents include claims covering at least three distinct delivery approaches (Antunes et al. 2013). Patent term extensions frequently target formulation patents rather than compound patents. Analysis of FDA Orange Book listings shows that 45% of patent term extensions granted in 2023 covered novel delivery technologies (Ledley and Cleary 2023; Paulekuhn et al. 2007). This trend reflects the growing commercial importance of formulation innovation (Mukherjee et al. 2020).

15.12 Challenges and Future Directions

15.12.1 Challenges in Biomaterials Research

15.12.1.1 Regulatory Hurdles

The biomaterials field faces significant regulatory challenges in moving new technologies from research to clinical use. These barriers include complex regulatory frameworks, which are especially difficult for combination products that combine biomaterials and therapeutic agents. For example, the FDA's 2020 guidance on drug-device combinations highlighted the difficulties in categorizing biomaterial-based delivery systems, as seen with Novartis' CAR-T cell therapy Kymriah, which employs biomaterial scaffolds (Sheridan 2017). The regulatory process prolonged development timelines and increased costs due to the need for additional safety and characterization studies. Ensuring manufacturing consistency and scalability is another significant regulatory challenge. The EMA's 2021 guidelines for advanced therapy medicinal products (ATMPs) emphasized the importance of rigorous quality control during manufacturing (Dario et al. 2020). Moderna encountered similar challenges when scaling up lipid nanoparticle systems for mRNA vaccines, where maintaining uniform particle size and stability was essential to regulatory approval. This demonstrated how production scalability could become a regulatory bottleneck (Hou et al. 2021). The lack of standardized testing protocols for new biomaterials further complicates the approval process. The ISO updated biocompatibility evaluation guidelines in 2022 (ISO 10993-23:2021), but many innovative biomaterials do not fit traditional testing frameworks. Researchers at MIT, for example, designed self-healing hydrogels but encountered regulatory challenges in developing new testing methodologies to meet these changing standards (Pellevoisin et al. 2022).

The regulatory requirements for long-term safety data have increased, particularly for degradation products and their biological effects. The FDA's 2023 biocompatibility guidance for implantable devices now requires extensive research into degradation products, which impacted Zimmer Biomet's resorbable bone scaffold (Wang et al. 2024; Hiranaka et al. 2023). Extended stability assessments were required to ensure that degradation products did not pose any safety risks, reflecting the increasing demand for long-term safety verification. Furthermore, regulatory harmonization remains a major issue globally. The International Medical Device Regulators Forum (IMDRF) has made some progress in aligning standards, but regional differences remain. For example, the requirements for demonstrating bioequivalence in biosimilar-device combinations differ between the FDA, the EMA, and Japan's PMDA (International Regulatory Co-operation 2016). These disparities complicate global market entry for biomaterial-based products and necessitate manufacturers navigating various compliance protocols. To address these challenges, several initiatives are underway, including standardized biomaterial characterization methods, AI-driven predictive toxicology, regulatory science associations to simplify approvals, and specialized pathways for biomaterial-inclusive combination products. These approaches aim to simplify the regulatory process for innovative biomaterials and promote efficient, safe clinical translation (Lui 2024).

15.12.1.2 Technical and Scientific Barriers

Despite recent advances, biomaterials research continues to face significant technical and scientific challenges, including precise control of material-tissue interactions at the molecular level. One significant challenge is managing immune responses; for example, even with advanced surface modifications, titanium implants frequently cause inflammation, as demonstrated by Zhang et al., necessitating ongoing efforts to develop "stealth" biomaterials that avoid immune detection (Wang et al. 2023c). Scaling laboratory production to industrial levels presents another challenge, particularly in maintaining quality in large-scale hydrogel production for drug delivery (Chen et al. 2023). Similarly, controlling degradation kinetics remains challenging, environmental factors can unpredictably alter degradation rates in PLGA scaffolds, affecting drug release and tissue regeneration. Mechanical property optimization presents yet another challenge, particularly in applications such as cardiovascular stents, where some authors identified difficulties in balancing flexibility, strength, and biocompatibility. Furthermore, limitations in current characterization techniques impede comprehensive in vivo tracking, authors also highlighted the need for improved imaging methods to monitor degradation and integration at microscopic levels (Zhang et al. 2022; Kim et al. 2021; Patel et al. 2021).

Looking ahead, promising strategies include advanced manufacturing technologies that use AI and machine learning to predict and control material properties at scale; the development of smart biomaterials with programmable biological responses, such as the self-healing hydrogels and improved imaging techniques that combine multiple modalities for better in vivo tracking (Chen et al. 2023; Meyer et al. 2023; Shokrani et al. 2023). Further directions include personalized biomaterials made possible by 3D printing and tailored to individual patient needs, as well as sustainable manufacturing methods that prioritize eco-friendly and biodegradable materials (Ramachandran et al. 2023). Addressing these challenges demands interdisciplinary collaboration across material science, biology, and clinical fields, building the platform for transformative biomaterial innovations.

15.12.2 Future Prospects

15.12.2.1 Emerging Technologies and Potential Breakthroughs

The future of biomaterial-based dosage forms in drug delivery is prime for transformative breakthroughs thanks to emerging technologies that are reshaping pharmaceutical development. Materials science advancements such as stimuli-responsive biomaterials, nanotechnology, and 3D bioprinting are paving the way for more efficient, patient-centered drug delivery systems (Zhu et al. 2022). Stimuli-responsive biomaterials, for example, respond to specific triggers such as pH, temperature, or light, allowing for controlled, localized drug release, lowering side effects and improving therapeutic outcomes (Shi et al. 2021). For cancer treatments, pH-sensitive hydrogels can selectively release drugs within acidic tumor environments, minimizing harm to healthy cells and increasing treatment precision. Nanotechnology has also enabled advancements in the development of nano-carriers that can cross biological barriers, evade immune responses, and deliver drugs directly to specific cells (Liu et al. 2022). Such technologies are critical for treating complex medical conditions such as cancer and neurological disorders, where targeted drug delivery can improve efficacy while reducing systemic toxicity.

Another promising innovation is 3D bioprinting, which is transforming the development of personalized, biomaterial-based drug delivery systems. This technology enables the development of dosage forms tailored to an individual's therapeutic requirements, such as customized implants with specific drug release profiles or biodegradable scaffolds which assist in tissue regeneration (Zhang et al. 2021). Recent patents have emerged on 3D-printed, antibiotic-infused implants that prevent infections while promoting bone healing, demonstrating their potential in personalized medicine. Artificial intelligence (AI) is another game changer, hastening biomaterial research by predicting material properties and optimizing design. AI applications in biomaterials can improve drug formulation by predicting interactions with biological systems and identifying optimal materials for drug delivery with greater precision. For example, AI has been used to create bioabsorbable stents with specific degradation rates that correspond to patient needs. These emerging technologies are driving patent activity and clinical trials, indicating increased commercial interest and market integration potential (Webster 2023). Regulatory frameworks will need to evolve to accommodate these advancements, ensuring the safety and efficacy of new biomaterial-based drug delivery systems. As these technologies advance, they promise a future in which biomaterials are safer, more effective, and customizable, paving the way for better patient outcomes and expanded applications in healthcare.

15.12.2.2 Predictions for the Evolution of Biomaterials

The future of biomaterials in drug delivery and medical applications appear promising, with a focus on therapeutic efficacy, safety, and personalization. Significant progress is expected in areas such as stimuli-responsive materials, nanotechnology, and sustainable production, all of which will transform the development of dosage forms and clinical applications to meet complex medical requirements (Ranakoti et al. 2023). One prominent trend involves "smart" biomaterials that respond to specific physiological stimuli, such as pH or temperature, allowing for targeted and controlled drug release. For example, pH-sensitive hydrogels are being studied in cancer therapy because they remain stable in the bloodstream and release their drug payload in the acidic environment of tumors, increasing efficacy while reducing systemic toxicity (Amukarimi et al. 2021). Similarly, temperature-sensitive hydrogels may offer precise diabetes treatment by releasing insulin in response to high blood glucose levels. Nanotechnology-enhanced biomaterials are expected to play a critical role in improving drug delivery and tissue engineering. For example, lipid-based nanoparticles have shown efficacy in mRNA vaccine delivery, as seen in COVID-19 vaccines, demonstrating the potential of nanoscale biomaterials for precise targeting and cellular uptake (Awan et al. 2022). Furthermore, biodegradable materials, such as the magnesium alloy MAGNEZIX®, can dissolve naturally in the body, avoiding the need for implant removal procedures (Biber et al. 2017). Sustainable biomaterials derived from renewable sources are also gaining popularity, as they address environmental concerns while still providing therapeutic benefits. Bio-derived materials, such as bacterial cellulose and alginate-based hydrogels, are being considered as potential replacements for synthetic polymers in a variety of medical applications, thereby reducing biomaterials' environmental impact (Wei et al. 2020b). AI is transforming biomaterial design by accelerating discovery and enabling personalized solutions tailored to patient needs, thereby improving healthcare outcomes and emphasizing the importance of adaptive regulations (Webster 2023).

References

ABG IP (n.d.) The Enlarged Board of Appeal of the EPO has decided that products obtained exclusively by an essentially biological process are not patentable

Adepu S, Ramakrishna S (2021) Controlled drug delivery systems: current status and future directions. Molecules 26:5905. https://doi.org/10.3390/molecules26195905

Adhikari J, Saha P, Sinha A (2018) Surface modification of metallic bone implants—polymer and polymer-assisted coating for bone in-growth. Fundam Biomater Metals (Elsevier):299–321. https://doi.org/10.1016/B978-0-08-102205-4.00014-3

Ahn W, Lee JH, Kim SR, Lee J, Lee EJ (2021) Designed protein- and peptide-based hydrogels for biomedical sciences. J Mater Chem B. https://doi.org/10.1039/d0tb02604b

Ajmal S, Athar Hashmi F, Imran I (2022) Recent progress in development and applications of biomaterials. Mater Today Proc. https://doi.org/10.1016/j.matpr.2022.04.233

Akbarzadeh A, Rezaei-Sadabady R, Davaran S, Joo SW, Zarghami N, Hanifehpour Y, Samiei M, Kouhi M, Nejati-Koshki K (2013) Liposome: classification, preparation, and applications. Nanoscale Res Lett 8:102. https://doi.org/10.1186/1556-276X-8-102

Aldhaher A, Shahabipour F, Shaito A, Al-Assaf S, Elnour AAM, Sallam EB, Teimourtash S, Elfadil AA (2023) 3D hydrogel/bioactive glass scaffolds in bone tissue engineering: status and future opportunities. Heliyon. https://doi.org/10.1016/j.heliyon.2023.e17050

Alemu D, Tafesse M, Mondal AK (2022) Mycelium-based composite: the future sustainable biomaterial. Int J Biomater 2022. https://doi.org/10.1155/2022/8401528

Al-Hashimi N, Begg N, Alany RG, Hassanin H, Elshaer A (2018) Oral modified release multiple-unit particulate systems: compressed pellets, microparticles and nanoparticles. Pharmaceutics 10:176. https://doi.org/10.3390/pharmaceutics10040176

Alhulaybi ZA (2023) Fabrication and characterization of poly(lactic acid)-based biopolymer for surgical sutures. ChemEngineering 7:98. https://doi.org/10.3390/chemengineering7050098

Allen LV (2010) Dosage form design: pharmaceutical and formulation considerations. In: Ansel's pharmaceutical dosage forms and drug delivery systems

Al-Shalawi FD, Mohamed Ariff AH, Jung D-W, Mohd Ariffin MKA, Seng Kim CL, Brabazon D, Al-Osaimi MO (2023) Biomaterials as implants in the orthopedic field for regenerative medicine: metal versus synthetic polymers. Polymers (Basel) 15:2601. https://doi.org/10.3390/polym15122601

Amukarimi S, Ramakrishna S, Mozafari M (2021) Smart biomaterials—a proposed definition and overview of the field. Curr Opin Biomed Eng. https://doi.org/10.1016/j.cobme.2021.100311

ANSI/AAMI/ISO 10993-1:2018 (2020) Biological evaluation of medical devices—Part 1: evaluation and testing within a risk management process. https://doi.org/10.2345/9781570207556.

Antunes A, Fierro I, Guerrante R, Mendes F, de Alencar MSM (2013) Trends in nanopharmaceutical patents. Int J Mol Sci. https://doi.org/10.3390/ijms14047016

Arcos D, Vallet-Regí M (2013) Bioceramics for drug delivery. Acta Mater 61:890–911. https://doi.org/10.1016/j.actamat.2012.10.039

Arcos D, Vallet-Regí M (2020) Substituted hydroxyapatite coatings of bone implants. J Mater Chem B 8:1781–1800. https://doi.org/10.1039/C9TB02710F

Assiri AA, Glover K, Mishra D, Waite D, Vora LK, Thakur RRS (2024) Block copolymer micelles as ocular drug delivery systems. Drug Discov Today 29:104098. https://doi.org/10.1016/j.drudis.2024.104098

Awan UA, Fatima Saeed R, Naeem M, Mumtaz S, Saleem Qazi A (2022) State of the art nanotechnology to combat current COVID-19 outbreak. Life Sci. https://doi.org/10.37185/lns.1.1.229

Baino F, Hamzehlou S, Kargozar S (2018) Bioactive glasses: where are we and where are we going? J Funct Biomater. https://doi.org/10.3390/jfb9010025

Baryakova TH, Pogostin BH, Langer R, McHugh KJ (2023) Overcoming barriers to patient adherence: the case for developing innovative drug delivery systems. Nat Rev Drug Discov 22:387–409. https://doi.org/10.1038/s41573-023-00670-0

Basu B, Dharamsi A, Makwana S, Makasana Y (2011) Prefilled syringes: an innovation in parenteral packaging. Int J Pharm Investig 1:200. https://doi.org/10.4103/2230-973X.93004

Bayer IS (2022) Advances in fibrin-based materials in wound repair: a review. Molecules 27:4504. https://doi.org/10.3390/molecules27144504

Bhat S, Kumar A (2013) Biomaterials and bioengineering tomorrow's healthcare. Biomatter 3. https://doi.org/10.4161/biom.24717

Biber R, Pauser J, Brem M, Bail HJ (2017) Bioabsorbable metal screws in traumatology: a promising innovation. Trauma Case Rep. https://doi.org/10.1016/j.tcr.2017.01.012

Boni R, Ali A, Shavandi A, Clarkson AN (2018) Current and novel polymeric biomaterials for neural tissue engineering. J Biomed Sci 25:90. https://doi.org/10.1186/s12929-018-0491-8

Bose S, Roy M, Bandyopadhyay A (2012) Recent advances in bone tissue engineering scaffolds. Trends Biotechnol. https://doi.org/10.1016/j.tibtech.2012.07.005

Brougher JT, Fazzolare DA (2014) USPTO guidance on patentable subject matter: impediment to biotechnology innovation? J Commer Biotechnol. https://doi.org/10.5912/jcb664

Burkus JK, Gornet MF, Dickman CA, Zdeblick TA (2002) Anterior lumbar interbody fusion using rhBMP-2 with tapered interbody cages. J Spinal Disord Tech. https://doi.org/10.1097/00024720-200210000-00001

Caixeta A, Campos CM, Felix C, Chieffo A, Capranzano P, Kawamoto H, Tamburino C, Diletti R, De Ribamar Costa J, Onuma Y, Van Geuns RJ, Bartorelli AL, Colombo A, Tamburino C, Serruys PW, Abizaid A (2019) Predictors of long-term adverse events after Absorb bioresorbable vascular scaffold implantation: a 1,933-patient pooled analysis from international registries. EuroIntervention. https://doi.org/10.4244/EIJ-D-16-00796

Caló E, Khutoryanskiy VV (2015) Biomedical applications of hydrogels: a review of patents and commercial products. Eur Polym J. https://doi.org/10.1016/j.eurpolymj.2014.11.024

Campochiaro PA, Marcus DM, Awh CC, Regillo C, Adamis AP, Bantseev V, Chiang Y, Ehrlich JS, Erickson S, Hanley WD, Horvath J, Maass KF, Singh N, Tang F, Barteselli G (2019) The port delivery system with Ranibizumab for Neovascular age-related macular degeneration: results from the randomized phase 2 Ladder clinical trial. Ophthalmology. https://doi.org/10.1016/j.ophtha.2019.03.036

Cartwright-Smith L (2014) Patenting genes: what does association for molecular pathology V. Myriad genetics mean for genetic testing and research? Public Health Rep 129:289–292. https://doi.org/10.1177/003335491412900311

Chattopadhyay S, Raines RT (2014) Collagen-based biomaterials for wound healing. Biopolymers 101:821–833. https://doi.org/10.1002/bip.22486

Che X, Zhao T, Hu J, Yang K, Ma N, Li A, Sun Q, Ding C, Ding Q (2024) Application of chitosan-based hydrogel in promoting wound healing: a review. Polymers (Basel) 16:344. https://doi.org/10.3390/polym16030344

Chen F-M, Liu X (2016) Advancing biomaterials of human origin for tissue engineering. Prog Polym Sci 53:86–168. https://doi.org/10.1016/j.progpolymsci.2015.02.004

Chen J, Yang J, Wang L, Zhang X, Heng BC, Wang DA, Ge Z (2021) Modified hyaluronic acid hydrogels with chemical groups that facilitate adhesion to host tissues enhance cartilage regeneration. Bioact Mater. https://doi.org/10.1016/j.bioactmat.2020.11.020

Chen B, Dong J, Ruelas M, Ye X, He J, Yao R, Fu Y, Liu Y, Hu J, Wu T, Zhou C, Li Y, Huang L, Zhang YS, Zhou J (2022) Artificial intelligence-assisted high-throughput screening of printing conditions of hydrogel architectures for accelerated diabetic wound healing. Adv Funct Mater. https://doi.org/10.1002/adfm.202201843

Chen H, Feng R, Xia T, Wen Z, Li Q, Qiu X, Huang B, Li Y (2023) Progress in surface modification of titanium implants by hydrogel coatings. Gels. https://doi.org/10.3390/gels9050423

Chong ETJ, Ng JW, Lee PC (2023) Classification and medical applications of biomaterials–a mini review. Bio Integr:10.15212/bioi-2022-0009

Collins LM, Dziak JJ, Kugler KC, Trail JB (2014) Factorial experiments: efficient tools for evaluation of intervention components. Am J Prev Med. https://doi.org/10.1016/j.amepre.2014.06.021

Cook-Deegan R, Heaney C (2010) Patents in genomics and human genetics. Annu Rev Genomics Hum Genet 11:383–425. https://doi.org/10.1146/annurev-genom-082509-141811

da Silva RYP, de Menezes DLB, da Oliveira VSA, Converti Á, de Lima AN (2023) Microparticles in the development and improvement of pharmaceutical formulations: an analysis of in vitro and in vivo studies. Int J Mol Sci 24:5441. https://doi.org/10.3390/ijms24065441

Dario P, Katrin B, Dominik B, Michael B (2020) Towards novel reimbursement models for expensive advanced therapy medicinal products (ATMPs). Swiss Med Wkly. https://doi.org/10.4414/smw.2020.20355

Davidson Hernandez ED, Reyes-Romero JR (2019) Characteristics of polymeric materials used in medicine. Mater Biomed Eng (Elsevier):479–506. https://doi.org/10.1016/B978-0-12-816874-5.00014-1

Davis R, Singh A, Jackson MJ, Coelho RT, Prakash D, Charalambous CP, Ahmed W, da Silva LRR, Lawrence AA (2022) A comprehensive review on metallic implant biomaterials and their subtractive manufacturing. Int J Adv Manuf Technol 120:1473–1530. https://doi.org/10.1007/s00170-022-08770-8

Davison NL, Barrère-de Groot F, Grijpma DW (2014) Degradation of biomaterials. Tissue Eng (Elsevier):177–215. https://doi.org/10.1016/B978-0-12-420145-3.00006-7

Demets DL, Furberg CD, Friedman LM (2006) Data monitoring in clinical trials: a case studies approach. https://doi.org/10.1007/0-387-30107-0

Deng X, Gould M, Ali MA (2022) A review of current advancements for wound healing: biomaterial applications and medical devices. J Biomed Mater Res B Appl Biomater. https://doi.org/10.1002/jbm.b.35086

Desai N, Rana D, Salave S, Gupta R, Patel P, Karunakaran B, Sharma A, Giri J, Benival D, Kommineni N (2023) Chitosan: a potential biopolymer in drug delivery and biomedical applications. Pharmaceutics. https://doi.org/10.3390/pharmaceutics15041313

Di Prima M, Coburn J, Hwang D, Kelly J, Khairuzzaman A, Ricles L (2016) Additively manufactured medical products—the FDA perspective. 3D Print Med. https://doi.org/10.1186/s41205-016-0005-9

DiMasi JA, Grabowski HG, Hansen RW (2016) Innovation in the pharmaceutical industry: new estimates of R&D costs. J Health Econ. https://doi.org/10.1016/j.jhealeco.2016.01.012

Drummond J, Tran P, Fary C (2015) Metal-on-metal hip arthroplasty: a review of adverse reactions and patient management. J Funct Biomater. https://doi.org/10.3390/jfb6030486

Du Y, Yu M, Ge J, Ma PX, Chen X, Lei B (2015) Development of a multifunctional platform based on strong, intrinsically Photoluminescent and antimicrobial silica-poly(citrates)-based hybrid biodegradable elastomers for bone regeneration. Adv Funct Mater. https://doi.org/10.1002/adfm.201501712

EBIN.PUB (n.d.) The role of intellectual property rights in biotechnology innovation 1847209807, 9781847209801

Elisetti SK, Arora V, Sharma RB (2023) Polymers for designing 3D printed pharmaceutical products. J Res Pharm. https://doi.org/10.29228/jrp.341

Elliott MJ, De Coppi P, Speggiorin S, Roebuck D, Butler CR, Samuel E, Crowley C, McLaren C, Fierens A, Vondrys D, Cochrane L, Jephson C, Janes S, Beaumont NJ, Cogan T, Bader A, Seifalian AM, Hsuan JJ, Lowdell MW, Birchall MA (2012) Stem-cell-based, tissue engineered tracheal replacement in a child: a 2-year follow-up study. Lancet. https://doi.org/10.1016/S0140-6736(12)60737-5

Ellis SG, Kereiakes DJ, Metzger DC, Caputo RP, Rizik DG, Teirstein PS, Litt MR, Kini A, Kabour A, Marx SO, Popma JJ, McGreevy R, Zhang Z, Simonton C, Stone GW (2015) Everolimus-eluting Bioresorbable scaffolds for coronary artery disease. N Engl J Med. https://doi.org/10.1056/nejmoa1509038

Elumalai A, Nayak Y, Ganapathy AK, Chen D, Tappa K, Jammalamadaka U, Bishop G, Ballard DH (2023) Reverse engineering and 3D printing of medical devices for drug delivery and drug-embedded anatomic implants. Polymers (Basel). https://doi.org/10.3390/polym15214306

European Parlament, Council of the European Union (2017) IVDR: regulation (EU) 2017/745 of the European Parliament and of the council of 5 April 2017 on medical devices. Official Journal of the European Union

Farshidfar N, Iravani S, Varma RS (2023) Alginate-based biomaterials in tissue engineering and regenerative medicine. Mar Drugs 21:189. https://doi.org/10.3390/md21030189

Fatimi A (2022) Chitosan-based hydrogels: patent analysis. In: IOCN 2022. MDPI, Basel, p 1. https://doi.org/10.3390/materproc2022009001

Feng H, Yu Z, Chu P (2006) Ion implantation of organisms. Mater Sci Eng R Rep 54:49–120. https://doi.org/10.1016/j.mser.2006.11.001

Food and Drug Administration (2018) Classify your medical device. https://www.fda.gov/medical-devices/overview-device-regulation/classify-your-medical-device

Food and Drug Administration (2019) Premarket approval. https://www.fda.gov/medical-devices/premarket-submissions/premarket-approval-pma

Food and Drug Administration (2020) 510(k) Clear. https://www.fda.gov/medical-devices/device-approvals-denials-and-clearances/510k-clearances

Food and Drug Administration (2021) Investigational Device Exemptions. https://www.fda.gov/medical-devices/investigational-device-exemption-ide/ide-approval-process

Frazer RQ, Byron RT, Osborne PB, West KP (2005) PMMA: an essential material in medicine and dentistry. J Long-Term Eff Med Implants 15:629–639. https://doi.org/10.1615/JLongTermEffMedImplants.v15.i6.60

Fu J, Su Y, Qin Y-X, Zheng Y, Wang Y, Zhu D (2020) Evolution of metallic cardiovascular stent materials: a comparative study among stainless steel, magnesium and zinc. Biomaterials 230:119641. https://doi.org/10.1016/j.biomaterials.2019.119641

Gadiya Y, Gribbon P, Hofmann-Apitius M, Zaliani A (2023) Pharmaceutical patent landscaping: a novel approach to understand patents from the drug discovery perspective. Artif Intell Life Sci. https://doi.org/10.1016/j.ailsci.2023.100069

García-Rey E, García-Cimbrelo E (2010) Polyethylene in total hip arthroplasty: half a century in the limelight. J Orthop Traumatol 11:67–72. https://doi.org/10.1007/s10195-010-0091-1

Gentile P, Chiono V, Carmagnola I, Hatton PV (2014) An overview of poly(lactic-co-glycolic) acid (PLGA)-based biomaterials for bone tissue engineering. Int J Mol Sci. https://doi.org/10.3390/ijms15033640

Georgieva D, Nikolova D, Vassileva E, Kostova B (2023) Chitosan-based nanoparticles for targeted nasal Galantamine delivery as a promising tool in Alzheimer's disease therapy. Pharmaceutics. https://doi.org/10.3390/pharmaceutics15030829

Geraili A, Xing M, Mequanint K (2021) Design and fabrication of drug-delivery systems toward adjustable release profiles for personalized treatment. View 2. https://doi.org/10.1002/VIW.20200126

Ghourichay MP, Kiaie SH, Nokhodchi A, Javadzadeh Y (2021) Formulation and quality control of orally disintegrating tablets (ODTs): recent advances and perspectives. Biomed Res Int 2021:1–12. https://doi.org/10.1155/2021/6618934

Gilbert PX, Walter C (2001) Patents and the human genome project – new claims for old? Trends Biotechnol 19:49–52. https://doi.org/10.1016/S0167-7799(00)01521-3

Guan Y, Du J, Torvik VI (2017) Geographical distribution of biomedical research in the USA and China. In: Proceedings of the 6th international workshop on mining scientific publication. ACM, New York, pp 40–45. https://doi.org/10.1145/3127526.3127534

Gubby H (2020) Is the patent system a barrier to inclusive prosperity? The biomedical perspective. Glob Policy. https://doi.org/10.1111/1758-5899.12730

Gul A, Papia E, Naimi-Akbar A, Ruud A, Vult von Steyern P (2024) Zirconia dental implants; the relationship between design and clinical outcome: a systematic review. J Dent 143:104903. https://doi.org/10.1016/j.jdent.2024.104903

Guo B, Ma PX (2014) Synthetic biodegradable functional polymers for tissue engineering: a brief review. Sci China Chem 57:490–500. https://doi.org/10.1007/s11426-014-5086-y

Gupta S, Kumar TP, Gowda DV (2020) Patent perspective on Orodispersible films. Recent Pat Drug Deliv Formul. https://doi.org/10.2174/1872211314666200904104022

Gurgula O (2020) Strategic patenting by pharmaceutical companies—should competition law intervene? IIC Int Rev Ind Prop Copyr Law. https://doi.org/10.1007/s40319-020-00985-0

Hakim LK, Yari A, Nikparto N, Mehraban SH, Cheperli S, Asadi A, Darehdor AA, Nezaminia S, Dortaj D, Nazari Y, Dehghan M, Hojjat P, Mohajeri M, Hasani Jebelli MS (2024) The current applications of nano and biomaterials in drug delivery of dental implant. BMC Oral Health 24:126. https://doi.org/10.1186/s12903-024-03911-9

Harrington JA, Hernandez-Guerrero TC, Basu B (2017) Early phase clinical trial designs—state of play and adapting for the future. Clin Oncol. https://doi.org/10.1016/j.clon.2017.10.005

Haynes R (2005) Media rights and intellectual property. Edinburgh University Press, p 162

Hernández-Melchor D, López-Bayghen E, Padilla-Viveros A (2023) The patent landscape in the field of stem cell therapy: closing the gap between research and clinic. F1000Research 11:997. https://doi.org/10.12688/f1000research.123799.4

Hicks CR, Crawford GJ, Lou X, Tan DT, Snibson GR, Sutton G, Downie N, Werner L, Chirila TV, Constable IJ (2003) Corneal replacement using a synthetic hydrogel cornea, AlphaCor™: device, preliminary outcomes and complications. Eye. https://doi.org/10.1038/sj.eye.6700333

Hiranaka T, Furumatsu T, Okazaki Y, Kintaka K, Higashihara N, Tamura M, Nakata E, Ozaki T (2023) Accuracy of the newly developed Zimmer Biomet Root Aiming guide in tibial tunnel creation compared with that of conventional guides, Asia-Pacific. J Sport Med Arthrosc Rehabil Technol. https://doi.org/10.1016/j.asmart.2023.03.001

Holzapfel BM, Reichert JC, Schantz JT, Gbureck U, Rackwitz L, Nöth U, Jakob F, Rudert M, Groll J, Hutmacher DW (2013) How smart do biomaterials need to be? A translational science and clinical point of view. Adv Drug Deliv Rev. https://doi.org/10.1016/j.addr.2012.07.009
Hosseini ES, Dervin S, Ganguly P, Dahiya R (2021) Biodegradable materials for sustainable health monitoring devices. ACS Appl Bio Mater 4:163–194. https://doi.org/10.1021/acsabm.0c01139
Hou X, Zaks T, Langer R, Dong Y (2021) Lipid nanoparticles for mRNA delivery. Nat Rev Mater. https://doi.org/10.1038/s41578-021-00358-0
Hou X, Zhang L, Zhou Z, Luo X, Wang T, Zhao X, Lu B, Chen F, Zheng L (2022) Calcium phosphate-based biomaterials for bone repair. J Funct Biomater 13:187. https://doi.org/10.3390/jfb13040187
Hu W, Yao X, Li Y, Li J, Zhang J, Zou Z, Kang F, Dong S (2023) Injectable hydrogel with selenium nanoparticles delivery for sustained glutathione peroxidase activation and enhanced osteoarthritis therapeutics. Mater Today Bio. https://doi.org/10.1016/j.mtbio.2023.100864
Huebsch N, Mooney DJ (2009) Inspiration and application in the evolution of biomaterials. Nature 462:426–432. https://doi.org/10.1038/nature08601
Hutmacher DW (2000) Scaffolds in tissue engineering bone and cartilage. Biomaterials. https://doi.org/10.1016/S0142-9612(00)00121-6
Hwang D, Ramsey JD, Kabanov AV (2020) Polymeric micelles for the delivery of poorly soluble drugs: from nanoformulation to clinical approval. Adv Drug Deliv Rev 156:80–118. https://doi.org/10.1016/j.addr.2020.09.009
Institute of Medicine (1999) Safety of silicone breast implants. The National Academies Press, Washington, DC. https://doi.org/10.17226/9602
International Organization for Standardization (2016) Medical devices – quality management systems – Requirements for regulatory purposes – DIN EN ISO 13485:2016. In: Competition and ownership in land passenger transport
International Regulatory Co-operation (2016) International Medical Device Regulators Forum (IMDRF). https://doi.org/10.1787/9789264244047-32-en
Ioele G, Chieffallo M, Occhiuzzi MA, De Luca M, Garofalo A, Ragno G, Grande F (2022) Anticancer drugs: recent strategies to improve stability profile, pharmacokinetic and pharmacodynamic properties. Molecules 27:5436. https://doi.org/10.3390/molecules27175436
Jaalouk JY, Dana P, Anesh G, Yoo DJ (2024) Dermatological reviews. Dermatol Rev:1–17. https://doi.org/10.1002/der2.244%0A
Janarthanan G, Shin HS, Kim I-G, Ji P, Chung E-J, Lee C, Noh I (2020) Self-crosslinking hyaluronic acid–carboxymethylcellulose hydrogel enhances multilayered 3D-printed construct shape integrity and mechanical stability for soft tissue engineering. Biofabrication 12:045026. https://doi.org/10.1088/1758-5090/aba2f7
Jarow JP, Baxley JH (2015) Medical devices: US medical device regulation. Urol Oncol Semin Orig Investig. https://doi.org/10.1016/j.urolonc.2014.10.004
Jin Y, Koh RH, Kim SH, Kim KM, Park GK, Hwang NS (2020) Injectable anti-inflammatory hyaluronic acid hydrogel for osteoarthritic cartilage repair. Mater Sci Eng C. https://doi.org/10.1016/j.msec.2020.111096
Jung H, Kim YY, Kim B, Nam H, Suh JG (2017) Improving glycemic control in model mice with type 2 diabetes by increasing superoxide dismutase (SOD) activity using silk fibroin hydrolysate (SFH). Biochem Biophys Res Commun. https://doi.org/10.1016/j.bbrc.2017.09.066
Karim K, Mandal A, Biswas N, Guha A, Chatterjee S, Behera M, Kuotsu K (2010) Niosome: a future of targeted drug delivery systems. J Adv Pharm Technol Res 1:374. https://doi.org/10.4103/0110-5558.76435
Kearney A, Linden K, Savage P, Menown IBA (2021) Advances in clinical cardiology 2020: a summary of key clinical trials. Adv Ther. https://doi.org/10.1007/s12325-021-01711-z
Kelly SH, Shores LS, Votaw NL, Collier JH (2017) Biomaterial strategies for generating therapeutic immune responses. Adv Drug Deliv Rev 114:3–18. https://doi.org/10.1016/j.addr.2017.04.009
Kereiakes DJ, Ellis SG, Metzger C, Caputo RP, Rizik DG, Teirstein PS, Litt MR, Kini A, Kabour A, Marx SO, Popma JJ, McGreevy R, Zhang Z, Simonton C, Stone GW (2017) 3-year clinical

outcomes with everolimus-eluting Bioresorbable coronary scaffolds: the ABSORB III trial. J Am Coll Cardiol. https://doi.org/10.1016/j.jacc.2017.10.010

Khunová V, Kováčová M, Olejniková P, Ondreáš F, Špitalský Z, Ghosal K, Berkeš D (2022) Antibacterial electrospun Polycaprolactone nanofibers reinforced by Halloysite nanotubes for tissue engineering. Polymers (Basel). https://doi.org/10.3390/polym14040746

Kim SM, Patel M, Patel R (2021) PLGA core-shell nano/microparticle delivery system for biomedical application. Polymers (Basel). https://doi.org/10.3390/polym13203471

Kostag M, El Seoud OA (2021) Sustainable biomaterials based on cellulose, chitin and chitosan composites – a review. Carbohydr Polym Technol Appl 2:100079. https://doi.org/10.1016/j.carpta.2021.100079

Kravanja KA, Finšgar M (2022) A review of techniques for the application of bioactive coatings on metal-based implants to achieve controlled release of active ingredients. Mater Des 217:110653. https://doi.org/10.1016/j.matdes.2022.110653

Kumar N, Alathur Ramakrishnan S, Lopez KG, Chin BZ, Devyapriya S, Kumar L, Baskar S, Vellayappan BA, Fuh JYH, Anantharajan SK (2022) Current trends and future scope in 3D printing for surgical management of spine pathologies. Bioprinting. https://doi.org/10.1016/j.bprint.2022.e00197

Kupikowska-Stobba B, Domagała J, Kasprzak MM (2024) Critical review of techniques for food emulsion characterization. Appl Sci 14:1069. https://doi.org/10.3390/app14031069

Ledley FD, Cleary EG (2023) NIH funding for patents that contribute to market exclusivity of drugs approved 2010–2019 and the public interest protections of Bayh-dole. PLoS One. https://doi.org/10.1371/journal.pone.0288447

Lee K, Silva EA, Mooney DJ (2011) Growth factor delivery-based tissue engineering: general approaches and a review of recent developments. J R Soc Interface. https://doi.org/10.1098/rsif.2010.0223

Levin B, Redmond SL, Rajkhowa R, Eikelboom RH, Marano RJ, Atlas MD (2010) Preliminary results of the application of a silk fibroin scaffold to otology. Otolaryngol – Head Neck Surg. https://doi.org/10.1016/j.otohns.2009.06.746

Li A, Bergan RC (2020) Clinical trial design: past, present, and future in the context of big data and precision medicine. Cancer. https://doi.org/10.1002/cncr.33205

Li S, Vert M (2002) Biodegradation of aliphatic polyesters. Degrad Polym. https://doi.org/10.1007/978-94-017-1217-0_5

Liang W, Zhou C, Zhang H, Bai J, Jiang B, Jiang C, Ming W, Zhang H, Long H, Huang X, Zhao J (2023) Recent advances in 3D printing of biodegradable metals for orthopaedic applications. J Biol Eng 17:56. https://doi.org/10.1186/s13036-023-00371-7

Liu B, Chen K (2024) Advances in hydrogel-based drug delivery systems. Gels 10:262. https://doi.org/10.3390/gels10040262

Liu Y, Ran Y, Ge Y, Raza F, Li S, Zafar H, Paiva-Santos AC, Yu C, Sun M, Zhu Y, Li F, Wu Y (2022) pH-sensitive peptide hydrogels as a combination drug delivery system for cancer treatment. Pharmaceutics. https://doi.org/10.3390/pharmaceutics14030652

Lombardo D, Calandra P, Pasqua L, Magazù S (2020) Self-assembly of organic nanomaterials and biomaterials: the bottom-up approach for functional nanostructures formation and advanced applications. Materials (Basel). https://doi.org/10.3390/ma13051048

Lui R (2024) Deus ex machina? The rise of artificial intelligence in toxicology. Chem Res Toxicol. https://doi.org/10.1021/acs.chemrestox.4c00050

Lüscher TF (2020) Clinical trials in interventional cardiology: a challenging necessity. Eur Heart J. https://doi.org/10.1093/eurheartj/ehaa585

Madan M, Bajaj A, Lewis S, Udupa N, Baig J (2009) In situ forming polymeric drug delivery systems. Indian J Pharm Sci 71:242. https://doi.org/10.4103/0250-474X.56015

Magill E, Demartis S, Gavini E, Permana AD, Thakur RRS, Adrianto MF, Waite D, Glover K, Picco CJ, Korelidou A, Detamornrat U, Vora LK, Li L, Anjani QK, Donnelly RF, Domínguez-Robles J, Larrañeta E (2023) Solid implantable devices for sustained drug delivery. Adv Drug Deliv Rev 199:114950. https://doi.org/10.1016/j.addr.2023.114950

Maisel WH (2004) Medical device regulation: an introduction for the practicing physician. Ann Intern Med. https://doi.org/10.7326/0003-4819-140-4-200402170-00012
Majee SB, Avlani D, Biswas GR (2017) Hpmc as capsule Shell material: physicochemical, pharmaceutical and biopharmaceutical properties. Int J Pharm Pharm Sci 9:1. https://doi.org/10.22159/ijpps.2017v9i10.20707
Makadia HK, Siegel SJ (2011) Poly lactic-co-glycolic acid (PLGA) as biodegradable controlled drug delivery carrier. Polymers (Basel) 3:1377–1397. https://doi.org/10.3390/polym3031377
Malathi S, Narayana Kalkura S, Balasubramanian S (2023) Polymeric nanoparticles for biomedical applications. Appl Multifunct Nanomater. https://doi.org/10.1016/B978-0-12-820557-0.00028-X
Markey N, Howitt B, El-Mansouri I, Schwartzenberg C, Kotova O, Meier C (2024) Clinical trials are becoming more complex: a machine learning analysis of data from over 16,000 trials. Sci Rep. https://doi.org/10.1038/s41598-024-53211-z
Meek RMD, Afolaranmi GA, Tettey J, Grant MH (2008) Release of chromium from Orthopaedic arthroplasties. Open Orthop J 2:10–18. https://doi.org/10.2174/1874325000802010010
Metwally WM, El-Habashy SE, El-Hosseiny LS, Essawy MM, Eltaher HM, El-Khordagui LK (2024) Bioinspired 3D-printed scaffold embedding DDAB-nano ZnO/nanofibrous microspheres for regenerative diabetic wound healing. Biofabrication. https://doi.org/10.1088/1758-5090/acfd60
Meyer TA, Ramirez C, Tamasi MJ, Gormley AJ (2023) A user's guide to machine learning for polymeric biomaterials. ACS Polym Au. https://doi.org/10.1021/acspolymersau.2c00037
Minssen T, Nordberg A (2015) The impact of "Broccoli II" and "Tomatoes II" on European patents in conventional breeding, GMOs, and synthetic biology: the grand finale of a juicy patents tale? Biotechnol Law Rep 34:81–98. https://doi.org/10.1089/blr.2015.29004
Mishchenko O, Yanovska A, Kosinov O, Maksymov D, Moskalenko R, Ramanavicius A, Pogorielov M (2023) Synthetic calcium–phosphate materials for bone grafting. Polymers (Basel) 15:3822. https://doi.org/10.3390/polym15183822
Mitchell MJ, Billingsley MM, Haley RM, Wechsler ME, Peppas NA, Langer R (2020) Engineering precision nanoparticles for drug delivery. Nat Rev Drug Discov 202(20):101–124. https://doi.org/10.1038/s41573-020-0090-8
Mobbs RJ, Coughlan M, Thompson R, Sutterlin CE, Phan K (2017) The utility of 3D printing for surgical planning and patient-specific implant design for complex spinal pathologies: case report. J Neurosurg Spine. https://doi.org/10.3171/2016.9.SPINE16371
Moorkens E, Vulto AG, Huys I (2020) An overview of patents on therapeutic monoclonal antibodies in Europe: are they a hurdle to biosimilar market entry? MAbs. https://doi.org/10.1080/19420862.2020.1743517
Moses JW, Leon MB, Popma JJ, Fitzgerald PJ, Holmes DR, O'Shaughnessy C, Caputo RP, Kereiakes DJ, Williams DO, Teirstein PS, Jaeger JL, Kuntz RE (2003) Sirolimus-eluting stents versus standard stents in patients with stenosis in a native coronary artery. N Engl J Med. https://doi.org/10.1056/nejmoa035071
Moura LIF, Dias AMA, Carvalho E, De Sousa HC (2013) Recent advances on the development of wound dressings for diabetic foot ulcer treatment – a review. Acta Biomater. https://doi.org/10.1016/j.actbio.2013.03.033
Mukherjee N, Adak A, Ghosh S (2020) Recent trends in the development of peptide and protein-based hydrogel therapeutics for the healing of CNS injury. Soft Matter. https://doi.org/10.1039/d0sm00885k
Narayanan G, Vernekar VN, Kuyinu EL, Laurencin CT (2016) Poly (lactic acid)-based biomaterials for orthopaedic regenerative engineering. Adv Drug Deliv Rev 107:247–276. https://doi.org/10.1016/j.addr.2016.04.015
Navarro M, Michiardi A, Castaño O, Planell JA (2008) Biomaterials in orthopaedics. J R Soc Interface. https://doi.org/10.1098/rsif.2008.0151
NCT05863221 (2023) A phase IV, open-label, randomized, active-controlled, single-center study of ZYNRELEF® for postoperative analgesia after laparoscopic sleeve gastrectomy. https://Clinicaltrials.Gov/Show/NCT05863221

New Developments in Biotechnology: Patenting Life – Special Report April 1989, Director (1989)

Nezhad-Mokhtari P, Ghorbani M, Roshangar L, Soleimani Rad J (2019) A review on the construction of hydrogel scaffolds by various chemically techniques for tissue engineering. Eur Polym J. https://doi.org/10.1016/j.eurpolymj.2019.05.004

Ni J, Shao R, Ung COL, Wang Y, Hu Y, Cai Y (2015) Valuation of pharmaceutical patents: a comprehensive analytical framework based on technological, commercial, and legal factors. J Pharm Innov. https://doi.org/10.1007/s12247-015-9225-6

Nielen MG, de Vries SA, Geijsen N (2013) European stem cell research in legal shackles. EMBO J 32:3107–3111. https://doi.org/10.1038/emboj.2013.249

Nogueira DMB, de Rosso MPO, Buchaim DV, Zangrando MSR, Buchaim RL (2024) Update on the use of 45S5 bioactive glass in the treatment of bone defects in regenerative medicine, world. J Orthop 15:204–214. https://doi.org/10.5312/wjo.v15.i3.204

Nsairat H, Khater D, Sayed U, Odeh F, Al Bawab A, Alshaer W (2022) Liposomes: structure, composition, types, and clinical applications. Heliyon 8:e09394. https://doi.org/10.1016/j.heliyon.2022.e09394

Olamide Omiyale B, Abiodun Rasheed A, Omoboyode Akinnusi R, Olumide Olugbade T (2023) Influence of mechanical properties of biomaterials on the reconstruction of biomedical parts via additive manufacturing techniques: an overview. In: Biotechnology – biosensors, biomaterials and tissue engineering. https://doi.org/10.5772/intechopen.104465

Oliver JA, Jiménez-Luna C, Melguizo C, Prados J, Caba O, Rodriguez-Serrano F, Alvarez PJ, Aranega A (2013) New biomarkers for colon cancer diagnosis: a review of recent patents. Recent Pat Biomark. https://doi.org/10.2174/2210309011303030003

Olorunsola EO, Udoh IE, Ekott MB, Alozie MF, Davies KG (2023) Biopharmaceutics and clinical outcomes of emerging dosage forms of insulin: a systematic review. Diabetes Epidemiol Manag. https://doi.org/10.1016/j.deman.2022.100120

Olsen A, Beall RF, Knox RP, Tu SS, Kesselheim AS, Feldman WB (2023) Patents and regulatory exclusivities on FDA-approved insulin products: a longitudinal database study, 1986–2019. PLoS Med. https://doi.org/10.1371/journal.pmed.1004309

Pallmann P, Bedding AW, Choodari-Oskooei B, Dimairo M, Flight L, Hampson LV, Holmes J, Mander AP, Odondi L, Sydes MR, Villar SS, Wason JMS, Weir CJ, Wheeler GM, Yap C, Jaki T (2018) Adaptive designs in clinical trials: why use them, and how to run and report them. BMC Med. https://doi.org/10.1186/s12916-018-1017-7

Palmese LL, Fan M, Scott RA, Tan H, Kiick KL (2020) Multi-stimuli-responsive, liposome-crosslinked poly(ethylene glycol) hydrogels for drug delivery. J Biomater Sci Polym Ed. https://doi.org/10.1080/09205063.2020.1855392

Pane J, Coloma PM, Verhamme KMC, Sturkenboom MCJM, Rebollo I (2017) Evaluating the safety profile of non-active implantable medical devices compared with medicines. Drug Saf. https://doi.org/10.1007/s40264-016-0474-1

Panova IG, Sudareva EA, Novoskoltseva OA, Spiridonov VV, Shtilman MI, Richtering W, Yaroslavov AA (2021) Temperature-induced unloading of liposomes bound to microgels. Colloids Surf A Physicochem Eng Asp. https://doi.org/10.1016/j.colsurfa.2021.127590

Park KJ, Hong SE, Do MS, Hyun CK (2002) Stimulation of insulin secretion by silk fibroin hydrolysate in streptozotocin-induced diabetic rats and db/db mice. Kor J Pharmacogn

Patel M, Jha A, Patel R (2021) Potential application of PLGA microsphere for tissue engineering. J Polym Res. https://doi.org/10.1007/s10965-021-02562-6

Patel D, Kuperkar K, Yusa S, Bahadur P (2023) Nanoscale self-assemblies from amphiphilic block copolymers as proficient templates in drug delivery. Drugs Drug Candidates 2:898–922. https://doi.org/10.3390/ddc2040045

Patrick DL, Burke LB, Powers JH, Scott JA, Rock EP, Dawisha S, O'Neill R, Kennedy DL (2007) Patient-reported outcomes to support medical product labeling claims: FDA perspective. Value Health. https://doi.org/10.1111/j.1524-4733.2007.00275.x

Patshina MV, Voroshilin RA, Osintsev AM (2021) Global biomaterials market: potential opportunities for raw materials of animal origin. Food Process Tech Technol. https://doi.org/10.21603/2074-9414-2021-2-270-289

Paulekuhn GS, Dressman JB, Saal C (2007) Trends in active pharmaceutical ingredient salt selection based on analysis of the orange book database. J Med Chem. https://doi.org/10.1021/jm701032y

PDF (n.d.) The role of the patent system in fostering innovation and entrepreneurship

Pellevoisin C, Coleman KP, Hoffmann S (2022) ISO 10993-23 In vitro irritation testing for medical devices: substantiating applicability to mild irritants and non-extractables. Toxicol In Vitro. https://doi.org/10.1016/j.tiv.2022.105371

Prete S, Dattilo M, Patitucci F, Pezzi G, Parisi OI, Puoci F (2023) Natural and synthetic polymeric biomaterials for application in wound management. J Funct Biomater 14:455. https://doi.org/10.3390/jfb14090455

Puluhulawa LE, Joni IM, Elamin KM, Mohammed AFA, Muchtaridi M, Wathoni N (2022) Chitosan–hyaluronic acid nanoparticles for active targeting in cancer therapy. Polymers (Basel). https://doi.org/10.3390/polym14163410

Rahman WA, Amenábar T, Hetaimish BM, Safir OA, Kuzyk PR, Gross AE (2016) Outcome of revision Total Hip Arthroplasty in Management of Failed Metal-on-Metal Hip Arthroplasty. J Arthroplast. https://doi.org/10.1016/j.arth.2016.04.033

Ramachandran MK, Kannan M, Velu R, Shanmugam P (2023) Design and development of 3D printing on bioinks and biomaterials for implants and tissue engineering. In: Digital design and manufacturing of medical devices and systems. https://doi.org/10.1007/978-981-99-7100-8_9

Ramakrishna S, Huang Z-M (2003) Biocomposites. In: Bioengineering, comprehensive structural integrity. Elsevier, pp 215–296. https://doi.org/10.1016/B0-08-043749-4/09064-9

Ramakrishna S, Mayer J, Wintermantel E, Leong KW (2001) Biomedical applications of polymer-composite materials: a review. Compos Sci Technol. https://doi.org/10.1016/S0266-3538(00)00241-4

Ranakoti L, Gangil B, Bhandari P, Singh T, Sharma S, Singh J, Singh S (2023) Promising role of Polylactic acid as an ingenious biomaterial in scaffolds, drug delivery, tissue engineering, and medical implants: research developments, and prospective applications. Molecules. https://doi.org/10.3390/molecules28020485

Ranganathan B, Miller C, Sinskey A (2018) Biocompatible synthetic and semi-synthetic polymers – a patent analysis. Pharm Nanotechnol. https://doi.org/10.2174/2211738505666171023152549

Reddy MSB, Ponnamma D, Choudhary R, Sadasivuni KK (2021) A comparative review of natural and synthetic biopolymer composite scaffolds. Polymers (Basel) 13:1105. https://doi.org/10.3390/polym13071105

Resnic FS, Normand S-LT (2012) Postmarketing surveillance of medical devices – filling in the gaps. N Engl J Med. https://doi.org/10.1056/nejmp1114865

Rizwan M, Yahya R, Hassan A, Yar M, Azzahari A, Selvanathan V, Sonsudin F, Abouloula C (2017) pH sensitive hydrogels in drug delivery: brief history, properties, swelling, and release mechanism, material selection and applications. Polymers (Basel) 9:137. https://doi.org/10.3390/polym9040137

Saha C, Bhattacharya S (2011) Intellectual property rights: an overview and implications in pharmaceutical industry. J Adv Pharm Technol Res 2:88. https://doi.org/10.4103/2231-4040.82952

Satchanska G, Davidova S, Petrov PD (2024) Natural and synthetic polymers for biomedical and environmental applications. Polymers (Basel) 16:1159. https://doi.org/10.3390/polym16081159

Schneider KH, Goldberg BJ, Hasturk O, Mu X, Dötzlhofer M, Eder G, Theodossiou S, Pichelkastner L, Riess P, Rohringer S, Kiss H, Teuschl-Woller AH, Fitzpatrick V, Enayati M, Podesser BK, Bergmeister H, Kaplan DL (2023) Silk fibroin, gelatin, and human placenta extracellular matrix-based composite hydrogels for 3D bioprinting and soft tissue engineering. Biomater Res 27. https://doi.org/10.1186/s40824-023-00431-5

Schuh JACL, Funk KA (2019) Compilation of international standards and regulatory guidance documents for evaluation of biomaterials, medical devices, and 3-D printed and regenerative medicine products. Toxicol Pathol. https://doi.org/10.1177/0192623318804121

Schulz KF, Grimes DA (2002) Generation of allocation sequences in randomised trials: chance, not choice. Lancet. https://doi.org/10.1016/S0140-6736(02)07683-3

Sedrakyan A, Paxton EW, Phillips C, Namba R, Funahashi T, Barber T, Sculco T, Padgett D, Wright T, Marinac-Dabic D (2011) The international consortium of Orthopaedic registries: overview and summary. J Bone Joint Surg Am. https://doi.org/10.2106/jbjs.k.01125

Seng D (2021) Intellectual property, information science, and quantitative legal analysis. In: Handbook of intellectual property research. https://doi.org/10.1093/oso/9780198826743.003.0030

Serrano DR, Kara A, Yuste I, Luciano FC, Ongoren B, Anaya BJ, Molina G, Diez L, Ramirez BI, Ramirez IO, Sánchez-Guirales SA, Fernández-García R, Bautista L, Ruiz HK, Lalatsa A (2023) 3D printing technologies in personalized medicine, nanomedicines, and biopharmaceuticals. Pharmaceutics 15:313. https://doi.org/10.3390/pharmaceutics15020313

Sheikh A, Md S, Alhakamy NA, Kesharwani P (2022) Recent development of aptamer conjugated chitosan nanoparticles as cancer therapeutics. Int J Pharm. https://doi.org/10.1016/j.ijpharm.2022.121751

Sheridan C (2017) First approval in sight for Novartis' CAR-T therapy after panel vote. Nat Biotechnol. https://doi.org/10.1038/nbt0817-691

Sherman RE, Anderson SA, Dal Pan GJ, Gray GW, Gross T, Hunter NL, LaVange L, Marinac-Dabic D, Marks PW, Robb MA, Shuren J, Temple R, Woodcock J, Yue LQ, Califf RM (2016) Real-world evidence — what is it and what can it tell us? N Engl J Med. https://doi.org/10.1056/nejmsb1609216

Shi H, Wang C, Ma Z (2021) Stimuli-responsive biomaterials for cardiac tissue engineering and dynamic mechanobiology. APL Bioeng. https://doi.org/10.1063/5.0025378

Shokrani H, Shokrani A, Seidi F, Kucińska-Lipka J, Makurat-Kasprolewicz B, Saeb MR, Ramakrishna S (2023) Artificial intelligence for biomedical engineering of polysaccharides: a short overview. Curr Opin Biomed Eng. https://doi.org/10.1016/j.cobme.2023.100463

Skallevold HE, Rokaya D, Khurshid Z, Zafar MS (2019) Bioactive glass applications in dentistry. Int J Mol Sci 20:5960. https://doi.org/10.3390/ijms20235960

Slavkova M, Tzankov B, Popova T, Voycheva C (2023) Gel formulations for topical treatment of skin cancer: a review. Gels 9:352. https://doi.org/10.3390/gels9050352

Smith TO, Hawker GA, Hunter DJ, March LM, Boers M, Shea BJ, Christensen R, Guillemin F, Terwee CB, Williamson PR, Dodd S, Roos EM, Loeser RF, Schnitzer TJ, Kloppenburg M, Neogi T, Ladel CH, Kalsi G, Kaiser U, Buttel TW, Ashford AE, Mobasheri A, Arden NK, Tennant A, Hochberg MC, De Wit M, Tugwell P, Conaghan PG (2019) The OMERACT-OARSI core domain set for measurement in clinical trials of hip and/or knee osteoarthritis. J Rheumatol. https://doi.org/10.3899/jrheum.181194

Sox HC, Greenfield S (2009) Comparative effectiveness research: a report from the institute of medicine. Ann Intern Med. https://doi.org/10.7326/0003-4819-151-3-200908040-00125

Spoială A, Ilie C-I, Ficai D, Ficai A (2023) Biomaterials:97–129. https://doi.org/10.1007/978-3-031-35832-6_4

Stapleton L, Zhu Y, Woo YJ, Appel E (2020) Engineered biomaterials for heart disease. Curr Opin Biotechnol 66:246–254. https://doi.org/10.1016/j.copbio.2020.08.008

Su J, Yang L, Sun Z, Zhan X (2024) Personalized drug therapy: innovative concept guided with Proteoformics. Mol Cell Proteomics 23:100737. https://doi.org/10.1016/j.mcpro.2024.100737

Suarato G, Li W, Meng Y (2016) Role of p H-responsiveness in the design of chitosan-based cancer nanotherapeutics: a review. Biointerphases. https://doi.org/10.1116/1.4944661

Sun J, Tan H (2013) Alginate-based biomaterials for regenerative medicine applications. Materials (Basel). https://doi.org/10.3390/ma6041285

Swider E, Koshkina O, Tel J, Cruz LJ, de Vries IJM, Srinivas M (2018) Customizing poly(lactic-co-glycolic acid) particles for biomedical applications. Acta Biomater 73:38–51. https://doi.org/10.1016/j.actbio.2018.04.006

Tabatabaei Hosseini BS, Meadows K, Gabriel V, Hu J, Kim K (2024) Biofabrication of cellulose-based hydrogels for advanced wound healing: a special emphasis on 3D bioprinting. Macromol Biosci. https://doi.org/10.1002/mabi.202300376

Tapscott DC, Wottowa C (2023) Orthopedic implant materials. StatPearls

Tavoni M, Dapporto M, Tampieri A, Sprio S (2021) Bioactive calcium phosphate-based composites for bone regeneration. J Compos Sci 5:227. https://doi.org/10.3390/jcs5090227

Thakur AK, Chellappan DK, Dua K, Mehta M, Satija S, Singh I (2020) Patented therapeutic drug delivery strategies for targeting pulmonary diseases. Expert Opin Ther Pat. https://doi.org/10.1080/13543776.2020.1741547

Thang NH, Chien TB, Cuong DX (2023) Polymer-based hydrogels applied in drug delivery: an overview. Gels 9:523. https://doi.org/10.3390/gels9070523

Timpe C, Stegemann S, Barrett A, Mujumdar S (2020) Challenges and opportunities to include patient-centric product design in industrial medicines development to improve therapeutic goals. Br J Clin Pharmacol. https://doi.org/10.1111/bcp.14388

Trenfield SJ, Basit AW (2020) Modified drug release: current strategies and novel technologies for oral drug delivery. Nanotechnol Oral Drug Deliv. https://doi.org/10.1016/B978-0-12-818038-9.00006-5

Tripathi AS, Zaki MEA, Al-Hussain SA, Dubey BK, Singh P, Rind L, Yadav RK (2023) Material matters: exploring the interplay between natural biomaterials and host immune system. Front Immunol 14. https://doi.org/10.3389/fimmu.2023.1269960

Trucillo P (2024) Biomaterials for drug delivery and human applications. Materials (Basel) 17:456. https://doi.org/10.3390/ma17020456

Tu Y, Chen N, Li C, Liu H, Zhu R, Chen S, Xiao Q, Liu J, Ramakrishna S, He L (2019) Advances in injectable self-healing biomedical hydrogels. Acta Biomater 90:1–20. https://doi.org/10.1016/j.actbio.2019.03.057

Turner BL, Twiddy J, Wilkins MD, Ramesh S, Kilgour KM, Domingos E, Nasrallah O, Menegatti S, Daniele MA (2023) Biodegradable elastomeric circuit boards from citric acid-based polyesters. npj Flexible Electron. https://doi.org/10.1038/s41528-023-00258-z

Vach Agocsova S, Culenova M, Birova I, Omanikova L, Moncmanova B, Danisovic L, Ziaran S, Bakos D, Alexy P (2023) Resorbable biomaterials used for 3D scaffolds in tissue engineering: a review. Materials (Basel) 16:4267. https://doi.org/10.3390/ma16124267

Valcourt DM, Dang MN, Day ES (2019) IR820-loaded PLGA nanoparticles for photothermal therapy of triple-negative breast cancer. J Biomed Mater Res A. https://doi.org/10.1002/jbm.a.36685

Vallittu PK (2018) Composite biomaterials: from lab to clinics. In: Translational oral health research. Springer International Publishing, Cham, pp 43–58. https://doi.org/10.1007/978-3-319-78205-8_5

Victor SP, Muthu J (2014) Polymer ceramic composite materials for orthopedic applications—relevance and need for mechanical match and bone regeneration. J Mechatronics 2:1–10. https://doi.org/10.1166/jom.2014.1030

Virmani T, Kumar G, Sharma A, Pathak K, Akhtar MS, Afzal O, Altamimi ASA (2023) Amelioration of cancer employing chitosan, its derivatives, and chitosan-based nanoparticles: recent updates. Polymers (Basel). https://doi.org/10.3390/polym15132928

Voskerician G (2017) Improving the hemocompatibility of biomedical composites, Hemocompatibility biomater. Clin Appl Blood-Biomater Interact:357–375. https://doi.org/10.1016/B978-0-08-100497-5.00010-0

Wang S, Noe RA, Wang Z-M (2014) Motivating knowledge sharing in knowledge management systems. J Manage 40:978–1009. https://doi.org/10.1177/0149206311412192

Wang C-S, Natale G, Virgilio N, Heuzey M-C (2016) Synergistic gelation of gelatin B with xanthan gum. Food Hydrocoll 60:374–383. https://doi.org/10.1016/j.foodhyd.2016.03.043

Wang M, Bai J, Shao K, Tang W, Zhao X, Lin D, Huang S, Chen C, Ding Z, Ye J (2021) Poly(vinyl alcohol) hydrogels: the old and new functional materials. Int J Polym Sci 2021:1–16. https://doi.org/10.1155/2021/2225426

Wang Z, Ma W, Fu X, Qi Y, Zhao Y, Zhang S (2023a) Development and applications of mRNA treatment based on lipid nanoparticles. Biotechnol Adv. https://doi.org/10.1016/j.biotechadv.2023.108130

Wang S, Chen Y, Guo J, Huang Q (2023b) Liposomes for tumor targeted therapy: a review. Int J Mol Sci 24:2643. https://doi.org/10.3390/ijms24032643

Wang S, Zhao X, Hsu Y, He Y, Wang F, Yang F, Yan F, Xia D, Liu Y (2023c) Surface modification of titanium implants with Mg-containing coatings to promote osseointegration. Acta Biomater. https://doi.org/10.1016/j.actbio.2023.07.048

Wang YT, Yang PC, Zhang YF, Sun JF (2024) Synthesis and clinical application of new drugs approved by FDA in 2023. Eur J Med Chem. https://doi.org/10.1016/j.ejmech.2024.116124

Wawrzyńska M, Arkowski J, Włodarczak A, Kopaczyńska M, Biały D (2018) Development of drug-eluting stents (DES). In: Functionalised cardiovascular stents. Elsevier, pp 45–56. https://doi.org/10.1016/B978-0-08-100496-8.00003-2

Webster TJ (2023) Artificial intelligence (AI) and biomaterials: a perfect BandAId™, Open Access Gov. https://doi.org/10.56367/oag-040-11059

Wei S, Ma J-X, Xu L, Gu X-S, Ma X-L (2020a) Biodegradable materials for bone defect repair. Mil Med Res 7:54. https://doi.org/10.1186/s40779-020-00280-6

Wei J, Wang B, Li Z, Wu Z, Zhang M, Sheng N, Liang Q, Wang H, Chen S (2020b) A 3D-printable TEMPO-oxidized bacterial cellulose/alginate hydrogel with enhanced stability via nanoclay incorporation. Carbohydr Polym. https://doi.org/10.1016/j.carbpol.2020.116207

White S (2018) What's my FDA medical device classification? Cortex Des

Williams DF (2008) On the mechanisms of biocompatibility. Biomaterials. https://doi.org/10.1016/j.biomaterials.2008.04.023

Williams HL (2017) How do patents affect research investments? Annu Rev Econ 9:441–469. https://doi.org/10.1146/annurev-economics-110216-100959

Wu Y-CJ, Lee P-J (2007) The use of patent analysis in assessing ITS innovations: US, Europe and Japan. Transp Res A Policy Pract 41:568–586. https://doi.org/10.1016/j.tra.2006.11.007

Wu Y, Yang Y, Lv X, Gao M, Gong X, Yao Q, Liu Y (2023) Nanoparticle-based combination therapy for ovarian cancer. Int J Nanomedicine 18:1965–1987. https://doi.org/10.2147/IJN.S394383

Xing H, Lee H, Luo L, Kyriakides TR (2020) Extracellular matrix-derived biomaterials in engineering cell function. Biotechnol Adv 42:107421. https://doi.org/10.1016/j.biotechadv.2019.107421

Xu C, Zhu D (2021) On conflicts between pharmaceutical patent protection and the right to life and health based on a stackelberg game. Int J Environ Res Public Health. https://doi.org/10.3390/ijerph18031119

Xu W, Ling P, Zhang T (2013) Polymeric micelles, a promising drug delivery system to enhance bioavailability of poorly water-soluble drugs. J Drug Deliv 2013:1–15. https://doi.org/10.1155/2013/340315

Xu G, Yu X, Zhang J, Sheng Y, Liu G, Tao W, Mei L (2016) Robust aptamer–polydopamine-functionalized M-PLGA–TPGS nanoparticles for targeted delivery of docetaxel and enhanced cervical cancer therapy. Int J Nanomedicine. https://doi.org/10.2147/IJN.S103513

Yang D (2023) Morgen stanleys acquisition of etrade: reasons, purposes and prospect. Adv Econ Manag Polit Sci. https://doi.org/10.54254/2754-1169/38/20231916

Yeung AWK, Torkamani A, Butte AJ, Glicksberg BS, Schuller B, Rodriguez B, Ting DSW, Bates D, Schaden E, Peng H, Willschke H, van der Laak J, Car J, Rahimi K, Celi LA, Banach M, Kletecka-Pulker M, Kimberger O, Eils R, Islam SMS, Wong ST, Wong TY, Gao W, Brunak S, Atanasov AG (2023) The promise of digital healthcare technologies. Front Public Health 11. https://doi.org/10.3389/fpubh.2023.1196596

Yusuf A, Almotairy ARZ, Henidi H, Alshehri OY, Aldughaim MS (2023) Nanoparticles as drug delivery systems: a review of the implication of nanoparticles' physicochemical properties on responses in biological systems. Polymers (Basel) 15:1596. https://doi.org/10.3390/polym15071596

Zhang G, Lu H, Mamidwar S, Wang M (2020) Composites. Biomater Sci (Elsevier):415–429. https://doi.org/10.1016/B978-0-12-816137-1.00029-5

Zhang B, Wang L, Song P, Pei X, Sun H, Wu L, Zhou C, Wang K, Fan Y, Zhang X (2021) 3D printed bone tissue regenerative PLA/HA scaffolds with comprehensive performance optimizations. Mater Des. https://doi.org/10.1016/j.matdes.2021.109490

Zhang W, Wang Q, Wang K, Ma H, Gong Y, Xu L, Zhang H (2022) Construction of an ornidazole/bFGF-loaded electrospun composite membrane with a core-shell structure for guided tissue regeneration. Mater Des. https://doi.org/10.1016/j.matdes.2022.110960

Zhao D, Witte F, Lu F, Wang J, Li J, Qin L (2017) Current status on clinical applications of magnesium-based orthopaedic implants: a review from clinical translational perspective. Biomaterials. https://doi.org/10.1016/j.biomaterials.2016.10.017

Zhong JX, Raghavan P, Desai TA (2023) Harnessing biomaterials for immunomodulatory-driven tissue engineering. Regener Eng Transl Med. https://doi.org/10.1007/s40883-022-00279-6

Zhu L, Zhang Y, Ji Y (2017) Fabricating poly(1,8-octanediol citrate) elastomer based fibrous mats via electrospinning for soft tissue engineering scaffold. J Mater Sci Mater Med. https://doi.org/10.1007/s10856-017-5906-7

Zhu W, Castro NJ, Shen YL, Zhang LG (2022) Nanotechnology and 3D/4D bioprinting for neural tissue regeneration. In: 3D bioprinting and nanotechnology in tissue engineering and regenerative medicine. https://doi.org/10.1016/B978-0-12-824552-1.00005-0

Zhu H, Liu F, Zhai X, Tong Z, Li H, Dong W, Wei W, Teng C (2023) Revisiting matrix hydrogel composed of gelatin and hyaluronic acid and its application in cartilage regeneration. Biochem Biophys Res Commun. https://doi.org/10.1016/j.bbrc.2023.09.060

Zothanpuii F, Ravindran R, Kanthiah S (2020) A review on stability testing guidelines of pharmaceutical products. Asian J Pharm Clin Res:3–9. https://doi.org/10.22159/ajpcr.2020.v13i10.38848

Regulatory Concern for Biomaterial and Its Toxicity

16

Shachi Patel, Swati Sharma, Jalpa Suthar, and Meghraj Suryawanshi

Abstract

Biomaterial is a material that has been created to take on a shape that may be employed alone or in conjunction with other sophisticated systems to manage interactions with biological systems to guide the course of any therapeutic or diagnostic technique. The state of art and potential of biomaterials, in biomedical engineering and pharmaceutical sciences is very well examined. These biomaterials are gradually being included into controlled drug delivery systems, tissue engineering, minimally invasive biomedical products, diagnostics, and artificial organs. Agencies ·like FDA, EMA, and American Society for Testing and Materials (ASTM) has consciously developing standards for biomaterial. However, the rapid and consistent development of biomaterial based products brings challenges to regulatory agencies for its approval. Biocompatibility, irritation and sensitization are of utmost concern in clinical adoption of biomaterial based products. Moreover, toxicological testing is a critical tool for studying the safety and biological effects of bio material. On the other hand, it is difficult to capture repeated and long-term toxicity. The application of regulatory science to biomaterials and associated medical products presents additional difficulties. Thus, biomaterials-based regulatory science system has been established to

S. Patel (✉) · S. Sharma
Department of Pharmaceutics, Ramanbhai Patel College of Pharmacy, Charotar University of Science and Technology (CHARUSAT), CHARUSAT campus, Changa, Gujarat, India
e-mail: shachikapatel.ph@charusat.ac.in

J. Suthar
Department of Pharmacology, Ramanbhai Patel College of Pharmacy, Charotar University of Science and Technology (CHARUSAT), CHARUSAT campus, Changa, Gujarat, India

M. Suryawanshi
Department of Pharmaceutics, Sandip Institute of Pharmaceutical Sciences (SIPS), Nashik, Maharashtra, India

Md. Faiyazuddin, M. Suryawanshi (eds.), *Design and Processing of Green Materials*, Biomaterials, Bioengineering and Sustainability 4,
https://doi.org/10.1007/978-3-031-91790-5_16

address safety concern and clinical needs. The chapter delve into the critical aspects of regulatory concerns associated with biomaterial toxicity, highlighting the importance of rigorous safety evaluations in the development and approval of biomaterials for clinical use.

Keywords

Bio materials · Bio compatibility · Regulatory science · Toxicity

16.1 Introduction

Biomaterials exemplify a pioneering paradigm shift in the field of pharmaceuticals and healthcare. These are categorised under those bioactive substances that can be engineered in a flexible form to have a direct or indirect interactions with the living creatures (Pires et al. 2023). The materials play an essential role in therapeutics and diagnostics by developing medical devices or pharmaceuticals. The biomaterial science is an interdisciplinary field connecting biology and engineering principles (Xiang and Chen 2020). The minimal side effects, self-healing nature and biomimetic behaviour have extended its potential beyond healthcare to diverse scientific domains (Trucillo 2024). Multiple requirements need to be considered while designing biomaterials, however biocompatibility is of utmost concern for fabrication and clinical application of biomaterials. A biocompatible material is able to demonstrate its purpose with no adverse effects to the cellular structure (Huzum et al. 2021). In some of the cases biodegradability is concerned letting biomaterials to shed off over the time which reduces the demand for surgery. Biomaterials are classified into metals, ceramics, or polymers. Materials like gold, ceria, silver, starch, cellulose, chitosan, alginate, collagen, silk-fibroin and others are used in pharmaceuticals and healthcare (Samir et al. 2022). A typical design is followed for designing a biomaterial (Ebrahimi 2020). These materials are used for drug delivery, medical devices, and other healthcare applications (Park et al. 2022; Lu et al. 2024).

Biomaterials have gained a unique visibility in the pharmaceutical sector. Current advances in biomaterials for controlled drug delivery have depicted substantial advances in medicine. The development of such dosage forms demands multidisciplinary collaborations between chemist, engineers and pharmacists (Fenton et al. 2018). Chitosan, silk based biomaterials, alginate, glycosaminoglycan, cellulose are few of the naturally occurring biomaterials used in designing advanced drug delivery (Troy et al. 2021). These are either used a scaffold or as a polymeric carrier for drug delivery. These biomaterials can be fabricated to incorporate the pharmaceutical compound and allow the controlled release of the pharmaceutical product. This approach allows prolonged release for chronic conditions. The biodegradability of the polymer undergoes time dependent erosion which results in sustained release of the medicament (Mehtani et al. 2019; Soni et al. 2022). A cisplatin loaded transmucosal drug delivery system demonstrated a good therapeutic efficiency against oral cancer in animals (Goldberg et al. 2022). In the study, chitosan loaded cisplatin nanoparticles demonstrated targeted release of medicament from the chitosan nanocarriers. Another study demonstrated stimuli responsive behaviour of docetaxel nanoparticles for breast cancer therapy. The study demonstrated the application of hyaluronic acid to target $CD44^+$ breast cancer cells. Similarly, biomaterials have also demonstrated its potential in various other chronic illness (Hsiao et al. 2019). Further biomaterials have also established its place as regenerative materials. These materials act as a scaffold and allowing the cells to proliferate. Regenerative medicine is one of most attracted field in biomedical engineering. Biomaterials support like an tissue's extracellular matrix, allowing the damaged cells to proliferate, function and heal naturally (Nii 2021). Biomaterials like glycosaminoglycan mimics the tissue's extracellular matrix and so are the effective carriers in stem cell encapsulation (Ashworth et al. 2021). A group of researcher established articular regeneration in rabbits by applying biomaterial loaded mesenchymal stem. The researchers developed a hydrogel carrier comprising of alginate, gelatin and chondroitin sulfate. Further the mesenchymal stem cells were loaded into the hydrogel. The hydrogel showed biocompatibility and supported cell proliferation and function. A novel cardiac patch was designed by a group of researchers from University of Pisa, Italy. These researchers developed alginate-gelatin sponge by applying avidin/biotin functionalization. The scaffold represented improved cardiac function (Frati et al. 2020). Further biomaterial based implants also support tissue engineering. Metals and ceramics are highly used materials specifically applied in orthopaedic implants (Hallab et al. 2020). These materials are established its specific properties like high flexibility, toughness, and most importantly biocompatibility. Magnesium, titanium based alloys have established its potential in bone regeneration (Uppal et al. 2022). Further biomaterial have established their role in diagnostics (Gervasoni et al. 2024). Owing to extensive application of biomaterials in tailor made drug delivery systems and medical devices, it is necessary to design multi-factorial design control processes to eradicate the unacceptable risk and achieve required benefits by validated biocompatibility, safety and biological performance (Feng et al. 2022). Though, the transformation of these biomaterials from bench to the bed-side remains challenging. As, biomaterials are not medical or pharmaceutical product, rather they are crucial components of it (Song et al. 2022b). Contemporary regulation only focused on final product approval, rather than single component (as in case of Biomaterial).

However, to ascertain safety and innovation of biomaterials across multidisciplinary field, agencies like International organization for standardization (ISO), or American Society for Testing materials (ASTM) (Eliaz and Metoki 2017) has constantly developing scientific approaches to characterise biomaterial.

The significant void between the scientific research and its commercial translation has led to the demand in designing new regulatory framework for biomaterial based product approval (Wang et al. 2022).

16.1.1 Overview of Regulatory Concern for Biomaterial

Biomaterials are part of formulation rather than a formulation itself, while essential parts of medicinal products biomaterials are not themselves medical items. Biomedical devices, medications, biologics, and combined medical products are categorized as medical products in this context and are all subject to different regulations. Biomaterials must be properly developed and constructed in accordance with the final product provisions and application criteria in order to abolish undesirable risks and simultaneously achieve desired benefits before they can become final medical products (Song et al. 2022b). The creation and therapeutic application of biomaterials with various origins on patients raise a multitude of ethical and legal concerns.

These includes questions on the procurement of the raw materials, their handling and processing, animal and human testing of the final products, marketing, and post-marketing biomaterial surveillance (Moharamzadeh 2017).

To guarantee the safety and compatibility of biomaterial with biological systems, biocompatibility testing is an indispensable part of the development and regulatory approval processes (Kanďárová and Pôbiš 2023). The assessments that are essential to this testing are the cytotoxicity, irritation, and sensitization tests, which are required practically for all biomaterial and medical devices containing such component that are being placed onto the market (Li et al. 2015).

Additional testing may be necessary depending on the proposed usage and final application of the biomaterial. Tragically, the adoption of alternate methods in biocompatibility testing in the medical field has been noticeably slower than in other industrial sectors (Hulsart-Billström et al. 2016).

Adherence to both domestic and global biocompatibility testing mandates is crucial for obtaining regulatory authorization and guaranteeing safe application. A number of international standards (refer Sect. 16.2.1) and regulations are strictly adhered by industry. Manufacturers must adhere to international standards and regulations that apply to their product. Testing prerequisites and final assessments have become somewhat unclear and confusing due to the lack of harmonization in the sector, despite numerous attempts. Developing preclinical in vivo testing standards and their subsequent clinical translation is difficult where newly developed materials were being constantly placed on to the market. Therefore, creating gold standard for established and newly developed biomaterial is cumbersome task for regulatory agencies.

16.2 Regulatory Landscape for Bio Material

Biomaterials are crucial to drug delivery systems since they are utilized as carriers—that is, inside biologics' packages—and then combined into vital parts of medical devices to assess, replace, cure, or alter a certain function, tissue, or organ (Razavi and Khandan 2017; Bernard et al. 2018a). Since, aegis healthcare industry have utilised various Natural, synthetic and hybrid biomaterial and therefore their character as raw materials and constituents of finished medicinal products, affect their safety and performance (Song et al. 2022b; Bhat and Kumar 2013). Hence, it is necessary to access biological compatibility before it is put on the market and used in clinical settings (Bernard et al. 2018b).

While essential parts of medicinal products, biomaterials are not themselves medical items. Entering a desired national or international market can be facilitated by understanding and using the appropriate testing standards during the pharmaceuticals development and testing phases.

Science of biocompatibility and biological response evaluation assessment have not advanced at the same rate. Most testing documents for pharmaceuticals and medical devices containing biomaterial are sourced from independent non-profit scientific bodies that rely on sales to fund the creation, manufacturing, and ongoing evaluation of testing documentation and standards; these are treated as standards rather than guidelines or recommendations that come from government agencies (Anderson 2016; Schuh and Funk 2019).

The written standards of several national standard organizations, national pharmacopeia's, the International Organization for Standardization (ISO), American Society for Testing and Materials (ASTM), and The Organisation for Economic Co-operation and Development (OECD) are crucial for understanding the biological testing, design of biomaterials and medical devices.

16.2.1 Overview of Key Regulatory Guidance on Biomaterial Testing

A. **ISO standards**

ISO is an alliance of standards organizations from more than 160 nations that collaborate to establish standards. Consequently, ISO 10993 is essentially used worldwide for biocompatibility testing and determination. There are 23 distinct sections that make up ISO 10993, and each is updated and maintained independently.

Amongst the series 10993, ISO 10993-1 is crucial as it contain algorithms to guidc study needs, summarles the rational of tests selection and help structural development plans for biological evaluation of related material or medical device (Anderson 2016; Schuh and Funk 2019). When doing biological evaluation with ISO 10993-1, experts must understand what is meant by "biocompatibility." It is the "the ability of a medical device or material to perform with an appropriate host response in a specific application".

According to ISO 10993-1, Biological examination of any material or medical device containing bio material can be done by examining historical clinical and preclinical data along with actual testing. In-case, material in specified role and physical form had noticeable safe history and adequate information on risk assessment of material, no testing is required. Moreover, Selection of *in-vitro* and *in-vivo* test depends upon careful consideration on intended usage. *In-vitro* test methods are highly recommended over *in-vivo* test, if methods of testing are validated, reliable, reproducible, reasonably and practically available.

Material characterization is prerequisite for biological testing, including physical, chemical characterization and performed as per ISO 10993-18 at product design stage before conducting biological evaluation.

The primary function of the ISO 10993-1 document is to offer a structural framework for biological evaluations in conjunction with the risk management techniques outlined in ISO 14971:2007. Additionally, it allows scientific advancement in the creation of *in-vitro* tissue models for toxicity in order to reduce the animal studies (Hampshire and Gilbert 2019).

B. **ASTM Standards**

In 1898, the ASTM was established as a non-profit organization with the goal of creating technical data and standards for materials, goods, systems, and services. Test procedures, guidelines, practices, requirements, classifications, and nomenclature are examples of ASTM standards. Over 10,000 standards have been published to date covering almost every aspect of material technology (Sekar et al. 2021).

The ASTM Committee F04 (Medical and Surgical Materials and Devices) was established in 1962. With 1050 members, the Committee now oversees approximately 338 standards that are listed in the ASTM Standards Annual Book (Committee F04 on Medical and Surgical Materials and Devices n.d.).

The resources, medical and surgical devices, orthopaedic devices, tissue-engineered goods, and computer-assisted surgical systems are the five main topics that are covered by the F04 technical subcommittees combined. The resources section covers requirements for materials including metals, polymers, and ceramics. It also covers requirements for basic data on test procedures, biocompatibility, and magnetic resonance imaging (ASTM n.d.-b).

Out of standards depicted in F04, Subcommittee F04.16 (Biocompatibility Test Methods) and F04.42 (Biomaterials and Biomolecules for tissue-engineered medical products TEMPs) were primordial in understanding testing requirement for bio material (Sekar et al. 2021).

Important F04 subparts were depicted below:

Under the guidance of F04.16, subcommittee F748-16 (Standard Practice for Selecting Generic Biological Test Methods for Materials and Devices), materials investigator choose the right steps to take while screening newly created or altered materials. These suggestions should be adjusted as needed because every material has different requirements, thus they shouldn't be the only analysis that's needed for a given material. Furthermore, these recommendations shouldn't be seen as the minimal requirement in any given circumstance (Amato 2015a, b; ASTM n.d.-a, n.d.-b).

Based on end-use applications, these biological test protocols can be applied to materials and medical devices intended for human use. Along with related topics like batch testing of production batches and pyrogen testing, the biological evaluation of materials and devices is covered. Standards suggest generalized test protocol using materials, final products, and extracts.

Subcommittee F04.42, provides standard guide and testing for many biomaterials like chitosan, sodium alginate, collagen, Porcine Fibrinogen and Hyaluronan for use in biomedicine and healthcare applications (Sekar et al. 2021).

By affecting cell activity and/or the release of bioactive compounds or medications, the physicochemical properties of raw material in developing regenerative medicines carry a large potential to alter product performance (ASTM 2016). In this regard, F2027-16 (Standard Guide for Characterization and Testing of Raw or Starting Materials for Tissue-Engineered Medical Products) can be used by material developer. This guideline outlines suggested parameters or first material characterizations to guarantee repeatability before they are fabricated into scaffolds for implantable biomedical products.

The criteria for good biomaterials are met by recently developed chitosan/alginate formulations that have exclusive a non-allergic response, antibacterial activity, oxygen permeability, biodegradation and an exclusive capability to heal faster (Dornish et al. 2001; Drewnowska et al. 2013). No regulatory body has yet to authorize these carriers for use in medical applications. Standard requirements for chitosan/alginate based salts suitable for use in pharmaceutical or biomedical applications are being researched by F2064-17 and F2103-18 ASTM F04 (Yaneva et al. 2020). These assessment of alginates/chitosan appropriate for usage in pharmaceutical, biomedical, or both applications, but not restricted to, TEMP's. This manual covers important factors that are pertinent to the alginates'/chitosan functionality, characterization, and purity. In this light, standardization of characterisation parameters and test procedures will enable alginate or chitosan functionality to match the demands of intended usage or final product (Dornish et al. 2001).

C. **OECD standards**

The most pertinent, globally accepted testing techniques used by government agencies, businesses, and autonomous laboratories to evaluate the safety of chemicals are compiled in the OECD guidelines. Their main applications are in safety testing regulations, followed by chemical notice and registration. The collection of test guidelines is updated frequently to stay up with scientific advancements and national regulatory requirements (ILibrary n.d.).

ISO 10993 series is most globally accepted standards, however with no experimental detail it states to refers the OECD guideline for testing (Wallin 1996; Elina Tenhunen 2024).

OECD provides series of guidelines for genetotoxicity, carcinogenicity testing under the broad heading of Testing of Chemicals, Section 4. OECD test guideline provides unique solution to genotoxicity testing for newly developed material or medical device, and hence it may not be useful for devices containing compounds that are previously known to be genotoxic (Azzarà et al. 2010).

For carcinogenicity testing OECD 451 or OECD 453 is applicable, where testing for reproductive toxicity may begin with OECD 421 to give a general idea of potential effects on development and/or reproduction. Positive test findings can help determine the need for and timing of more testing and provides preliminary hazard assessment. If more testing is thought to be required, it must be done in compliance with OECD 414, OECD 415, or OECD 416, as applicable.

Besides that, for regulatory decision-making, the OECD Mutual Acceptance of Data (MAD) system has generated, it offers a standard foundation for national authorities to work together and prevent the development of non-tariff trade obstacles. Important feature of MAD includes avoidance of duplication as test conducted on particular materials will be valid to all 38 OECD members as well as the 7 non-member countries that adhere to MAD (The Mutual Acceptance of Data (MAD) System n.d.; Visser n.d.). Testing must follow OECD guidelines for test procedures (OECD Test Guidelines) and OECD Principles of Good Laboratory Practice (GLP) for data quality and integrity in order to comply with the OECD MAD system (Rasmussen et al. 2019).

D. **USFDA Regulation**

Since the 1938 enactment of the Food, Drug, and Cosmetic Act (FD&C Act), regulation governing the marketing of biomaterial products has undergone significant change. From a framework standpoint, the regulation of biomaterials is similar to that of other medical devices and biologic products; however, from an application standpoint, there are certain differences. Thus, it is essential to comprehend the main regulatory procedures associated with device commercialization in order to establish a grasp of those procedures for biomaterials (Mueller et al. 1991).

FD&C Act was amended by the US Congress to include the Medical Device Amendments (MDA), the first significant part of medical device regulation in the country. According to the manufacturer, the FDA does not approve biomaterials; instead, it approves them for specific uses in the form of devices (Amato 2015a).

The FDA assigns its biomaterial based product approval via three regulatory bodies like Center for drug evaluation and research (CDER), Center for device and radiological health (CDRH) and Center for biologics and evaluation research (CBER) (Pashuck and Stevens 2012). The selection of such FDA centre for regulatory approval of biomaterial based products remains challenging in terms of time and expense required for the approval process.

On realising widespread applicability of ISO 10993 series, FDA recently published commentaries on use of ISO 10993, part 1, which included the characterizing and testing of material within a risk management procedure, specifically targeting to industry and FDA staff in September, 2023.These document serve the view of FDA on ISO 10993-1, as well as other biocompatibility standards (e.g., additional ISO 10993 series of standards, OECD, USP ASTM, ICH) (U.S. Food and Drug Administration 2023). However, ISO 10993 include generalized methods with numerous options, which sometimes limit to address assessment of results or include acceptance criteria (U.S. Food and Drug Administration 2023). The FDA recommends using the risk management approach as the foundation for any biomaterial component, such as medical devices, for evaluating biological properties, as shown in ISO 10993-1. As part of the risk assessment, the finished product or component needs to be assessed. The Agency has neither cleared nor certified any specific materials used in the production of medical devices. Hence, the risk evaluation should consider the device's materials, their processing, production techniques (such sterilization), and the residue from any manufacturing aids employed in the process.

The FDA suggests that a thorough report be supplied for each test carried out in order to strengthen the biomaterial regulatory science. Additionally, the "Accreditation Scheme for Conformity Assessment " (ASCA) being applied for the testing (FDA/CDRH 2023).

Contemporary to that, "National Medical Products Administration of China" (NMPA) propelled "Regulatory Science Action Plan (RSAP)" in 2019 for the first time. Furthermore, the People's Republic of China's Ministry of Science and Technology launched "National Key Research & Development Programs", which included the projects on "Medical Equipment and Biomaterials (2021–2025)"and "Research and Development of Biomaterials and Repair and Replacement of Tissues and Organs (2016–2020)". Regulatory science investigation on new biomaterials used in medical devices was a component of both major initiatives. Additionally, on April-2021, the "Biomaterials Innovation Cooperation Platform," which was started by the NMPA's "Centre for Medical Device Evaluation" (CMDE), was enacted (Song et al. 2022a).

16.3 Regulatory Challenges in Integration of Biomaterial in Pharmaceutical Products

As discussed earlier, biomaterials are essential parts of medicinal products, biomaterials are not themselves medical items. Hence, all of them are considered medical products in this context and are subject to regulatory agency scrutinization. Premarket assessment and post market surveillances are the two phases of the biomaterials evaluation process. The two biggest biomaterials markets, the US and the EU, use rather different approaches to biomaterials evaluation (Song et al. 2022b).

Before starting "first-in-man" clinical trial for a biomaterial derived from tissue, a product safety and function data package needs to satisfy regulatory bodies, institutional review boards, and investigators that the product's potential benefits exceed its purported risks. Preclinical testing for drugs and medical devices is based on an advanced framework: study designs, diagnostics, statistical plans, and animal models essentially from a "cookbook" that makes the risk/benefit analysis process easier for all reviewers. Such a manual does not exist for biomaterials including biologic elements. Consequently, innovators are need to create their own preclinical research, and each study's unique design prompts a (justifiably) cautious regulatory scrutiny (Masaeli et al. 2019).

To address important challenges and offer deep insights into the various forms and ways that biomaterials interact with surrounding tissues, an inclusive taxonomy of biomaterials is necessary (Song et al. 2022b). Although the terms "medical-grade materials" or "implantable materials" are frequently used in scholarly papers, none of these has a defined meaning nor set of guidelines.

The conditions for approval of a particular biomaterial vary throughout regulatory systems. International standards like ISO, ASTM and OECD has been utilized by different manufacturer and regulator for biological assessment. As discussed in regulatory framework section ISO 10993 guidelines, highlights standards that serve as a simple outline to aid manufacturers in selecting appropriate tests for biological assessment and are not a checklist to be followed methodically. These standards provide extensive range of biological tests depending on the exposure conditions. (O'Brien et al. 2022). Hence, manufacture needs to navigate the suitable test methods that establish safety along with efficacy for biomaterial.

Emerging advances in biomaterials and medical devices present new challenges for biocompatibility assessment such as nanoparticles, tissue engineered medical devices and functional materials. While new approaches of delivering therapeutic moiety will be very beneficial to patients' health and quality of life, they also carry a larger risk than traditional therapy (Halamoda-Kenzaoui et al. 2019). Therefore, novel biomaterials' new standards for terms, classification, specifications, test procedures, and practice—created by research in regulatory science—may also make it easier to produce and use pharmaceuticals based on biomaterials.

The intricacy of next-generation biomaterials and their influence on commercialization provide the biggest regulatory challenges to their clinical translation. Biomaterials' long-term potential is in their combination with drugs, which includes biologics, small-molecule medications, and other materials or technologies, that offer transformative therapies, which puts regulators under pressure to reconsider their long-standing approaches that are firmly based in single-platform technology products (Prestwich et al. 2012).

16.4 Safety Concerns and Toxicological Testing of Biomaterials

In recent years, biomaterials are frequently utilized to treat multiple diseases or restore damaged tissues; however, there is a danger of infection, inflammation, and ultimately rejection while using them. Novel materials with antibacterial and anti-inflammatory properties are being explored to tackle these concerns. Before any clinical applications are considered, the aforementioned materials must undergo extensive preclinical testing (Helmus et al. 2008).

Multiple requirements need to be considered while incorporating biomaterials in medical products. However, biocompatibility is of utmost concern for fabrication and clinical application of biomaterials. A biocompatible material is able to demonstrate its potential with no adverse effects to the biological systems (Huzum et al. 2021).

Bioinertia is just one aspect of biocompatibility; other aspects include biofunction and biostability. This fundamental idea is highly dependent on the material's characteristics, how it interacts with the biological environment (Huzum et al. 2021;

Cvrček and Horáková 2018; Rahmati et al. 2020). The "Big Three" assessments—cytotoxicity, irritation, and sensitization testing—are essential to this testing and must be carried out. Testing for biocompatibility is a crucial step in the regulatory approval processes that guarantees products' safety and suitability for use with biological systems (Kanďárová and Pôbiš 2023).

When it comes to biomaterial testing, both *in-vivo* and *in-vitro* assessment techniques are essential for determining the performance, safety, and efficacy of biomaterials meant for medical application. Each Biomaterial are tested in controlled laboratory settings away from living things in the in vitro research.

While researchers are working to create new *in-vitro* tissues for drug screening and tissue replacement, in vitro models for evaluating novel biomaterials are incredibly rare and relatively new. After *in-vitro* testing is finished, *in-vivo* biological testing according to the intended usage can be carried out. These tests can include skin irritation testing, hemocompatibility testing, and implantation testing. (Frisch et al. 2023).

Nonetheless, before accessing biomaterial for *in-vitro* and *in-vivo* testing, material characterization and analysis of a device's components are conducted. The amount of necessary biological testing can frequently be decreased with the use of material characterization data. To show that an evaluated material or device is equivalent to one that has been commercialized, characterization is essential (Bhat and Kumar 2013). Such material characterization before accessing biocompatibility testing is depicted in Fig. 16.1.

While producing biocompatibility results, accuracy and reproducibility of test remain paramount concern to the manufacturer which can be attained through proper sample preparation. To guarantee safety and efficacy of biological material in medical applications, accurate, consistent, and reliable biocompatibility test findings must be obtained through correct sample preparation as depicted in ISO 10993-12, which includes detail about the sample preparation and reference materials. In addition to that general USP monograph and specialized sample preparation according to ASTM also utilized to ensure data uniformity for regulatory submission (Wallin 1998; Bernard et al. 2018b).

16.4.1 Testing of Biomaterials and Sample Preparation

Prerequisite step in biomaterial testing involves collection of sample in aseptic environment using sterilized tools and containers to avoid contamination. While sample collection procedure, appropriate conditions should be sustaining to ascertain the properties (e.g., temperature, humidity).

Testing of biomaterials involves several crucial steps to ensure the integrity and reliability of the results. Prior to biocompatibility testing, the biomaterials from the medical devices in final form is removed using a cell culture medium in tiny tubes or a semi-physiological media (saline solution, cottonseed or sesame seed oil). After that, they are preliminary incubated at 37 °C for 24–72 h. In addition, resultant extract solutions are used in the biological experiments after being decanted into

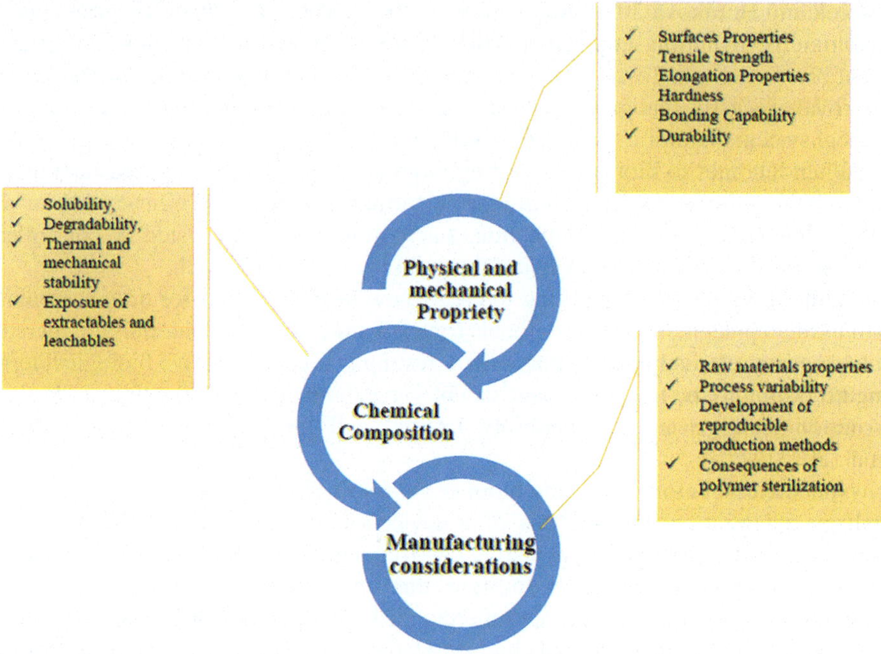

Fig. 16.1 Prerequisite for biocompatibility test

sterile glass tubes (Assad and Jackson 2019). The process of extraction is quite delicate. It is only possible to extract certain compounds from a material's surface. The primary drawbacks of this method are its low extraction yield (less than 10%), the material's brief contact with cells or tissue fluids in comparison to *in-vivo* conditions, and the extraction medium's inappropriateness for organic hydrophobic substances. Utilizing an extraction medium with dimethyl sulphoxide concentrations as low as <0.5% enables the solubilization of both hydrophilic and hydrophobic compounds (Bruinink and Luginbuehl 2011). Excessive force that alters material properties should be avoided. Depending on the testing requirements, surfaces might need to be polished, coated, or treated to mimic physiological conditions (Navarro et al. 2008).

16.4.2 *In-Vitro* Testing

Here we summarize the preclinical techniques and models that are currently available for biological in vitro assessment of biomaterials. We address the current issues in this field and outline various biological tests that are carried out in the biocompatibility assessment process, which is a fundamental stage in the development of novel materials. This chapter delivers an overview of the biological safety rules for biomaterials and the latest developments for ensuring those requirements. The optimal procedures for biological evaluation including biocompatibility and in vitro testing of biomaterials are thoroughly described in this piece of content.

16.4.2.1 *In-Vitro* Cytotoxicity Testing

Using cell culture-based techniques, cytotoxicity tests are carried out as the initial stage of material biocompatibility screening. Cytotoxicity assays offer both qualitative and quantitative assessments of a biomaterial's possible risk. A positive result necessitates additional research as it could be an early indication of a possible biological risk associated. Combining several assays can yield a more thorough understanding of the cytotoxicity mechanism that causes cell death (Nguyen et al. 2020; Mohd Shafiee et al. 2021).

According to ISO 10993-5, material and come into contact via three different methods:

(a) Direct contact; (b) indirect contact using an agar layer or a filter diffusion test; and (c) extract methods, which involve applying extraction media to cells under physiological or heightened circumstances prior to cytotoxicity assessment. While indirect contact techniques enable the investigator to identify compounds that may be leachable and potentially harmful, direct contact methods are designed to assess the cytotoxicity of the entire system (Bernard et al. 2018b; ISO 2009).

Various direct and indirect methods are useful to determine cytotoxicity according to ISO 10993-5 is depicted in Table 16.1.

16.4.2.2 *In-Vitro* Genotoxicity/Carcinogenicity/Reproductive Toxicity Testing of Biomaterial

ISO 10993-3:1992 represents evaluation of genotoxicity, carcinogenicity and reproductive toxicity guideline can serve as basis to check various toxicity.

16.4.2.3 Tests for Genotoxicity

Tests for genotoxicity identify gene mutations, alterations in the number and structure of chromosomes, and other gene toxicities brought on by biomaterials, medical devices, or their extracts.

All devices or material used in extracorporeal blood-contacting circuits are required to undergo genotoxicity assessments, even if the contact lasts for less than 24 h. If authenticated literature as well as chemical analysis of material depict safe genotoxic profile there is no need to conduct genotoxicity further (U.S. Food and Drug Administration 2023).

Two *in-vitro* tests and an optional third *in-vivo* test should be carried out since no single test is able to identify every genotoxin. Various *in-vitro* test was depicted in Table 16.1.

16.4.2.4 Test for Carcinogenicity

When evaluating the carcinogenicity potential of devices with long-term contact—that is, exposure lasting longer than 30 days—a risk assessment is typically used. This includes implants and devices that communicate externally, as well as devices

Table 16.1 List of *in-vitro* biocompatibility studies for biomaterials as per regulatory standards

Sr No	Test Name	Procedure	References
1	**Cytotoxicity testing**		
1.1	MTT cytotoxicity test	Required cells are seeded onto 96-well plates to create a semi-confluent monolayer and cultured for 24 h They are exposed to various concentrations of the test substance for 24-h The formazan production at each treatment concentration is measured and compared with the results from control cultures % growth inhibition for each therapy is calculated	ISO 10993-5 (Inayat-Hussain et al. 2008)
1.2	Neutral red uptake (NRU) assay	BALB/c 3T3 cells are cultivated in 96-well plates for 24 h to create a semi-confluent monolayer They are exposed to various concentrations of the test substance NRU is calculated for every treatment concentration after a 24-h exposure and is contrasted with that found in control cultures	ISO 10993-5 (Inayat-Hussain et al. 2008)
1.3	Colony formation cytotoxicity test	V79 cells are seeded into six-well plates and maintained in culture for 24 h to start growing in a logarithmic phase They are then exposed to the test compound over a range of concentrations and incubated for 6 days to make colonies Colonies are fixed with methanol, stained with Giemsa solution, and counted	ISO 10993-5 (Inayat-Hussain et al. 2008)
Note: For 1.2 and 1.3 following should be checked If the extract has a cytotoxic impact % growth inhibition is computed for each treatment The concentration that results in a 50% reduction in NRU, or the IC50, is determined			
1.4	**Test by indirect contact**		
	Agar diffusion assay	A test piece is placed on an agar layer that covers a confluent monolayer of cells The thin agar layer containing the test sample may allow toxic compounds to seep out, killing or damaging nearby monolayer cells	ISO 10993-5 (Inayat-Hussain et al. 2008)
1.5	Test by direct contact	Pipette a known aliquot of the continuously agitated cell suspension Gently rotate the cells horizontally to distribute them uniformly Allow culture to reach subconfluency, incubate at 37 ± 1 °C in air, with or without carbon dioxide depending on the buffer system that was selected for the culture medium	ISO 10993-5 (Inayat-Hussain et al. 2008)

(continued)

Table 16.1 (continued)

Sr No	Test Name	Procedure	References
2	**Genotoxicity testing**		
2.1	In vitro test		
2.1.1	Bacterial gene mutation assay	Salmonella typhimurium and Escherichia coli strains that have been modified to detect frameshift mutations and all potential single base pair modifications are used in this test	OECD 471 (Doak et al. 2012)
2.1.2	The mouse lymphoma gene mutation	Detects the broadest set of genotoxic mechanisms associated with carcinogenic activity;	OECD 490 (Vaman et al. 2012)
2.1.3	*In-vitro* chromosomal aberration (CA) assay	In vitro chromosomal aberration (CA) assay	OECD 473 (Vaman et al. 2012)
2.1.4	*In-vitro* micronucleus assay	Detection of both clastogens and aneugens and it can simultaneosly detect mitotic delay, apoptosis, chromosome breakage, chromosome loss and non-disjunction	OECD 487 (Corvi et al. 2008)
3	**Hemocompatibility**		
3.1	Devices with direct contact of circulating blood	Direct and indirect (extract) methods for material/surface-mediated hemolysis: Under specific static conditions, test and control material specimens or extracts are exposed to blood, and the amount of released haemoglobin is quantified Comparisons are drawn between the test specimens and control specimens that were subjected to identical procedures	ASTM F756-17 (ASTM F 756-17 2017)
3.2	Complement activation	Evaluating SC5b-9 fragment activation using a recognized ELISA test procedure	(Cambiaghi 2018)
		Procedure A describes exposure of solid materials to a standard lot of human serum, using a 0.1-mL serum/13 × 100-mm disposable test tube. Cellulose acetate powders and fibers are used as examples of test materials Procedure B describes assaying the exposed serum for significant functional whole complement depletion as compared to control samples	ASTM F1984 (Cambiaghi 2018)
3.3	Thrombogenicity	Assessment of platelets (e.g., adhesion, activation), and the coagulation system [e.g., Thrombin-Antithrombin Complex (TAT), Partial Thromboplastin Time (PTT)	(U.S. Food and Drug Administration 2023)

(continued)

Table 16.1 (continued)

Sr No	Test Name	Procedure	References
4	Sterility Testing and Ethylene oxide sterilization residuals	Ensures the biomaterial is free from microbial contamination that could cause infection The standard specifies allowable limits for residual ethylene oxide (EO) and ethylene chlorohydrin (ECH) in individual EO-sterilized medical devices, procedures for the measurement of EO and ECH, and methods for determining compliance so that devices may be released	ISO 11737-2, ISO 10993-7
5	Immunological testing	Assesses the immune response triggered by the biomaterial, including inflammation and immune cell activation	ISO 10993-20:2006 (Bernard et al. 2018b)
6	Biocompatibility/ Cytocompatibility evaluation	Evaluates the ability of the biomaterial to stimulate specific cellular responses, such as promoting cell attachment or tissue growth These methods specify the incubation of cultures cells in contact with a device and/ or extracts of a device either directly or through diffusion	ISO 10993-5 (Rivki et al. n.d.)

that come into contact with compromised or breached surfaces (such as wound healing) (U.S. Food and Drug Administration 2023; Organization 2014).

16.4.2.5 Reproductive Toxicity Testing of Biomaterial (U.S. Food and Drug Administration 2023; Reifschneider 2017)

Reproductive and developmental toxicity be evaluated in order to ascertain the potential properties of bio materials and/or their extracts on embryonic development (teratogenicity), prenatal and early postnatal development, and reproductive function as defined in ISO 10993-1. Furthermore, devices intended for use on reproductive organs, medicinal devices that can be used on pregnant women, and innovative materials all need to take developmental and reproductive toxicity into consideration.

16.4.2.6 Hemocompatibility (U.S. Food and Drug Administration 2023; Inayat-Hussain et al. 2008)

Common tests to determine hemocompatibility include thrombogenicity, complement activation, and haemolysis. Detail test is illustrated in Table 16.1.

16.4.3 *In-Vivo* Testing

The physiological milieu is reactive, degradative, kinetic, and dynamic, therefore predicting the long-term *in-vivo* response just from fast screening using in-vitro cell culture models is insufficient. (Basu 2017). Additionally, *in-vivo* testing may be

Table 16.2 List of *in-vivo* biocompatibility studies for biomaterials as per regulatory standards

Sr No	Test Name	Procedure	References
2.1	Implantation studies	Involve surgically implanting the biomaterial into animal models (e.g., rodents, rabbits, or mice) to observe its integration with surrounding tissues, stability, and long-term effects This may also be applied to medical devices that are intended to be used topically in clinical indications where the surface or lining may have been breached, in order to evaluate local tissue responses	ISO 10993-6:2016 (Bernard et al. 2018b)
2.2	Tests for irritation and delayed-type hypersensitivity	The standard describes the procedure for the assessment of medical devices and their constituent materials with regard to their potential to produce irritation and skin sensitization. ISO 10993-10:2010 includes: Pretest considerations for irritation in silico and in vitro methods for dermal exposure; Details of in vivo (irritation and sensitization) test	ISO 10993-10
2.3	***In vivo* cytogenetics assay**		
		Bone marrow micronucleus (MN) Assay or	OECD 474 (U.S. Food and Drug Administration 2023)
		Bone marrow chromosomal aberration (CA) Assay or	OECD 475(U.S. Food and Drug Administration 2023)
		Peripheral blood MN Assay	OECD 474(U.S. Food and Drug Administration 2023)

required to assess the safety of biomaterials and/or medical devices. While some of these *in-vivo* tests like genotoxicity responses—are required to validate or counter results from in vitro assays, other toxicological endpoints might only be ascertained by in vivo testing (De Jong et al. 2019). This reduces the requirement to evaluate multiple aspects, including genotoxicity, epigenetics, carcinogenicity, and immunogenicity, to enhance the material's clinical acceptability. Important *in-vivo* testing for biomaterial is depicted in Table 16.2.

Although the biocompatibility data from both *in-vitro* and *in-vivo* studies can be used to infer a newer material's safety, the material's exact future won't be known until it is utilized clinically amongst patients. Thus, to guarantee a longer existence of material in biological system —it must finally be proved in an ethically sound manner, a comprehensive toxicity study at every stage of development, from *in-vitro* to *in-vivo*. Hence it is necessary to conduct clinical trial.

16.5 Toxicological Testing

Absorption, diffusion, and metabolism of leachable from the device or its components to regions of the body result into systemic toxicity (Strickland et al. 2018). These effects are generalized effects as well on specific organs and organ systems. These tests assess the potential harmful effects of biomaterials on biological systems over short-term (acute) and long-term (chronic) exposure.

According to ISO 10993-1 testing criteria, the length of the toxicity study (subacute, subchronic, and chronic) should correspond with the length of time the device was exposed too.

16.5.1 Types of Toxicity Testing

16.5.1.1 Acute Systemic Toxicity

When a chemical is exposed to a single dose at any moment during the first 72 h following a single, repeated, or constant exposure of a test sample for 24 h, acute systemic toxicity testing determines the possibility that the chemical will have negative effects far from the site of entry (Strickland et al. 2018; De Jong et al. 2019). General information on potential health risks associated with acute exposure through the proposed clinical route is provided by acute systemic toxicity.

16.5.1.2 Subacute/Subchronic Toxicity

Devices with long term exposure (>30 days) also need to address subacute and subchronic toxicity, while devices with protracted exposure (>24 h to ≤30 days) just need to address subacute toxicity studies (De Jong et al. 2019). Subchronic toxicity is the capacity of a drug to produce particular negative effects over prolonged exposure. Evaluating this attribute involves determining a drug's lethal dosage (Zhang et al. 2019). The term "subacute systemic toxicity" refers to adverse effects that follow repeated or constant exposure that lasts between 24 h and 28 days. While, adverse effects that rise during recurrent or continuous administration of a test sample throughout an animal's life span are referred to as subchronic systemic toxicity (90 days in rodent) (De Jong et al. 2019).

16.5.1.3 Chronic Toxicity (De Jong et al. 2019)

Adverse effects that develop after a test sample is administered repeatedly or continuously for a significant portion of life span are known as chronic toxicity. This is generally last between 6 and 12 months for rodents.

16.5.2 Challenges for Toxicity Testing (Anderson et al. 2008; Williams 2008; Anderson and Shive 2012; Ratner 2015)

Long-term and repeated toxicity testing of biomaterials presents unique challenges that arise from their complex interactions with biological systems.

- Biomaterials can elicit immune responses, leading to inflammation, fibrosis, or other adverse reactions. Predicting and assessing these responses over the long term is challenging. The potential for materials to be known as foreign by the body's immune system requires careful evaluation.
- Many biomaterials are designed to degrade over time, releasing harmful degradation products. The toxicity of these degradation products must be assessed, which can be complex due to varying degradation rates and by-products. Understanding the long-term effects of degradation products on the body is difficult, requiring extensive and prolonged studies.
- *In-vitro* models may not fully replicate the *in-vivo* environment, leading to discrepancies in toxicity results. Choosing appropriate animal models that closely mimic human physiological responses to biomaterials is challenging.
- Long-term studies involving animals raise ethical concerns about their welfare, particularly given the potential for chronic suffering. Ethical issues also arise in the context of human trials, especially for implants and other long-term applications.
- Regulatory standards for biomaterials are continuously evolving, and keeping up with these changes is a challenge. Different countries have diverse regulatory requirements, complicating international research and product approval.
- Long-term toxicity testing is expensive due to the need for prolonged observation periods, advanced technologies, and comprehensive data analysis.

Reproducing long-term studies is challenging due to the time and cost involved, raising concerns about the reliability of the data.

16.6 Diverse Clinical Aspects of Biomaterials

16.6.1 Biomaterials in Medical Devices

The wide range of uses of biomaterials in medical devices and their profound influence on patient health outcomes have rarely documented in research. The biomaterials have demonstrated its potential in the ocular based products. TopCEM Dual Treatment Resin cement, which contains a paste of catalyst and base, is the most advanced and adaptable dental dual cure (chemical and/or light cure) luting cement available. It is applied to a variety of dental restorations as a permanent cement. The gadget is essentially the same as the reference devices, which have been sold legally for many years without experiencing any negative side effects in patients. (USFDA 2024b).

Osteostimulative calcium-phosphorous sodium-silicate granules, ranging in size from 0.5–0.8 mm to 1.0–2.0 mm, constitute Bonalive® Orthopedics granules, a sterile medical device. It is an implant designed to close any bone gaps or holes in the skeletal system, such as those in the pelvis and extremities. These osseous defect, which are not inherent to the integrity of the bony structure, can be the result of surgically induced defects, traumatic bone injuries, benign bone cysts and tumors, or both. When the body heals, these granules resorb and are replaced with bone. (Devices@FDA 2023). Moreover, biomaterials are also used as scaffold for fabrication of allogenic cell graft. USFDA approved GINTUIT, an allogenic cultured keratinocyte and fibroblast in bovine collagen. It is fabricated to treat mucogingival conditions in adults (USFDA n.d.).

16.6.2 Biomaterials in Pharmaceutical Application

Biomaterials are acceptable for fabricating medical devices however, it is still facing recognition issues in the pharmaceutical formulations. The prime issue with the biomaterial based pharmaceuticals is its source, impurities or heavy metal contamination. There are few of the biomaterials which has been into the clinical trials however, still not accepted by the regulatory agencies. A biomaterial named Chitosan has been approved as dietary supplement. A clinical trial study (NCT03650023) demonstrated the applications of chitosan oligosaccharide in diabetes. The study revealed that the chitosan oligosaccharide consumption effectively reduces the blood glucose level helping in controlling diabetes (Jeong et al. 2019).

Moreover, biomaterials are also used in dermal fillers. Polymers like hyaluronic acid, polylactic acid are been accepted as wrinkle fillers (USFDA 2020). FDA has approved intradermal injection of hyaluronic acid in combination with lidocaine to reduce rejuvenate facial skin and remove wrinkles. The product was named Skinvive was launched by Juvederm (USFDA 2022). Additionally, a case report published by Stolic et al., showed the use of hyaluronic acid in surgical lips deformity. The study demonstrated a case of 40-year-old woman who had a surgery of cleft lip, primary and secondary palate. The study showed remarkable effect of hyaluronic acid on reducing the post-operative scars (Stolic et al. 2015). Moreover, Zahi Badran from University of Sharjah initiated a clinical trial (NCT05682833) on the application of hyaluronic acid and chlorhexidine based gel for the treatment of Periodontitis (NCT05682833). The results depicted significant improvement in the clinical parameters of patients (Diehl et al. 2022). Further, hyaluronic acid is a principle ligand to the CD44 receptor which is a highly expressed cell surface receptor to the cancer cell (Misra et al. 2015). However, none of the anticancer formulation has been designed including hyaluronic acid as an adjuvant.

Further, biomaterials have also contributed significantly in the wound healing. MaxiCel is the chitosan based wound dressing. The dressing is marketed by Axio Biosolutions Pvt. Ltd. HemCon, Chitopack are few other UDFDA approved chitosan dressings (USFDA 2024a). Numerous biomaterial based formulations like hydrogel, foams, films have showed remarkable effects in wound healing (Ansari and Darvishi 2024). However, facing significant challenges in getting accepted to the market. The major place where pharmaceutical scientists and bioengineers face issue is the lack of clarity in the available documents. The natural polymers are accepted in the inactive ingredient guide, however these natural polymers based products are not accepted by the regulatory agencies due to its source or impurity guidelines. Hence proper, compiled and stringent document dedicated to approval of biomaterial based products will lead to a better documentation and study plan for a biomaterial based product which will in turn led to approval.

16.7 Future Prospect and Conclusion

Although biomedical technologies and medicine have made tremendous progress in the laboratory, it is frequently disappointing when these advances are not translated to meet patients' unmet clinical needs (Ebrahimi 2020). Since, procedure used to assess medical devices is followed while evaluating biomaterials. For this reason, biomaterials and medical devices are interchangeable terms in many regulatory procedures (Masaeli et al. 2019). Additionally, the conditions for approval of a particular biomaterial vary throughout regulatory systems, different rules for a given product make the process of developing biomaterials expensive and time-consuming. There isn't a mutual classification for the materials found in regulations and hence, no uniform method that extends to all biomaterial classes. Therefore, confusion and inefficiency in assessment processes are mostly triggered by residual inconsistencies in evaluation techniques (Reeve and Baldrick 2017; Masaeli et al. 2019).

Encouraging greater collaboration between the regulatory bodies and the sector is one possible option. In order to change the dynamics, this collaboration involve the industry supplying additional test materials, concerning regulators initially in the developmental activities and validation phase of product (Kanďárová and Pôbiš 2023).

In addition to that, innovation and regulatory acceptance are substantially aided by the cross-sector harmonization that can be achieved between different industry branches, scientific fields, and regulatory agencies. While, ISO plays a crucial role in guaranteeing this universality. Starting a formal conversation between national parties, similar to the OECD-MAD approach, would be a crucial tactic. In order to match standards as well as practices, these enable a more unified implementation of innovative testing procedures, such a discourse is necessary (Kanďárová and Pôbiš 2023; Myers et al. 2017).

Unrestricted access to information is an essential part of global harmonization efforts. There is an abundance of scientific literature that explains the various types of biomaterials, as well as their features and applications. Accumulating and developing worldwide data standards to compile information and findings from scientific publications that have been published in the past are currently in progress that discuss how various materials interact with living tissues in an effective way to expedite and improve the financial benefits of biomaterials evaluations. Through eliminating duplication, knowledge from earlier research might accelerate current investigations. In order to promote cooperation, stimulate innovation, and guarantee that all parties involved in the harmonization process are informed and actively involved, it is imperative that data, research findings, and methodological developments be transparently shared (Kanďárová and Pôbiš 2023; Helmus et al. 2008).

Regulatory agencies in the biomaterials sector need to adopt novel testing methodologies and personalized strategies in order to stay up with the high pace of innovation in this area. Ultimately, a balanced strategy that promotes patient safety and encourages innovation is necessary for the successful integration of biomaterials into clinical practice. Through the resolution of crucial regulatory issues and the on-going upgrading of standards, the biomedical community may fully utilize biomaterials to transform healthcare and improve the quality of life for patients globally.

References

Amato SF (2015a) Clinical development and endpoint strategies for biomaterials and medical devices. In: Regulatory affairs for biomaterials and medical devices. Woodhead Publishing Limited. https://doi.org/10.1533/9780857099204.47

Amato SF (2015b) Regulatory strategies for biomaterials and medical devices in the USA: classification, design, and risk analysis. In: Regulatory affairs for biomaterials and medical devices. Elsevier, pp 27–46. https://doi.org/10.1533/9780857099204.27

Anderson JM (2016) Future challenges in the in vitro and in vivo evaluation of biomaterial biocompatibility. Regener Biomater 3(2):73–77. https://doi.org/10.1093/RB/RBW001

Anderson JM, Shive MS (2012) Biodegradation and biocompatibility of PLA and PLGA microspheres. Adv Drug Deliv Rev 64(SUPPL):72–82. https://doi.org/10.1016/j.addr.2012.09.004

Anderson JM, Rodriguez A, Chang DT (2008) Foreign body reaction to biomaterials. Semin Immunol 20(2):86–100. https://doi.org/10.1016/j.smim.2007.11.004

Ansari M, Darvishi A (2024) A review of the current state of natural biomaterials in wound healing applications. Front Bioeng Biotechnol 12(March):1–26. https://doi.org/10.3389/fbioe.2024.1309541

Ashworth S et al (2021) Chondroitin sulfate as a potential modulator of the stem cell niche in cornea. Front Cell Dev Biol 8(January):1–10. https://doi.org/10.3389/fcell.2020.567358

Assad M, Jackson N (2019) Biocompatibility evaluation of orthopedic biomaterials and medical devices: a review of safety and efficacy models. In: Encyclopedia of biomedical engineering. Elsevier. https://doi.org/10.1016/B978-0-12-801238-3.11104-3

ASTM (2016) Standard guide for characterization and testing of raw or starting materials for tissue-engineered medical products. https://doi.org/10.1520/F2027-16

ASTM (n.d.-a) Standard practice for selecting generic biological test methods for materials and devices. https://doi.org/10.1520/F0748-16

ASTM (n.d.-b) International Committee F04 on Medical and Surgical Materials and Devices. Available at: https://mcsdocs.astm.org/committee-documents/F04_Fact_Sheet_2021.pdf

ASTM F 756-17 (2017) Standard practice for assessment of hemolytic properties of materials. In: Annual book of ASTM, p 6

Azzarà A et al (2010) Increased level of DNA damage in some organs of obese Zucker rats by c-H2AX analysis. Environ Mol Mutagen 405(April):391–405. https://doi.org/10.1002/em.22115

Basu B (2017) Probing toxicity of biomaterials and biocompatibility assessment. https://doi.org/10.1007/978-981-10-3059-8_9

Bernard M et al (2018a) Biocompatibility of polymer-based biomaterials and medical devices-regulations,: in vitro screening and risk-management. Biomater Sci 6(8):2025–2053. https://doi.org/10.1039/c8bm00518d

Bernard M et al (2018b) Biocompatibility of polymer-based biomaterials and medical devices-regulations, In vitro screening and risk-management. Biomater Sci 6(8):2025–2053. https://doi.org/10.1039/c8bm00518d

Bhat S, Kumar A (2013) Biomaterials and bioengineering tomorrow's healthcare. Biomatter 3(3). https://doi.org/10.4161/biom.24717

Bruinink A, Luginbuehl R (2011) Evaluation of biocompatibility using in vitro methods: interpretation and limitations. Adv Biochem Eng Biotechnol 126:117–152. https://doi.org/10.1007/10_2011_111

Cambiaghi A (2018) Biological evaluation of medical devices as an essential part of the risk management process: updates and challenges of. Eurofins [Preprint]

Committee F04 on Medical and Surgical Materials and Devices (n.d.) Available at: https://www.astm.org/committee-f04

Corvi R et al (2008) ECVAM retrospective validation of in vitro micronucleus test (MNT). Mutagenesis 23(4):271–283. https://doi.org/10.1093/mutage/gen010

Cvrček L, Horáková M (2018) Plasma modified polymeric materials for implant applications. In: Non-thermal plasma technology for polymeric materials: applications in composites, nanostructured materials, and biomedical fields. https://doi.org/10.1016/B978-0-12-813152-7.00014-7

De Jong WH, Carraway JW, Geertsma RE (2019) In vivo and in vitro testing for the biological safety evaluation of biomaterials and medical devices. In: Biocompatibility and performance of medical devices, 2nd edn. Elsevier Ltd. https://doi.org/10.1016/B978-0-08-102643-4.00007-0

Devices@FDA (2023) Filler, bone void, calcium compound. CDRH

Diehl D et al (2022) Adjunctive application of hyaluronic acid in combination with a sodium hypochlorite gel for non-surgical treatment of residual pockets reduces the need for periodontal surgery—retrospective analysis of a clinical case series. Materials 15(19). https://doi.org/10.3390/ma15196508

Doak SH et al (2012) In vitro genotoxicity testing strategy for nanomaterials and the adaptation of current OECD guidelines. Mutat Res Genet Toxicol Environ Mutagen 745(1–2):104–111. https://doi.org/10.1016/j.mrgentox.2011.09.013

Dornish M, Kaplan D, Skaugrud Ø (2001) Standards and guidelines for biopolymers in tissue-engineered medical products: ASTM alginate and chitosan standard guides. Ann N Y Acad Sci 944(388):388–397. https://doi.org/10.1111/j.1749-6632.2001.tb03850.x

Drewnowska O et al (2013) Chitosan-a promising biomaterial in veterinary medicine. Pol J Vet Sci 16(4):843–848. https://doi.org/10.2478/pjvs-2013-0119

Ebrahimi M (2020) Standardization and regulation of biomaterials. LTD, Handbook of biomaterials biocompatibility. https://doi.org/10.1016/B978-0-08-102967-1.00013-X

Eliaz N, Metoki N (2017) Calcium phosphate bioceramics: a review of their history, structure, properties, coating technologies and biomedical applications. Materials 10(4). https://doi.org/10.3390/ma10040334

ElinaTenhunen(2024)Alternativestoanimaltestinginthedevelopmentofmedicaldevices.Availableat: https://measurlabs.com/blog/animal-testing-alternatives-in-medical-device-biocompatibility/

FDA/CDRH (2023) Accreditation Scheme for Conformity Assessment (ASCA), Fda Cdrh. Available at: https://blog.johner-institute.com/systems-engineering/asca/

Feng F et al (2022) Translation of a spinal bone cement product from bench to bedside'. Bioact Mater 10(July 2021):345–354. https://doi.org/10.1016/j.bioactmat.2021.08.011

Fenton OS et al (2018) Advances in biomaterials for drug delivery. Adv Mater 30(29):1–29. https://doi.org/10.1002/adma.201705328

Frati C et al (2020) Reinforced alginate/gelatin sponges functionalized by avidin/biotin-binding strategy: a novel cardiac patch. J Biomater Appl 34(7):975–987. https://doi.org/10.1177/0885328219886029

Frisch E et al (2023) Preclinical in vitro evaluation of implantable materials: conventional approaches, new models and future directions. Front Bioeng Biotechnol 11(July):1–15. https://doi.org/10.3389/fbioe.2023.1193204

Gervasoni S et al (2024) A human-scale clinically ready electromagnetic navigation system for magnetically responsive biomaterials and medical devices. Adv Mater 2310701:1–20. https://doi.org/10.1002/adma.202310701

Goldberg M et al (2022) A nanoengineered topical transmucosal cisplatin delivery system induces anti-tumor response in animal models and patients with oral cancer. Nat Commun 13(1). https://doi.org/10.1038/s41467-022-31859-3

Halamoda-Kenzaoui B et al (2019) Mapping of the available standards against the regulatory needs for nanomedicines. Wiley Interdiscip Rev Nanomed Nanobiotechnol 11(1):1–17. https://doi.org/10.1002/wnan.1531
Hallab NJ, Jacobs J, Oshua J (2020) Orthopedic applications. In: Biomaterials science, 4th edn. Elsevier. https://doi.org/10.1016/B978-0-12-816137-1.00071-4
Hampshire VA, Gilbert SH (2019) Refinement, reduction, and replacement (3R) strategies in preclinical testing of medical devices. Toxicol Pathol 47(3):329–338. https://doi.org/10.1177/0192623318797289
Helmus MN, Gibbons DF, Cebon D (2008) Biocompatibility: meeting a key functional requirement of next-generation medical devices. Toxicol Pathol 36(1):70–80. https://doi.org/10.1177/0192623307310949
Hsiao MY et al (2019) Drug-loaded hyaluronic acid hydrogel as a sustained-release regimen with dual effects in early intervention of tendinopathy. Sci Rep 9(1):1–9. https://doi.org/10.1038/s41598-019-41410-y
Hulsart-Billström G et al (2016) A surprisingly poor correlation between in vitro and in vivo testing of biomaterials for bone regeneration: results of a multicentre analysis. Eur Cells Mater 31(June):312–322. https://doi.org/10.22203/eCM.v031a20
Huzum B et al (2021) Biocompatibility assessment of biomaterials used in orthopedic devices: an overview (review). Exp Ther Med 22(5):1–9. https://doi.org/10.3892/etm.2021.10750
ILibrary O (n.d.) OECD guidelines for the testing of chemicals, section 4. https://doi.org/10.1787/20745788
Inayat-Hussain S, Rajab NF, Siew EL (2008) In vitro testing of biomaterials toxicity and biocompatibility. In: Cellular response to biomaterials. Woodhead Publishing Limited. https://doi.org/10.1533/9781845695477.3.508
ISO (2009) Biological evaluation of medical devices – Part 5: tests for in vitro cytotoxicity. ISO
Jeong S et al (2019) Chitosan oligosaccharide (GO2KA1) improves postprandial glycemic response in subjects with impaired glucose tolerance and impaired fasting glucose and in healthy subjects: a crossover, randomized controlled trial. Nutr Diabetes 9(1):31. https://doi.org/10.1038/s41387-019-0099-4
Kanďárová H, Pôbiš P (2023) The "Big Three" in biocompatibility testing of medical devices: implementation of alternatives to animal experimentation—are we there yet? Front Toxicol 5(January):1–9. https://doi.org/10.3389/ftox.2023.1337468
Li W, Zhou J, Xu Y (2015) Study of the in vitro cytotoxicity testing of medical devices. Biomed Rep 3(5):617–620. https://doi.org/10.3892/br.2015.481
Lu XJFC et al (2024) Chemical and structural engineering of gelatin-based delivery systems for therapeutic applications: a review. Biomacromolecules 25(2):564–589. https://doi.org/10.1021/acs.biomac.3c01021
Masaeli R, Zandsalimi K, Tayebi L (2019) Biomaterials evaluation: conceptual refinements and practical reforms. Ther Innov Regul Sci 53(1):120–127. https://doi.org/10.1177/2168479018774320
Mehtani D et al (2019) Biomaterials for sustained and controlled delivery of small drug molecules. In: Biomaterials and bionanotechnology. Elsevier Inc. https://doi.org/10.1016/B978-0-12-814427-5.00004-4
Misra S et al (2015) Interactions between hyaluronan and its receptors (CD44, RHAMM) regulate the activities of inflammation and cancer. Front Immunol 6(May). https://doi.org/10.3389/fimmu.2015.00201
Moharamzadeh K (2017) Processing and preservation of biomaterials and regulatory issues. In: Biomaterials for oral and dental tissue engineering. Elsevier, pp 141–153. https://doi.org/10.1016/B978-0-08-100961-1.00009-8
Mohd Shafiee MA, Muhamad Asri MA, Syed Alwi SS (2021) Review on the in vitro cytotoxicity assessment in accordance to the international organization for standardization (ISO). Malays J Med Health Sci 17(2):261–269
Mueller E, Kammula R, Marlowe D (1991) Regulation of "biomaterials" and medical devices. MRS Bull 16(9):39–41. https://doi.org/10.1557/S0883769400056037

Myers DK et al (2017) From in vivo to in vitro: the medical device testing paradigm shift. Altex 34(4):479–500. https://doi.org/10.14573/altex.1608081

Navarro M et al (2008) Biomaterials in orthopaedics. J R Soc Interface 5(27):1137–1158. https://doi.org/10.1098/rsif.2008.0151

Nguyen PAH et al (2020) How to select the appropriate method(s) of cytotoxicity analysis of mammalian cells at biointerfaces: a tutorial. Biointerphases 15(3):031201. https://doi.org/10.1116/6.0000136

Nii T (2021) Biomaterial-assisted regenerative medicine, pp 1–18

O'Brien MT et al (2022) Scientific and regulatory policy committee points to consider for medical device implant site evaluation in nonclinical studies. Toxicol Pathol 50(4):512–530. https://doi.org/10.1177/01926233221103202

Organization W (2014) Health International agency for research on cancer. IARC monographs on the evaluation of carcinogenic risks to humans. Available at: https://monographs.iarc.who.int/wp-content/uploads/2018/08/14-002.pdf

Park S, Kim J, Lee C (2022) Injectable rapidly dissolving needle-type gelatin implant capable of delivering high concentrations of H2O2 through intratumoral injection. Biomed Pharmacother 156:113910. https://doi.org/10.1016/j.biopha.2022.113910

Pashuck ET, Stevens MM (2012) Designing regenerative biomedical therapies for the clinic. Biomaterials 4(160):1–12

Pires PC et al (2023) Polymer-based biomaterials for pharmaceutical and biomedical applications: a focus on topical drug administration. Eur Polym J 187(February):111868. https://doi.org/10.1016/j.eurpolymj.2023.111868

Prestwich GD, Bhatia S, Breuer CK, Dahl SLM, Mason C, McFarland R, McQuillan DJ, Sackner-Bernstein J, Schox J et al (2012) What is the greatest regulatory challenge in the translation of biomaterials to the clinic? Sci Transl Med. https://doi.org/10.1126/scitranslmed.3004915

Rahmati M et al (2020) Biological responses to physicochemical properties of biomaterial surface. Chem Soc Rev 49(15):5178–5224. https://doi.org/10.1039/d0cs00103a

Rasmussen K et al (2019) Developing OECD test guidelines for regulatory testing of nanomaterials to ensure mutual acceptance of test data. Regul Toxicol Pharmacol 104(February):74–83. https://doi.org/10.1016/j.yrtph.2019.02.008

Ratner BD (2015) The biocompatibility of implant materials. In: Host response to biomaterials: the impact of host response on biomaterial selection. Elsevier Inc. https://doi.org/10.1016/B978-0-12-800196-7.00003-7

Razavi M, Khandan A (2017) Safety, regulatory issues, long-term biotoxicity, and the processing environment. In: Nanobiomaterials science, development and evaluation. Elsevier. https://doi.org/10.1016/B978-0-08-100963-5.00014-8

Reeve L, Baldrick P (2017) Biocompatibility assessments for medical devices–evolving regulatory considerations. Expert Rev Med Devices 14(2):161–167. https://doi.org/10.1080/17434440.2017.1280392

Reifschneider A (2017) The new U. S. FDA regulations on biocompatibility and reprocessing for medical devices, pp 9–13. Available at: http://dgra.de/deutsch/studiengang/master-thesis/2017-Anna-Reifschneider-The-new-U.S.-FDA-regulations-on-biocompatibility-and-reprocessin

Rivki M et al (n.d.) Biomaterial & biocompatibility testing laboratory, (112)

Samir A et al (2022) Recent advances in biodegradable polymers for sustainable applications. npj Mater Degrad 6(1). https://doi.org/10.1038/s41529-022-00277-7

Schuh JACL, Funk KA (2019) Compilation of international standards and regulatory guidance documents for evaluation of biomaterials, medical devices, and 3-D printed and regenerative medicine products. Toxicol Pathol 47(3):344–357. https://doi.org/10.1177/0192623318804121

Sekar MP et al (2021) Current standards and ethical landscape of engineered tissues—3D bioprinting perspective. J Tissue Eng 12. https://doi.org/10.1177/20417314211027677

Song X, Hu M et al (2022a) Advancing medical device regulatory reforms for innovation, translation and industry development in China. J Orthop Transl 37(October):89–93. https://doi.org/10.1016/j.jot.2022.09.015

Song X, Tang Z et al (2022b) Biomaterials and regulatory science. J Mater Sci Technol 128:221–227. https://doi.org/10.1016/j.jmst.2022.04.018
Soni SS et al (2022) Sustained release of drug-loaded nanoparticles from injectable hydrogels enables long-term control of macrophage phenotype. Biomater Sci 10(24):6951–6967. https://doi.org/10.1039/d2bm01113a
Stolic D et al (2015) The surgical lips deformity corrected with hyaluronic fillers: a case report. Open Access Maced J Med Sci 3(3):423–425. https://doi.org/10.3889/oamjms.2015.067
Strickland J, Clippinger AJ, Brown J et al (2018) Status of acute systemic toxicity testing requirements and data uses by U.S. regulatory agencies. Regul Toxicol Pharmacol 94:2018–2196. https://doi.org/10.1016/j.yrtph.2018.01.022
The Mutual Acceptance of Data (MAD) System (n.d.) Available at: https://www.oecd.org/en/topics/sub-issues/testing-of-chemicals/mutual-acceptance-of-data-system.html
Troy E et al (2021) Nature-based biomaterials and their application in biomedicine. Polymers 13(19):1–37. https://doi.org/10.3390/polym13193321
Trucillo P (2024) Biomaterials for drug delivery and human applications. Materials 17(2). https://doi.org/10.3390/ma17020456
U.S. Food & Drug Administration (2023) Use of International Standard ISO 10993-1, "Biological evaluation of medical devices – part 1: evaluation and testing within a risk management process". U.S. Department of Health and Human Services Food and Drug Administration, (301), pp 1–68
Uppal G et al (2022) Magnesium based implants for functional bone tissue regeneration – a review. J Magnesium Alloys 10(2):356–386. https://doi.org/10.1016/j.jma.2021.08.017
USFDA (2020) FDA-approved dermal fillers. Available at: https://www.fda.gov/medical-devices/aesthetic-cosmetic-devices/fda-approved-dermal-fillers#approved
USFDA (2022) SKINVIVE by JUVÉDERM – P110033/S059. Available at: https://www.fda.gov/medical-devices/recently-approved-devices/skinvive-juvederm-p110033s059
USFDA (2024a) Hemostatic wound dressing without thrombin or other biologics. USFDA
USFDA (2024b) Material, tooth shade, resin. Available at: https://www.accessdata.fda.gov/scripts/cdrh/devicesatfda/index.cfm?db=pmn&id=K233082
USFDA (n.d.) GINTUIT (Allogeneic cultured keratinocytes and fibroblasts in bovine collagen)
Vaman VSA et al (2012) Assessment of micronuclei and chromosomal anomalies of five biocompatible materials in mice. Toxicol Environ Chem 94(6):1225–1237. https://doi.org/10.1080/02772248.2012.694999
Visser R (n.d.) 14. Mutual recognition in the testing of chemicals through the OECD, pp 415–438
Wallin RF (1996) Improving biocompatibility standards for the global market. Available at: https://www.mddionline.com/regulatory-quality/improving-biocompatibility-standards-for-the-global-market
Wallin RF (1998) A practical guide to ISO 10993-12: sample preparation and reference materials. Medical Device & Diagnostic Industry Magazine MDDI. Available at: https://www.mddionline.com/materials/a-practical-guide-to-iso-10993-12-sample-preparation-and-reference-materials
Wang L et al (2022) Key considerations on the development of biodegradable biomaterials for clinical translation of medical devices: with cartilage repair products as an example. Bioact Mater 9(August 2021):332–342. https://doi.org/10.1016/j.bioactmat.2021.07.031
Williams DF (2008) On the mechanisms of biocompatibility. Biomaterials 29:2941–2953. https://doi.org/10.1016/j.biomaterials.2008.04.023
Xiang H, Chen Y (2020) Materdicine: interdiscipline of materials and medicine. Viewpoints 1(3):1–29. https://doi.org/10.1002/VIW.20200016
Yaneva Z et al (2020) The 21st century revival of chitosan in service to bio-organic chemistry. Biotechnol Biotechnol Equip 34(1):221–237. https://doi.org/10.1080/13102818.2020.1731333
Zhang Z et al (2019) Systematic acute and subchronic toxicity evaluation of polysaccharide-protein complex-functionalized selenium nanoparticles with anticancer potency. Biomater Sci 7(12):5112–5123. https://doi.org/10.1039/c9bm01104h

The manufacturer's authorised representative in the EU is Springer Nature Customer Service Centre GmbH, Europaplatz 3, 69115 Heidelberg, Germany. If you have any concerns regarding our products, please contact ProductSafety@springernature.com

Printed and bound by CPI Group (UK) Ltd, Croydon, CR0 4YY
30/09/2026
02225501-0013